Advanced Coal Preparation and Beyond

Advanced Coal Preparation and Beyond

CO_2 Capture and Utilization

S. Komar Kawatra

CRC Press
Taylor & Francis Group
Boca Raton London New York

CRC Press is an imprint of the
Taylor & Francis Group, an **informa** business

CRC Press
Taylor & Francis Group
6000 Broken Sound Parkway NW, Suite 300
Boca Raton, FL 33487-2742

Printed on acid-free paper

International Standard Book Number-13: 978-0-367-22884-2 (Hardback)

Visit the Taylor & Francis Web site at
http://www.taylorandfrancis.com

and the CRC Press Web site at
http://www.crcpress.com

Contents

Preface

This book is an update of the book *Coal Desulfurization: High Efficiency Preparation Methods*, which was written when controlling the sulfur content of coal was very important. The scene has since changed: today the primary concern with coal has become controlling carbon dioxide (CO_2) emissions.

Coal has historically been linked to many different pollutants. However, what if these pollutants could be turned into opportunities to extract additional value from the coal? Technologies are in development to transform CO_2 into valuable compounds. If these technologies can be refined to the extent that they can be cost effective, then coal becomes a valuable carbon source in addition to a source of energy. The amount of CO_2 available from coal is staggering, and capturing and utilizing this CO_2 presents a tremendous opportunity.

U.S. coal production in 2017 was down 33% from the peak production of 1,054.8 million metric tons in 2006. In 2016, 30% of electric power was produced using coal as a fuel. In the author's opinion coal will not disappear, especially if pollution-free energy could be produced from coal or the pollutants can be transformed into valuable products. The United States alone has reserves on the order of 475 billion short tons, and the world reserves are estimated at around 1.03 trillion metric tons. At current use rates, these reserves are expected to outlast oil and gas reserves by over double – 134 years versus approximately 50 each for oil and gas.

It has been believed that the implementation of CO_2 capture and sequestration in a coal power plant would increase the footprint by 25%, but this is not necessarily true. A steel plant in Louisiana has added a CO_2 scrubber and completely covered the associated energy cost with unused waste heat already available in the plant.

The ideal scenario would be to remove the pollutants from coal and use them to create additional profit by selling them or valuable derivatives. Although CO_2 is the most ubiquitous pollutant, it has also been found that rare earths can be concentrated in the coal ashes. The rare earths are critical to many aspects of our everyday lives and the defense and security of our nations and economies.

The primary audience for this book is intended to be practicing coal preparation engineers who need complete information about all the coal preparation technologies that are available to them now, or that may be available in the future. It will also be valuable for researchers who need details and comparative data for "cutting-edge" technologies that are still under development. The main emphasis is on physical separation processes for desulfurization, but this book also covers a diverse range of topics relevant to coal processing including dust suppression, CO_2 capture, CO_2 utilization opportunities, rare earth recovery opportunities, and chemical and biological processing methods. We have also included details of processes and techniques that but were subsequently abandoned, with the hope that this will help engineers and researchers avoid repeating past mistakes.

I would like to thank several of my graduate students including Victor Claremboux, Howard Haselhuhn, Scott Moffatt, and Jacob McDonald.

S. Komar Kawatra
Houghton, Michigan
August 2, 2019

Author

Dr S. Komar Kawatra obtained an MS in Physics from the University of Poona, India and a PhD in Metallurgy from the University of Queensland, Australia. At present he is a Professor of Chemical Engineering at Michigan Technological University (Michigan Tech). His research focuses on the optimization of mineral processing operations and reusing industrial wastes.

His research experience includes work at the Bhabha Atomic Research Center, Trombay, Bombay, India; Julius Kruttschnitt Mineral Research Centre, Brisbane, Australia; and Canada Centre for Mineral and Energy Technology, Ottawa; the National Energy Technology Laboratory, USA; the University of Alberta; and Michigan Tech.

Dr Kawatra has received several awards from the Society for Mining, Metallurgy and Exploration including the Distinguished Member Award, the Taggart Award, the Gaudin Award, and the Presidential Citation. He has also received the Robert H. Richards Award and the Frank F. Aplan Award from the American Institute of Mining, Metallurgy, and Petroleum Engineers, the Michigan Tech Research Award and Graduate Student Mentor Award, the IEEE Certificate of Service, and the Michigan Association of Governing Boards Distinguished Faculty Member Award. He completed the Fulbright Scholarship Program in 2012.

Dr Kawatra has received several patents stemming from his research. One notable patent is for the E-Iron Nugget Process (U.S. Patent No. 7,632,330), which produces iron using environmentally benign reducing agents. This process was independently found to be commercially viable by Purdue University. Carbontec Energy Corporation recently formed E-Iron International, LLC to construct a 300,000 metric tons/year plant in Burns Harbor, IN, which will sell to a U.S. market which currently imports 4 million metric tons/year of pig iron.

Dr Kawatra is the Editor-in-Chief of the *Journal of Mineral Processing and Extractive Metallurgy Review*, published by Taylor & Francis.

1 Introduction

There has been increasing worldwide interest in reducing industrial emissions to the atmosphere. Coal is under particular scrutiny, as when it is burned, it emits pollutants such as carbon dioxide (CO_2), sulfur oxides (SO_2 and SO_3), and nitrogen oxides (NO_x) in addition to particulate matter. These gases are precursors to climate change and acid rain, which has a harmful effect on precipitation, oceans, lakes, and vegetation. Over 36 billion metric tons of CO_2 are emitted around the world each year (Janssens-Maenhout et al., 2017), and the total worldwide SO_2 emissions from coal utilization are approximately 100 million metric tons/year (Crippa et al., 2018), of which the emissions arising from power generation are about 55 million metric tons/year. These SO_2 emissions can be reduced by coal cleaning to remove sulfur from coal before it is burned (Couch, 1995).

To reduce acid rain problems, the U.S. Clean Air Act amendments of 1990 require that SO_2 emissions be reduced to 2.5 lb/million British Thermal Units (BTU) (1.08 g/MJ) by 1995 and to 1.2 lb/million BTU (0.52 g/MJ) by 2000 (Hoddinot, 1992), where pounds of SO_2 per million BTU or grams of SO_2 per megajoule is calculated from proximate analysis of the coal by using one of the following relationships (Cavallaro et al., 1991b):

$$\text{lb } SO_2/\text{million BTU} = 20{,}000 \times \frac{(\text{Total Sulfur, wt\%})}{(\text{Heating value, BTU/lb})} \tag{1.1}$$

$$\text{g } SO_2/\text{MJ} = 20 \times \frac{(\text{Total Sulfur, wt\%})}{(\text{Heating value, MJ/kg})} \tag{1.2}$$

Environmental legislation to limit sulfur emissions is already in place in many countries and is under consideration in most others. Since coal is an international commodity, all coal producers will either have to be able to meet these standards in the near future, or will have to take them into account when deciding how to continue to market high-sulfur coals.

The total international export of coal in 2016 amounted to 1.19 billion tons (IEA, 2018). Of coal produced in the world, approximately 45% was burned to generate steam (IEA, 2018). The production is widely dispersed, with numerous countries mining significant amounts of coal, as shown in Table 1.1 (Hoddinot, 1992, 1994; BGR, 2017). The expected long-term trend is for increasing demand for low-sulfur coals with high heating value, in order to minimize the sulfur emitted per unit of energy. Each of these coal producers will need to ensure that their products can meet stringent sulfur emissions standards if they wish to keep their share of the market, and therefore precombustion desulfurization technologies will be of great interest. Additionally, many countries are already implementing or strongly considering implementing emissions regulations on carbon dioxide. In the United States, many

TABLE 1.1
Coal Output for Selected Countries (millions of tons) (BGR, 2017)

	2013		2014		2015		2016	
Country	**Hard**	**Lignite**	**Hard**	**Lignite**	**Hard**	**Lignite**	**Hard**	**Lignite**
China	3,602	147	3,495	145	3,423	140	3,423	140
United States	823	70	535	72	749	65	749	66
Commonwealth of Independent States	477	85	461	83	443	84	443	85
Poland	77	66	73	64	73	63	73	60
Germany	8	183	8	178	7	178	7	172
India	566	44	609	48	640	44	252	45
South Africa	256	—	261	—	252	—	254	—
Australia	411	60	441	58	441	61	444	60
United Kingdom	13	—	12	—	9	—	9	—
Czech/Slovakia	9	41	8	38	8	38	8	39
Serbia	—	40	—	30	—	37	—	38
Canada	60	9	61	9	51	8	52	9
Greece	—	54	—	50	—	46	—	32
North Korea	32	7	34	7	34	7	34	7
Turkey	—	58	—	63	—	56	—	57
Romania	—	25	—	24	—	25	—	23
Bulgaria	—	27	—	31	—	31	—	31
Colombia	85.5	—	89	—	86	—	86	—
Hungary	—	10	—	10	—	9	—	9
Thailand	—	18	—	18	—	15	—	17
Indonesia	430	65	411	60	402	60	402	60
Mexico	16	—	16	—	16	—	12	—
Other	120	46	120	38	112	38	117	40
World	6,984	1,054	6,933	1,025	6,743	1,012	6,291	990

utility companies are switching to low-sulfur western coals in spite of increased transportation costs and lower heating values, because it is more cost-effective than trying to meet air pollution standards while burning higher-sulfur coals from the eastern and Midwestern United States.

There are several methods of controlling carbon dioxide and sulfur oxide emissions. Some of the most industrially useful methods are (Cavallaro et al., 1991a)

- Physical removal of pyritic sulfur from coal before combustion.
- Use of fluidized-bed combustion, with limestone ($CaCO_3$) added directly to the combustor to absorb sulfur oxides during combustion.
- Use of flue-gas scrubbers to remove sulfur oxides from the plant gas emissions after combustion.
- Use of flue-gas scrubbers to remove carbon dioxide from the plant gas emissions after combustion.

- Conversion of the coal to a low-sulfur clean fuel by gasification or liquefaction, possibly while simultaneously capturing carbon dioxide.
- Fuel switching, and blending of high-sulfur coals with low-sulfur coals.

These methods are not typically terribly expensive. Carbon capture and sulfur technologies are both well-developed technologies and can be implemented quite effectively on most plants. However, especially carbon dioxide, the equipment required to perform these separations has a considerable footprint on the plant and a nonnegligible energy cost. These energy costs in particular have slowed industry investment in carbon capture technologies, as the primary use case for carbon dioxide captured this way has an energy cost on the order of 25% of a typical coal-fired power facility's entire output (Rochelle, 2009). However, in reality, the capture aspect of carbon capture is a very well-developed, very efficient technology – the majority of the energy cost really is in the subsequent storage, use, and transportation of the carbon dioxide.

In addition, there is a great opportunity for new power plants, whether ultimately coal-fired or via other processes. For example, integrated gasification cycles require a complete reconsideration of the materials used throughout the entire plant but make up for this additional cost in large efficiency gains and easy sulfur and carbon dioxide separation. We are approaching a point where there is only so much technology that can be retrofitted onto existing plants, whether due to size and space constraints or due to material decisions made long before these technologies were even conceived. Thus, new facilities are in a prime position to take full advantage of every aspect of these technologies.

1.1 NEED FOR CARBON DIOXIDE CAPTURE

Carbon dioxide is a greenhouse gas that is emitted when coal is burned. From simple stoichiometry, it can be shown that for every kilogram of carbon burned, 3.67 kg of carbon dioxide is formed. As carbon is a key component of coal, burning coal necessarily leads to the formation of carbon dioxide. This is unavoidable.

For this reason, the emission of carbon dioxide is under consideration for regulation in many countries. While the formation of carbon dioxide while burning coal is unavoidable, the release of it is not. As such, to avoid introducing excess carbon dioxide into the atmosphere, technology has been developed to capture it. Climate change is at least an issue that is receiving regulatory attention, but the theory and consequences predicted by it are also supported by numerous independent scientists.

Coal, however, is a readily available, easy to use, and considerably dense energy source. It is one of the most suitable energy sources for rapid application in developing countries. If we wish to use coal responsibly however, we must consider its impact on the environment. For the sake of having an energy infrastructure at all, along with a functional economy, the threat posed by uncontrolled climate change cannot be swept under the rug. Carbon dioxide capture is a key step to allowing access to the energy resource that coal represents while simultaneously mitigating the risks it poses. That is the need for carbon dioxide capture.

TABLE 1.2
Typical Sulfur Value Ranges for Coals from Various Countries (Monticello and Finnerty, 1985)

Region	Range of Total Sulfur Values, %
Belgium	0.5–4.5
France	0.8–1.4
Great Britain	1.0–3.6
Germany (Western)	1.3–1.5
India (Assam province)	6.0–8.0
The Netherlands	1.0–3.0
U.S. (Eastern)	0.2–7.0
U.S. (Western)	0.2–1.0

1.2 NEED FOR SULFUR REMOVAL

Sulfur levels in coal are highly variable worldwide, as shown in Table 1.2. The sulfur contents of coals range from approximately 0.5% to greater than 11% (Monticello and Finnerty, 1985). While some coals can be sold directly from the mine, others require a good deal of cleaning before they can meet environmental requirements. There are several advantages to removing the sulfur from coal before burning, such as

- being able to sell the coal to customers regardless of whether they have installed pollution control measures;
- reducing the load on flue-gas scrubbers (and hence reducing the amount of scrubber sludge which must be landfilled); and,
- for some processes, being able to remove much of the ash while removing the sulfur.

To use any desulfurization process effectively, it is first necessary to understand the forms that sulfur can take in coal, determine how these forms are affected by various processes, and consider how the variations in coal properties influence the processes that can be used to treat it effectively.

1.3 FORMS OF SULFUR IN COAL

The removal of sulfur from coal is greatly complicated by the complex chemistry of sulfur. It can occur as oxidized sulfates, as inorganic sulfides, as a wide range of organic compounds, and in some cases even as elemental sulfur and polysulfides. Because sulfur can form all of these compounds with nearly equal ease, it is not uncommon for most of the possible forms to be present in a single coal. Each type of compound presents its own particular separation difficulties. The two most important forms are pyritic and organic sulfur, and the amounts and relative proportions of these two forms for average U.S. coals are given in Table 1.3 (Cavallaro et al., 1976). Normally, the pyritic sulfur is the dominant form, but it is not uncommon for

TABLE 1.3
Average Amounts of Various Sulfur Forms in U.S. Coals, by Region (Cavallaro et al., 1976)

Region	% Pyritic Sulfur	% Total Sulfur	% Organic Sulfur
Northern Appalachian	2.01	3.01	1.00
Southern Appalachian	0.37	1.04	0.67
Alabama	0.69	1.33	0.64
Eastern Midwest	2.29	3.92	1.63
Western Midwest	3.58	5.25	1.67
Western	—	0.68	—
Overall U.S.	1.91	3.02	1.11

the organic sulfur to be dominant. For example, some Yugoslavian coals contain more than 11% sulfur, with most of it present as organic compounds (Monticello and Finnerty, 1985).

Sulfur-bearing minerals, despite their low proportions in coal compared to other mineral matter such as clay minerals, are much more important because of the environmental impact of sulfur emissions. Mineral matter in coal has frequently been classified into two groups: (1) inherent mineral matter and (2) extraneous mineral matter. The inherent mineral matter is defined as the minerals that have their origin in the organic constituents of the plants from which the coal was derived. It is generally present in very small amount, usually not more than 2% and sometimes less than 1%. It does not exist as a separate entity in the coal, but is so finely and intimately mixed with the coal substance that it may be considered a structural part of the coal. Because of the small amounts of inherent mineral matter present in coals, very little of the sulfur-bearing minerals are from this source. The majority of the sulfur-bearing minerals are extraneous matter, which arise as a result of contamination of the coal by outside sources.

1.3.1 Sources of Sulfur in Coal

The wide variations in the sulfur contents of coal are a result of differences in their deposition environments. The single largest factor is whether the original peat bed that became the coal seam was in contact with seawater when it formed. Seawater is rich in dissolved sulfate, mostly $MgSO_4$. This sulfate is converted by bacteria to hydrogen sulfide, as shown in Figure 1.1. Depending on the conditions in the peat during its deposition, this hydrogen sulfide then could react in a number of different ways to produce the various sulfur forms found in coals. If the peat formed in a freshwater environment, then there was no source of external sulfur, and the only sulfur is what was originally present in the plants that formed the peat. Coals that were in intermittent contact with seawater during their formation will obviously have intermediate sulfur levels, and can vary a great deal in sulfur content over quite short distances. Since it is common for one part of a coal seam to have formed in

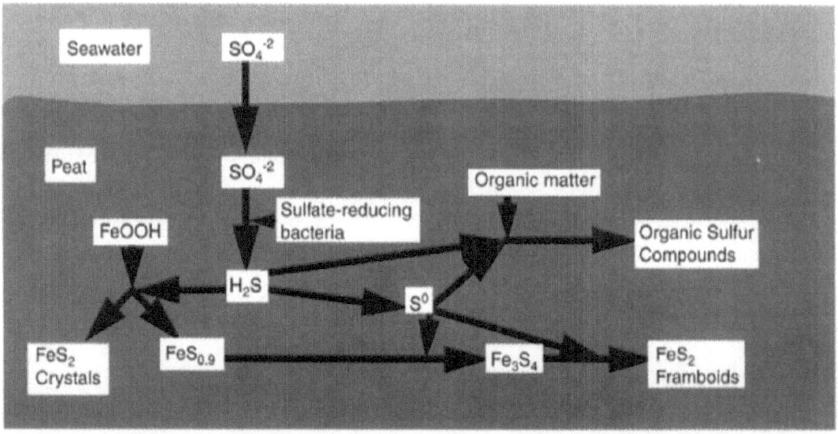

FIGURE 1.1 Formation of high-sulfur coals. In marine peats, the dissolved sulfate in the seawater diffuses into the peat and is converted to hydrogen sulfide by the action of bacteria. This hydrogen sulfide can then react in numerous ways to form pyrite (FeS_2) and a number of sulfur-bearing organic compounds. The relative proportions of these compounds are controlled by the precise chemistry of the peat. When the peat becomes converted to coal, it then has a high sulfur content, with the sulfur present as either pyritic or organic compounds. Other forms of sulfur, such as sulfates and elemental sulfur, form as weathering products when the coal is exposed to air. Freshwater peats contain little sulfur other than what was originally present in the plants that formed them, and so produce low-sulfur coal (Chou, 1990).

freshwater while another part formed in seawater, the variation in sulfur content in a given coal seam can be considerable.

1.3.2 Pyritic Sulfur

There are many known compounds of iron and sulfur, as shown in Table 1.4. The forms that are generally accepted as occurring in coals in significant quantities are pyrite and marcasite, which together are frequently referred to as "pyritic sulfur." Pyritic sulfur occurs in coal as discrete inclusions of iron disulfide and generally accounts for approximately 30%–70% of the total sulfur in U.S. coals. Pyrite is the most commonly reported sulfide mineral and is ubiquitous in coal, although marcasite has often been found in lesser amounts (Gray et al., 1963). Pyrite and marcasite both have the chemical composition FeS_2, but they are dimorphs, which are minerals that are identical in their chemical composition, but different in crystalline form. Pyrite has a cubic crystal structure, while marcasite crystals are orthorhombic. The two can be conveniently distinguished by X-ray diffraction methods. It is believed that marcasite forms when the original coal deposition environment was acidic (Rimmer and Davis, 1986).

The physical characteristics of pyrite present in coal are given in Appendix 1 (Hucko and Gala, 1990). From this table, it can be seen that pyrite inclusions can take many forms and can have particle sizes ranging from 0.1 to 1,000 μm. These inclusions can in principle be broken free of the surrounding coal and physically

TABLE 1.4
Known Forms of Iron Sulfide Which Can Potentially Appear in Coal (Rossi, 1993)

Chemical Formula	S/Fe Atom Ratio	Fe, wt%	Name
F_eS	1.00	63.5	Troilite
$Fe_{11}S_{12}$	1.09	61.5	
$Fe_{10}S_{11}$	1.10	61.3	Hexagonal pyrrhotites
Fe_9S_{10}	1.11	61.0	
Fe_7S_8	1.14	60.4	Monoclinic pyrrhotite
Fe_9S_{11}	1.22	58.8	Smythite
Fe_3S_4	1.33	56.6	Gregite
Fe_2S_3	1.50	53.7	$-\gamma$ FeS
FeS_2	2.00	46.5	Pyrite, marcasite

separated. The high density of these minerals makes the separation of coal and pyrite particles straightforward, provided that they are coarser than about 500 μm. Pyrite and marcasite can also be oxidized to soluble sulfates by a number of oxidizing agents, with marcasite being much more reactive than pyrite (Hurlbut and Klein, 1977), and so these minerals are also very amenable to oxidative leaching. The morphology of individual pyrite particles varies widely depending on how they were deposited and on the characteristics of the surrounding coal. They range from massive nodules ("sulfur balls") to minuscule inclusions such as submicron single crystals, framboids (clusters of submicron crystals, resembling a raspberry when seen with a microscope), and replacements of plant tissue. As a result, the surface chemistry, density, and degree of locking to coal vary considerably, which complicates the removal of the smallest pyritic sulfur particles. Because of these factors, coal pyrite does not behave in the same manner as pyrite from hard-rock mineral deposits; therefore, processes that work well for separating pure pyrite will often not work for coal pyrite.

1.3.3 Organic Sulfur

The organic sulfur in coal generally accounts for approximately 30%–70% of the total sulfur in U.S. coals and is composed of a variety of compounds where sulfur atoms are directly bonded to carbon, which in turn are part of the coal structure. Physical separation of organic sulfur is impractical, since the organic sulfur is approximately uniformly distributed throughout the coal, and so there are no high-organic-sulfur inclusions to break free and remove. A number of chemical processes that can remove organic sulfur have been developed, and some with great effectiveness, although reagent costs are a serious consideration.

Organic sulfur can occur in coal in any of the four basic types of structure: thiols, sulfides, disulfides, and thiophenes (Morrison, 1981). Examples of these structures are as follows, where R and R′ are unspecified aliphatic hydrocarbons.

1.3.3.1 Thiols

Thiols are the sulfur-based analogs of alcohols and include both aliphatic compounds (mercaptans) and aromatic compounds (thiophenols):

1.3.3.2 Sulfides

The sulfide form of organic sulfur includes the thioethers:

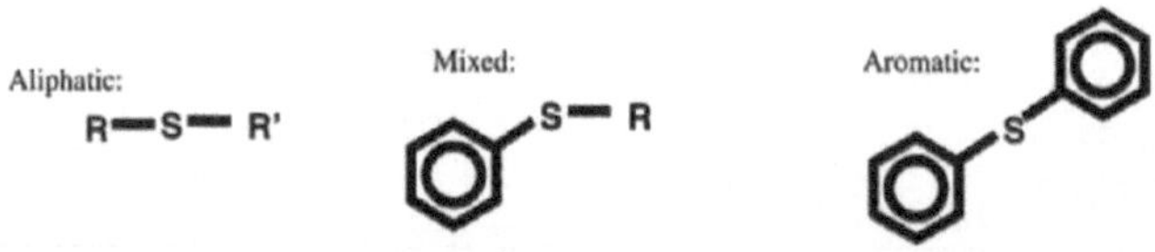

1.3.3.3 Disulfides

Disulfides have a structure superficially similar to sulfides, except that they contain two sulfur atoms instead of one:

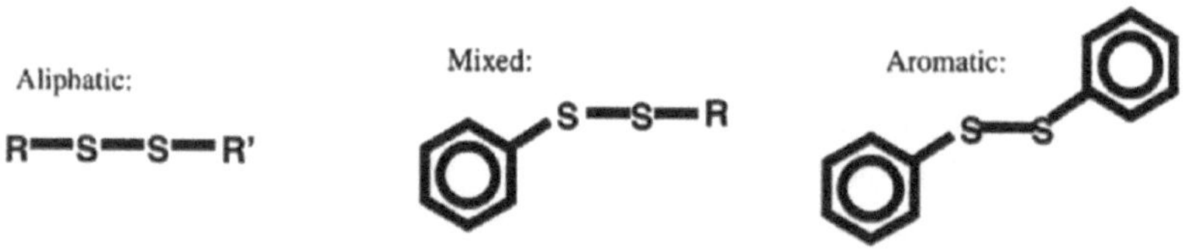

1.3.3.4 Thiophenes

These are compounds that include the thiophene ring, such as dibenzothiophene:

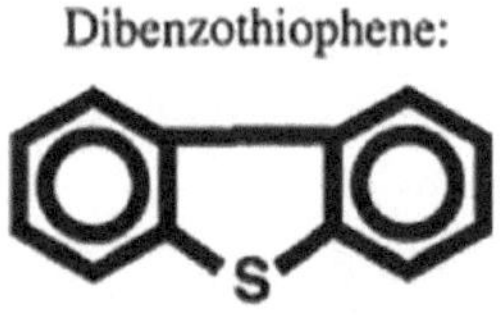

1.3.3.5 Occurrence of the Forms of Organic Sulfur

There is evidence to suggest that lignites and high-volatile bituminous coals have a higher content of thiols than is seen in low-volatile coals. It has also been reported that as the rank of the coal increases, the proportion of the organic sulfur that is in the highly stable thiophenic form also increases (Morrison, 1981).

1.3.4 Sulfate Sulfur

In pristine, unoxidized coal and sulfate salts such as gypsum are present in small quantities (typically less than 0.01%), and so are not a major source of sulfur. Oxidized coals, however, frequently contain a good deal of iron sulfates, which form by oxidation of pyrite and marcasite. Marcasite, which is particularly reactive, readily oxidizes in air, while pyrite oxidizes more slowly unless it is in an acidic environment and certain iron-oxidizing bacteria are present. Since the sulfates are mostly soluble in water, they are almost completely removed by most coal-washing operations and are not considered to be a serious source of sulfur in the clean coal.

1.3.5 Elemental Sulfur

When coal oxidizes, the pyritic sulfur sometimes forms elemental sulfur and sulfur-rich polymers on its surface. This forms most readily during oxidation of marcasite (Hurlbut and Klein, 1977), but can also be formed by partial oxidation of pyrite (Stock and Wolny, 1990), apparently by the action of iron-oxidizing bacteria (Chou, 1990). Elemental sulfur is not found in freshly mined coals, but only in coals that have oxidized to some extent. It is an intermediate stage in the oxidation to iron sulfate, and as a result the absolute level of elemental sulfur is quite low. The significance is that the elemental sulfur forms on the surface of the pyrite and marcasite and has a powerful effect on the surface chemistries of these minerals. Like coal, elemental sulfur is a naturally hydrophobic compound, and it therefore makes the surfaces of pyrite and marcasite to more closely resemble the surface of coal. Since the most effective processes for physically separating very fine particles are based on surface chemistry differences, elemental sulfur can be a serious problem for physical separation of very fine pyritic sulfur from coal.

1.4 PROPERTIES OF COAL

Since coal is not a tightly defined substance, the properties of coals can vary considerably depending on how they were deposited, what types of organic matter were present in the original bog, and the extent of alteration since deposition (rank). The largest variation is in the surface chemistry, with coals ranging from strongly hydrophobic to hydrophilic depending on rank, volatiles content, and degree of surface oxidation. There is also variation in the density, with the specific gravities of various coal components ranging from 1.2 to 1.6 g/cm^3 (1,200 to 1,600 kg/m^3) in bituminous coals (Dyrcacz et al., 1984). For extremely high-rank coals, the density can approach that of graphite, 2.6 g/cm^3 (2,600 kg/m^3). The electrical conductivity of coal can also vary considerably. This variability in physical and chemical properties causes problems for physical separation processes, since they depend on differences in some physical property to separate coal from its contaminants. The density variation is not a great problem for sulfur removal, since the pyrite inclusions are considerably denser than even the densest coals, but the surface chemistry variation is much more troublesome. Separation of very fine particles is most effectively carried out by surface-chemistry-based processes, yet many coals (particularly low-rank and oxidized coals) are not treatable by these methods, because their properties are insufficiently different from those of the contaminant minerals, particularly pyrite.

1.5 LIBERATION

To use a physical separation to remove sulfur from coal, the sulfur must be in a form that can be broken free of the coal (liberated). This generally limits physical separations to removing sulfate sulfur and pyritic sulfur, because organic sulfur is spread throughout the structure of the coal and cannot be liberated by crushing or grinding the coal.

To liberate the pyrite so that it can be separated, the coal must be crushed to a size that is smaller than the particle size of the pyrite inclusions. This is simple if the pyrite is present as reasonably coarse particles, but separations become very difficult when particles are smaller than approximately 10 μm. Such fine particles are commonly referred to in the coal industry as "slimes," and they cannot be treated by conventional coal-cleaning technologies. The degree of liberation that is practical to achieve is therefore limited by the ability of the separation process in use to handle the ultrafine particles.

Even when the coal is crushed to a size finer than the pyrite inclusions, there will still be a substantial number of particles that are partly coal and partly pyrite. These "locked particles" can be a significant source of sulfur in the clean coal, because the attached coal can make them behave as if they are coal particles. Particular care is therefore needed to avoid recovering locked particles.

1.6 ADVANCED COAL CLEANING

The technologies commonly referred to as advanced coal cleaning methods are those that are capable of producing superclean or ultraclean coals. Superclean coals are defined as those containing less than 3% ash by weight, and ultraclean coals are defined as containing less than 1% ash. Advanced coal cleaning methods are primarily associated with size reduction of the coal to below 100 μm, and sometimes to as little as 5 or 10 μm.

The differences in terminology used by various researchers and manufacturers have led to some confusion about what exactly is meant by "advanced methods" (Couch, 1991). In this book, the term "advanced coal cleaning methods" is used to describe any of the following three types of separation processes:

- Physical separations carried out on material where the particle size has been reduced until the mineral matter is almost completely liberated from the coal (Couch, 1991);
- Chemical or biological methods; and
- Modifications to conventional coal-cleaning technologies that improve their effectiveness.

1.7 RELATIONSHIPS FOR CALCULATING SULFUR REMOVAL PERFORMANCE

The following standard definitions and formulas are used for quantifying the separation of impurities such as sulfur from coal (Kaiser Engineers, 1989).

1.7.1 Yield

The yield is the proportion of the total raw coal weight that reports to the clean-coal product. Yield is used only for total coal weight, and the term "recovery" (defined below) is used for the proportion of specific components that report to the clean coal.

1.7.2 Combustibles

The combustibles value is the weight proportion of the coal that burns away when the material is ashed using ASTM International's D3174 procedure. Combustibles are calculated as follows:

$$\text{Combustibles} = 100 - \%\ \text{Ash} \tag{1.3}$$

where the ash is calculated on a moisture-free basis.

1.7.3 Sulfur Reduction

Sulfur reduction is the percentage change in the amount of sulfur in the coal when the coal is cleaned. It is calculated using the following equation:

$$\%\ \text{Sulfur Reduction} = \frac{(\text{Raw Coal S content}) - \text{Clean Coal S content}}{(\text{Raw Coal S content})} \times 100 \tag{1.4}$$

where the sulfur content is expressed in pounds of SO_2 per million BTU or grams of SO_2 per megajoule. So, for example, if the coal originally contained 4.0 lb SO_2/million BTU (1.72 g SO_2/MJ), and a cleaned product produced from it contained only 0.5 lb SO_2/million BTU (0.22 g SO_2/MJ), the % sulfur reduction would be 87.5%. The sulfur reduction value does not take into account the actual clean coal yield, and is of somewhat limited usefulness.

1.7.4 Sulfur Removal

Sulfur removal is the weight percentage of the sulfur that was originally in the coal, which is removed by the coal-cleaning process. This is calculated as follows:

$$\%\ \text{Sulfur Removal} = \frac{(100 - \%\ \text{Yield}) \times (\text{Refuse}\ \%\ \text{S})}{100 \times (\text{Raw Coal}\ \%\ \text{S})} \times 100 \tag{1.5}$$

The % sulfur is expressed as weight percent, and the yield is the proportion of the total weight of raw coal that is extracted as clean coal.

1.7.5 Desirables Recovery

Desirables recovery is the proportion of the valuable portion of the coal (combustible material or heating value) that is recovered in the clean product. It is calculated from the formula:

$$\%\ \text{Recovery} = \frac{\%\ \text{Yield} \times (\text{Clean-Coal Value})}{100 \times (\text{Raw-Coal Value})} \times 100 \tag{1.6}$$

The raw-coal and clean-coal values are typically either weight % combustibles or energy units such as BTU/lb or MJ/kg.

1.7.6 Separation Efficiency

It is often helpful to have a single number that can be used for comparing various separations. The separation efficiency can be used for this purpose. The separation efficiency can be defined in any of the following ways, depending on what impurity is being removed:

$$\text{Separation Efficiency} = \%\ \text{Desirables Recovery} - (100 - \%\ \text{Ash Removal}) \quad (1.7)$$

$$\text{Separation Efficiency} = \frac{\%\ \text{Desirables Recovery}}{-(100 - \%\ \text{Total Sulfur Removal})} \quad (1.8)$$

$$\text{Separation Efficiency} = \frac{\%\ \text{Desirables Recovery}}{-(100 - \%\ \text{Pyritic Sulfur Removal})} \quad (1.9)$$

The larger the separation efficiency number is, the better the separation performance. For a perfect separation, the value of the separation efficiency will be 100%.

1.8 SUMMARY

The sulfur content of coal and the carbon dioxide released by burning coal are the greatest obstacles to environmentally responsible coal utilization. While the sulfur-bearing minerals and compounds in coal make up a much lower proportion of the coal than the ash-forming minerals, they are an important problem because of the environmental impact of sulfur oxide emissions. Most coals need some degree of sulfur removal to meet the steadily tightening clean-air standards. However, carbon dioxide is an intrinsic result of burning coal, and preventing its release into the atmosphere requires capturing it.

Depending on the deposition environment of the coal, the sulfur content can vary greatly from place to place. Removing the sulfur from coal is complicated by the numerous forms that it can take, with most of the sulfur being in the form of pyrite or sulfur-bearing organic compounds. Since each form differs in its physical properties and its degree of dispersion in the coal, different methods will be needed to remove each form most effectively. Coal does not have a fixed composition and varies a good deal in its properties. Therefore, processes that can easily remove sulfur from one coal may work poorly or not at all for another.

REFERENCES

BGR (2017), "BGR Energy Study: Data and Developments Concerning German and Global Energy Suppliers," *BGR*, pp. 140–141, 148–149.

Cavallaro, J.A., Deurbrouck, A.W., Killmeyer, R.P., Fuchs, W. and Jacobsen, P.S. (1991a), "Sulfur and Ash Reduction Potential and Selected Chemical and Physical Properties of United States Coals," U.S. Department of Energy, Pittsburgh Energy Technology Center, Report No. DOE/PETC/TR-91/1, February 1991.

Cavallaro, J.A., Deurbrouck, A.W., Killmeyer, R.P., Fuchs, W. and Jacobsen, P.S. (1991b), "Sulfur and Ash Reduction Potential and Selected Chemical and Physical Properties of United States Coals," U.S. Department of Energy, Pittsburgh Energy Technology Center, Report No. DOE/PETC/TR-91/2, June 1991.

Cavallaro, J.A., Johnston, M.T. and Deurbrouck, A.W. (1976), "Sulfur Reduction Potential of U.S. Coals: A Revised Report of Investigations," EPA-600/2-76-091, Bureau of Mines RI 8118.

Chou, C.L. (1990), "Geochemistry of Sulfur in Coal," *Geochemistry of Sulfur in Fossil Fuels*, ACS Symposium Series No. 429, American Chemical Society, Washington, DC, pp. 30–52.

Couch, G.R. (1991), "Advanced Coal Cleaning Technology," IEA Coal Research, Report No. IEACR/44, pp. 1–96.

Couch, G.R. (1995), "Power from Coal—Where to Remove Impurities," IEA Coal Research, Report No. IEACR/82g, pp. 1–87.

Crippa, M., Guizzardi, D., Muntean, M., Schaaf, E., Dentener, F., van Aardenne, J. A., Monni, S., Doering, U., Olivier, J. G. J., Pagliari, V. and Janssens-Maenhout, G. (2018), "Gridded Emissions of Air Pollutants for the Period 1970–2012 within EDGAR v4.3.2," *Earth System Science Data*, Vol. 10, 1987–2013, doi: 10.5194/essd-10-1987-2018. Accessed at http://edgar.jrc.ec.europa.eu/overview.php?v=432 (doi: 10.2904/JRC_DATASET_EDGAR).

Dyrcacz, G.R., Bloomquist, C.A.A., Ruscic, L. and Horwitz, E.P. (1984), "Variations in Properties of Coal Macerals Elucidated by Density Gradient Separation," *Chemistry and Characterization of Coal Macerals*, ACS Symposium Series No. 252 (Winans and Crelling, eds.), American Chemical Society, Washington, DC, pp. 65–78.

Gray, R.J., Schapiro, N. and Coe, G.D. (1963), "Distribution and Forms of Sulfur in a High Volatile Pittsburgh Seam Coal," *Transactions of the Society of Mining Engineers*, Vol. 226, pp. 113–121.

Hoddinot, P.J. (1992), *Coal*, Metals and Minerals Annual Review, Mining Journal Ltd, London, pp. 118–128.

Hoddinot, P.J. (1994), *Coal*, Metals and Minerals Annual Review, Mining Journal Ltd, London, pp. 105–114.

Hucko, R.E. and Gala, H. Be (1990), "Surface Science of Coal Preparation," U.S. Department of Energy, Pittsburgh Energy Technology Center, DOE/PETC/TR-90/2 (DE90007161).

Hurlbut, C.S. and Klein, C. (1977), *Manual of Mineralogy* (*after J.D. Dana*), 19th edition, John Wiley & Sons, New York.

IEA (2018). "World Energy Balances, 2018," *International Energy Agency.* Accessed at www.iea.org/statistics/?country=WORLD&year=2016&category=Coal&indicator=CoalImportsExports&mode=table&dataTable=BALANCES#.

Janssens-Maenhout, G., Crippa, M., Guizzardi, D., Muntean, M., Schaaf, E., Olivier, J.G.J., Peters, J.A.H.W. and Schure, K.M. (2017). Fossil CO_2 and GHG Emissions of All World Countries, EUR 28766 EN, Publications Office of the European Union, Luxembourg, ISBN 978-92-79-73207-2, doi: 10.2760/709792, JRC107877. Accessed at http://edgar.jrc.ec.europa.eu/overview.php?v=CO2andGHG1970-2016

Kaiser Engineers (1989), Formula Definitions for Project, "Engineering Development of Advanced Physical Fine Coal Cleaning Technologies—Froth Flotation," U.S. DOE-PETC, DE-AC-88PC88881.

Monticello, D.J. and Finnerty, W.R. (1985), "Microbial Desulfurization of Fossil Fuels," *Annual Review of Microbiology*, Vol. 39, pp. 371–389.

Morrison, G.F. (1981), "Chemical Desulfurization of Coal," IEA Coal Research, Report No. ICTIS/TR15, pp. 1–72.

Rimmer, S.M. and Davis, A. (1986), "Geologic Controls on the Inorganic Composition of Lower Kittanning Coal," *Mineral Matter and Ash in Coal*, ACS Symposium Series No. 301 (Vorres, ed.), American Chemical Society, Washington, DC, pp. 41–52.

Rochelle, G.T. (2009). "Amine Scrubbing for CO_2 Capture," *Science*, Vol. 325, pp. 1652–1654.

Rossi, G. (1993), "Biodepyritization of Coal—Achievements and Problems," *Fuel*, Vol. 72, No. 12, pp. 1581–1592.

Stock, L.M. and Wolny, R. (1990), "Elemental Sulfur in Bituminous Coals," *Geochemistry of Sulfur in Fossil Fuels*, ACS Symposium Series No. 429 (Orr and White, eds.), American Chemical Society, Washington, DC, pp. 241–248.

2 Analytical Methods

The design of any emissions control process, whether sulfur or carbon dioxide, for a particular coal requires the knowledge of how much sulfur and carbon is in the coal. In particular, it is also necessary to understand what specific forms of sulfur are present (pyritic, elemental, or organic). Sulfur analysis is also needed to monitor the quality of the coal being produced, and the efficiency of the processing can be greatly improved if real-time, on-line sulfur analysis is used for control of the process. Several analytical techniques are available or under development, which can be used effectively for these purposes.

It must be remembered that analytical results are only as good as the quality of the sample. Care must always be taken to collect samples that are representative of the entire lot, using correct sampling practices (Pitard, 1993, 2019). Once a representative sample has been collected, care must be taken to keep the sample well mixed to avoid segregation effects and to ensure that the coarsest particles are small enough that random variations in the composition of analysis samples can be made smaller than the random errors in the analysis itself.

In routine analytical work, it is important to have a standard reference material (SRM) that has a composition similar to the material being studied, and which has been thoroughly analyzed to a high degree of accuracy. Such SRMs are available from standards organizations, such as the National Institute of Standards and Technology (NIST), Washington, DC (USA). These standards are crucial for calibration and periodic checks of most analytical techniques, and particularly for the various types of instrumental analysis.

Analysis for sulfur is not included in the Proximate Analysis procedure typically used for routine analyses of coal, which includes moisture, volatile matter, fixed carbon, and ash. The standard Ultimate Analysis procedure does include sulfur analysis, and consists of determinations of carbon, hydrogen, nitrogen, total sulfur, and oxygen.

2.1 CHEMICAL METHODS

These are techniques that are normally used by analytical laboratories. The most important methods are those that have been approved by the American Society for Testing and Materials (ASTM), which are generally accepted and used by both the producers and users of coal. It should be noted that some recent studies have suggested that the ASTM methods for determining forms of sulfur may not be fully selective for the forms that they purport to measure (Bottrell et al., 1994). As a result, if any new techniques are developed for determining sulfur forms, they may not fully agree with results obtained by the ASTM methods.

2.1.1 ASTM Standard Methods[1]

The ASTM standard methods for determining sulfur in coal are D4239 for measuring total sulfur (ASTM, 2018a) and D2492 and D8214 for determining the quantities of sulfur present as sulfates, pyrite, and organic sulfur (ASTM, 2012a, 2018b). In these methods, the total sulfur and sulfate sulfur are determined directly, while the pyritic sulfur is determined by measuring the quantity of pyritic iron present. Organic sulfur is determined by subtracting the quantity of sulfate sulfur and pyritic sulfur from the total sulfur. These standard methods are fully detailed by ASTM, and copies of the standards should be obtained before attempting the procedures.

2.1.1.1 Total Sulfur

There are two techniques that have been approved by ASTM for measurement of the total sulfur content. These two techniques are high-temperature combustion methods as described in D4239. The high-temperature combustion methods are rapid, requiring only a few minutes for completion, but depend on specific equipment and the availability of reference samples.

2.1.1.1.1 High-Temperature Combustion Methods (ASTM D4239)

In each of these methods, the coal is first burned in a tube furnace either at 1,150°C in the presence of tungsten (vi) oxide (WO_3) or at 1,350°C or higher in a stream of oxygen, to convert the sulfur completely into gaseous oxides. The quantity of these oxides in the gaseous combustion products is then determined by infrared absorption as follows:

The gas stream is filtered and dehydrated with anhydrous magnesium perchlorate. The dry gas then passes through a cell which measures the infrared absorption at the frequency which is most strongly absorbed by sulfur dioxide.

High-temperature combustion methods are used by the various commercially available automated sulfur analyzers. They are very rapid and do not require a high level of operator skill to use; however, they are empirical and require careful calibration with an SRM before the results are meaningful. The SRMs used are coal samples that have had their sulfur content analyzed precisely by other means, such as the Eschka method. 2,5-Di(5-tertbutylbenzoxazol-2-yl)thiophene (BBOT) can also be used as a certified sulfur reference with a sulfur content of 7.47%. In recent years, research has also been carried out to produce SRMs using ion chromatography and isotope dilution thermal ionization mass spectrometry. Both methods can produce standards that have the sulfur determined to a high precision (Kelly et al., 1994; Thomas, 1995).

2.1.1.2 Sulfate Sulfur (ASTM D2492, D8214)

Sulfate sulfur is mainly from oxidation of pyrite and consists of iron sulfates, which are soluble in hydrochloric acid. In Standard D2492, the sulfate sulfur is first dissolved from the weighed analysis sample of coal by boiling for 30 min in an open

[1] Current copies of individual standards can be obtained directly from ASTM, 1916 Race Street, Philadelphia, PA, 19103-1187.

beaker with 4.8 M hydrochloric acid. In addition to the sulfate, this dissolves the nonpyritic iron, certain alkaline ash-forming minerals, and in some cases a portion of the coal. The solution of sulfate is filtered to remove the solids, and the solids are saved for use in the pyritic sulfur determination (Section 2.1.1.3). The sulfate is then precipitated from the solution using barium chloride. The resulting barium sulfate is filtered, ashed, and weighed in order to determine the quantity of sulfate sulfur. Alternatively, the dissolved sulfate in hydrochloric acid solution can be mixed with a noninterfering standard and analyzed with inductively coupled plasma atomic emission spectroscopy (ICP-AES), as in D8214 (ASTM, 2018b).

Since iron oxides, hydroxides, and sulfates are soluble in hydrochloric acid, the determination of sulfate sulfur removes the nonpyritic iron from the coal. Since the nonpyritic iron would interfere with the determination of pyritic sulfur, this analysis prepares the coal sample for the pyrite analysis.

2.1.1.3 Pyritic Sulfur (ASTM D2492, D8214)

Determination of pyritic sulfur is based on the fact that pyrite does not dissolve in hydrochloric acid, but does dissolve in nitric acid. In the sulfate sulfur analysis (Section 2.1.1.2), the sulfate sulfur and the nonpyritic iron is dissolved by HCl leaching, and so the only iron remaining in the coal is the iron contained in pyrite. To determine the quantity of pyrite, the solids remaining after the sulfate sulfur determination are leached with nitric acid solution either by boiling for 30 min or by standing overnight at room temperature to dissolve the pyrite. In addition to pyrite, the nitric acid dissolves some of the organic material from the coal and can also dissolve the more loosely bonded organic sulfur, such as thiols. The solids are then removed by filtration, leaving a solution that contains the iron that was present in the pyrite. The solution is then analyzed by atomic absorption spectroscopy to determine the quantity of pyritic iron. The amount of pyritic sulfur originally present is calculated from the quantity of iron by assuming that the pyrite was present as stoichiometric FeS_2. The amount of sulfur can also be determined with ICP-AES spectroscopy after mixture with a noninterfering standard, as in D8214 (ASTM, 2018b).

An alternative method given by ASTM D2492, which does not require the use of nitric acid, is to incinerate the solids remaining after sulfate extraction by slowly heating to 700°C–750°C in a muffle furnace, ashing the coal without allowing it to burst into flame. While the sulfur from the pyrite will be lost by combustion, the pyritic iron will remain in the ash as iron oxides, which are soluble in hydrochloric acid and can be used to calculate the quantity of pyrite that was originally present. The ash is digested in a hot HCl solution to dissolve the iron, filtered, and the quantity of iron is determined by atomic absorption spectroscopy.

The quantity of sulfur from the pyrite is not measured directly because the procedures for dissolving pyrite also dissolve a variable amount of the organic sulfur. Instead, the quantity of iron from the pyrite is measured, and the content of pyritic sulfur is calculated from the stoichiometric ratio of sulfur to iron in pyrite (FeS_2, S/Fe ratio = 1.148). The analysis assumes that there are negligible amounts of iron monosulfides and sulfide minerals other than pyrite. This assumption is generally valid for most coals, but is not valid for other carbonaceous materials, such as coke (ASTM, 2012a).

A potential source of error in this analysis is incomplete dissolution of submicron pyrite particles, even though the coal is ground to finer than 250 μm before analysis. The ultrafine pyrite particles can be completely enclosed by coal so that they are not in contact with the leaching solutions and so never dissolve in the nitric acid. When this happens, the sulfur in the ultrafine pyrite is recorded as being part of the organic sulfur. This effect can be minimized by grinding the coal to even finer sizes before analysis, but the problem is difficult to eliminate completely. It has been recommended that samples for pyrite determination be pulverized to finer than 45 μm to give satisfactory results (Stiller et al., 1990).

2.1.1.4 Organic Sulfur

In the ASTM procedure D2492, organic sulfur is not measured directly. Instead, it is calculated from the values for the other forms of sulfur, as follows:

$$\%\ \text{Organic Sulfur} = \%\ \text{Total Sulfur} - (\%\ \text{Sulfate Sulfur} + \%\ \text{Pyritic Sulfur}) \quad (2.1)$$

As a result, the analytical errors for each of the earlier analyses all accumulate in the organic sulfur value, making it the least accurate determination.

An alternate method for determining sulfate, sulfide, pyritic, and organic sulfur in a single determination by stepwise selective oxidation has been developed, which avoids to some extent the difference problem mentioned above (McGowan and Markuszewski, 1987, 1988; Ailey-Trent et al., 1993). This scheme sequentially removes sulfate, pyritic sulfur, and organic sulfur by boiling the coal in perchloric acid ($HClO_4$) at 120°C, 155°C, and 205°C, respectively, and measuring the quantity of sulfate produced at each stage by turbidimetric measurement. The process is unfortunately complex, and the use of perchloric acid produces a potential hazard (Calkins, 1994).

Several other researchers have also attempted to develop methods that would reduce the errors by measuring the organic sulfur content directly (Riley et al., 1990), but none have yet been approved by ASTM. One such method (Riley et al., 1990) was to first combine 6 g of coal with 120 mL of 2M nitric acid (HNO_3) solution and boil for 30 min to completely dissolve the sulfates and pyrite, leaving only organic sulfur in the coal. The solids were then filtered and washed with hot deionized water and dried at 110°C under nitrogen for 3 h. The quantities of moisture, ash, and total sulfur remaining in the dried, extracted coal were then determined, and the sulfur content was reported on a dry, mineral-matter-free basis. It was assumed that the nitric acid leach removed all of the inorganic sulfur, leaving the organic sulfur in the extracted coal. The organic sulfur values determined by this technique were generally lower than those obtained with the ASTM procedure, but the precision of the results for the direct-determination method was reported to be much better. It should be noted that certain forms of organic sulfur are soluble in nitric acid and so will not be included with the organic sulfur that is determined by this method. This partial dissolution of certain forms of organic sulfur accounts for the method giving lower organic sulfur values than the ASTM procedure. A critical analysis of methods for direct measurement of organic sulfur in coal is given by Davidson (1994).

2.1.2 Elemental Sulfur

Elemental sulfur is not determined by any standard ASTM method. The standard ASTM procedure for determining forms of sulfur tends to include the elemental sulfur with the organic sulfur, since it is not detected by the determination of either the sulfate sulfur (because it is not soluble in HCl solutions) or the pyritic sulfur (because it does not include iron). Since elemental sulfur is an oxidation product, it is not an important factor in pristine coals, although it can be important in oxidized coals.

The procedures being developed for determining elemental sulfur in coal start with dissolving the elemental sulfur in a suitable solvent, such as hexane, ethanol, acetone, benzene, cyclohexane, or tetrachloroethylene. The quantity of sulfur dissolved is then determined by another method such as high performance liquid chromatography (HPLC) (Stock and Wolny, 1990; Thomas, 1995). It is also possible to determine whether an elemental sulfur layer has begun to form on coal pyrite by using cyclic voltammetry (Tao et al., 1994); however, this only indicates the presence of elemental sulfur, and it is not useful for determining the actual quantity.

2.1.3 Organic Sulfur Forms

The nature of organic sulfur-bearing compounds in coal is difficult to determine in detail for any given coal, but methods for removing organic sulfur from coal would benefit from determining the types and quantities of sulfur compounds present in the coal (Winans and Neill, 1990). The techniques used to date include mass spectrometry (Winans and Neill, 1990), isotope dilution thermal ionization mass spectrometry (Kelly et al., 1994), nuclear magnetic resonance (NMR) analysis combined with controlled oxidation (Palmer et al., 1990), isothermal flash pyrolysis (Torres-Ordonez et al., 1990), and many others. However, there are no routine, low-temperature, unambiguous methods for characterization of organic sulfur in coal (Palmer et al., 1990).

2.2 SPECTROSCOPIC METHODS

Spectroscopic methods determine the quantity of sulfur based on how the coal interacts with electromagnetic and accelerated-particle radiation. This includes a broad range of techniques, ranging from specialized research tools to methods that are suitable for on-stream analyzers.

2.2.1 X-Ray Spectroscopy

There are two methods for using X-rays to determine the amount of sulfur in coal. X-ray fluorescence provides a means for measuring the total quantity of sulfur atoms in the coal, while X-ray diffraction is a method for determining the quantity of specific sulfur-bearing minerals, such as pyrite or various sulfates.

2.2.1.1 Fluorescence

In X-ray fluorescence, the sample being analyzed is irradiated with high-energy X-rays or γ-rays. These X-rays are absorbed by the electrons surrounding the atoms, pushing them into high-energy states. The electrons then return to their original energy, and emit fluorescent X-rays that have specific energies that are characteristic of the particular atom. By measuring the energy spectrum of the fluorescent X-rays, the types and quantities of atoms present can be measured. The spectrum can be determined by using wavelength-dispersive analyzers, solid-state energy-dispersive analyzers, gas proportional counters, or scintillation counters, with the selection of detector based on the degree of accuracy needed, and the cost of the detector (Dos Santos and Conde, 1994). The limitation on measuring elements by this method is that elements with a low atomic number emit low-energy fluorescent X-rays, which are easily absorbed by air and are therefore difficult to detect. It is also important that the X-rays being used to excite the fluorescence have a different energy than the fluorescent X-rays, to avoid confusing the two spectra.

Sulfur is a fairly low atomic number element, and so its fluorescent X-rays are of low energy and rapidly absorbed by air. Special care is therefore needed to detect its fluorescence spectrum. If the path from the sample to the X-ray detector is either through vacuum or through a gas with a low X-ray absorption cross-section (such as helium), then the sulfur fluorescence can be detected.

A related technique is sulfur K-edge X-ray absorption spectroscopy (George et al., 1990), which measures the changes in the spectrum of an X-ray beam that are caused by absorption of X-rays by the coal sample. This can provide a measure of the total sulfur, and the spectra can be analyzed to determine the X-ray absorption near-edge structure (XANES), which can be matched with model compounds to determine what compounds contain sulfur (George et al., 1990; Huffman et al., 1995; Gorbaty et al., 1990).

2.2.1.2 Diffraction

X-ray diffraction is used to determine the crystalline structures of solids by measuring the distances between the planes of atoms in the crystals. This is accomplished by slowly rotating the sample in an X-ray beam that has a known wavelength and measuring the angles at which the beam is diffracted by the sample. The diffraction angles and X-ray wavelength are then used to calculate the interatomic spacings, which are characteristic of the particular crystal and can be used to identify it. This method can therefore be used to determine how much pyrite is contained in a sample, as well as the amounts of other crystalline materials that contain sulfur. Sulfur that exists in a noncrystalline form (such as organic sulfur) cannot be detected by this means.

2.2.1.3 X-Ray Photoelectron Spectroscopy (XPS)

In this method, a sample is irradiated with monochromatic X-rays. When the X-rays strike the sample, they cause electrons to be ejected. The energies of the ejected electrons are then measured, and the differences between the energy of the incident X-ray photons and the electron energies are determined. This difference is equal to the binding energy that the electron had in the sample. Since electron binding energies are characteristic for each element and are also diagnostic of the chemical form, X-ray photoelectron spectroscopy (XPS) is a powerful tool for chemical analysis. The technique is sometimes also referred to as electron spectroscopy for chemical analysis, or ESCA.

The most important feature of XPS is that the energies of the emitted electrons are low (less than 1.5 keV), and so they have very little penetrating power. As a result, only the electrons emitted from the surface layers of the sample can reach the detector, with any electrons ejected below the surface being reabsorbed before they can escape. XPS is therefore only sensitive to surface compositions, and is ideally suited for analyses of coal to determine the extent of surface phenomena such as oxidation and weathering (Putnis, 1992). This is particularly valuable for measuring the formation of elemental sulfur on the coal surface. Weitzsacker and Gardella (1992) have used this technique for analysis of the Argonne Premium coals, and other investigators have used XPS for the study of lignite lithotypes (Fiedler and Bendler, 1992) and sulfur forms in Illinois No. 6 and Rasa coals (Kelemen et al., 1990).

2.2.2 Mössbauer Spectroscopy

Mössbauer spectroscopy differs from the other spectrographic techniques discussed, in that it uses the absorption of γ-rays by the nucleus of the target atom, rather than the absorption by the electron shells. The atomic nuclei are the heavy cores of atoms and are generally considered to be composed of protons and neutrons. The number of protons in the nucleus is the atomic number of the atom, and this value determines the chemical properties of the atom. The Mössbauer effect is concerned with the emission of γ-rays by a radioactive nucleus and the subsequent reabsorption of these γ-rays by other nuclei of the same type as the emitting atom, as shown in Figure 2.1. For this absorption to occur, the γ-ray must match the resonance of the absorbing atom very closely, and so the technique is extremely selective.

Mössbauer spectroscopy is used to detect the iron atoms contained in the pyrite for the determination of pyrite in coal. The radioisotope source for this application is cobalt-57 (^{57}Co), which decays to produce iron-57 (^{57}Fe) in its excited state. The excited ^{57}Fe then rapidly decays, emitting its characteristic γ-ray. Since the target

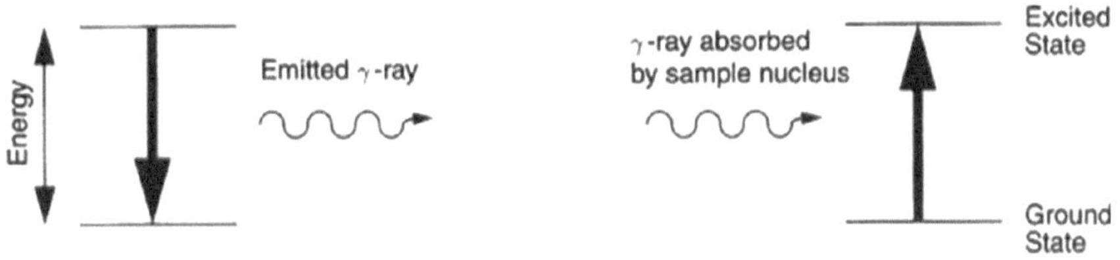

FIGURE 2.1 Schematic representation of the events occurring in Mössbauer spectroscopy. The horizontal lines represent the nuclear energy states. When the source atom undergoes radioactive decay, it goes from the excited state to the ground state, emitting a γ-ray. The γ-ray is subsequently absorbed by the second nucleus, raising it to an excited state. This absorption can be detected either by the decreased transmission of γ-rays through the absorber or by the subsequent decay of the absorber nucleus from the excited state.

nucleus must precisely resonate with the γ-ray to absorb it, the radiation will only be absorbed by the ^{57}Fe nuclei in the sample. Thus, this technique is sensitive only to the iron contained in the coal. By measuring the quantity of absorbed γ-rays, the quantity of iron (and therefore pyrite) in the sample can be precisely determined. Eissa et al. (1994) used Mössbauer spectroscopy to study the oxidation of pyrite during the combustion of Egyptian Maghara coal, and Weng (1994) used it to monitor the removal of pyrite by a microwave treatment/magnetic separation process. Stiller et al. (1990) used Mössbauer spectroscopy to precisely determine the quantities of iron in coal samples, so that they could evaluate the effectiveness of the standard ASTM method for pyrite determination.

This measurement is not interfered with by the presence of other types of atoms, and so it is not affected by the type of coal, presence of sulfur, or other interferences that cause trouble for other analysis techniques (Huffman et al., 1987; Gibb, 1976). In many respects, the γ-ray fluorescence used in Mössbauer spectroscopy is similar to X-ray fluorescence; however, the resonance in γ-ray fluorescence is much sharper, because the energy of the γ-ray photons is much larger than for X-ray photons (Tsai, 1982).

2.2.3 Infrared Spectroscopy

The infrared spectrum is the part of the electromagnetic spectrum that has wavelengths in the range of 0.7–1,000 μm, which is longer than those of visible light (400–700 nm), but shorter than those of microwaves (greater than 1 mm). Infrared spectroscopy is used for studying the structure of coal, because it can be used to identify specific functional groups such as C–H, C=O, and O–H. Fourier transform infrared spectroscopy (FTIR) is a particularly popular variation of this technique. The details of coal structure that have been determined by this means are extensively discussed in the literature (Atar and Hendrickson, 1982). However, these techniques have only limited application for the direct study of sulfur in coal, because they are limited to surface analyses, but they are well suited for studies of surface phenomena such as oxidation and weathering.

FTIR spectroscopy is very useful to characterize the hydrophobic/hydrophilic balance at a coal surface in terms of chemical composition. This technique is a powerful tool for analysis of surface compositions and has been used to characterize the hydrophobicity of coals of different ranks (Ye et al., 1988). It is therefore a useful analytical technique for desulfurization processes that are based on surface chemistry, such as froth flotation and oil agglomeration. FTIR has also been used to monitor the behavior of sulfur in coals during pyrolysis (Shao et al., 1994).

2.2.4 Raman Spectroscopy

When light is scattered by molecules, the scattered light changes slightly in frequency due to transfer of energy to the molecules. In Raman spectroscopy, this frequency change is used to determine the characteristics of the molecule. If the frequency of the incident light is (ν_0) and that of the scattered light is (ν_r), then the frequency shift is

$$\nu_r - \nu_0 = \Delta\nu \quad (2.2)$$

which is referred to as the Raman frequency for the molecule causing photon scattering (Woodward, 1967). Raman spectroscopy is used for analyses of coal structure but has only limited application for the study of sulfur forms in coal.

2.2.5 Nuclear Magnetic Resonance

NMR provides a rapid and nondestructive determination of the total hydrogen content and hydrogen distribution among the various chemical components present in coal. This technique does not involve high-energy ionizing radiation. Instead, it measures the interactions of atomic nuclei with a strong magnetic field, by probing with electromagnetic radiation at radio frequencies. As a result, use of NMR does not represent a radiation hazard.

NMR spectroscopy is useful for determination of structural variations in coal macerals. Studies using NMR to analyze sulfur in coal are limited, because NMR analysis is not highly sensitive to sulfur or sulfur compounds.

2.2.6 Auger Electron Spectroscopy

Auger spectroscopy is similar to X-ray fluorescence because the sample is excited by electron, photon, or ion bombardment and gives off an emission that is characteristic of the elements involved. In Auger spectroscopy, however, the characteristic emissions are of electrons with specific energies. The emission of Auger electrons is produced when an atom is ionized in one of its core-level electron orbitals. The core-level vacancy is immediately filled by another electron, releasing energy. The released energy can be in the form of fluorescent X-rays, or it can be transferred to another electron, which is ejected from the atom with a specific energy. The energy of this Auger electron is characteristic of the source atom and independent of the energy of the excitation radiation. For Auger electrons to be produced, the atom must have at least two electron energy states and three electrons; therefore, hydrogen and helium cannot emit Auger electrons.

Auger electrons cannot travel through a great thickness of material, and so Auger spectroscopy is a surface analysis technique, limited to depths of 1–10 Å (Goldstein et al., 1977). This technique also requires a very high vacuum (10^{-9} torr or better) for high-precision work, which is difficult to achieve with materials such as coal that outgas in vacuum. Auger spectroscopy is therefore only used to a limited extent in coal characterization.

2.2.7 Secondary Ion Mass Spectroscopy

In secondary ion mass spectrometry (SIMS), a solid sample is bombarded with energetic ions in a vacuum. The ions penetrate the sample and dissipate their energy to atoms in the sample through inelastic collisions. The energy from these collisions is sometimes sufficient to eject atoms from the sample, in a phenomenon known as "sputtering." A fraction of the sputtered atoms are ionized so that they can be collected and analyzed using a mass spectrometer, which can detect all elements, including hydrogen. SIMS has been used to study the organic and inorganic sulfur contents of a variety of coals. It is also possible to use SIMS in conjunction with electron microprobe analysis, allowing it to be applied to individual phases in the coal. Hou et al. (1995) used this technique to determine the in situ organic and inorganic composition of coal in conjunction with traditional petrographic analysis.

2.3 MICROSCOPIC TECHNIQUES

2.3.1 Optical Microscopy

Optical microscopy has long been used for macerai and mineral analysis of coals. In polished specimens, viewed under reflected light, it is relatively straightforward to distinguish between the various coal macerals and mineral matter. Pyrite is a particularly easy mineral to distinguish by this means because it is highly reflective and generally yellow in color. It is generally not sufficient, however, to simply note the presence of pyrite. Some means is needed to quantify how much pyrite is present compared to other types of particles.

A common method for determining the relative abundances of pyrite and other particle types is point counting, which determines the frequency of occurrence of each phase of interest at random positions in the specimen. The random positions can be selected by using a grid reticule in the ocular of the microscope, with the points examined being the points that lie at the intersections of the grid, as shown in Figure 2.2. Alternatively, a point-counting stage and cross-hair reticule can be used, where the stage moves the specimen by small, preset increments and the points are taken at the cross-hair intersection after each move. In either method, the identity of the phase present at the point directly under the intersection is determined and tabulated. This is repeated for a statistically significant population to give the desired degree of accuracy, generally between 300 and 1,000 points. The number of points that fall onto a particular phase is proportional to the volume fraction of that phase in the specimen, and so the point counts can be used to calculate volume percentages of each phase measured.

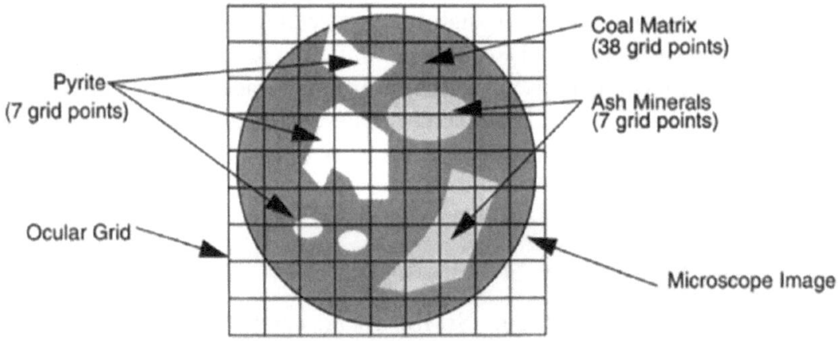

FIGURE 2.2 Illustration of the ocular grid point-counting technique for determining volume fractions of pyrite and other mineral grains in coal. Grains that do not fall under at least one grid intersection do not contribute to the point count.

This procedure is quite labor-intensive, and it must always be kept in mind that the specimen under the microscope is generally too small to be a properly representative sample of the parent lot of coal. Results of microscopic analysis should therefore only be considered to be qualitative indications of the characteristics of a coal. Also, the resolution of optical microscopy is limited, and ultrafine pyrite will not necessarily be seen.

2.3.2 Electron Microprobe Analysis

An electron microprobe can be used to determine the forms and quantities of sulfur atoms over very small distances, on the order of 1 μm or less (Ge and Wert, 1990). When it is used in combination with coal petrography, the sulfur content in individual coal macerais can be determined.

The microprobe uses X-ray fluorescence to determine the elemental composition of the spot being analyzed. In X-ray fluorescence, the individual atoms will emit their characteristic X-ray spectrum whenever the atom is excited by a sufficient amount of energy. The nature of the incident radiation that excites the atom is irrelevant. Instead of using X-rays to excite the atoms to fluoresce, the microprobe uses an electron beam to irradiate the sample. Although the electrons do not excite fluorescence as efficiently as incident X-rays would, the electron beam can be tightly focused so that only the particular spot of interest is excited (Putnis, 1992).

The electron microprobe can be used to determine the sulfur concentrations at specific points and to find out how the sulfur atoms are distributed. In coal, the sulfur atoms are clustered into discrete clumps if they are mineral sulfur, while the atoms that are uniformly distributed through the coal are organic sulfur. The iron atom distribution can also be determined, and the points that emit both iron and sulfur X-rays can be identified as pyrite (Tsai, 1982). Electron imaging and electron diffraction can be used in conjunction with X-ray fluorescence to determine the structures of the parts of the sample that contain sulfur (Ge and Wert, 1990).

The combination of coal petrography with microprobe analysis as described above determines the organic and pyritic sulfur contents in coal macerai groups (Harris et al., 1977; Tsai, 1982). First, coal petrography is used to determine the volume fractions of macerai groups and pyrite in coal. The weight fractions of the macerai groups and pyrite are then obtained by multiplying the volume fractions by the appropriate density factors. The sulfur X-ray intensities for individual macerai groups and pyrite are then measured using the microprobe, and the intensities are multiplied by the weight fractions of each constituent. These values are then compared with suitable sulfur standards to determine the sulfur content of each maceral.

A related technique is proton-induced X-ray emission (PIXE) microanalysis, which has been shown to be capable of making standardless determinations of trace element contents of coals (Hickmott and Baldridge, 1995). This technique uses a proton beam to excite fluorescent X-rays from the sample, and is reported to be considerably more accurate and sensitive than electron microprobe analysis. PIXE requires a particle accelerator as the source of the proton beam, and so facilities capable of this work are limited.

Microprobe analysis is mainly a research tool because only a very small amount of the coal can be analyzed. It is not practical to use such microscopy techniques for routine analyses of bulk coal because of the impossibility of obtaining a representative sample at such a small size.

2.3.3 Image Analysis

Image analysis is a method of quantitatively evaluating features such as size, shape, or area fraction from a digital image. The digital image consists of pixels (picture elements) of intensities from 0 (black) to 255 (white). Any type of image that illustrates features of interest as differences in brightness can be processed, whether it is optical, electron, or X-rays. Automated image analysis systems can be programmed to acquire, process, and evaluate images with a minimal amount of operator input. In addition, automated image analysis systems are frequently integrated with other analytical tools, such as an EDS (energy-dispersive spectroscopy) analyzer on a scanning electron microscope, to provide additional data so that the images can be correlated with physical measurements.

The user sets a range of gray-level intensities for the image analyzer to process, and excludes all others. Usually this is accomplished by producing a binary image of the features of interest, although some systems process the gray-level digital images directly. Once the image analyzer identifies and processes a feature, any additional integrated systems will be directed to the location of that feature for additional analysis.

Characterization of coals using automated image analysis is most effectively accomplished using electron imaging, sometimes in conjunction with X-ray analysis. In a backscattered electron image, concentrations of heavy elements appear bright, and concentrations of light elements appear dark. For coals, the coal macerals will appear dark, clay and silicate contaminants will appear gray, and pyrite will be bright in backscattered electron images. The concentrations, grain sizes, grain shapes, and degree of locking of any or all of the minerals present can be evaluated by specifying its corresponding range of gray-level intensities on the digital image. A chemical composition of each feature is obtained with the X-ray analysis. Thus, each particle on the image can be measured and analyzed independently. This makes the image analysis useful for determining the size to which coal must be ground to liberate the pyritic sulfur (Irdi and Rohar, 1988; Irdi et al., 1988).

An example of the application of image analysis to measure the size distribution of pyrite grains is shown in Figure 2.3 (Kramer, 1995, Personal Communication). For this analysis, the coal was mounted in epoxy and polished to 1 μm. Optical microscopy was first used to carry out manual point counting using the method described in Section 2.3.1., with 441 points examined. The coal was determined to contain 75.3% vitrinite, 1.62% inertinite, 5.2% exinite, and 3.4% pyrite. However, optical point counting could not provide a measurement of the grain size distribution of the pyrite. To make this determination, the polished specimen was carbon-coated and examined using a scanning electron microscope. It was then imaged using backscattered electrons, which gives excellent contrast between high-atomic-number grains (pyrite) and low-atomic-number grains (coal and clays). In the photograph shown in Figure 2.3a, the bright white features are pyrite framboids, while the clays are light gray.

(a)

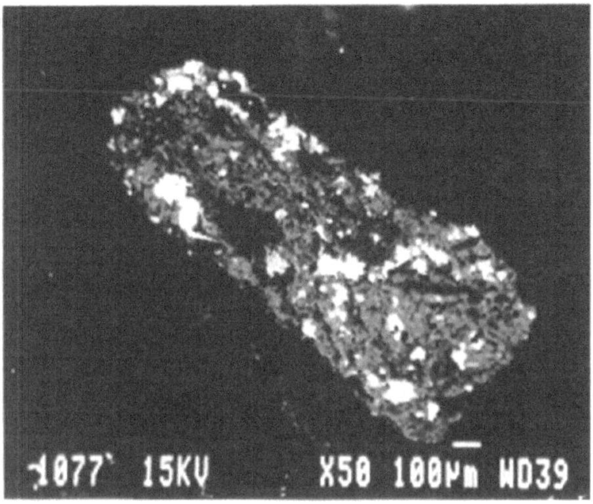

(b)

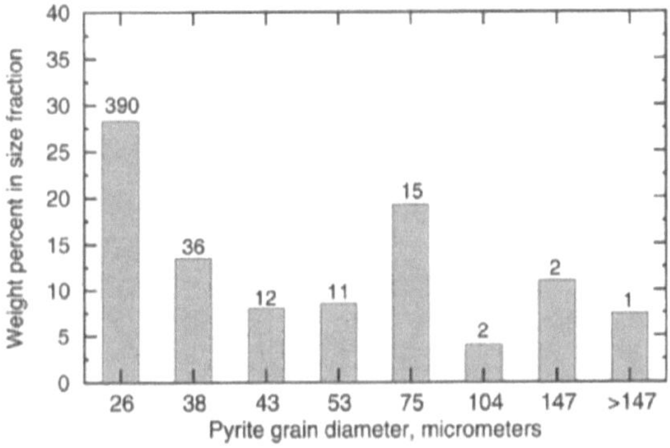

FIGURE 2.3 Example of the application of image analysis to determining pyrite grain size distribution. (a) Illinois No. 6 coal particle, backscattered electron image, 50× magnification. (b) Weight distribution of pyrite grains in individual particle size fractions, as determined by image analysis. The numbers above each bar are the number of individual particles found in each size fraction.

For pyrite sizing, binary images were produced from ten frames of backscattered electron images, and 469 pyrite particles were sized using the image analyzer. The sizing data generated was then read into a spreadsheet, and the diameters of particles were grouped into size ranges representing mesh sizes from –26 pm to +147 μm. The particle areas corresponding to the diameters of the particles in each size fraction were also calculated and expressed as a percentage of the total area, which is directly proportional to the weight percent. Since pyrite was the only phase measured, the sum of the individual weight fractions is equal to 100%. The results of this analysis are plotted in Figure 2.3b. It was found that over 80% of the individual pyrite grains were smaller than 26 pm and that these particles accounted for 28%

of the pyrite present in the coal. From this analysis, it can be concluded that grinding the pyrite to 26 pm will leave approximately 28% of the total pyrite in locked pyrite–coal grains.

A similar analysis was carried out for an Upper Freeport coal, which was found to contain 74.2% vitrinite, 7.2% inertinite, 12.5% exinite, and 7% pyrite by optical point counting (343 points). Image analysis determined that over 40% of the individual pyrite grains were finer than 500 mesh, but because of the presence of a number of very large pyrite grains, these fine pyrite particles accounted for only 4% of the total pyrite. It was shown that 88% of the total pyrite was coarser than 100 mesh, and would therefore be relatively easy to liberate.

2.4 OTHER TECHNIQUES

2.4.1 Float–Sink Analysis

Float–sink analysis is used to determine the washability of coal by density-based separations. ASTM Standard D4371 sets forth a standardized method for carrying out this analysis (ASTM, 2012b). The basic principle of the test is to suspend the coal in a series of progressively denser liquids, and collect and analyze the material that floats and sinks in each liquid. This produces a distribution of coal particles by density. Since pyrite is much denser than coal, the pyritic sulfur tends to concentrate in the high-density fractions.

Formerly the most common liquids used for this analysis were typically commercially available organic liquids, with specific gravities ranging up to nearly 3.0. Commonly used liquids, and their specific gravities, were petroleum spirit (0.7), white spirit (0.77), naphtha (0.79), toluene (0.86), perchloroethylene (1.6), methylene bromide (2.49), bromoform (2.9), and tetrabromoethane (2.96). By mixing these liquids, any desired specific gravity between 2.96 and 0.7 can be obtained. These liquids are toxic and care must be taken in their use. Because of the toxicity concerns with traditional heavy liquids, there has been a good deal of interest in the development of nontoxic heavy liquids, and the organic heavy liquids are being phased out of use.

The most successful of these alternatives are the metatungstate salts (sodium, lithium, and ammonium metatungstate) and cesium chloride. These salts can be dissolved in water to give a wide range of specific gravities. The metatungstate salts can reach specific gravities as high as 3.1, while cesium chloride is effective up to a specific gravity of approximately 1.8 before its viscosity becomes high enough to interfere with the separation. The drawbacks of these salts are, first, that they are extremely expensive (several hundred dollars per liter), and second, that they can react with other soluble species from the coal, forming insoluble precipitates (Suardini, 1995). Additionally, the high viscosity caused by the sodium metatungstate species in practice leads to slower filtration rates and less efficient separations.

Thus, the heavy liquids are summarized below:

- Organic heavy liquids: toxicity concerns, but nonreactive with coal compounds.
    - Petroleum spirit, white spirit, naphtha, toluene, perchloroethylene, methylene bromide, bromoform, and tetrabromomethane, among others.

- Tungsten-based heavy liquids: expensive, potentially reactive with impurities in coal, but considerably less toxic.
  - Ammonium, lithium, and sodium metatungstate solutions.

There are two common procedures for running a series of float–sink analyses of a coal. In the incremental procedure, a single sample is taken and subjected to float–sink analysis, usually at the lowest density to be tested. The sink product is then taken and treated at the next lowest specific gravity in the series. This is repeated for each density to be tested, and the products are the material that falls into a specific density fraction. In the cumulative procedure, a set of identical coal samples are used, each of which is independently subjected to float–sink analysis at a single density. The products of this procedure represent the total amount of material that will float and sink at each density. The incremental approach is often preferred because the results are more intuitively obvious, but the cumulative approach is less prone to sampling errors and procedural errors because each sample is only subjected to float–sink analysis once (Zhou et al., 1993).

Float–sink washability analyses are of critical importance in evaluating the performance of physical coal desulfurization techniques. The float–sink results are generally considered to represent the best possible separation that can be made by a physical separation of a given coal at a given particle size. If float–sink analysis does not produce a low-density product that is low in sulfur, then there are no low-sulfur particles available in the coal at that particle size. This means that it will not be possible for any physical desulfurization technology to produce a low-sulfur product from that coal. A sink–float analysis will therefore give a good indication of how closely a physical desulfurization process is approaching ideal performance.

The method used for the float–sink test depends on the particle size of the coal. Coal coarser than 2.36 mm can be suspended in a vat of heavy liquid, and once the particles have had time to float or sink, the float product can be skimmed off with a wire-mesh strainer. Finer particles, between 2.36 mm and 75 µm, make use of specially designed separatory flasks or funnels, which can segregate the float and sink products. For very fine particles (less than 75 µm), an ASTM standard method is not available. This is a continuing problem in coal analysis, and studies are ongoing to develop a satisfactory method for fine-particle analyses.

A comparative laboratory study was conducted by the Pittsburgh Energy Technology Center to study the problem of float–sink analysis of very finely ground coals. Wide variations were found in the float–sink results reported by various laboratories, even though they were testing identical coal samples. The disagreements between laboratories became more pronounced as the particle size of the coal was reduced, and at the finest sizes (finer than 14 µm), the results were dependent on the type of heavy liquid that was used (Killmeyer et al., 1992). With the current trend in coal desulfurization, where coals are being ground to progressively finer sizes to liberate the pyrite, it is of particular importance that the fine-particle float–sink procedures be improved. A useful procedure will need to overcome the following problems with fine particles: (1) agglomeration of fine particles, particularly when the agglomerates contain both coal and fine mineral matter and (2) slow separation, because of the low settling rates of fine particles.

The agglomeration problem can be reduced by using ultrasonic conditioning to break up agglomerates (Zhou et al., 1993; Suardini, 1995). It should be kept in mind that for soft or friable coals, the ultrasonic treatment can cause the coals to break to finer sizes, changing the float–sink results (Suardini, 1995). Dispersants, such as Aerosol OT-1OO (sulfobutanedioic acid 1,4-bis-(2-ethylhexyl) ester sodium salt) and Brij-35 (polyoxyethylene (23) lauryl ether), were also found to be helpful.

A useful technique for increasing the fine particle settling rate is centrifuging the particle/heavy-liquid suspension (Palowitch and Nasiatka, 1961). Workers in the area recommend the use of a refrigerated centrifuge, so that the temperature can be controlled at 25°C, and use centrifuge speeds that provide over 500 times the force of gravity (Zhou et al., 1993; Suardini, 1995). Once the suspension has been centrifuged, some care is needed to remove the float product from the centrifuge tube. In one method, floats can be separated from the sinks by freezing the bottom of the centrifuge tube in liquid nitrogen. The floats product can then be poured off, leaving the sinks frozen in the bottom of the tube. Another technique is to use a suction hose to carefully remove the float product without disturbing and resuspending the sinks product (Zhou et al., 1993).

The most suitable heavy liquids for laboratory work are organic liquids (such as the Certigrav® liquids sold by Interstate Chemical Co.) or cesium chloride (CsCl) solutions. Metatungstate salts have also been proposed, but these have a tendency to react with calcium that dissolves from certain coals, making a calcium metatungstate precipitate that is difficult to deal with. It has been found that for coals coarser than 26 μm, and for separating densities higher than 1.8, the Certigrav solutions give the best results. For coals finer than 26 μm and for separating densities less than 1.8, the cesium chloride produces the most efficient separation (Suardini, 1995).

Based on the factors discussed above, and on the results of an extensive test program, the following procedures have recently been proposed for float–sink analysis of fine coals (Suardini, 1995):

- General Float–Sink Procedure:
  - Use a refrigerated centrifuge, at 25°C, capable of centrifuging 600 mL bottles at 500 × gravity.
  - Process a maximum of 100 g of coal per 600 mL bottle.
  - Run the procedure incrementally, starting at the lowest specific gravity of separation, and reprocess the sinks product at progressively higher densities.
  - Maintain the specific gravity of the media at ±0.001 specific gravity units.
  - Dry bituminous coals at 40°C–65°C and sub-bituminous or lignite coals at 25°C–40°C, to prevent size degradation during drying.
- Organic Liquid Procedure, for all 150 × 26 μm size fractions:
  - Add Aerosol OT-100 surfactant to the organic liquid, at a rate of 2% of the coal weight for fresh liquids and 1% of the coal weight for recycled liquids.
  - Agitate coal/heavy-liquid mixture for 3 min, with no ultrasonic treatment.
  - Centrifuge for 10 min, allow the rotor to slow without braking.

  - After filtering the solids from the heavy liquid, wash the solids with 3 parts ethanol/1 part coal to remove the surfactant (Aerosol OT-100 contains sulfur, and could interfere with sulfur determinations).
- Cesium Chloride Procedure, for –26 μm bituminous coals, specific gravities less than 1.8:
  - Add Brij-35 surfactant to the CsCl solution, at a rate of 8% of the coal weight for fresh liquids, and 2% of the coal weight for recycled liquids.
  - Agitate coal/heavy liquid mixture for 2 min, followed by 1 min of ultrasonic treatment and a further 1 min of final agitation.
  - Centrifuge for 15 min and allow the rotor to slow without braking.
  - After filtering out the heavy liquid, wash the coal with 5 parts hot water, followed by 5 parts cold water, and finish with 3 parts ethanol to remove residual CsCl and Brij-35.
  - Readjust the pH of recycled heavy liquid to 6.0–7.0 with HCl or NaOH additions.
- Organic Liquid Procedure, for –26 μm sub-bituminous and lignite coals, or for –26 μm bituminous coals at specific gravities greater than 1.8:
  - Add Aerosol OT-100 surfactant to the organic liquid, at a rate of 2% of the coal weight for fresh liquids and 1% of the coal weight for recycled liquids.
  - Agitate coal/heavy liquid mixture for 2 min, followed by 1 min of ultrasonic treatment, and a further 1 min of final agitation.
  - Centrifuge for 15 min, allow the rotor to slow without braking.
  - After filtering the solids from the heavy liquid, wash the solids with 3 parts ethanol/1 part coal to remove the surfactant (Aerosol OT-100 contains sulfur, and could interfere with sulfur determinations).

2.4.2 Thermal Analysis

When coal is heated, the various sulfur-bearing compounds release their sulfur into a gaseous form at different temperatures and at different rates. By heating the coal at controlled rates and under controlled (oxidizing or reducing) atmospheres, and by measuring the rate at which the sulfur is released, it is possible to determine both how much total sulfur was present and the types of compounds that originally contained it. Several methods are available for doing this, including programmed temperature oxidation (PTO), programmed temperature hydrodesulfurization (PTH), programmed temperature reduction (PTR), programmed temperature pyrolysis (PTP), and flash pyrolysis–gas chromatography–mass spectroscopy (Boudou, 1990; Bakel et al., 1990; Garcia-Labiano et al., 1995; Gryglewicz, 1995). These methods are used as routine laboratory test methods for coals, kerogens, and heavy oil characterization.

PTR provides good discrimination between the various types of sulfur-bearing groups in coal. This technique is based on the fact that when coal is heated in a hydrogen atmosphere, all of the sulfur groups can be converted to hydrogen sulfide gas (LaCount et al., 1987). Since each type of sulfur-bearing group converts to hydrogen sulfide over a particular temperature range, a plot of temperature versus hydrogen sulfide evolution can be used to determine how much of each type of sulfur-bearing group was present.

2.4.3 Chromatographic Analysis

If the coal structure is broken down enough to convert the sulfur-bearing compounds into soluble or volatile forms, then the individual compounds can be separated, concentrated, and measured quantitatively, using chromatographic methods.

In general, chromatography is a class of physical separation methods, in which the components to be separated are distributed between two mutually immiscible phases. One of these phases consists of a stationary bed of large surface area (the stationary phase), and the other phase is a fluid (the moving phase) that percolates along or through the stationary bed (Keulemans, 1959). The sample components are initially suspended in the moving phase, and the different components of the sample undergo a series of exchanges between the two phases. Components with minimal interactions with the stationary phase migrate through the bed most rapidly and are eluted from the column first, while components that interact more strongly with the stationary phase migrate through more slowly. The stationary bed can be either a solid or a liquid, and the moving phase can be either a liquid or a gas, with the various chromatographic techniques classified as in Table 2.1. A schematic of the basic process is shown in Figure 2.4 (Day and Underwood, 1986).

In a typical chromatographic analysis, a pulse of sample is injected into the moving phase at the beginning of the column. The eluted material is then collected as a function of time and analyzed by any convenient method for the species in question. Individual components are eluted from the column at different times, with their concentrations ideally being in the form of Gaussian-shaped peaks, in order of increasing interaction with the stationary phase (Dean, 1995). A chromatogram will therefore have a form similar to that shown in Figure 2.5.

Chromatographic analysis is a massive field of broad application, and a complete discussion is far beyond the scope of this book. The important point is that chromatographic methods can be used to separate sulfur species after they have been extracted from coal. The separation can be made based on chemical affinities, solubilities in various solvents, and diffusion rates. Once separated, the species can be identified by a wide range of detector types, including mass spectrometers, electrochemical methods, or spectrographic methods.

While chromatography can be used for coal analysis, these methods are by their nature only suited for analysis of liquid or gaseous materials. The difficulty

TABLE 2.1
General Classification of Chromatographic Methods (Day and Underwood, 1986)

	Moving Phase	
Stationary Phase	**Gas**	**Liquid**
Solid	Gas–Solid Chromatography (GSC)	Ion-exchange chromatography, HPLC,
Liquid	Gas–Liquid Chromatography (GLC)	Partition chromatography using silica-gel columns

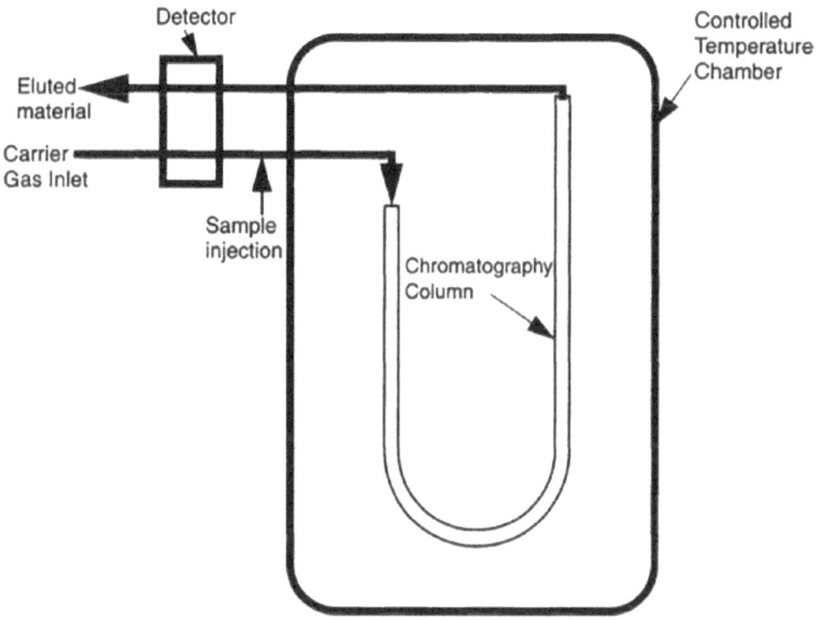

FIGURE 2.4 Schematic diagram of a gas–liquid chromatograph. The liquid stationary phase is supported by a high-surface-area packing in the chromatography column. Other types of chromatographs have similar configurations.

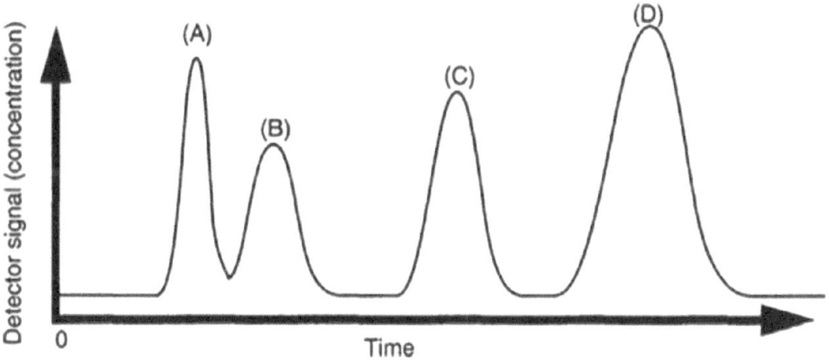

FIGURE 2.5 Ideal form of a chromatogram, with the sample injected into the moving phase at time $t = 0$. Components are considered to be separated when the difference in residence times is large enough that the peaks do not overlap. In this chromatogram, component A has the least affinity for the stationary phase, and component D has the greatest affinity. Components A and B are incompletely separated, while C and D are completely separated.

in coal chromatography is that the sulfur-bearing compounds must be first converted to a mobile form, either in a liquid solvent or in a gas. Since the main reason for concern over organic sulfur forms in coal is that they are difficult to extract from coal, it is evident that the mobilization step is the most difficult part of coal chromatography. The methods that have been used to date for organic sulfur mobilization include

- Supercritical extraction with alcohol/NaOH mixtures (Yurum et al., 1990);
- Supercritical extraction with carbon dioxide (Louie et al., 1994);
- Extraction with hot perchloroethylene (Buchanan et al., 1993; Lee and Fullerton, 1992).

2.5 ON-LINE ANALYSIS

Coal processing plants generally do not include on-line composition analyzers that can be used for process control, although such equipment is being developed and is gradually being adopted, as the units become less expensive and more plant operators are convinced that they will give significant improvements in plant performance and product quality. Particular care must be taken to ensure that the material that passes through the composition analyzer is representative of the whole stream, which is very difficult to do reliably. Many on-line analysis installations have failed because of poor sampling systems (Pitard, 1993, 2019). On-line analyses for coal processes have been developed and tried in laboratories and plant operations (Kawatra, 1976b, 1979, 1980, 1985; Kawatra et al., 1979, 1984; Laurila and Kawatra, 1985).

2.5.1 X-Ray Fluorescence

To determine sulfur content of coal on-line, a very rapid analysis is needed. One suitable analysis technique is X-ray fluorescence, with the sulfur content (and optionally the pyrite content) determined by measurement of the fluorescent X-rays emitted by sulfur and iron (Gorlov and Onishchenko, 1994; Alaya et al., 1994; Tsarenko et al., 1994). The capability to measure sulfur and iron fluorescence can readily be added to ash analyzers that use X-ray backscatter to measure ash content, with little additional cost. The only requirements are that the radiation path should be through a helium atmosphere to minimize absorption of the low-energy sulfur fluorescence and that the coal passing through the sensor is regularly sampled for laboratory analysis to determine the sulfur content, so that the analyzer can be calibrated to deal with the particular coal being processed.

2.5.2 Prompt Gamma Neutron Activation Analysis

2.5.2.1 Principle of Operation

Neutron activation analysis is an important analytical tool with many applications. If a material is bombarded with neutrons, the neutrons are captured or scattered by the atomic nuclei in the sample. When this occurs, energy is released in the form of γ-rays. These γ-rays can be classified as "prompt" if they are emitted within 10^{-12} s of the neutron capture, and as "delayed" if they are emitted much later, when the nucleus undergoes radioactive decay. Since very few atoms emit delayed γ-rays that are useful in elemental analysis, most work has concentrated on the use of prompt γ-rays. These γ-rays arise from two predominant mechanisms: thermal neutron capture and inelastic scattering of fast neutrons. The individual elements can be identified by the characteristic energies of the γ-rays that they emit (Kawatra, 1987; Kawatra and DeLa'O, 1995).

Prompt gamma neutron activation analysis (PGNAA) has been developed commercially for the bulk analysis of coal and cement. It does not give a direct bulk measurement of the ash content, but it gives measurements of the individual elements from which the ash content can be inferred. It also gives a direct measurement of the total sulfur content, although it does not distinguish between pyritic, organic, and sulfate sulfur. In this technique, a known volume of coal is exposed to a beam of neutrons from a neutron-emitting radioisotope, such as californium-252 (^{252}Cf). These neutrons enter the coal and are slowed by collisions with the hydrogen nuclei, until they slow enough that they can be captured by the nuclei of the various elements. When a nucleus captures a neutron, it is in an "excited" state due to the energy that it gained from the capture, as shown in Figure 2.6. The excited nucleus then rapidly emits a γ-ray, which can be detected by a scintillation detector. A nucleus of a particular element will emit a series of γ-rays that are unique to that element, and so determining the emitted radiation spectrum provides a means for identifying the elements present (Page, 1991; Zumberge, 1987).

Each element has a particular ability to capture thermal neutrons, called its thermal neutron capture cross-section. This is defined as the area presented by a nucleus to oncoming radiation and is measured in barns (1 barn = 10^{-28} m^2). Elements that have large cross-sections will provide a good target for thermal neutrons and will produce a strong γ-ray signal. Some elements have very low cross-sections and are correspondingly more difficult to detect by this means. The capture cross-sections for the most common elements in coal are given in Table 2.2. It can be seen from this table that sulfur has an intermediate cross-section, indicating that it is reasonably easy to detect by this means. It should be noted that chlorine has an unusually high cross-section, and as a result it produces a very pronounced peak in the γ-ray spectrum even when it is present in low quantities. In practice, it has been found that when the chlorine concentration is higher than 0.5% by weight, the emissions from the chlorine are so intense that they swamp the spectrum from the other elements, which impairs the accuracy of the measurement.

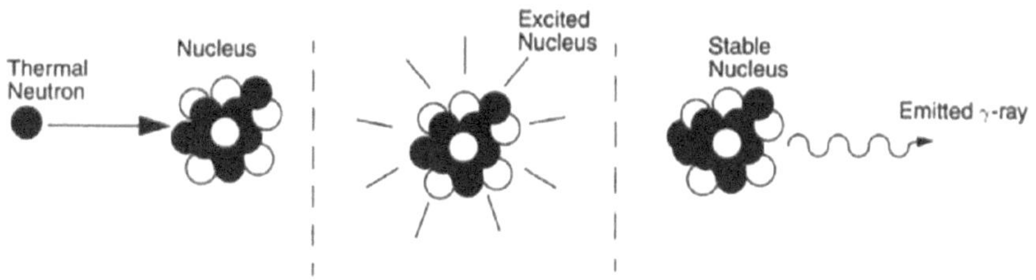

FIGURE 2.6 Emission of γ-rays in prompt neutron activation analysis.

TABLE 2.2
Thermal Neutron Capture Cross-Sections for the Most Common Elements in Coal (Clayton et al., 1983; Mughabghab, 2006)

Element	Neutron Capture Cross-Section, Barns
Aluminum	0.231
Carbon	0.0035
Chlorine[a]	33.2
Hydrogen	0.3326
Iron	2.59
Nitrogen	0.080
Oxygen	0.00019
Potassium	2.1
Silicon	0.172
Sodium	0.517
Sulfur	0.51

[a] Chlorine cross-section is averages of chlorine-35 (43.6) and chlorine-37 (0.433) cross-sections.

2.5.2.2 Calibration

To produce quantitative measurements of the elemental concentrations, the PGNAA must be calibrated so that the intensity of the emitted γ-rays can be correlated with the concentrations. In the first stage of calibration, a block of either a pure plastic or a coal of known composition is used. The block is placed inside the analyzer and exposed to the neutron source so that a base level of γ-ray emission can be determined. The block is then accurately doped with a series of known quantities of a particular element, and the resulting γ-ray spectra with varying intensities are used to produce a calibration equation for that element. This is repeated for each element to be measured, producing a spectra library for the instrument to use.

Once the instrument is calibrated, it can be used to analyze an unknown sample, producing a γ-ray spectrum. By matching peaks in the measured spectrum with the spectrum library, the unit can then identify the elements present. The computer then determines what relative concentrations the elements must have in order to give the observed peak intensities.

This method works well when all of the elements have roughly similar thermal neutron capture cross-sections. However, if an element with a high capture cross-section (such as chlorine) is present in significant amounts, it will emit a disproportionately large quantity of γ-rays. The spectrum from such an element can overwhelm the other spectra, making calibration very difficult. In general, existing instruments do not work well for high-chlorine coals.

The calibration must also be fine-tuned to the coal that the instrument will be used to analyze. To do this, coal samples are collected from the site where the analyzer will be installed, and they are carefully analyzed to determine how much of each element of interest they contain. The analyzed coal samples are then run through the instrument, and the measured analyses are compared with the analyses produced by the PGNAA device. Differences between the two analyses are then corrected by adjusting the calibration equations, bringing the PGNAA into agreement with what is actually present. This procedure must be repeated any time the composition of the coal changes.

2.5.2.3 Equipment Design

A typical PGNAA consists of a large housing surrounding the radiation source, the γ-ray detector, and the associated electronics. Thorough shielding of the neutron source is critical, and typically two tiers of shielding are used. The first tier is a lead or bismuth shield surrounding the source. This absorbs the γ-rays emitted by the radioisotope, while letting the neutrons through to reach the sample. The second tier of shielding surrounds the source shield, the detector, and the portion of the coal under analysis. This shield is made up of materials that are rich in hydrogen (such as high-density polyethylene, water, or polyester resin), which are effective for slowing the neutrons. The second-tier shield is often loaded with boron, which is extremely effective for capturing neutrons once they have been slowed down by interactions with hydrogen. The second shield tier is mainly intended to protect plant personnel from being irradiated by neutrons.

A schematic of a typical sensor assembly is shown in Figure 2.7. Coal flows down through the chute, through the measurement volume where it is bombarded by neutrons and emits γ-rays. The radiation from the coal is then received by the detector, and the data is sent to an external computer for analysis. The computer then calculates the elemental concentration and displays the results with updates several times a minute.

2.5.2.4 Applications

PGNAA was first applied to coal analysis at the Morgantown Energy Technology Center in the early 1970s. This work showed that it was practical to determine sulfur by this means. Subsequent work was funded by Electric Power Research Institute (EPRI)

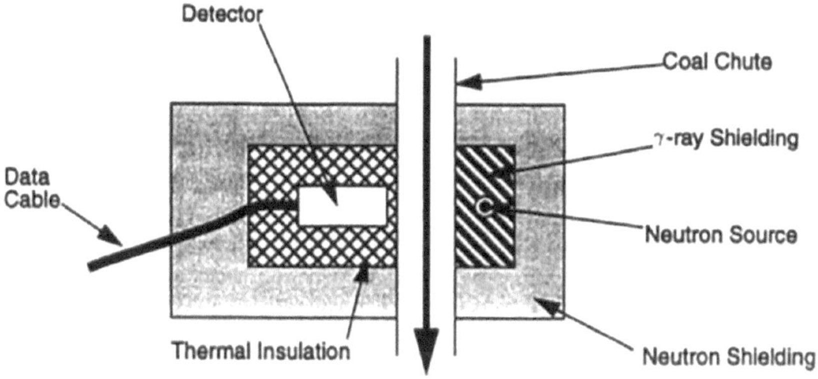

FIGURE 2.7 Sensor assembly for a PGNAA unit.

in the late 1970s, which led to a number of instruments. Currently, there are two major manufacturers of on-line PGNAA units: Thermo Gamma-Metrics and Scantech.

- *Thermo Gamma-Metrics.* The Thermo Gamma-Metrics coal analyzer (Model 3612C) was introduced in 1984 and could directly analyze coal flows up to 500 tons/h. In 1986, the smaller Model 1812C was introduced, with a capacity of 100 tons/h. In addition, the company has developed a coal blending system that uses a PGNAA analyzer to infer the analyses of up to three source coals and control the quality of the blended product (Woodward, 1988, 1989a,b).
- *Coalscan 9500X.* Coalscan is a brand of an on-line coal measurement system developed by Scantech, an Australian company that manufactures on-line ash meters. In addition to the PGNAA system, this unit also incorporates an ash monitor using a dual-energy γ-ray transmission technique, and a microwave moisture meter.
- *Other units.* The UK Harwell laboratory has developed a prototype PGNAA analyzer, which uses an array of 13 sodium iodide detectors in place of the single detector used by other systems. This arrangement allows greater resolution of the γ-ray spectrum without a significant loss in efficiency (Warmald, 1989a,b).

The basic price of these analyzers is typically under $500,000, with additional costs for additional equipment such as supplementary analyzers and sampling systems. In some cases, the rapid on-line analysis that these units produce is beneficial enough that the payback period is quite short. For sulfur concentrations in the coal between 0.5% and 2.5%, the sulfur can be determined by PGNAA with a 2σ precision of 0.1%–0.2% sulfur (Page, 1991).

2.5.3 On-Line Ash Analysis in Coal Slurries

Ash analysis is the determination of noncombustible mineral matter, especially alumina, silica, and iron compounds, present in a coal. This material is largely excess weight, which will be transferred to the fly ash or the bottom ash after combustion, and presents no particular benefit to the combustion process. It is often desirable to remove these ashes during coal cleaning, but doing so effectively requires a method to measure their presence. It is also very important to have an accurate understanding of solids and ash content in coal slurries when designing and controlling flotation cells, jigs, and heavy media separators.

Measuring the ash compounds is a matter of determining the percent of ash and solids in slurries. By combining these two measurements, it is possible to determine the ash content of the coal at a given location in the process.

The percent ash can be determined by the Rayleigh and Compton scattering of X-rays passing through the material. Rayleigh scattering, also known as elastic scattering, does not reduce the energy of the incoming photons. Compton scattering, also known as inelastic scattering, does reduce the energy of the incoming photons. If a photon from a 10 keV source strikes a molecule and is detected at 10 keV, then it underwent Rayleigh scattering.

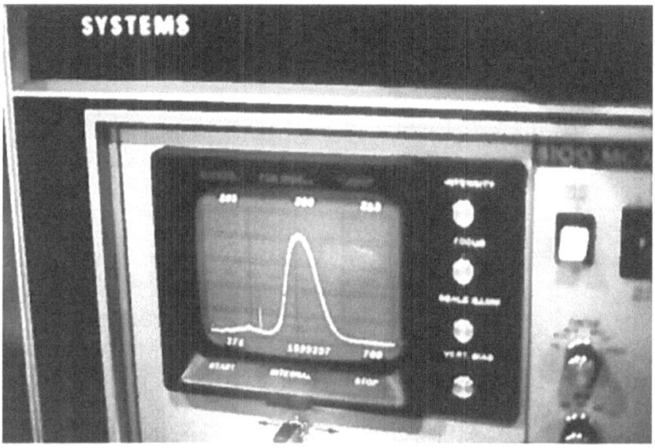

FIGURE 2.8 Measurement of Rayleigh scattering (large peak on the right) and Compton scattering (small peak on the left) on an oscilloscope at the Banning mine in Pennsylvania.

In Figure 2.8, the Rayleigh and Compton scattering peaks from a Cd-109 source are shown. The small peak on the left is the Compton scattering peak, which is nearly negligible in this figure. The peak on the right is the Rayleigh scattering peak. Common radiation sources for this measurement would include Cd-109 which primarily emits X-rays at 22.1 keV and Pu-238 which primarily emits X-rays between 11.6 and 20.7 keV. The area under the peak in Figure 2.8 is correlated with the total ash content of the coal material.

To determine the percent solids in the slurry, a gamma density gauge can be used. Originally gamma density gauges were built for heavier materials such as chalcopyrite. Coal is a much lighter material, and therefore it is important to use a lower energy radiation source. Otherwise, the transmission path lengths required for sufficient absorption and scattering to take place become very large. If the path length is too short (not enough absorption) or too long (too much absorption), the sensitivity of a gamma density gauge decreases considerably (Kawatra, 1976a). For chalcopyrite ($CuFeS_2$), a typical source would be Cs-137 (emitting at 662 keV). For coal, it is preferable to use a Gd-153 source with emissions at 41, 97, and 103 keV (Kawatra, 1976b). The γ-rays at 97 and 103 keV are more useful than the 41 keV X-rays for coal, but the X-rays can be blocked with a lead shield. This model can be fit using a linear model, such as Equation 2.3.

$$\%\ \text{Ash} = a_0 + a_1 N_1 + a_2 N_2 \tag{2.3}$$

where a_0, a_1, and a_2 are fitting parameters, and N_1 and N_2 are sensor inputs. This setup is outlined in Figure 2.9.

Pyrite minerals can be separately detersmined in this setup with the use of a Cd-109 source to trigger X-ray fluorescence in iron. The fluorescent X-rays from iron allowed for the independent characterization of iron in the sample, allowing for pyrite to be accounted for (Kawatra, 1993). This information can be incorporated into Equation 2.3 directly as an additional parameter.

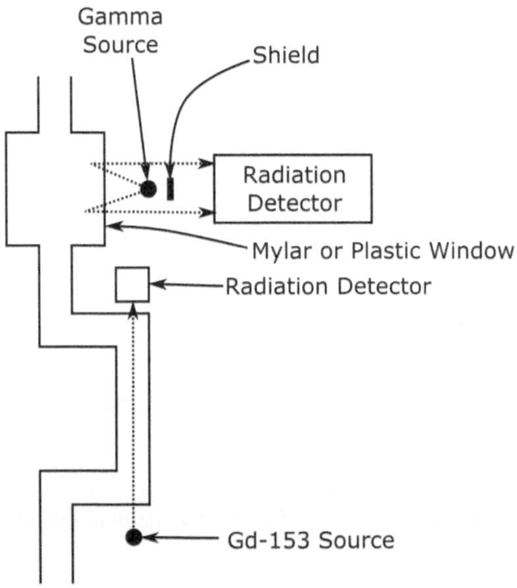

FIGURE 2.9 Example arrangement of X-ray backscattering sensor and gamma density gauge for on-line coal slurry ash analysis.

2.6 EMISSIONS MONITORING

In addition to measuring the sulfur content of a coal before it enters a power plant, it is also important to be able to measure the sulfur content of the gases leaving the power plant. This has historically been done by collecting gas samples directly and analyzing them in the laboratory to determine sulfur content. Increasingly stringent regulations, however, are making it more important to monitor the sulfur emissions frequently, rapidly, and with high accuracy; therefore, improved methods are under development for this purpose. Regulatory agencies have been demanding remote-sensing methods so that they can measure the sulfur content of stack-gas plumes even when they do not have access to the inside of the plant. Two remote-sensing methods under development are

- FTIR of stack-gas plumes, using the sky background as the IR source (Chaffin et al., 1995).
- Differential absorption lidar, using scattered laser light to detect sulfur compounds (Cunningham et al., 1994).

2.7 SUMMARY

A number of techniques are available for determining the amount of sulfur in coal, and routine analyses using ASTM-approved procedures can be used to measure total sulfur, pyritic sulfur, sulfate sulfur, and organic sulfur. This is sufficient for most purposes, but there is still great difficulty in determining the precise compounds in coal that contain sulfur. This is not a serious problem for most physical separations, which are mostly concerned with removing pyritic sulfur; however, chemical and biological techniques for removing organic sulfur require much more advanced analytical techniques that are still being developed.

Analytical procedures such as float–sink analysis are also needed that can determine the maximum possible separation performance of a physical sulfur separation, and rapid on-line analysis techniques are crucial for controlling desulfurization processes on an industrial scale.

REFERENCES

Ailey-Trent, K.S., McGowan, C.W., Lachowicz, E. and Markuszewski, R. (1993), "The Effect of Time, Additives, and Method of Measurement on Sulfate during the Selective Oxidation of Coal with Perchloric Acid," *Fuel*, Vol. 72, pp. 1197–1201.

Alaya, J.M., Buergo, M.A. and Xiberta, J. (1994), "Use of Energy Dispersive X-ray Fluorescence (EDXRF) as an Approximate Method of Analyzing Ash and Sulfur Content in the Coals of the Asturias, Spain," *Nuclear Geophysics*, Vol. 8, No. 1, pp. 99–102.

ASTM (2012a), "Standard Test Method for Forms of Sulfur in Coal," *Annual Book of ASTM Standards*, Vol. 5.06, Standard No. D2492-02, American Society for Testing and Materials, Philadelphia.

ASTM (2012b), "Standard Test Method for Determining the Washability Characteristics of Coal," *Annual Book of ASTM Standards*, Vol. 5.06, Standard No. D4371-06, American Society for Testing and Materials, Philadelphia.

ASTM (2018a), "Standard Test Method for Sulfur in the Analysis Sample of Coal and Coke Using High-Temperature Tube Furnace Combustion," *Annual Book of ASTM Standards*, Vol. 5.06, Standard No. D4239-18e1, American Society for Testing and Materials, Philadelphia.

ASTM (2018b), "Standard Test Method for Forms of Sulfur in Coal by Inductively Coupled Plasma Atomic Emission Spectroscopy," *Annual Book of ASTM Standards*, Vol. 5.06, Standard No. D8214-18, American Society for Testing and Materials, Philadelphia.

Atar, A.T. and Hendrickson, G.G. (1982), "Functional Groups and Heteroatoms in Coal," *Coal Structure* (Meyers, ed.), Academic Press, New York, pp. 132–192.

Bakel, A.J., Philp, R.P. and Galvez-Sinibaldi, A. (1990), "Characterization of Organosulfur Compounds in Oklahoma Coals by Pyrolysis-Gas Chromatography," *Geochemistry of Sulfur in Fossil Fuels*, ACS Symposium Series No. 429 (Orr and White, eds.), American Chemical Society, Washington, DC, pp. 326–344.

Bottrell, S.H., Louie, P.K.K., Timpe, R.C. and Hawthorne, S.B. (1994), "Use of Stable Sulfur Isotope Ratio Analysis to Assess Selectivity of Chemical Analyses and Extractions of Forms of Sulfur in Coal," *Fuel*, Vol. 73, No. 10, pp. 1578–1582.

Boudou, J.-P. (1990), "Coal Desulfurization by Programmed-Temperature Pyrolysis and Oxidation," *Geochemistry of Sulfur in Fossil Fuels*, ACS Symposium Series No. 429 (Orr and White, eds.), American Chemical Society, Washington, DC, pp. 345–364.

Buchanan, D.H., Coombs, K.J., Murphy, P.M. and Chaven, C. (1993), "Convenient Method for the Quantitative Determination of Elemental Sulfur in Coal by HPLC Analysis of Perchloroethylene Extracts," *Energy and Fuels*, Vol. 7, No. 2, pp. 219–221.

Calkins, W.H. (1994), "The Chemical Forms of Sulfur in Coal—A Review," *Fuel*, Vol. 73, No. 4, pp. 475–484.

Chaffin, C.T., Marchall, T.L., Combs, R.J., Knapp, R.B., Kroutil, R.T., Fateley, W.G. and Hammaker, R.M. (1995), "Passive Fourier Transform Infrared (FTIR) Monitoring of SO_2 in Plumes: A Comparison of Remote Passive Spectra of an Actual Emission Spectra Collected with a Heatable Cell," *Proceedings of SPIE—the International Society for Optical Engineering*, Vol. 2365, pp. 302–313.

Clayton, C.G., Hassan, A.M. and Warmald, M.R. (1983), "Multi-Element Analysis of Coal During Borehole Logging by Measurement of Prompt Gamma-Rays from Thermal Neutron Capture," *International Journal of Applied Radiation and Isotopes*, Vol. 34, No. l, pp. 83–93.

Cunningham, D.L., Pence, W.H. and Moody, S.E. (1994), "Mobile SO_2 and NO_2 DIAL Lidar System for Enforcement Use," *Proceedings of SPIE—the International Society for Optical Engineering*, Vol. 2112, pp. 229–238.

Davidson, R.M. (1994), "Quantifying Organic Sulfur in Coal—A Review," *Fuel*, Vol. 73, No. 7, pp. 988–1005.

Day, R.A. Jr. and Underwood, A.L. (1986), *Quantitative Analysis*, 5th Edition, Prentice-Hall, Englewood Cliffs, NJ.

Dean, J.A. (1995), *Analytical Chemistry Handbook*, McGraw-Hill, New York.

Dos Santos, J.M.F. and Conde, C.A.N. (1994), "Application of Gas Proportional Scintillation counters to the Analysis of Sulfur in Coal," *Nuclear Geophysics*, Vol. 8, No. 1, pp. 103–106.

Eissa, N.A., Ahmed, A.A., Yaseen, O.M. and Ahmed, M.A. (1994), "Moessbauer Study of Iron Phases Produced during the Combustion of Egyptian Maghara Coal," *Arab Journal of Nuclear Science and Its Applications*, Vol. 27, No. 2, pp. 23–32.

Fiedler, R. and Bendler, D. (1992), "ESCA Investigation on Schleenhain Lignite Lithotypes and Hydrogenation Residues," *Fuel*, Vol. 47, No. 4, pp. 381–388.

Garcia-Labiano, F., Hampartsoumian, E. and Williams, A. (1995), "Determination of Sulfur Release and Its Kinetics in Rapid Pyrolysis of Coal," *Fuel*, Vol. 74, No. 7, pp. 1072–1079.

Ge, E. and Wert, C. (1990), "Spatial Variation of Organic Sulfur in Coal," *Geochemistry of Sulfur in Fossil Fuels*, ACS Symposium Series No. 429 (Orr and White, eds.), American Chemical Society, Washington, DC, pp. 316–325.

George, G.N., Gorbaty, M.L. and Kelemen, S.R. (1990), "Sulfur K-Edge X-ray Absorption Spectroscopy of Petroleum Asphaltenes and Model Compounds," *Geochemistry of Sulfur in Fossil Fuels*, ACS Symposium Series No. 429 (Orr and White, eds.), American Chemical Society, Washington, DC, pp. 220–230.

Gibb, T.C. (1976), *Principles of Mössbauer Spectroscopy*, John Wiley and Sons, New York, p. 254.

Goldstein, J.I., Yakowitz, H., Newbury, D.E., Lifshin, E., Colby, J.W. and Coleman, J.R. (1977), *Practical Scanning Electron Microscopy*, Plenum Press, New York, pp. 87–91.

Gorbaty, M.L., George, G.N. and Kelemen, S.R. (1990), "Direct Determination and Quantification of Sulphur Forms in Heavy Petroleum and Coals. 2. The Sulphur K Edge X-ray Absorption Spectroscopy Approach," *Fuel*, Vol. 69, No. 8, pp. 945–949.

Gorlov, Yu.I. and Onishchenko, A.M. (1994), "Simple Method of Automating Control of Coal Quality," *Koks i Khimiya* (*Coke and Chemistry*), No. 8, pp. 25–27.

Gryglewicz, G. (1995), "Sulfur Transformations during Pyrolysis of a High Sulfur Polish Coking Coal," *Fuel*, Vol. 74, No. 3, pp. 356–361.

Harris, L.A., Yust, C.S. and Crouse, R.S. (1977), "Direct Determination of Pyritic and Organic Sulfur by Combined Coal Petrography and Microprobe Analysis (CPMA)—A Feasibility Study," *Fuel*, Vol. 56, pp. 456–457.

Hickmott, D.D. and Baldridge, W.S. (1995), "Application of PIXE Microanalysis to Macerals and Sulfides from the Lower Kittanning Coal of Western Pennsylvania," *Economic Geology and the Bulletin of the Society of Economic Geologists*, Vol. 90, No. 2, pp. 246–254.

Hou, X., Ren, D., Mao, H., Lei, J., Jin, K., Chu, P.K., Reich, F. and Wayne, D.H. (1995), "Application of Imaging TOF-SIMS to the Study of Some Coal Macérais," *International Journal of Coal Geology*, Vol. 27, No. 1, pp. 23–32.

Huffman, G.P., Huggins, F.E., Shah, N., Bhattarcharyya, D., Pugmire, R.J., Davis, B., Lytle, F.W. and Greegor, R.B. (1987), "Investigation of the Atomic and Physical Structure of Organic and Inorganic Sulfur in Coal," *Processing and Utilization of High Sulfur Coals II* (Chugh and Caudle, eds.), Elsevier, Amsterdam, pp. 3–12.

Huffman, G.P., Shah, N., Hugins, F.E., Stock, L.M., Chatterjee, K., Kilbane II, J.J., Chou, M.-I.M. and Buchanan, D.H. (1995), "Sulfur Speciation of Desulfurized Coals by XANES Spectroscopy," *Fuel*, Vol. 74, No. 4, pp. 549–555.

Irdi, G.A. and Rohar, P.C. (1988), "Validation of Image Analysis and Sample Pellet Preparation Methods for Estimating Pyrite Particle Size Distribution in Crushed Coals," Pittsburgh Energy Technology Center, U.S. DOE, Report No. DOE/PETC/TR-88/3 (DE88010319).

Irdi, G.A. Rohar, P.C. and Chaparro, L.F. (1988), "The Development of an Automated Image Analysis Method to Estimate Pyrite Particle Liberation in Crushed Coals," Pittsburgh Energy Technology Center, U.S. DOE, Report No. DOE/PETC/TR-88/4 (DE88014876).

Kawatra, S.K. (1976a). "The Influence of Length of Flow Cell and Strength of Source on the Performance of a Gamma Density Gauge," *International Journal of Mineral Processing*, Vol. 3, pp. 167–174.

Kawatra, S.K. (1976b). "The Use of a Gd-153 Gamma Ray Density Gauge for Coal Slurries in an On-Line X-Ray Fluorescence System," *Canadian Journal of Spectroscopy*, Vol. 21, No. 4, pp. 97–100.

Kawatra, S.K., Dalton, J.L. and Hill, W.A. (1979). "A Gamma Density Gauge for Coal Cleaning Processes," *American Nuclear Society Transactions*, Vol. 32.

Kawatra, S.K. (1980). "On-Line Characterization of Coal Slurries," Argonne National Laboratory, Publication No. ANL-80-62.

Kawatra, S.K., Seitz, R.A. and Suardini, P.J. (1984), "The Control of Coal Flotation Circuits," Chapter 23, *Control '84* (Herbst, George, and Sastry, eds.), Society of Mining Engineers, New York, pp. 225–233.

Kawatra, S.K. (1985), "The Development and Plant Trials of an Ash Analyzer for Control of a Coal Flotation Circuit," *Proceedings of the 15th International Mineral Processing Congress*, Vol. 3, Cannes, France, pp. 176–188.

Kawatra, S.K. (1987), "Instrumentation and Control of Fine Coal Processing," Chapter 12, *Fine Coal Processing* (Mishra and Klimpel, eds.), Noyes Publications, Park Ridge, NJ, pp. 294–328.

Kawatra, S.K. (1993), "The Theory and Development of a Michigan Tech/Outokumpu Slurry Ash Analyzer," Chapter 28, *Emerging Computer Techniques for the Minerals Industry* (Scheiner and Stanley, eds.), Society for Mining, Metallurgy, and Exploration, Littleton, CO, pp. 259–270.

Kawatra, S.K. and DeLa'O, K.A. (1995), "On-Line Sensors and Process Control for Coal Preparation Plants," Chapter 25, *Industrial Practice of Fine Coal Processing* (Klimpel and Luckie, eds.), Society for Mining, Metallurgy, and Exploration, Inc., Littleton, CO, pp. 257–274.

Kelemen, S.R., George, G.N. and Gorbaty, M.L. (1990), "Direct Determination and Quantification of Sulphur Forms in Heavy Petroleum and Coals. 1. The X-ray Photoelectron Spectroscopy (XPS) Approach," *Fuel*, Vol. 69, No. 8, pp. 939–944.

Kelly, W.R., Paulsen, P.J., Murphy, K.E., Vocke Jr., R.D. and Chen, L.-T. (1994), "Determination of Sulfur in Fossil Fuels by Isotope Dilution Thermal Ionization Mass Spectrometry," *Analytical Chemistry*, Vol. 66, No. 15, pp. 2505–2513.

Keulemans, A.I.M. (1959), *Gas Chromatography*, Reinhold Publishing, New York.

Killmeyer, R.P., Hucko, R.E. and Jacobsen, P.S. (1992), "Centrifugal Float-Sink Testing of Fine Coal: An Interlaboratory Test Program," *Coal Preparation*, Vol. 10, pp. 107–118.

LaCount, R.B., Anderson, R. R., Friedman, S. and Blaustein, B. D. (1987), "Sulfur in Coal by Programmed Temperature Oxidation," *Fuel*, Vol. 66, No. 7, pp. 903–913.

Laurila, M.J. and Kawatra, S.K. (1985), "On-Line Instrumentation for Control of a Coal Preparation Facility," *Instrumentation in the Mining and Metallurgy Industries*, Vol. 12, The Instrument Society of America, pp. 81–88.

Lee, S. and Fullerton, K.L. (1992), "Characterization of Desulfurization Extracts from Midwestern U.S. Coals," *Fuel Science and Technology International*, Vol. 10, No. 7, pp. 1137–1159.

Louie, P.K.K., Timpe, R.C., Hawthorne, S.B. and Miller, D.J. (1994), "Sulfur Removal from Coal by Analytical-Scale Supercritical Fluid Extraction (SFE) Under Pyrolysis Conditions," *Fuel*, Vol. 73, No. 7, pp. 1173–1178.

McGowan, C.W. and Markuszewski, R. (1987), "Fate of Sulfur Compounds in Coal During Oxidative Dissolution in Perchloric Acid," *Fuel Processing Technology*, Vol. 17, No. 1, pp. 29–41.

McGowan, C.W. and Markuszewski, R. (1988), "Direct Determination of Sulfate, Sulfide, Pyritic, and Organic Sulfur in a Single Sample of Coal by Selective, Step-Wise Oxidation with Perchloric Acid," *Fuel*, Vol. 67, No. 8, pp. 1091–1095.

Mughabghab, S.F. (2006), "*Atlas of Neutron Resonances: Resonance Parameters and Thermal Cross Sections. Z= 1–100*," Elsevier, Amsterdam.

Page, D. (1991), "Methods for On-Line Coal Quality Measurement," *Mine & Quarry*, Vol. 20, pp. 13–16.

Palmer, S.R., Hippo, E.J., Kruge, M.A. and Crelling, J.C. (1990), "Characterization of Organic Sulfur Compounds in Coals and Coal Macerals," *Geochemistry of Sulfur in Fossil Fuels*, ACS Symposium Series No. 429 (Orr and White, eds.), American Chemical Society, Washington, DC, pp. 296–315.

Palowitch, E.R. and Nasiatka, T.M. (1961), "Using a Centrifuge for Float-and-Sink Testing Fine Coal," Report of Investigations RI 5741, U.S. Department of the Interior, Bureau of Mines.

Pitard, F.F. (1993), *Pierre Gy's Sampling Theory and Sampling Practice: Heterogeneity, Sampling Correctness, and Statistical Process Control*, CRC Press, Boca Raton, FL.

Pitard, F.F. (2019), *Theory of Sampling and Sampling Practice*, 3rd Edition, Chapman and Hall/CRC, New York.

Putnis, A. (1992), *Introduction to Mineral Sciences*, Cambridge University Press, Cambridge, MA.

Riley, J.T., Ruba, G.M. and Lee, C.C. (1990), "Direct Determination of Total Organic Sulfur in Coal," *Geochemistry of Sulfur in Fossil Fuels*, ACS Symposium Series No. 429 (Orr and White, eds.), American Chemical Society, Washington, DC, pp. 231–240.

Shao, D., Hutchinson, E.J., Heidbrink, L, Pan, W.-P. and Chou, C.-L. (1994), "Behavior of Sulfur during Coal Pyrolysis," *Journal of Analytical and Applied Pyrolysis*, Vol. 30, pp. 91–100.

Stiller, A.H., Renton, J.J. and Montano, P.A. (1990), "Comparison of Methods for the Determination of the Pyritic Sulfur Content of Coal," *Mining Science and Technology*, Vol. 11, No. 2, pp. 153–156.

Stock, L.M. and Wolny, R. (1990), "Elemental Sulfur in Bituminous Coals," *Geochemistry of Sulfur in Fossil Fuels*, ACS Symposium Series No. 429 (Orr and White, eds.), American Chemical Society, Washington, DC, pp. 241–248.

Suardini, P.J. (1995), "Improving Fine-Coal Washability Procedures," *High Efficiency Coal Preparation* (Kawatra, ed.), Society for Mining, Metallurgy, and Exploration, Littleton, CO, pp. 119–128.

Tao, D.P., Li, Y.Q., Richardson, P.E. and Yoon, R.-H. (1994), "Incipient Oxidation of Pyrite," *Colloids and Surfaces A: Physicochemical and Engineering Aspects*, Vol. 93, pp. 229–239.

Thomas, J.V. (1995), "Coal," *Analytical Chemistry*, Vol. 67, No. 12, pp. 317R–319R.

Torres-Ordonez, R.J., Calkins, W.H. and Klein, M.T. (1990), "Distribution of Organic- Sulfur-Containing Structures in High Organic Sulfur Coals," *Geochemistry of Sulfur in Fossil Fuels*, ACS Symposium Series No. 429 (Orr and White, eds.), American Chemical Society, Washington, DC, pp. 287–295.

Tsai, S.C. (1982), *Fundamentals of Coal Beneficiaron and Utilization*, Elsevier, New York, pp. 375.

Tsarenko, P.L., Zubova, T.V., Glinskii, V.G., Demyanets, L.N. (1994), "The Use of X-Ray Spectral Quantometers for the Analytical Control of Products of Metallurgical Plants," *Koks i Khimiya* (*Coke and Chemistry*), No. 10, pp. 39–40.

Warmald, M.R. (1989a), "Pair Spectrometer NaI(Tl) Array for Neutron-Induced Prompt Gamma-Ray Analysis," *Nuclear Geophysics*, Vol. 3, No. 4, pp. 373–380.

Warmald, M.R. (1989b), "A Bulk Materials Analyzer Using Pair and Compton-Suppressed Gamma-Ray Spectrometry," *Nuclear Geophysics*, Vol. 3, No. 4, pp. 461–466.

Weitzsacker, C.L. and Gardella, J.A. (1992), "Quantitative Electron Spectroscopic Analysis of Argonne Premium Coals," *Analytical Chemistry*, Vol. 64, No. 9, pp. 1068–1075.

Weng, S. (1994), "Study of Influence of Irradiation Time on the Microwave-Magnetic Desulfurization of Raw Coal by Mossbauer Method," *He Jishu* (*Nuclear Techniques*), Vol. 17, No. 7, pp. 437–442.

Winans, R.E. and Neill, P.H. (1990), "Multiple-Heteroatom-Containing Sulfur Compounds in a High Sulfur Coal," *Geochemistry of Sulfur in Fossil Fuels*, ACS Symposium Series No. 429 (Orr and White, eds.), American Chemical Society, Washington, DC, pp. 249–260.

Woodward, L.A. (1967), "General Introduction," *Raman Spectroscopy* (Szymanski, ed.), Plenum Press, New York, pp. 1–43.

Woodward, R.C. (1988), "On-Line Coal Analysis," *Paper presented at the American Mining Convention*, Denver, CO.

Woodward, R.C. (1989a), "Automated Blending with an On-Line Analyzer," *Proceedings of the 6th International Coal Preparation Exhibit and Conference*, Lexington, Kentucky, pp. 257–266.

Woodward, R.C. (1989b), "Automated Blending with an On-Line Analyzer," *Paper presented at the EPRI Conference on Utility Applications of On-Line Coal Analysis*, Pittsburgh, Pennsylvania.

Ye, Y., Jin, R. and Miller, J.D. (1988), "Thermal Treatment of Low-Rank Coal and Its Relationship to Flotation Response," *Coal Preparation*, Vol. 6, pp. 1–16.

Yurum, Y., Osyoru, H., Gulce, H. and Tugluhan, A. (1990), "Supercritical Extraction and Desulfurization of Beypazari Lignite by Ethyl Alcohol/NaOH Treatment," *Fuel Science and Technology International*, Vol. 8, No. 7, pp. 699–718.

Zhou, R., Arnold, B., Chander, S., Hogg, R. and Polat, M. (1993), "Problems in Sink-Float Analysis of Fine Coal," *Processing and Utilization of High-Sulfur Coals V* (Parekh and Groppo, eds.), Elsevier, Amsterdam, pp. 189–205.

Zumberge, J.F. (1987), "Measurements of Coal Quality by Prompt Gamma Neutron Activation Analysis," *Journal of Coal Quality*, Vol. 6, pp. 120–123.

3 Precombustion/ Postcombustion Desulfurization

There are two basic options for controlling sulfur emissions from a coal-fired power plant. The first is to remove the sulfur from the coal before it is burned (precombustion), and the second is to "scrub" the sulfur oxides from the combustion gases after the coal is burned (postcombustion) (Davies, 1989). Each of these approaches has its own advantages and limitations, and it is generally best to be able to use both types of processes to remove the greatest proportion of sulfur at the lowest cost.

3.1 PRECOMBUSTION DESULFURIZATION

Precombustion processes include all of the various physical and chemical methods for removing sulfur from coal, which will be discussed later in this book. This also includes the option of switching to a low-sulfur coal, either to use alone or to blend with high-sulfur coal to reduce the total sulfur content per British Thermal Units (BTU) (Harrison and Stallard, 1992). Low-sulfur U.S. coals tend to be low-rank western sub-bituminous or lignite coals, and switching to a lower-sulfur coal is not always a practical solution, for the following reasons (Makansi, 1993a; Skorupska, 1993):

- Low-rank coals are prone to spontaneous combustion and require extensive fire prevention and suppression equipment.
- Coal pulverizers are designed for coals of specific grindability and moisture content, and switching coals can cause a degradation in pulverizer performance.
- Changes in ash composition will change the slagging, fouling, and corrosion in the boiler and can increase maintenance requirements or harm boiler performance.
- Increases in feed coal ash content per BTU delivered can strain the capacity of the ash handling system. Also, many low-sulfur coals produce ashes that will cement together when exposed to moisture.
- Low-rank coals often do not flow readily from bins and hoppers, because of their higher moisture content. Chute angles, bin designs, conveyor capacities, etc. may have to be altered.
- The lower heat content per ton of low-rank coals will require increases in the amount of supplied air, possibly requiring changes in the blowers.

Because of these factors, many plants will prefer to desulfurize the coals they are currently using, rather than trying to switch to a low-sulfur coal.

3.1.1 Benefits of Precombustion Desulfurization

There are several significant benefits of processing coal to remove sulfur before combustion, as follows (Couch, 1995):

- Precombustion sulfur removal also generally removes the various ash-forming minerals, resulting in a higher-quality fuel.
- If the sulfur and mineral matter are removed and disposed of at the mine site, the disposal costs can be much lower at the mine than they would be at the power plant, often by as much as a factor of ten.
- The removal of mineral matter increases the heating value of a ton of coal, which reduces the shipping cost.
- Coals that are near-compliance before preparation can be reduced in sulfur content until their sulfur emissions per unit energy content are low enough to make them compliance coals.

However, precombustion coal desulfurization also has a number of limitations, which prevent these methods from being a complete solution to sulfur emission problems (Couch, 1995; Vernon and Jones, 1993):

- Many high-sulfur coals cannot be made into compliance coals using precombustion treatment because they contain much of their sulfur in forms that cannot be economically removed.
- The effectiveness of precombustion desulfurization depends on the characteristics of the coal, and so no one method is universally suitable for all coals.
- Customers are often not willing to pay a premium price for lower-sulfur coals, making it often difficult to justify expensive precombustion desulfurization processes.
- Some precombustion treatments are best performed on-site at the power plant, immediately before combustion. Utility companies, however, have traditionally preferred not to carry out such processing at their plants and have left it to the coal producers to do the coal preparation.
- The removal of organic sulfur remains an exceedingly difficult challenge, so coals containing a comparatively large amount of organic sulfur may not be very amenable to precombustion sulfur removal. Chemical removal of sulfur in the precombustion step is largely a laboratory curiosity (Xia and Xie, 2017), with some of the more promising methodologies involving oxidative removal with ionic liquids achieving only 16.76% organic sulfur removal (Wang et al., 2019) or oxidative leaching with sodium hypochlorite (Li and Cho, 2005). Neither of these technologies has been implemented in an operating plant.

In general, precombustion desulfurization is most effective for coals with a large proportion of their sulfur as coarse pyrite (such as northern hemisphere Carboniferous coals); for coals with sulfur contents that are near compliance and can be made into compliance coals by removing a portion of the sulfur; for coals that contain a great deal of ash or sulfur, whose quality can be markedly improved by precombustion processing; and for producing premium coals for markets other than coal combustion.

3.1.2 Processing Costs

The cost of precombustion coal desulfurization is highly variable, depending on the process used, the nature of the coal being processed, and the degree of sulfur removal needed. Conventional coal cleaning of coal fractions coarser than approximately 0.635 cm (1/4 in) costs $2–$3/metric ton of coal, while cleaning of the finer fractions can cost up to $10/metric ton. More advanced desulfurization processes are more expensive, at approximately $15–$30/metric ton (Couch, 1995). On average, coal cleaning adds $2–$12/metric ton to the cost of coal. This is offset by a reduction of transportation costs by as much as 20% (due to the increased calorific value of clean coal) and by reductions in coal handling, storage, pulverizing, and boiler maintenance costs (due to removal of bulky, abrasive, and slagging ash minerals) (Elliott, 1992).

Precombustion coal cleaning permits smaller, and therefore, less expensive, flue-gas desulfurization equipment. Any reduction of the quantity of sulfur in the coal will decrease the quantity of sulfur-absorbing reagents required for flue-gas desulfurization and will reduce the amount of waste disposal required. Cleaned coal can improve the power station heat rate by reducing the amount of power that must be dedicated to flue-gas handling systems and by allowing lower air heater exit temperatures, thus increasing boiler efficiency. Removal of ash during cleaning also improves the quality of coal, and pilot-scale combustion tests have shown that boiler efficiency can be significantly improved by cleaning coal prior to combustion, as shown in Table 3.1 (Cichanowicz and Harrison, 1989). Since deep cleaning of coal requires very fine grinding to liberate the mineral matter, there is interest in developing economical methods for micronizing coal (Jones, 1994; Anonymous, 1990a).

TABLE 3.1
The Effects of Cleaning a Coal on the Operation of a Boiler, as an Example

	Run-of-Mine Coal	Medium-Cleaned Coal	Deep-Cleaned Coal
% Moisture	1.7	1.7	1.6
% Sulfur	3.8	3.7	2.0
% Ash	23.5	7.1	3.5
Heating value, MJ/kg	23.38	31.03	32.66
Flue gas SO_3 concentration, ppm	7	4	3
Air heater exit temperature, °C	136	—	120
% Boiler efficiency	88.4	—	90.1
% Flue gas volume reduction	—	—	6

The cleaned coals have a higher heating value per kilogram of coal, contain less sulfur, result in fewer flue gas emissions overall, and result in higher boiler efficiencies (Cichanowicz and Harrison, 1989).

The value of desulfurization before combustion depends on the cost of removing a ton of SO_2 after combustion. This cost is variable, depending on equipment, operating, and maintenance costs, and is estimated to be in the range of \$275–\$1,650/metric ton of SO_2 removed (Barnett and Feeley, 1992). The cost of many precombustion coal desulfurization processes falls within this range. For example, studies supported by the U.S. Department of Energy have estimated that advanced column flotation can remove the equivalent of 1 metric ton of SO_2 from moderate- to high-sulfur bituminous coals at a cost of \$275/metric ton SO_2, while selective agglomeration with heavy oil costs \$325/metric ton SO_2 (Barnett and Feeley, 1992).

3.2 POSTCOMBUSTION DESULFURIZATION

In postcombustion desulfurization, the sulfur oxides that are produced during coal combustion are removed from the combustion gases using an absorptive chemical, such as calcium oxide. Since nitrogen oxides are also produced during combustion, it is possible to develop combined desulfurization methods that can remove both sulfur dioxides and nitrogen oxides (Anonymous, 1990b). Postcombustion absorbers capture the sulfur in a solid form, so that it can be disposed of. There are two basic approaches to capturing the sulfur: (1) divert the combustion gases through a scrubber unit that is separate from the combustor, or (2) design the combustor so that the sulfur absorbent can be injected directly along with the fuel.

3.2.1 Scrubber Technologies

Scrubbers are devices that receive the gases from the coal burner and selectively remove the sulfur oxides from the gases. The advantage is that they can be used to retrofit existing plants (Ellison, 1991). There are several types of scrubbers available, and many different chemical reactions have been used for the extraction of sulfur oxides, as shown in Table 3.2.

3.2.1.1 Wet Scrubbers

Wet scrubbers use a slurry or solution of a sulfur absorbent in water, which is generally contacted with the flue gases using a scrubber tower, such as that shown in Figure 3.1. Flue gases rise through the tower through a falling spray of absorbent. The absorbent spray removes the sulfur oxides from the gases and collects them in the base of the tower, where it is removed. The mist eliminator at the top of the column is to prevent fine droplets of absorbent from being carried up the stack, where they could cause corrosion or deposition problems in the stack or be released as particulate pollution.

3.2.1.1.1 Limestone and Lime Absorbents

The most common absorbents are lime (calcium hydroxide) and limestone (calcium carbonate) slurries. Limestone is the preferred absorbent in many modern scrubbers, because of its low cost compared to lime and other absorbents (Bryan et al., 1993; Dettmer, 1994). Lime is also used, however, because of its higher reactivity, which allows it to absorb sulfur more rapidly. This makes it possible to use smaller scrubbers

TABLE 3.2
Absorbents That Have Been Studied for Removal of Sulfur Oxides from Coal Combustion Gases

Absorbent	Regenerable?	Product	References
$CaCO_3$	No	$CaSO_3$ or $CaSO_4$	
$Ca(OH)_2$	No	$CaSO_3$ or $CaSO_4$	Couch (1995)
CaO	No	$CaSO_3$ or $CaSO_4$	
MgO	Yes. Heat to release SO_2	SO_2	Burnett and Wells (1982)
Na_2SO_3	Yes. Heat to release SO_2	SO_2	Couch (1995)
$CeO_2 + Al_2O_3$	Yes. Heat to release SO_2	SO_2	Hedges and Yeh (1992)
$Ca(OH)_2$ + fly-ash	No	Calcium silicate sulfates	Kind et al. (1994)
$CuO + MnO_2$	Yes. Heat with H_2 at 200°C–560°C	H_2S	Bjöernbom et al. (1995)
$Ca(OH)_2$ + methanol	No	$CaSO_3$ or $CaSO_4$	Withum and Yoon (1989)
CaO + MgO	No	$CaSO_3$ and $MgSO_3$	Makansi (1993a)
$NaHCO_3$	Yes. React with $Ca(OH)_2$	$CaSO_4$	Valencia (1982)
Gaseous oxidation	(Nothing to regenerate)	H_2SO_4	Durrani (1994)
$CaCO_3$ + NaCl	No	$CaSO_3$ or $CaSO_4$	Bulewicz and Janicka (1990)
Chalk	No	$CaSO_3$ or $CaSO_4$	Dennis and Hayhurst (1989)
Cement flue dust	No	Calcium silicate sulfates	
Alkali + Al_2O_3	Yes. Heat with CO at 700°C–800°C	COS, S, SO_2	Gavalas et al. (1987)
Ca-Mg acetate	No	$CaSO_3$ or $CaSO_4$	Levendis et al. (1993)

Some are low-cost throw-away absorbents, while others are regenerable and convert sulfur oxides into marketable products.

to treat a given quantity of gas when lime is the absorbent. A set of proposed reactions for the collection of sulfur dioxide when lime or limestone is used are as follows (Fellman and Cheremisinoff, 1993):

- Sulfur Dioxide Hydration:

$$SO_2(g) + H_2O(l) \rightarrow H_2SO_3(aq) \tag{3.1}$$

$$H_2SO_3(aq) \rightarrow H^+(aq) + HSO_3^-(aq) \tag{3.2}$$

- Lime Reactions:

$$Ca(OH)_2(s) \rightarrow Ca^{2+}(aq) + {}_2OH^-(aq) \tag{3.3}$$

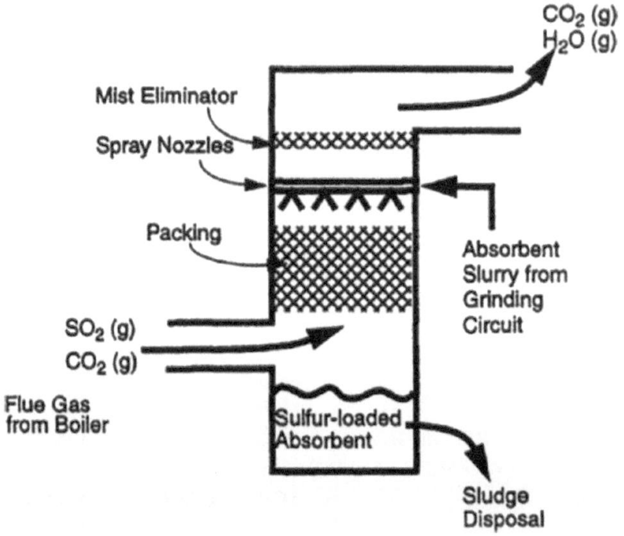

FIGURE 3.1 Basic schematic of a wet-scrubber column. Absorbent percolates down through the packing, while the flue gases flow upwards. The most common SO_2 absorbents are limestone (calcium carbonate), lime (calcium hydroxide), and "magnesium-enhanced" lime. The sulfur-bearing sludge for some scrubbers is market-grade gypsum, but for other scrubbers, it is a waste product that must be landfilled.

$$Ca^{2+}(aq) + H_2SO_3^-(aq) + \frac{1}{2}H_2O(l) \rightarrow CaSO_3 \cdot \frac{1}{2}H_2O(s) + \frac{1}{2}H_2O(l) \tag{3.4}$$

$$H^+(aq) + OH^-(aq) \rightarrow H_2O(l) \tag{3.5}$$

$$\text{Overall}: Ca(OH)_2(s) + SO_2(g) \rightarrow CaSO_3 \cdot \frac{1}{2}H_2O(s) + \frac{1}{2}H_2O(l) \tag{3.6}$$

- Limestone Reactions:

$$H^+(aq) + CaCO_3(s) \rightarrow Ca^{2+}(aq) + HCO_3^-(aq) \tag{3.7}$$

$$Ca^{2+}(aq) + HSO_3^-(aq) + \frac{1}{2}H_2O(l) \rightarrow CaSO_3 \cdot \frac{1}{2}H_2O(s) + H^+(aq) \tag{3.8}$$

$$H^+(aq) + HCO_3^-(aq) \rightarrow H_2CO_3(aq) \tag{3.9}$$

$$H_2CO_3(aq) \rightarrow CO_2(g) + H_2O(l) \tag{3.10}$$

$$\text{Overall}: CaCO_3(s) + SO_2(g) + \frac{1}{2}H_2O(l) \rightarrow CaSO_3 \cdot \frac{1}{2}H_2O(s) + CO_2(g) \tag{3.11}$$

The solid product from each of these sets of reactions is primarily calcium sulfite hemihydrate $CaSO_3 \cdot \frac{1}{2}H_2O$, which has been confirmed by X-ray diffraction analysis of scrubber sludges (Kawatra and Eisele, 1993; Kawatra et al., 1995). A similar set of reactions collect sulfur trioxide (SO_3) from the flue gases, forming gypsum ($CaSO_4 \cdot 2H_2O$) as the solid product, but under normal boiler conditions, sulfur trioxide makes up only about 0.5% of the total sulfur oxides, so its removal is less important than the removal of sulfur dioxide (Fellman and Cheremisinoff, 1993; Eisele and Kawatra, 1995).

A serious problem occurs in many wet scrubbers if sulfur dioxide is partially oxidized to sulfur trioxide. In this case, the main precipitate is calcium sulfite hemihydrate, with up to 15% calcium sulfate in solid solution in the sulfite particles. If more than 15% is as calcium sulfate, then it can no longer precipitate with the sulfite crystals, and instead it precipitates as separate crystals of gypsum. There is a shortage of gypsum seed crystals in the slurry in this situation, and much of the gypsum crystallizes on the walls of the scrubber. This can rapidly plug the scrubber, and must be avoided. Many plants prevent this problem by adding thiosulfate $S_2O_4^{-2}$ to the scrubber slurry as a reducing agent. This prevents the oxidation of sulfur dioxide and eliminates the formation of gypsum and the buildup of gypsum scale in the scrubber. A second solution to the plugging problem is to completely oxidize the calcium sulfite to gypsum, which provides more seed crystals for the gypsum and also prevents plugging.

When the solid sludge is removed from the scrubber as unoxidized calcium sulfite, as is done in many older scrubbers, it has no market value and must be disposed of by landfilling. The more advanced scrubbers include an oxidation step, which converts the sulfite to sulfate, and the solid product is then gypsum (Dettmer, 1994). If it is sufficiently pure, this synthetic gypsum can be marketed to make plaster, wallboard, cement, and other construction products. The major barrier to widespread marketing of scrubber sludge is that it contains a number of impurities and is of uneven quality, which makes it unattractive for most purposes in its raw form. Potential users are therefore not eager to purchase the material, even with significant price breaks (Ellison and Hammer, 1988; Van der Brugghen and Koppins-Odink, 1989). Typical quality requirements for gypsum for use in wallboard manufacture are given in Table 3.3. For the gypsum to be salable, it should meet or exceed these requirements.

A complete circuit for an advanced scrubber is shown in Figure 3.2, which includes oxidation of the sludge to form gypsum (Makansi, 1993a). In this circuit, limestone is first reduced to a fine particle size by a grinding mill, producing a slurry. The slurry is then added to the absorber tank and pumped into the scrubber tower. A portion of the descending absorbent is diverted back to the absorber tank, which provides more time for the sulfur dioxide and limestone to react. The remaining absorbent collects in the base of the tower, where it is oxidized by injected air while being recirculated in the lower portion of the scrubber. A portion of the absorbent is continuously drawn off to a hydrocyclone, which separates the gypsum particles from the absorbent slurry, and returns the liquid to the scrubber.

The most efficient wet-scrubber technology, from the standpoint of sulfur removal efficiency and equipment size, is the magnesium-enhanced lime process. This type of scrubber uses lime that contains up to 12% magnesium, which increases the absorption capacity of the lime to approximately 10–15 times that of the limestone

TABLE 3.3
Impurity Limits for Gypsum for Use in Plaster or Wallboard Manufacture (Shoop et al., 1996)

Impurity	Maximum % by Weight
Fe_2O_3	1.5
SiO_2	1.0
MgO	0.1
K_2O	0.1
Na_2O	0.04
Cl	0.01
CO_3	1.5
SO_2	0.25
Moisture	8.0

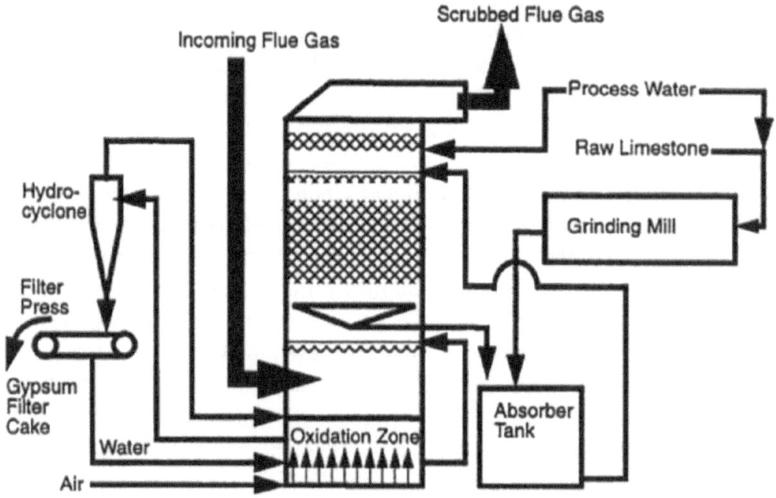

FIGURE 3.2 Circuit for a limestone scrubber, with oxidation of the solids to gypsum. The absorber tank simplifies control of the process. Air is injected into the oxidation zone to oxidize the calcium sulfite to gypsum. Solid gypsum particles are removed from the absorbent slurry using a hydrocyclone separator.

scrubbers described previously. The principal advantage is that the magnesium-enhanced lime is soluble enough that the SO_2 removal is governed by the degree of gas–liquid contact in the scrubber, and not by the degree of absorbent dissolution as is the case with limestone. A disadvantage is that the magnesium-enhanced lime is comparatively expensive because it must be calcined by heating before use. Also, the magnesium specifically inhibits the formation of gypsum. This inhibition not only helps to prevent plugging and scaling but also results in the sludge being an unmarketable sulfite sludge instead of gypsum. Finally, the magnesium content of the sludge is too high for use in most synthetic gypsum markets.

3.2.1.1.2 Other Absorbents

In addition to lime and limestone, a number of other absorbents have been used to improve the efficiency of sulfur removal or to recover the sulfur in a marketable form while regenerating the absorbent. Next-generation scrubbers are therefore under development to improve the efficiency and reduce the quantity of unmarketable waste products (Feeney, 1995; Anonymous, 1992a,b). Several of the scrubber technologies that use other absorbents are as follows:

- *Dual-alkali process*: In this process, the absorption of sulfur dioxide is first carried out using a solution of a sodium alkali, such as NaOH, Na_2CO_3, or Na_2SO_3. Since these are all very soluble in water, they can absorb the sulfur dioxide very rapidly and completely and can be easily oxidized afterwards. Also, the absorbent is a clear liquid rather than a slurry, and so the problems with scaling and plugging of the scrubber are much reduced (Valencia, 1982). The oxidized sulfur-bearing alkali is then circulated to a vessel, where it is reacted with lime or limestone, which precipitates the sulfur as calcium sulfate and regenerates the sodium alkali. A flow diagram of the process is shown in Figure 3.3. The dual-alkali process is reported to be stable and resistant to disturbances, and to be capable of removing more than 99% of the sulfur dioxide from flue gases (Valencia, 1982; Hodges et al., 1992).

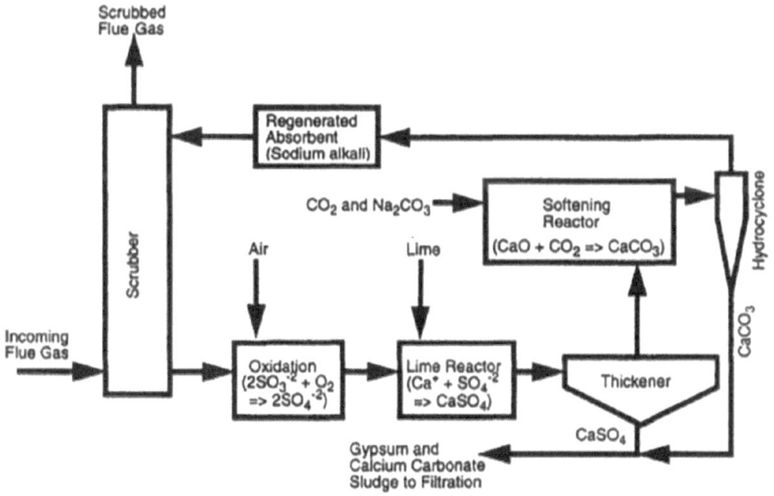

FIGURE 3.3 Flow diagram of the dual-alkali scrubber process, using lime to regenerate the sodium alkali. The clarified liquid from the thickener contains dissolved calcium sulfate, which would make calcium carbonate scale if it were allowed to contact the carbon dioxide in the flue gas. It is therefore precipitated by injecting carbon dioxide in the softening reactor, and the resulting calcium carbonate is removed by the hydrocyclone. Additional sodium alkali is also added at this point, to make up sodium alkali losses to the sludge.

- *Wellman-Lord process*: This is a regenerable-sorbent process, producing SO_2 gas, which can be sold for industrial uses. It uses a solution of sodium sulfite (Na_2SO_3), which absorbs SO_2 and becomes a sodium bisulfite solution ($NaHSO_3$). The sodium bisulfite is then decomposed in a forced circulation evaporator, releasing the SO_2 at sufficiently high concentration to be compressed and sold as SO_2 gas or used for producing elemental sulfur or sulfuric acid (Couch, 1995).
- *Magnesium oxide process*: The magnesium oxide slurry is used to collect SO_2, and the resulting magnesium sulfite is thermally treated to release the SO_2 and regenerate the absorbent, as shown in Figure 3.4. Like the Wellman–Lord process, this process is relatively complex and has a capital cost about 14% higher than limestone scrubbers (Burnett and Wells, 1982). It is therefore only economically viable when there is a reliable market for the by-products (Vernon and Jones, 1993).

3.2.1.2 Spray-Dry Scrubbers

Spray-dry scrubbers are an alternative to conventional wet scrubbers. In this type of scrubber, an alkaline slurry or solution is sprayed in fine droplets into a reaction vessel, along with the flue gas. The droplets rapidly react with the sulfur dioxide while drying to a fine powder of sulfite salts. This powder is entrained in the gas stream, and is carried to a dust precipitator where it is collected, as shown in Figure 3.5. Most of the sulfur dioxide is collected in liquid-phase reactions while the droplets are drying, but 10%–15% additional sulfur dioxide can be absorbed in gas/solid reactions as the absorbent powder is swept through the ductwork and particulate collector. These are cocurrent devices; therefore, the limestone utilization and sulfur removal efficiency are inherently lower than for countercurrent devices such as wet scrubbers. Partial recycle of the sorbent is often used to improve the sorbent utilization.

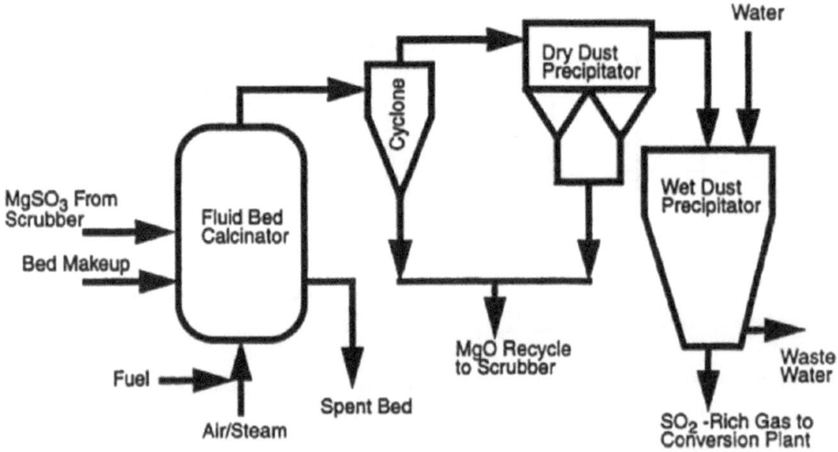

FIGURE 3.4 Magnesium oxide regeneration and sulfur dioxide recovery for magnesium oxide scrubbers (Chiang and Cobb, 1993).

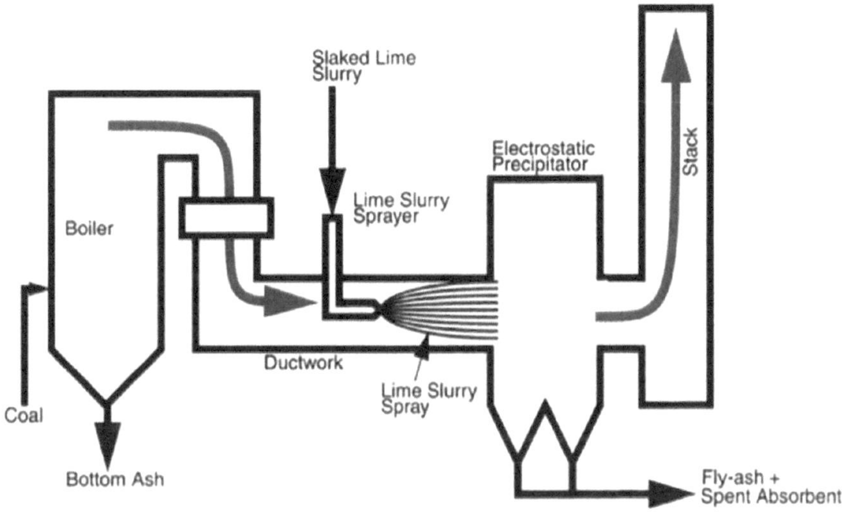

FIGURE 3.5 Basic configuration for a single-stage spray-dry sulfur absorber, with no recycle of absorbent.

It is typical to install the spray-dryer before the plant fly-ash collector, so that the existing dust control equipment can be used to collect the used absorbent. Slaked lime ($Ca(OH)_2$) is the most common absorbent, although sodium carbonate (Na_2CO_3) is used in some plants. Spray-dryers have also been used with regenerable magnesium oxide absorbent (Burnett and Wells, 1982).

Spray-dryers are simpler and more compact than conventional wet scrubbers and have a lower capital and operating cost. They also do not produce large quantities of wastewater, and the spent absorbent is dry, eliminating the need for thickening and filtration of the sludge. If the same dust precipitator is used for both the fly-ash and the spray-dryer product, the mixture of fly-ash and spent absorbent that they produce is unmarketable and must be disposed of. Also, they require more expensive absorbents than conventional wet scrubbers. They are most suitable for retrofitting small plants that burn medium-sulfur coals, where capital costs and space restrictions are more of a consideration (Vernon and Jones, 1993).

3.2.1.3 Venturi Scrubbers

Venturi scrubbers are mainly used for collecting fine particulates (such as fly-ash) from gas streams, but they have also been adapted for absorption of sulfur dioxide (Brady and Legatski, 1993). These units are mechanically very simple, consisting of a reducing inlet with liquid sprays, a narrow throat where the gas/liquid contact occurs, and an expanding region, as shown in Figure 3.6. The flue gases are injected into the venturi at the contracting inlet, along with an absorbent, such as lime slurry. The gas accelerates to high speed as it enters the throat and atomizes the absorbent, providing good gas/liquid contact. The gas and atomized liquid then expands and slows in the expander region, and it is then diverted to a mist eliminator to separate the liquid droplets from the scrubbed gases.

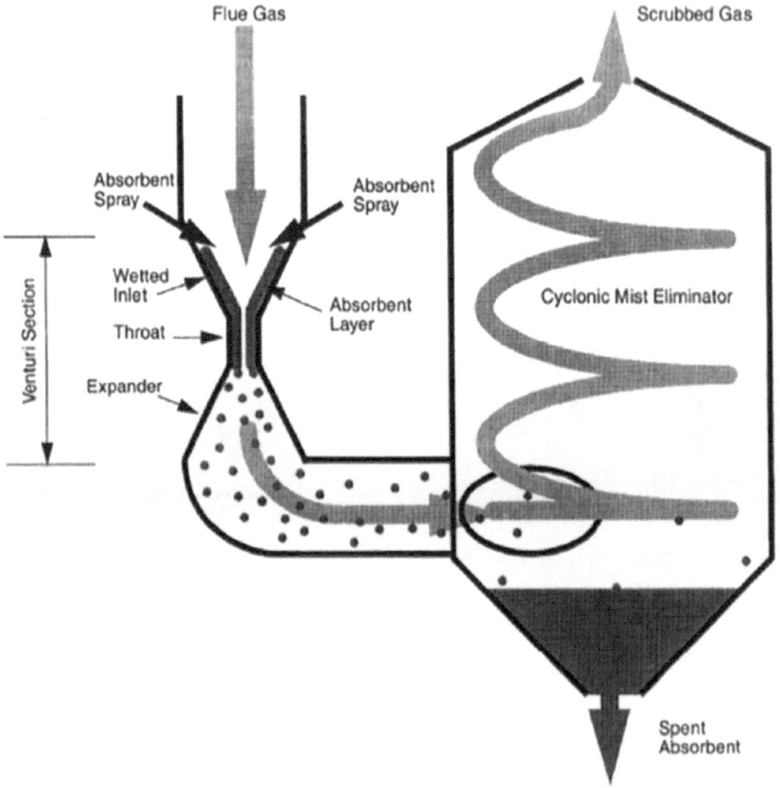

FIGURE 3.6 Schematic diagram of a venturi scrubber, with a cyclonic mist eliminator.

Venturi scrubbers are lower in capital cost than other types of scrubbers because they are mechanically simple, but they have a high energy consumption because of the need for pressurizing the gas to force it through. They also double as a fly-ash collection device, and so there is no need for separate scrubbers and fly-ash collectors when these units are used (Brady and Legatski, 1993).

Since venturi scrubbers are cocurrent devices, with the flue gas and absorbent both traveling in the same direction, they cannot remove sulfur dioxide as completely as countercurrent devices, such as wet-scrubber towers (Porter, 1984; Edwards, 1984). They also produce a wet mixture of fly-ash and alkaline absorbent, which is unmarketable, and can form a cement-like substance upon disposal.

3.2.1.4 Dry Sorbent Injection

Dry sorbent injection is similar to the use of spray-dryers, except that the sorbent is injected as a dry powder rather than as an atomized slurry (Yeh et al., 1982; Stouffer et al., 1993). The most common sorbent is hydrated lime, but other sorbents can also be used. The sorbent is usually injected directly into the existing ductwork, so the amount of space required is negligible compared to other flue-gas desulfurization processes. This makes dry sorbent injection a very low-cost option. Unfortunately, the reactivity of dry absorbents is much lower than for absorbent slurries or solutions, so dry sorbent injection is only suitable for applications where less than about 70% of the sulfur dioxide needs to be removed from the flue gases (Vernon and Jones, 1993).

In cases where hydrated lime does not remove enough of the sulfur dioxide, but economics make more efficient absorbents impractical, a two-stage absorbent injections scheme can be used, as shown in Figure 3.7. Here, the relatively low-cost calcium hydroxide is used to remove the bulk of the sulfur dioxide, which is then followed by a spray of more effective (but higher-cost) sodium bicarbonate. In addition to further reduction of the sulfur dioxide content, the sodium bicarbonate spray also reduces the content of nitrogen oxides.

It is also possible to use limestone in dry sorbent injection, as is done in the Limestone Injection Multistage Burner (LIMB) system (Figure 3.8). In this system, pulverized limestone is injected into the boiler directly, where the temperature is high enough to flash-calcine the $CaCO_3$ to CaO. The CaO dust is carried off with the flue gases until the temperature drops enough for $CaSO_3$ to become stable. The CaO then captures the SO_2, and the resulting $CaSO_3$ is removed by the electrostatic precipitators, along with the fly-ash. If necessary, a second, wet SO_2 scrubber is used to finish the sulfur removal (Chiang and Cobb, 1993).

A related technology, the SO_x–NO_x–Rox Box™, or SNRB, also destroys nitrogen oxides while removing sulfur dioxide (Makansi, 1992). This unit is a replacement for electrostatic precipitators and is installed in the flue-gas stream between the economizer and the combustion air heater, where the flue gases are still hot. An alkali is injected as sulfur absorbent, along with anhydrous ammonia. The gases then enter the high-temperature catalytic baghouse, which consists of catalyst-impregnated ceramic filter "bags," as shown in Figure 3.9. The ceramic filters capture and remove the sulfur-loaded absorbent, and the catalyst in the filters catalyzes the reduction of the nitrogen oxides by the ammonia, producing nitrogen gas and water.

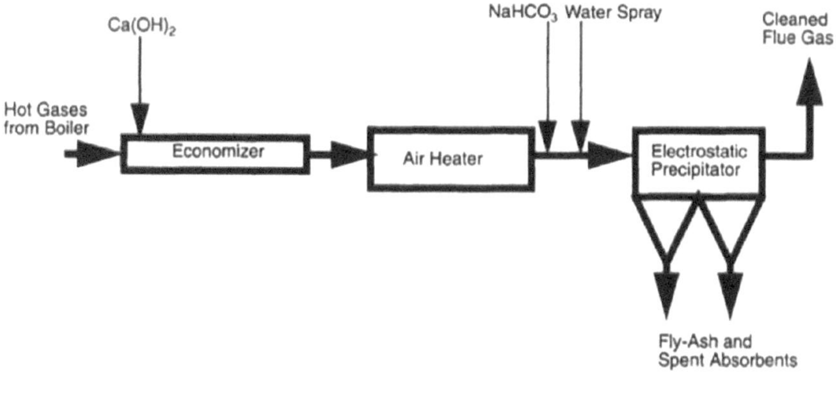

FIGURE 3.7 Integrated dry injection process, utilizing both calcium hydroxide and sodium bicarbonate to reduce sulfur dioxide and nitrogen oxide emissions (Helfritch et al., 1992). Calcium hydroxide is added to the hot flue gases before they are cooled in the economizer and the combustion air heater, while sodium bicarbonate is added to the cooled gases before they enter the electrostatic precipitator.

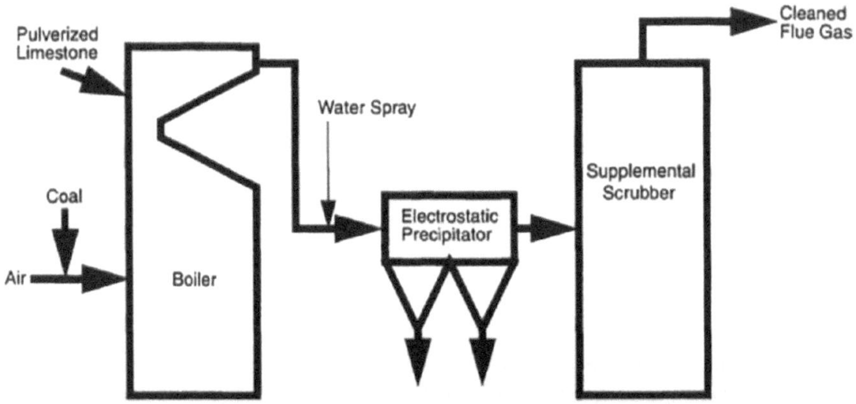

FIGURE 3.8 The basic integrated LIMB dry sorbent injection system.

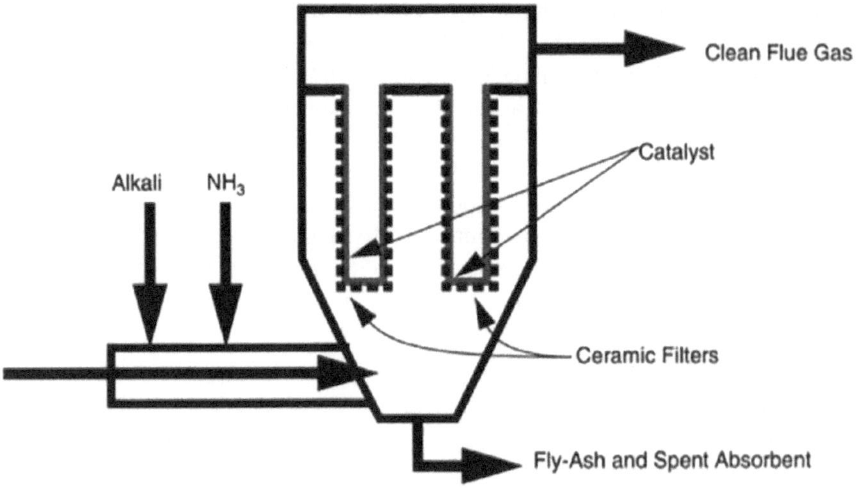

FIGURE 3.9 Schematic of the SNRB Catalytic Baghouse (Kudlac et al., 1992).

3.2.1.5 SNOX System

The SNOX system is designed for removing both sulfur oxides and nitrogen oxides from flue gases and is unusual in that it does not use an alkali as an absorbent to collect the sulfur dioxide. Instead, it oxidizes the sulfur dioxide to sulfur trioxide and uses a special condenser to collect the sulfur trioxide as marketable sulfuric acid. This is combined with catalytic destruction of nitrogen oxides with ammonia, producing the overall circuit shown in Figure 3.10. The system is reported to be capable of more than 90% removal of both SO_2 and NO_x, while producing sulfuric acid at 93%–95% concentration (Durrani, 1994).

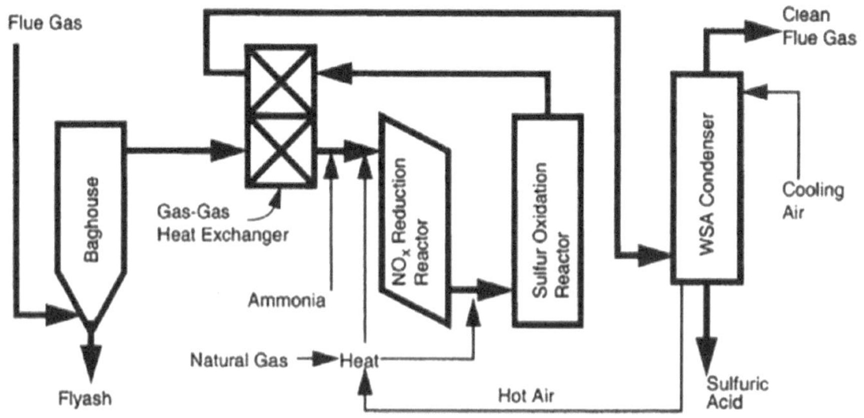

FIGURE 3.10 Diagram of the SNOX system for collection of sulfuric acid from flue gases, while simultaneously destroying nitrogen oxides (Durrani, 1994).

3.2.2 In-Combustor Technologies

These technologies are designed to minimize the release of sulfur from the combustion step, rather than attempting to capture the sulfur after it has already reached the flue gases. Such techniques have been under development for many years, but the only ones that are actually being used commercially are fluid-bed combustion (FBC) and integrated gasification/combined-cycle combustion (Vernon and Jones, 1993).

3.2.2.1 Fluid-Bed Combustion (FBC)

FBCs burn pulverized coal mixed with a bed material (such as sand mixed with a sulfur absorbent), which is fluidized by air being blown upward through the bed. If a limestone-bearing bed is used, it will absorb the sulfur dioxide as the coal burns, and so the sulfur will not be released into the flue gases. The limestone forms a solid sulfate, which is discarded along with the combustor ash. This is possible because the operating temperature of an FBC is 800°C–900°C, which is low enough for the calcium sulfate to be stable. These combustors can capture up to 90% of the sulfur dioxide, with a calcium/sulfur ratio of approximately 2. Other sulfur absorbents that have been used in this application are chalk, dolomite, and high-calcium waste dust from cement manufacture (Bulewicz et al., 1986; Dennis and Hayhurst, 1989). Some work has been done to treat the ash produced by FBCs with water or water vapor to reactivate it, so that it can be recycled to the combustor and improve the absorbent utilization (Schmal, 1985). It has also been reported that salts such as NaCl improve the effectiveness of limestone-based sulfur absorbents in this application (Bulewicz and Janicka, 1990).

Several variations of FBC technology are used, including atmospheric FBC (AFBC), circulating FBC (CFBC), and pressurized FBC (PFBC).

3.2.2.1.1 Atmospheric Fluid-Bed Combustion

In AFBC, the combustion is carried out at atmospheric pressure, using a combustor such as that shown schematically in Figure 3.11. Air is blown upward through the combustor and bubbles through the bed, fluidizing it. To function properly, these combustors require that the bed material be coarse enough not to be entrained in the air-flow and carried out of the combustor. The fluidized bed provides good contact between the air, coal, and sulfur absorbent, and the sulfur is largely collected from the gases before they leave the bed.

These units are mainly used in small-scale industrial boilers. To capture a large proportion of sulfur dioxide, they require a deep bubbling bed and a higher sorbent use than for other types of FBC. It is reported that few combustors of this type have been able to routinely achieve 90% SO_2 removal, and most installations have had to switch to lower-sulfur coals to meet their emissions targets.

3.2.2.1.2 Circulating Fluid Bed Combustion

A CFBC system differs from other FBCs in that the bed particles are fine enough that a significant proportion of them are entrained into the gas stream. As a result, these units do not have a well-defined bed surface. A cyclone is used with this type of combustor to return these particles to the combustion zone, as shown in Figure 3.12.

The design of CFBC systems makes it possible to use higher air velocities, giving more complete combustion. The recirculation of solids to the bed improves combustion efficiency further and ensures good contact between sulfur absorbent and combustion gases (Neathery, 1996). These units can achieve 90% sulfur removal with a calcium/sulfur ratio of only 1.5, which is a considerable improvement over the performance of AFBC combustors.

CFBCs are becoming larger as the technology matures, and in recent years, numerous installations have been commissioned (Makansi, 1993b). The largest unit

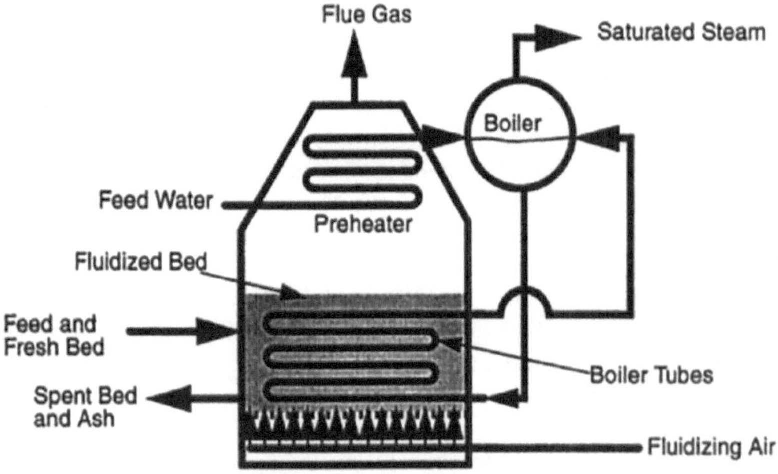

FIGURE 3.11 Atmospheric fluidized-bed combustor (AFBC).

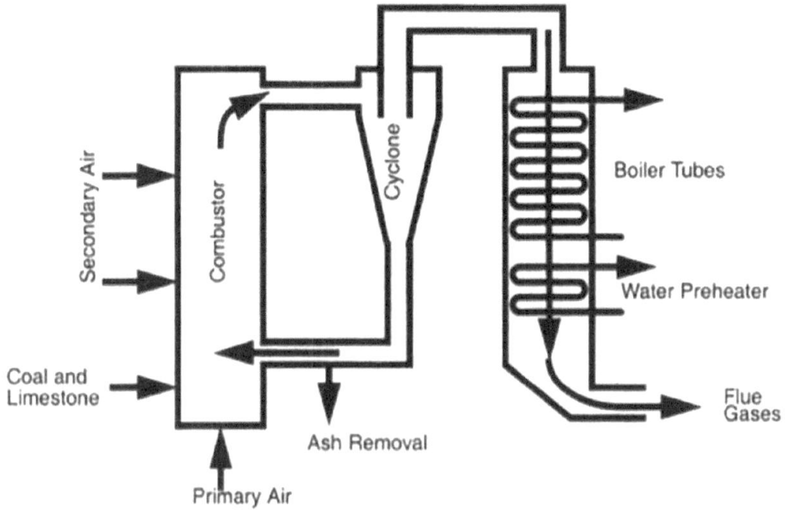

FIGURE 3.12 Circulating fluidized-bed combustor (CFBC).

in operation in 1993 could produce 175 MW of electrical power. These units also have considerable fuel flexibility, burning fuels ranging from bituminous coal to peat (Vernon and Jones, 1993).

3.2.2.1.3 Pressurized Fluid Bed Combustion

Unlike AFBC and CFBC, a PFBC uses an enclosed, pressurized combustion chamber. The advantage of this is that the increased pressure allows more oxygen to be supplied, and the unit can therefore achieve higher combustion intensities. The combustor can also be operated under conditions that evolve combustible gases (combined-cycle operation), which can be used to power a gas turbine. This turbine is then used both to compress the air for the combustor and supply additional power, as shown in Figure 3.13.

The benefit of a PFBC is that the increased combustion intensity and combined-cycle operation makes it possible to build combustors that are less than half the size of a conventional boiler of the same capacity (Maude, 1993). A disadvantage is that it is necessary to make provisions to force coal and sorbent across the pressure boundary, and similar provisions must be made to withdraw the ash. One approach is to crush and grind the coal and absorbent to a fine particle size and combine with water to make a paste. This paste is then pumped into the combustor using concrete pumps. Other combustor designs can inject dry feed. Another disadvantage is that increasing pressure can reduce the sulfur capacity of the absorbent (Bulewicz et al., 1986).

PFBC units have shown very high sulfur removal rates over a wide range of feed sulfur contents. In commercial-scale plants, over 90% SO_2 removal has been achieved at a calcium/sulfur ratio of 2 with coals containing as little as 0.65% sulfur. A 97% SO_2 removal was achieved with a coal containing 7% sulfur (EPRI, 1992).

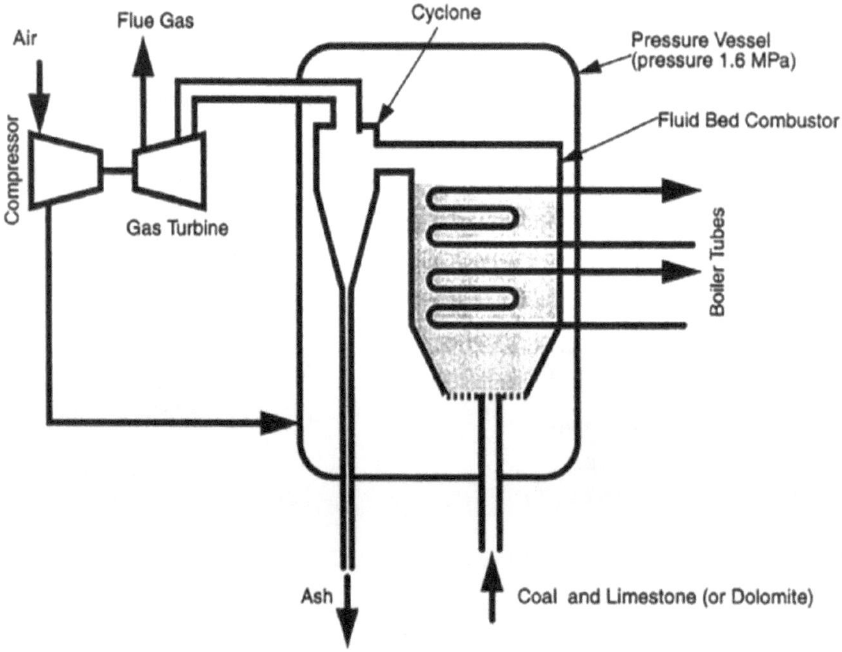

FIGURE 3.13 Pressurized fluidized-bed combustor (PFBC).

3.2.2.2 Integrated Gasification Combined Cycle

Integrated gasification combined cycle (IGCC) systems do not burn the coal directly; instead, they convert the coal into combustible gases, which are then burned to produce power as shown in Figure 3.14. There are four major steps in the process:

- Formation of fuel gas by reaction of coal with high-temperature steam and air (or oxygen).
- Purification of the gas to remove sulfur and particulates.
- Combustion of the clean gas in a gas turbine to produce electricity.
- Recovery of heat from the gas turbine to boil water for a conventional steam turbine.

The most critical step of the process is the coal gasification. In the gasification step, the coal reacts with steam and oxygen according to the following formulas (Mahagaokar and Krewinghaus, 1993):

$$C + O_2 \rightarrow CO_2 \tag{3.12}$$

$$C + \frac{1}{2}O_2 \rightarrow CO \tag{3.13}$$

$$CO + H_2O \rightarrow CO_2 + H_2 \tag{3.14}$$

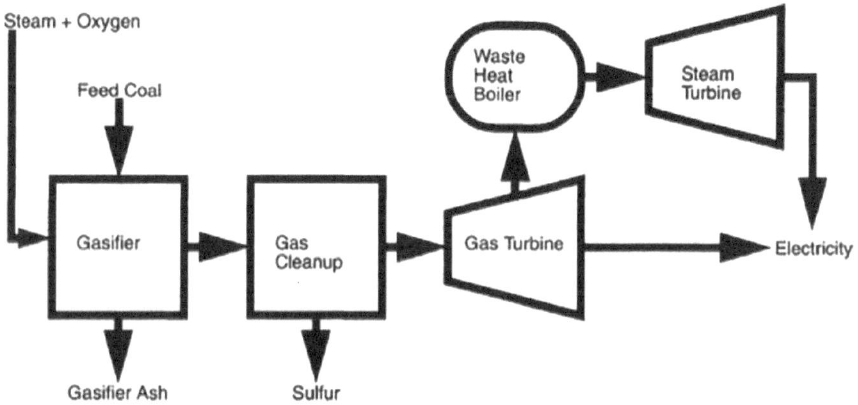

FIGURE 3.14 Basic components of an integrated gasification/combined-cycle process (Maude, 1993).

$$C + H_2O \rightarrow CO_2 + H_2 \tag{3.15}$$

$$C + CO_2 \rightarrow 2CO \tag{3.16}$$

$$CO + 3H_2 \rightarrow CH_4 + H_2O \tag{3.17}$$

$$C + 2H_2 \rightarrow CH_4 \tag{3.18}$$

These reactions can be carried out using any of the types of reactor shown in Figure 3.15, each of which are best suited for different types of coal.

In the moving-bed gasifier, steam and oxygen travel upward through a bed of coal in the size range of 5cm×0.6cm. Relatively coarse coal is needed, so that it will not be carried away by the gas stream. As the coal reacts, it breaks down and descends through the gasifier, until finally the remaining ash discharges through the bottom grate. This type of gasifier therefore has a countercurrent flow of the coal and the gas stream, allowing low gasification temperatures, low oxygen requirements, and a high methane content in the product. Moving-bed gasifiers, however, cannot handle fines and perform poorly with caking coals. A slagging version of this gasifier has been developed by Lurgi/British Coal, with improved ability to handle caking coals and fines.

Fluidized-bed gasifiers are similar to FBCs, with a well-mixed bed of coal reacting with the gases at a temperature below the ash fusion temperature to prevent it from becoming sticky and forming lumps. Fines are captured from the off-gases and recycled to improve carbon utilization. These units operate at high enough temperatures to gasify unreactive high-rank coals.

Entrained-flow gasifiers are plug-flow systems, with fine coal particles being carried along by the steam/oxygen stream. They operate at higher temperatures than the other types of gasifiers, which ensures that carbon conversion will be very rapid and the ash can be collected as a molten slag. Residence times are very short, allowing a high capacity.

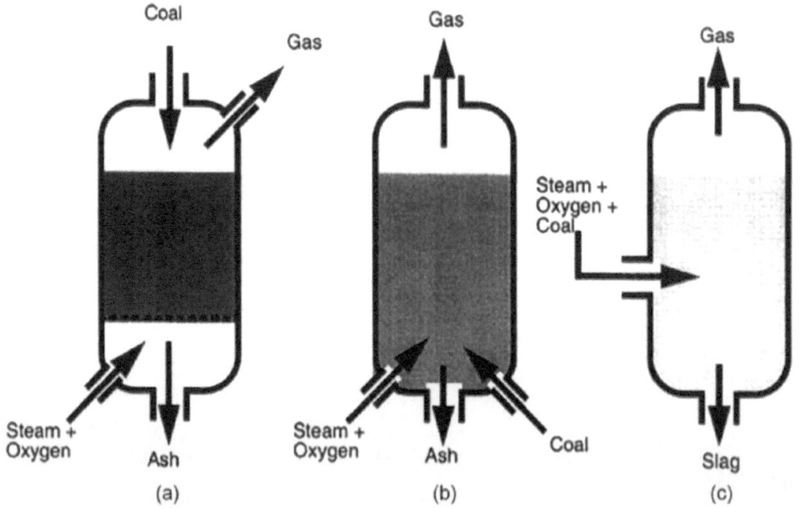

FIGURE 3.15 Types of coal gasifier in use: (a) moving-bed gasifier, (b) fluidized-bed gasifier, and (c) entrained-flow gasifier (Mahagaokar and Krewinghaus, 1993).

The gas cleanup step uses technologies that have already been developed for natural gas production and can readily remove over 99% of the sulfur, which is largely present as H_2S (Khare et al., 1995). Gasification can be accomplished using moving bed, fluidized bed, or entrained-flow systems, with the entrained-flow technique being preferred in work done to date. A number of large-scale demonstration projects are underway and have shown the practicality of the process.

The main advantages of the IGCC process over other coal combustion techniques are as follows:

- IGCC produces lower quantities of solid residues, and the gasifier ash is a vitreous material that is easily disposed of.
- Combined-cycle operation has a potentially higher operating efficiency than furnace/boiler systems because of the ability to use high-efficiency gas turbines.
- The desulfurization stage can produce a salable sulfur by-product, reducing the waste disposal requirements still further.
- Desulfurization can be much more complete than is possible using other types of combustion.

3.3 RELATIVE COSTS OF SULFUR CONTROL PROCESSES

3.3.1 Postcombustion Sulfur Removal Costs

A comparison of the benefits and disadvantages of the various types of precombustion and postcombustion desulfurization technologies is given in Appendix 2. In considering the cost of each desulfurization technique, it is necessary to consider factors such as fuel quality, waste disposal costs, degree of sulfur removal needed to meet standards, and availability of sorbent chemicals. Overall, the choice of which process or processes to use is site-specific, and there is not yet a single "best" solution for all applications.

A comparison of the estimated sulfur removal ability and costs for various post-combustion processes is given in Table 3.4. Clearly, there is considerable variation in costs, depending on factors such as the size of the plant, fraction of time that the plant operates at full capacity, ability to retrofit existing facilities, differences in fuel prices, sulfur content of the coal, and ability of existing equipment such as electrostatic precipitators to cope with changes in the process. In general, fuel-switching is currently the cheapest option at the current price difference of $10/ton between high-sulfur and low-sulfur coals. However, as low-sulfur coal prices increase, this will become less economical. Dry processes are the next most economical option, but they may not be able to meet emission control standards. The various types of scrubbers have broadly similar costs, and the choice will depend on the specific plant under consideration. It appears from Table 3.4 that IGCC and fluidized-bed technologies are drastically more expensive than the other choices, but it should be remembered that the comparison is between retrofitting existing plants (for the scrubbers) and building new plants (for the IGCC and fluidized bed). The price differential between the two types of control processes is much less for completely new plants, or for repowering existing plants (Vernon and Jones, 1993).

TABLE 3.4
Cost Comparison of Various SO_2 Control Options (White and Maibodi, 1991)

		Capital Costs, $/kW			Operating Costs, mills/kWh			
Technology	**% SO_2 Removal**	**Low**	**Base**	**High**	**Low**	**Base**	**High**	**Most Sensitive Parameters**
Fuel-switching/blending	2%–80%	20	28	30	3	6	13	% S, CF, FPD, SCA
Lime/limestone flue-gas desulfurization	90%	120	240	520	5	16	150	MW, RF, CF, % S
Lime spray drying with existing ESP	76%	70	170	540	3	10	130	MW, RF, CF, % S, SCA
Lime spray drying with new fabric filter	86%	140	240	620	5	13	150	MW, RF, CF, % S
Integrated gasification combined cycle	95%	1,710	2,100	2,800	44	91	605	MW, CF, heat rate
Atmospheric fluid-bed combustion	90%	1,360	1,680	2,250	40	80	480	MW, CF, heat rate
Dry sorbent injection	70%	25	50	110	2	6	40	MW, CF, % S, SCA

The ranges of both the operating costs and the capital costs for typical installations of each type are given. The "Most Sensitive Parameters" column lists the factors that are of most importance in determining whether a given technology should be used in a particular situation.

CF, capacity factor; FPD, fuel price differential; MW, size of plant (megawatts); % S, fuel sulfur content; RF, retrofit factor; SCA, specific collection area of electrostatic precipitator (ESP).

3.3.2 Benefits of Combining Precombustion and Postcombustion Desulfurization

In general, the performance of a postcombustion desulfurization process will be improved if the sulfur content of the feed coal is reduced. A lower-sulfur feed provides the following benefits:

- Gas-scrubbing processes require less absorbent.
- A lower percentage of the total SO_2 can be removed and still reach emission targets, allowing the less-efficient, lower-cost technologies to be used.
- The quantity of desulfurization waste that must be disposed of is reduced.

It is therefore beneficial to pretreat the coal to remove as much sulfur as is practical before combustion, so that the expense of postcombustion desulfurization can be reduced. Combined with the other benefits of coal cleaning, it is evident that precombustion coal treatment is valuable even when it is not sufficient to completely desulfurize the coal by itself.

3.4 SUMMARY

A wide range of postcombustion technologies are available for reducing the sulfur emissions from coal-fired power plants. These include both wet and dry scrubbers, which either produce sulfur-bearing wastes or marketable by-products, and new types of combustors, which capture the sulfur before it ever reaches the flue gases. In general, the less-expensive techniques are also the least effective for reducing sulfur emissions. The choice of which process to use depends on the quantity of sulfur in the fuel and on the emissions target, as well as many other factors such as plant size, whether it is a retrofit of an existing plant or a new plant, availability of low-sulfur fuel, cost of waste disposal, and availability of markets for by-products. For example:

- Fuel-switching is the best choice if a suitable fuel is available at a low cost and can be burned with minimal modification of existing equipment.
- Precombustion coal cleaning is most useful if it is performed at the mine, where waste disposal costs are likely to be lower than at the power plant. It will improve the overall quality of the coal and reduce shipping costs, and will be useful regardless of the additional sulfur removal measures being taken during or after combustion.
- Spray-dryers or dry sorbent injection are a good choice if there is little space in the plant for installing new equipment, if there is no market for by-products, and if the plant emissions can be brought into regulatory compliance with relatively modest reductions in sulfur emissions.
- Wet scrubbers are well-suited for plants with significant room for expansion, and which need to remove a large proportion of the sulfur from their emissions. Scrubbers with throw-away absorbents are the best choice when waste disposal is cheap, while those with regenerable absorbents and/or

marketable by-products are the best choice when waste disposal is expensive and markets for by-products are nearby.

- FBCs are a good choice for new installations with small- to medium-size boilers or for replacement of existing boilers, provided that waste disposal is cheap.
- IGCC is a good choice for large, new facilities being built in industrialized areas, where by-product markets are available, emissions regulations are stringent, and waste disposal is expensive.

Precombustion and postcombustion technologies should not be considered to be in competition. Rather, they are complementary technologies that should be used together for maximum benefit. Many precombustion desulfurization processes are low in cost, but cannot remove all of the sulfur, while postcombustion treatment can remove all of the sulfur, but their costs increase as the coal sulfur content increases. By using both types of processes together, the sulfur emissions can be reduced the maximum amount at the minimum cost.

REFERENCES

Anonymous (1990a), "Latest Micronized-Coal Mills Consume Less Energy, Cost Less," *Power*, Vol. 134, pp. 39–44.

Anonymous (1990b), "Will Combined SO_2/NO_x Processes Find a Niche in the Market?" *Power*, Vol. 134, pp. 26–28.

Anonymous (1992a), "DOE Kicks Off Clean Air Program to Boost Performance of Scrubbers," *Fossil Energy Review*, U.S. Department of Energy, pp. 18–19.

Anonymous (1992b), "Southern Company Begins Testing 2nd Generation Clean Coal Scrubber," *Fossil Energy Review*, U.S. Department of Energy, pp. 28–29, 39.

Barnett, W.P. and Feeley III, T.J. (1992), "Status of Advanced Coal Cleaning as a Compliance Technology," *Mining Engineering*, Vol. 44, No. 10, pp. 1225–1230.

Bjöernbom, E.N., Druesne, S., Zwinkels, M.F.M. and Järås, S.G. (1995), "Study on the Regeneration of Copper-Manganese Sorbent for Removal of Sulfur Dioxide from Flue Gases," *Industrial Engineering Chemistry Research*, Vol. 34, No. 5, pp. 1853–1858.

Brady, J.D. and Legatski, L.K. (1993), "Venturi Scrubbers," Chapter 11, *Air Pollution Control and Design for Industry* (Cheremisinoff, ed.), Marcel Dekker, New York, pp. 339–357.

Bryan, R.R., Smith, A.A. and Farmer, C. (1993), "Texas Plant Demonstrates Viability of Coal Option," *Power*, Vol. 137, pp. 57–64.

Bulewicz, E.M., Kandefer, S. and Jurys, C. (1986), "Desulphurization during the Fluidized Combustion of Coal Using Calcium-Based Sorbents at Pressures Up to 600 kPa," *Journal of the Institute of Energy*, Vol. 59, pp. 188–195.

Bulewicz, E.M. and Janicka, E. (1990), "Catalytic Effect of NaCl on Flue-Gas Desulphurization by Limestone-Based Sorbents during the FB Combustion of Coal," *Journal of the Institute of Energy*, Vol. 63, pp. 124–130.

Burnett, T.A. and Wells, W.L. (1982), "Conceptual Design and Economics of and Improved Magnesium Oxide Flue Gas Desulfurization Process," *Flue Gas Desulfurization* (Hudson and Wells, eds.), American Chemical Society, Washington, DC, pp. 381–411.

Chiang, S.-H. and Cobb Jr., J.T. (1993), "Coal Conversion Processes (Desulfurization)," *Kirk-Othmer Encyclopedia of Chemical Technology*, 4th Edition, Vol. 6, John Wiley & Sons, Hoboken, NJ, pp. 511–540.

Cichanowicz, J.E. and Harrison, C.D. (1989), "The Role of Coal Cleaning in Integrated Environmental Control," *Reducing Power Plant Emissions by Controlling Coal Quality*, Electric Power Research Institute, EPRI-GS-6281, pp. 7/1–7/24.

Couch, G.R. (1995), Power from Coal—Where to Remove Impurities? IEA Coal Research, London, Report No. IEACR/82.

Davies, G. (1989), "Getting the Sulphur out of Combustion," *The Chemical Engineer*, No. 461, p. 111.

Dennis, J.S. and Hayhurst, A.N. (1989), "Alternative Sorbents for Flue-Gas Desulphurization, Especially in Fluidized-Bed Combustors," *Journal of the Institute of Energy*, Vol. 62, pp. 202–207.

Dettmer, R. (1994), "Sans Sulphur: The Drax FGD Project," *IEE Review*, Vol. 40, No. 2, pp. 88–89.

Durrani, S.M. (1994), "The SNOX Process: A Success Story," *Environmental Science and Technology*, Vol. 28, No. 2, pp. 88A–90A.

Edwards, W.M. (1984), "Mass Transfer and Gas Absorption," *Perry's Chemical Engineers' Handbook*, 6th Edition (Green and Maloney, eds.), McGraw-Hill, New York City, NY, pp. 14/39.

Eisele, T.C. and Kawatra, S.K. (1995), "Separation of the Components of Flue Gas Scrubber Sludge by Froth Flotation," *Proceedings of the 19th International Mineral Processing Congress*, Society of Mining, Metallurgy, and Exploration, Littleton, CO, Vol. 4, Chapter 33, pp. 163–166.

Elliott, T.C. (1992), "Coal Handling and Preparation," *Power*, Vol. 136, pp. 17–32.

Ellison, W. and Hammer, E.L. (1988), "FGD Gypsum Use Penetrates U.S. Wallboard Industry," *Power*, Vol. 132, No. 2, pp. 29–33.

Ellison, W. (1991), "Today's FGD Systems Satisfy Retrofit Needs for 1990s," *Power*, Vol. 135, pp. 101–106.

EPRI (1992), *AFPS Developments*, Issue E, Electric Power Research Institute, Palo Alto, CA.

Feeney, S. (1995), "Upgrade Scrubbers to Improve Performance," *Power*, Vol. 139, pp. 32–37.

Fellman, R.T. and Cheremisinoff, P.N. (1993), "Lime/Limestone Scrubbing for SO_2 Removal," Chapter 12, *Air Pollution Control and Design for Industry* (Cheremisinoff, ed.), Marcel Dekker, New York, pp. 339–357.

Gavalas, G.R., Edelstein, S., Flytzani-Stephanopoulous, M. and Weston, T.A. (1987), "Alkali-Alumina Sorbents for High-Temperature Removal of SO_2," *AIChE Journal*, Vol. 33, No. 2, pp. 258–266.

Harrison, C.D. and Stallard, G.S. (1992), "The Coal Quality Expert: Introduction to the Acid Rain Advisor," *Mining Engineering*, Vol. 44, No. 10, pp. 1257–1261.

Hedges, S.W. and Yeh, J.T. (1992), "Kinetics of Sulfur Dioxide Uptake on Supported Cerium Oxide Sorbents," *Environmental Progress*, Vol. 11, No. 2, pp. 98–103.

Helfritch, D., Bortz, S., Beittel, R., Bergman, P. and Toole-O'Neil, B. (1992), "Combined SO_2 and NO_x Removal by Means of Dry Sorbent Injection," *Environmental Progress*, Vol. 11, No. 1, pp. 7–10.

Hodges, G.J., Roset, G.K., Woodland, L.R. and Stevenson, J.A. (1992), "Dual-Alkali Scrubbing at Stillwater Mining Company," *Mining Engineering*, Vol. 44, No. 10, pp. 1269–1271.

Jones, C. (1994), "Utility Converts from Gas to Micronized Coal," *Power*, Vol. 138, pp. 67–68.

Kawatra, S.K. and Eisele, T.C. (1993), "Separation of Flue-Gas Scrubber Sludge into Marketable Products," First Quarterly Technical Progress Report, U.S. Department of Energy, Pittsburgh Energy Technology Center, Contract No. DE-FG22-93PC93214.

Kawatra, S.K., Eisele, T.C. and Banerjee, D.D. (1995), "Recovery of Gypsum and Limestone from Scrubber Sludge by Water-Only Cyclone and Conventional Froth Flotation," *New Remediation Technology in the Changing Environmental Arena* (Scheiner, et al., eds), Society for Mining, Metallurgy, and Exploration, Inc., Littleton, CO, pp. 99–104.

Khare, G.P., Delzer, G.A., Kubicek, D.H. and Greenwood, G.J. (1995), "Hot Gas Desulfurization with Phillips Z-Sorb Sorbent in Moving Bed and Fluidized Bed Reactors," *Environmental Progress*, Vol. 14, No. 3, pp. 146–150.

Kind, K.K., Wasserman, P.D. and Rochelle, G.T. (1994), "Effects of Salts on Preparation and Use of Calcium Silicates for Flue Gas Desulfurization," *Environmental Science and Technology*, Vol. 28, No. 2, pp. 277–283.

Kudlac, G.A., Farthing, G.A., Szymanski, T. and Corbett, R. (1992), "SNRB Catalytic Baghouse Laboratory Pilot Testing," *Environmental Progress*, Vol. 11, No. 1, pp. 33–38.

Levendis, Y.A., Zhu, W., Wise, D.L. and Simons, G.A. (1993), "Effectiveness of Calcium Magnesium Acetate as an SO_x Sorbent in Coal Combustion," *AIChE Journal*, Vol. 39, No. 5, pp. 761–773.

Li, W. and Cho, E.U. (2005), "Coal Desulfurization with Sodium Hypochlorite," *Energy & Fuels*, Vol. 19, pp. 499–507.

Mahagaokar, U. and Krewinghaus, A.B. (1993), "Coal Conversion Processes (Gasification)," *Kirk-Othmer Encyclopedia of Chemical Technology*, 4th Edition, Vol. 6, John Wiley & Sons, Hoboken, NJ, pp. 541–568.

Makansi, J. (1992), "DOE's Clean Coal Program: What Industry Has Learned," *Power*, Vol. 136, pp. 56–61.

Makansi, J. (1993a), "Controlling SO_2 Emissions," *Power*, Vol. 137, pp. 23–56.

Makansi, J. (1993b), "CFB Technology Injects Life into Dry Scrubbing," *Power*, Vol. 137, pp. 79–81.

Maude, C. (1993), Advanced Power Generation—A Comparative Study of Design Options for Coal, IEA Coal Research, London, Report No. IEACR/55.

Neathery, J.K. (1996), "Model for Flue-Gas Desulfurization in a Circulating Dry Scrubber," *AIChE Journal*, Vol. 42, No. 1, pp. 259–268.

Porter, H.F. (1984), "Solids Drying and Gas-Solid Systems," *Perry's Chemical Engineers' Handbook*, 6th Edition (Green and Maloney, eds.), McGraw-Hill, New York City, NY, pp. 20/93.

Schmal, D. (1985), "Improving the Action of Sulfur Sorbents in the Fluidized-Bed Combustion of Coal," *Industrial Engineering Chemistry Process Design and Development*, Vol. 24, No. 1, pp. 72–77.

Shoop, K.J., Blystone, S.S. and Kawatra, S.K. (1996), "Zeta Potential Measurements of the Components of Wet Flue-Gas Scrubber Sludge," *Presented at the 1996 SME Annual Meeting*, Phoenix, AZ, Preprint No. 96–99.

Skorupska, N.M. (1993), Coal Specifications—Impact on Power Station Performance, IEA Coal Research, London, Report No. IEACR/52.

Stouffer, M.R., Rosenhoover, W.A. and Withum, J.A. (1993), "Advanced Coolside Desulfurization Process," *Environmental Progress*, Vol. 12, No. 2, pp. 133–139.

Valencia, J.A. (1982), "The Limestone Dual Alkali Process for Flue Gas Desulfurization," *Flue Gas Desulfurization* (Hudson and Wells, eds.), American Chemical Society, Washington, DC, pp. 325–347.

Van der Brugghen, F.W. and Koppins-Odink, J.M. (1989), "Rue Gas Cleaning in Power Stations in the Netherlands," *Proceedings of the First Combined Flue Gas Desulfurization and Dry SO_2 Control Symposium*, St. Louis, MO, Electric Power Research Institute, paper 1–5, pp. 1/69 to 1/94.

Vernon, J.L. and Jones, T. (1993), Sulphur and Coal, IEA Coal Research, London, Report No. IEACR/57.

Wang, L., Jin, G. and Xu, Y. (2019), "Desulfurization of Coal Using Four Ionic Liquids with $[HSO_4]^-$," *Fuel*, Vol. 236, pp. 1181–1190.

White, D.M. and Maibodi, M. (1991), "Assessment of Control Technologies for Reducing Emissions of SO_2 and NO_x from Existing Coal-Fired Utility Boilers," Project Summary, Environmental Protection Agency, Report No. EPA/600/S7-90/018.

Withum, J.A. and Yoon, H. (1989), "Treatment of Hydrated Lime with Methanol for InDuct Desulfurization Sorbent Improvement," *Environmental Science and Technology*, Vol. 23, No. 7, pp. 821–827.

Xia, W. and Xie, G. (2017), "A Technological Review of Developments in Chemical-Related Desulfurization of Coal in the Past Decade," *International Journal of Mineral Processing*, Vol. 161, pp. 65–71.

Yeh, J.T., Demski, R.J. and Joubert, J.I. (1982), "Control of SO_2 Emissions by Dry Sorbent Injection," *Flue Gas Desulfurization* (Hudson and Wells, eds.), American Chemical Society, Washington, DC, pp. 349–368.

4 Heavy-Media Separation

All of the coal-cleaning devices that are in current industrial use, with the exception of froth flotation, use density-based concentration methods. These devices clean coal based upon the differences in specific gravity of coal and mineral matter, such as shale, clay, and silica. Coal cleaning is a nearly ideal application for separations based on particle density, because of the large specific gravity difference between coal and the associated minerals. The specific gravity of coal is approximately 1.5, while shale, clays, and silicates have a specific gravity of about 2.6, and pyrite and marcasite have a specific gravity of nearly 5.0. Even though the densities of each component type can vary over a significant range (see Table 4.1), the wide separation in densities prevents the ranges from overlapping. In particular, gravity separation is well-suited for removing pyritic sulfur from coal, provided that the pyrite liberates at a coarse enough size for gravity separations to work well.

In a typical coal preparation plant, raw coal is crushed to about 100–200 mm top particle size. It is then screened into coarse, intermediate, and fine particle sizes. Crushing the coal liberates some of the pyritic sulfur, with the degree of liberation increasing as the particle size decreases. The effectiveness of gravity separation devices will decrease with decreasing particle size, because as particles become smaller, they are more influenced by viscous drag forces. At the same time, the gravitational force becomes less important because of the decreasing mass of small particles. When particles are coarser than about 0.5 mm, however, density-based separations are extremely effective and provide a number of high-capacity, low-cost methods for removing the portion of the sulfur, which liberates from the coal at a fairly coarse size. As a result, these methods are almost universally used for cleaning the coarse and intermediate particle size ranges. The processes that provide the sharpest separation based on density are the heavy-media techniques described in this chapter. Sulfur removal using other density-based separations, including jigs,

TABLE 4.1
Specific Gravity Ranges of Coal and of Common Contaminant Minerals When Found Associated with Coal

Component	Low SG	High SG
Coal	1.23	1.72
Shale, clays, and silicates	2.0	2.6
Pyrite and marcasite	3.3	4.8

SG, specific gravity.

spirals, and concentrating tables, is discussed in Chapter 5. The purpose of these chapters is not to give an exhaustive list and description of the various gravity concentrating methods but to concentrate on their ability to desulfurize coal. Detailed descriptions of these devices are available in the literature (Leonard, 1991; Wills, 1992; Burt, 1984).

4.1 DETERMINATION OF SULFUR-REMOVAL ABILITY OF DENSITY SEPARATORS

The performance of devices that remove impurities from coal based on their densities is rarely reported in terms of percent British Thermal Units (BTU) recovery or percent sulfur reduction. Instead, it is normal to specify a separation efficiency for the particular separator and to report the yield achieved with a particular coal (Fonseca, 1995). This is done because the laboratory sink–float test provides a convenient tool for determining how a particular coal would behave in an ideal density separation (Browning, 1961), and there are procedures that can determine how close the operation of a particular density separator is to ideal performance. It is possible to predict reasonably accurately how well a particular coal will be cleaned by a specific separator.

A useful tool for evaluating the performance of a heavy-media separator is the partition curve, or Tromp curve. This curve is constructed by first carrying out sink–float analyses of the separator products. This data is then used to calculate the fraction of the total material in each density fraction that reported to the separator sinks product. This fraction is the partition coefficient for the given density. When the partition coefficients are plotted versus density, they define the partition curve. Examples of such curves for heavy-media separators are shown in Figure 4.1.

The simplest method for creating these curves would be to plot the individual points and then to draw a smooth curve through them by hand, however, for many data sets the resulting curve is subjective, and it will vary depending on the judgment of the person drawing it (Wizzard et al., 1983). In these cases, it is helpful to use a standard formula for the curve, using a curve-fitting computer program to calculate the adjustable parameters. If this is done, then the curve-fitting can be standardized and made reproducible. A number of such formulas have been developed over the years, which are each suited for particular types of separators (Terra, 1954; Peng, 1983; Klima and Lucide, 1986; Wizzard et al., 1983; Rao and Rao, 1975; Gottfried, 1978; Meloy, 1982; Peng and Luckie, 1991). It is generally best to use an equation with a small number of adjustable parameters when possible, since this makes the curve-fitting operation simpler. This results in values for the adjustable parameters that are tightly constrained and are more likely to be physically meaningful.

The following equation, which was originally used for hydrocyclone efficiency curves (Lynch, 1977), can be curve-fitted to symmetrical partition curves (such as those produced by heavy-media separators), using any of the several standard nonlinear fitting algorithms (Press et al., 1988; Dennis et al., 1981):

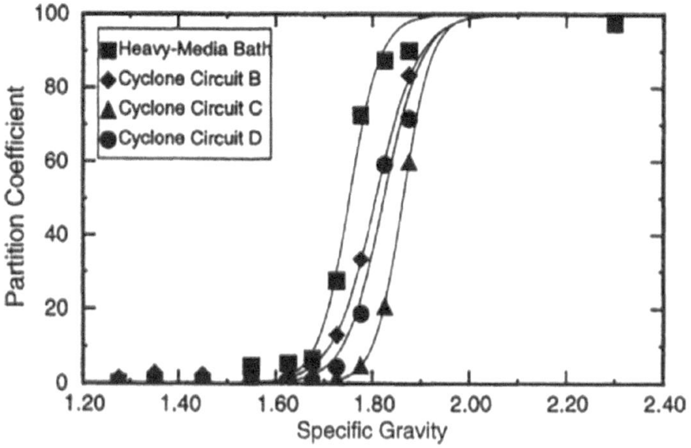

FIGURE 4.1 Partition curves for a heavy-media bath and three heavy-media cyclone circuits in an operating plant (Kawatra et al., 1995). The partition coefficient is the percent of the feed material at a particular specific gravity that reports to the sink product. These curves can be used in conjunction with a sink–float analysis to determine how well a particular separator will desulfurize a given coal.

$$y = \frac{\left(e^{\alpha x} - 1\right)}{\left(e^{\alpha x} + e^{\alpha} - 2\right)} \tag{4.1}$$

where

$x = \rho/\rho_{50}$
y = fraction of material of the given density that reports to the sink product
ρ = particle density
ρ_{50} = density at which 50% of the particles report to the sink product
α = sharpness parameter.

The values of α and ρ_{50} for each circuit provide a quantitative measure of the performance, where increasing values of α indicate an increasingly sharp separation, and ρ_{50} is a measure of the density at which the separation is actually being made.

The standard measure of separation accuracy, which is widely used in the coal industry, is the Ecart Probability, or E_p. This is defined as

$$E_p = \frac{\rho_{75} - \rho_{25}}{2}, \tag{4.2}$$

where ρ_{75} and ρ_{25} are the densities on the partition curve where 75% and 25% of the particles, respectively, would report to the floats product. The Ecart Probability is the inverse of the slope of the efficiency curve in its middle region, where it is approximately linear. The quality of the separation improves as the E_p value decreases. For a perfect separation, the value of E_p would be zero.

The value of E_p can be calculated from Lynch's model, using a formula that was derived as follows (Kawatra et al., 1995). First, Equation 4.1 was solved for x, to give

$$x = \left(\frac{1}{\alpha}\right)\ln\left(\frac{y\left(2 - e^{\alpha}\right) - 1}{y - 1}\right) \tag{4.3}$$

By substituting $y = 0.75$ and $y = 0.25$ into Equation 4.3, we obtained

$$x_{75} = \left(\frac{1}{\alpha}\right)\ln\left(3e^{\alpha} - 2\right) \tag{4.4}$$

$$x_{25} = \frac{1}{\alpha}\ln\left(\frac{2 + e^{\alpha}}{3}\right) \tag{4.5}$$

Since $x = \rho/\rho_{50}$, and therefore, $\rho = x\,\rho_{50}$:

$$E_p = \frac{x_{75}\rho_{50} - x_{25}\rho_{50}}{2} = \left(\frac{\rho_{50}}{2}\right)\left(x_{75} - x_{25}\right) \tag{4.6}$$

By substituting Equations 4.4 and 4.5 into Equation 4.6 and simplifying, we obtained the exact solution for calculating the Ecart Probability from Lynch's model for the partition curve:

$$E_p = \left(\frac{\rho_{50}}{2\alpha}\right)\ln\left(\frac{9e^{\alpha} - 6}{2 + e^{\alpha}}\right) \tag{4.7}$$

If the separation is efficient, the value of α from Lynch's equation will be large. For large α, the value of e^{α} becomes so large that the effects of subtracting 6 and adding 2 are negligible, and so Equation 4.7 can be greatly simplified to

$$E_p = \left(\frac{\rho_{50}}{2\alpha}\right)\ln(9) = (1.0986)\left(\frac{\rho_{50}}{\alpha}\right) \tag{4.8}$$

Whenever α is greater than 5, the difference between Equations 4.7 and 4.8 is less than 1%, and when $\alpha > 27$, the difference between the two equations is less than the round-off error of most calculators and computers (10^{-9}). For comparison, the typical partition curves shown in Figure 4.1 all had values of α greater than 40, so for these cases, Equation 4.8 is just as accurate as Equation 4.7.

Using Equations 4.1 and 4.8, it is straightforward to estimate how efficiently a separator with given values of ρ_{50}, α, and E_p will remove sulfur from a particular coal. Given the sink–float analysis of a particular feed coal, these parameters can be used in Equation 4.1 to calculate the amount of material from each density fraction that will report to the clean product. The weights and sulfur contents can then be summed to give the predicted sulfur content of the clean coal. This provides a convenient way to compare the sulfur-removal abilities of separators whose ρ_{50} and E_p values are known.

An example analysis of a plant, using the described procedure, is given in the next section.

4.2 CASE STUDY: ANALYSIS OF SULFUR REMOVAL AT THE MEIGS PLANT

4.2.1 Plant Description

A simplified flowsheet of the plant is given in Figure 4.2 (Dilley, 1993, Personal Communication). The plant used heavy-media separation to clean all of the coal coarser than 0.6 mm (28 mesh). The heavy media was a slurry of finely ground magnetite in water. After being reduced to pass 152 mm (6 in) by a rotary breaker, the feed was screened to remove a 152 × 16 mm (6 in × 5/8 in) coarse fraction. The material finer than 16 mm (5/8 in) was then further screened at 0.6 mm (28 mesh)

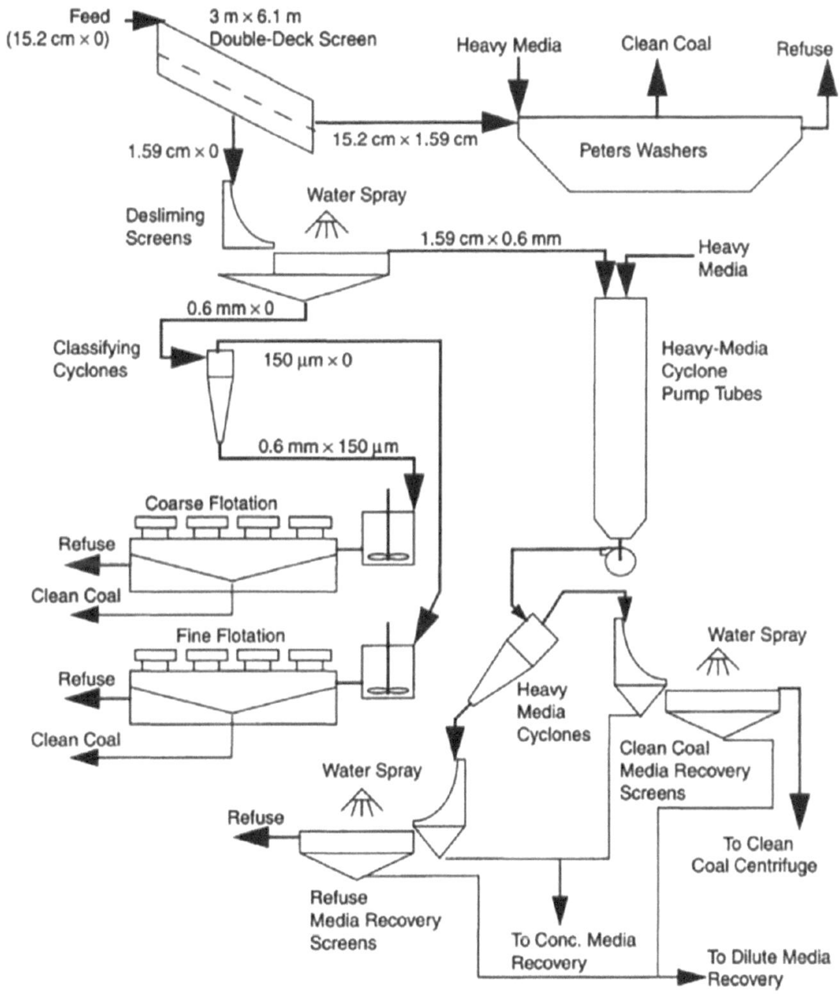

FIGURE 4.2 Simplified flow diagram of the Meigs coal-cleaning plant, not including the reclamation circuits for heavy media and water (Dilley, 1993, Personal Communication).

TABLE 4.2
Specifications of Krebs 61 cm (24 in) Heavy-Media Cyclone, Model 24b-20FA, Used in the Meigs Plant (Dilley, 1993, Personal Communication)

Inlet Area	290 cm² (45 in²)
Overflow diameter	28 cm (11 in)
Underflow diameter	15.9 cm (6.25 in)
Cone angle	0.174 radians (10°)
Flow rating	7570 L/min @ 103 kPa (2000 gpm @ 15 psi)

using a sieve bend and vibrating screen to produce a 16 × 0.6 mm (5/8 in× 28 mesh) intermediate product. The 0.6 mm (–28 mesh) material was then separated by a classifying cyclone into 600 × 150 µm (28 mesh × 100 mesh) and 150 × 0 µm (–100 mesh) size fractions that were treated in two separate froth-flotation circuits, which are not discussed here.

The 152 × 16 mm (6 in × 5/8 in) material was cleaned using two Peters washers, which are heavy-media baths. The rated capacity of these units was 454 metric tons/h (500 short tons/h), or 907 metric tons/h (1,000 tons/h) for the two units. The feed rate to the entire plant was 1,868 metric tons/h (2,059 tons/h), of which approximately 40%, or 748 metric tons/h (824 tons/h), reported to the Peters washers.

The 16 × 0.6 mm (5/8 in × 28 mesh) coal accounted for approximately 45% of the plant feed and was treated by heavy-media cyclones. Normally three of the four available cyclone circuits were operating at any given time. Each circuit consisted of two 61 cm (24 in) diameter heavy-media cyclones, rated at 136 metric tons/h (150 tons/h) each. The six operating cyclones received 841 metric tons/h (927 tons/h), or 141 metric tons/h (155 tons/h) each. The specifications of the cyclones were as shown in Table 4.2.

The correct media specific gravity targets were 1.70 in the heavy-media baths and 1.75 in the heavy-media cyclones. Actual separating gravities were 1.75 in the heavy-media baths and 1.82–1.88 in the heavy-media cyclones.

4.2.2 Feed Characteristics

The coal was mined from the Clarion 4A seam, and was a high-volatile bituminous coal containing approximately 1.05% organic sulfur and 0.06% sulfate sulfur, with 3.5%–4.6% pyritic sulfur. The composition of the feeds to the heavy-media bath and to the three heavy-media cyclone circuits described is given in Table 4.3, with the feed size distributions given in Table 4.4. The washability of each of the feed-coal streams was determined by heavy-liquid sink–float analysis, and the weight and sulfur content in each density fraction is given in Table 4.5. From this data, it is seen that it is possible to reject a greater proportion of the sulfur from the heavy-media cyclone feeds than from the heavy-media bath feed, because the heavy-media cyclone feeds contain a greater amount of low-density, low-sulfur material. This is

TABLE 4.3
Heavy-Media Bath and Heavy-Media Cyclone Feed Coal Composition at the Meigs Plant (Dilley, 1993, Personal Communication)

	Feed Size Range, mm	% Ash	% Sulfur	BTU/lb
Heavy-media bath	152 × 16	46.27	5.63	6,950
Heavy-media cyclone circuit B	16 × 0.6	40.97	4.68	7,801
Heavy-media cyclone circuit C	16 × 0.6	45.56	4.54	7,149
Heavy-media cyclone circuit D	16 × 0.6	44.41	4.74	7,271

TABLE 4.4
Heavy-Media Bath and Heavy-Media Cyclone Feed Coal Size Distributions at the Meigs Plant (Cumulative Percent Passing Each Size)

Size	Heavy-Media Bath	Heavy-Media Cyclone Circuit B	Heavy-Media Cyclone Circuit C	Heavy-Media Cyclone Circuit D
152 mm (6 in)	100.00	100.00	100.00	100.00
16 mm (5/8 in)	15.00	93.76	97.17	90.12
9.5 mm (3/8 in)	—	69.56	75.07	64.38
6.3 mm (1/4 in)	4.23	50.98	54.59	47.71
2.36 mm (8 mesh)	—	24.39	27.15	26.17
0.6 mm (28 mesh)	—	2.21	3.74	5.31

Since the cyclone feeds have a finer size distribution, the liberation of coal and pyrite is higher than for the bath feed, so there is a larger quantity of low-sulfur, low-density coal that can be recovered (Dilley, 1993, Personal Communication).

a result of the finer size distribution of the heavy-media cyclone feeds, which allows better liberation of the pyrite from the coal.

4.2.3 Performance Summary

A summary of the separation results for the four circuits studied is given in Table 4.6. As would be expected from the sink/float results in Table 4.5, the heavy-media cyclone products (Table 4.6) are all lower in sulfur than the product from the heavy-media bath, and the % weight yields and BTU recoveries are higher, due to better liberation at the finer particle sizes where the heavy-media cyclones operate.

While the data in Table 4.6 is useful for determining whether the product being produced is meeting specifications, it does not give any indication of whether the performance of the circuit can be improved significantly. The procedure outlined in Section 4.1 is useful for calculating the sharpness of the separation being made by the separator.

TABLE 4.5
Sink/Float Washability Results for the Feeds to the Peters Washer (Heavy-Media Bath) and Heavy-Media Cyclones in the Meigs Plant (Dilley,1993, Personal Communication)

	Heavy-Media Bath		Heavy-Media Cyclone Circuit B		Heavy-Media Cyclone Circuit C		Heavy-Media Cyclone Circuit D	
SG Range	**% wt**	**% S**	**% wt**	**% S**	**% wt**	**% S**	**% wt**	**% S**
−1.3	15.81	2.70	33.19	2.55	29.86	2.58	28.34	2.65
+1.3/−1.4	17.65	3.99	13.73	3.52	12.04	3.56	14.34	3.37
+1.4/−1.5	4.89	5.60	4.79	5.48	4.62	4.74	4.39	4.95
+1.5/−1.6	2.35	6.72	2.12	6.08	2.45	5.83	2.34	5.91
+1.6/−1.65	0.78	5.65	0.57	5.88	0.73	6.12	0.94	5.66
+1.65/−1.7	1.08	6.02	0.71	5.31	0.79	5.56	0.91	5.14
+1.7/−1.75	2.00	8.82	0.92	6.71	1.02	7.93	1.12	7.53
+1.75/−1.8	5.79	9.13	2.51	8.31	2.43	8.96	2.42	8.91
+1.8/−1.85	3.03	8.83	2.12	9.84	1.51	8.66	2.00	9.42
+1.85/−1.9	1.45	9.09	1.78	8.07	1.31	8.25	1.20	8.41
+1.9	45.16	6.33	37.56	6.02	43.24	5.44	42.00	5.85
Total	99.99	5.63	100.00	4.68	100.00	4.54	100.00	4.74

SG, specific gravity.

TABLE 4.6
Plant Performance Summary, Clean Coal Products, for the Meigs Plant

	% Weight Recovery (Yield)	% BTU Recovery	% Ash	% S	% Pyritic Sulfur	% Pyritic Sulfur Removed	BTU/lb (MJ/kg)
Heavy-media bath	45.18	71.48	14.99	4.28	3.23	68.1	12,090 (28.05)
Heavy-media cyclone circuit B	60.00	86.40	12.47	3.80	2.75	54.5	12,429 (28.84)
Heavy-media cyclone circuit C	56.32	86.64	12.09	3.80	2.75	55.6	12,533 (29.08)
Heavy-media cyclone circuit D	56.67	85.81	12.92	3.93	2.88	55.8	12,361 (28.68)

The pyritic sulfur contents were estimated from the total sulfur contents, assuming that all products contained 1.05% organic sulfur (Dilley, 1993, Personal Communication).

4.2.4 Performance Curves

A partition curve is a means for measuring the separation efficiency of a separator. Partition curves for the plant being studied were calculated based on sink/float analyses of the separator products. The partition data calculated for these tests is plotted in Figure 4.1, along with the best-fit curves for each circuit, as determined by nonlinear regression, using Equation 4.1. The values and the uncertainties of the calculated parameters are given in Table 4.7. Two general-purpose curve-fitting programs were used (Sherrod, 1992; Temple University, 1993), which gave identical results even though they used different curve-fitting algorithms (Dennis et al., 1981; Press et al., 1988). Lynch's model fits the central portion of the partition curve very well and deviates only a small amount near the ends. If the deviations at the ends were large, this would be an indication that there were mechanical problems with the separator, which was allowing either high-density or low-density particles to be misplaced into the wrong product.

To show the effect of changes in the density of separation (ρ_{50}) and the separation sharpness parameter (α), Equation 4.1 was used to calculate the % clean coal yield and the % pyritic sulfur that would be expected from the Peters Washer and Heavy-Media Cyclone Circuit B feeds as ρ_{50} and α were varied. The results of these calculations are given in Table 4.8. The pyritic sulfur contents of the products were estimated based on the assumption that all products contained 1.05% organic sulfur.

Comparing the actual plant results from Tables 4.6 and 4.7 with the predicted results in Table 4.8, it is seen that the predicted results show lower sulfur contents and higher sulfur removals than was actually seen in the plant. This difference is due to the small deviations of Lynch's model from the actual data near the ends of

TABLE 4.7
Results of Fitting Lynch's Equation to the Meigs Plant Partition Data, and Using the Parameters α and ρ_{50} to Calculate the Ecart Probability Number, E_p, Using Equation 4.8

	A	ρ_{50}	E_p
Heavy-media bath	54.87 ± 5.93	1.752 ± 0.004	0.0351
Heavy-media cyclone circuit B	41.28 ± 1.66	1.807 ± 0.002	0.0480
Heavy-media cyclone circuit C	64.07 ± 1.64	1.8633 ± 0.0007	0.0319
Heavy-media cyclone circuit D	46.25 ± 5.12	1.824 ± 0.005	0.0433

Increasing values of α indicate an increasingly sharp separation, the ρ_{50} value increases as the separating density of the separator increases, and the value of E_p approaches zero as the separation becomes sharper and correspondingly separation efficiency improves (Kawatra et al., 1995). All of these separations are relatively efficient, but circuit C has the best separation performance having the highest a value and the lowest E_p value.

TABLE 4.8
Calculation of the Predicted Performance of the Peters Washer (Heavy-Media Bath) and Heavy-Media Cyclone Circuit B at the Meigs Plant, Based on the Sink/Float Analyses in Table 4.5, as the Specific Gravity of Separation (ρ_{50}) and Sharpness Parameter (α) Are Varied

			Heavy-Media Bath (Feed 4.58% Py. S)			Heavy-Media Cyclone Circuit B (Feed 3.63% Py. S)		
ρ_{50}	α	E_p	% Yield	% Pyritic S	% Pyrite Removal	% Yield	% Pyritic S	% Pyrite Removal
1.70	60	0.031	42.31	2.89	73.30	54.82	2.20	66.78
	55	0.034	42.37	2.91	73.08	54.84	2.20	66.76
	45	0.042	42.54	2.94	72.69	54.87	2.20	66.74
	35	0.053	42.77	2.98	72.17	54.90	2.22	66.42
	20	0.093	42.77	3.11	70.96	54.33	2.26	66.17
1.75	60	0.032	44.34	3.09	70.09	55.89	2.27	65.05
	55	**0.035**	**44.45**	**3.11**	**69.82**	55.94	2.27	65.02
	45	0.043	44.71	3.14	69.35	56.05	2.29	64.64
	35	0.055	45.00	3.19	68.66	56.17	2.31	64.26
	20	0.096	44.99	3.28	67.78	55.84	2.36	63.70
1.80	60	0.033	47.62	3.42	64.44	57.50	2.40	61.98
	55	0.036	47.64	3.42	64.43	57.53	2.41	61.80
	45	0.044	47.67	3.43	64.30	57.60	2.42	61.60
	35	0.056	47.67	3.44	64.20	57.65	2.43	61.41
	20	0.099	47.13	3.45	64.50	57.24	2.45	61.37
1.81	**41**	**0.048**	48.28	3.49	63.21	**57.96**	**2.45**	**60.88**
1.85	60	0.034	51.06	3.72	58.53	59.44	2.57	57.92
	55	0.037	50.95	3.71	58.73	59.41	2.57	57.94
	45	0.045	50.67	3.69	59.18	59.33	2.57	58.00
	35	0.058	50.28	3.66	59.82	59.19	2.56	58.26
	20	0.102	49.07	3.60	61.43	58.50	2.54	59.07

Values corresponding to the actual plant performance are highlighted in bold (Kawatra et al., 1995).

the partition curve. Since the ends of the curve represent either very low-ash coal or very high-ash reject, a very small deviation in the curve fit will result in a disproportionately large error in the predicted values. The particles that fall in the ends of the curve represent misplaced particles; therefore, the difference between the curve of Lynch's model and the actual data shows how much room for improvement there is in the performance of a particular separator.

From Table 4.8, it can be seen that increasing the ρ_{50} increases both the clean coal yield and the % sulfur in the clean coal, as would be expected, since a higher separating density will carry more locked coal/pyrite particles into the float product. There are also changes in yield and product quality when α is varied, but the changes are very small because these feeds do not contain a great deal of near-gravity material. This shows that in this plant, the separation performance is good enough that the product quality is determined more by the characteristics of the feed than by the performance of the plant. Increasing the sharpness of the separation will have a small, but potentially significant, effect on the plant performance for this feed.

4.3 PARTITION CURVES FOR OTHER DENSITY SEPARATORS

Using partition curves, the desulfurization abilities of various types of density separators can be easily compared, as shown in Figure 4.3. From these curves, it is seen that heavy-media separators give the sharpest separation and are therefore the most effective of these processes for sulfur removal. The other processes shown are less effective, but are generally much cheaper to install and operate. They are most suitable for coals where the sulfur is well-liberated and for coals that have relatively little material with a density near the separating density. It has also been determined in laboratory tests that heavy-media separation can produce a sharper separation than froth flotation (Gayle and Smelley, 1960).

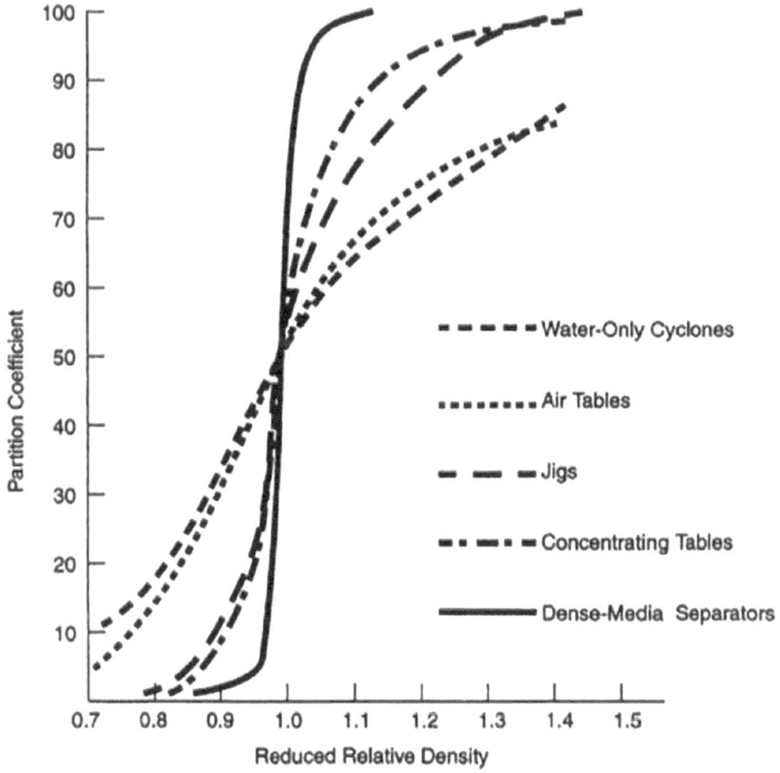

FIGURE 4.3 Comparative partition curves for various types of density separators. It should be kept in mind that the nature of the curves will also change as the particle size changes.

4.4 PRINCIPLES OF HEAVY-MEDIA SEPARATION

Heavy-media separation is the most straightforward technique for separating particles based on their densities. These processes work by immersing the coal in a fluid that has a density between the density of coal and of the ash. The coal then floats in the fluid, while the ash sinks. This is the same basic principle as the standard laboratory sink–float test that is used for determining coal washability (ASTM, 1994). Heavy-media separators have the following advantages over other coal-cleaning processes:

- Can make a sharp separation, even when there is a significant amount of material with a density near the separating density;
- Density can be controlled to within ±0.005 specific gravity units;
- Can be used for coals ranging from several centimeters in diameter down to 150 μm or less;
- Allow the specific gravity of separation to be changed freely, so that the coal product can be adjusted according to the feed quality or market requirements;
- Can easily adapt to changes in the quality and quantity of the feed.

These advantages are partially offset by the following disadvantages:

- High capital costs, because of the need for collecting and recycling the heavy media;
- High operating costs, because of media losses and the cost of operating the media recycle circuit;
- High maintenance costs when an abrasive media is used;
- System start-up problems if the plant goes down unexpectedly, and media is allowed to settle out and plug pumps, sumps, and lines.

Like other gravity separation devices, the ability of heavy-media separators to remove sulfur is limited to removing the pyrite particles which are liberated at a size coarser than the operating limit of the machine. The various heavy-media baths, which have a lower size limit of approximately 0.635 cm (1/4 in), are therefore limited to removing the coarser pyrite lumps. Heavy-media cyclones are somewhat more effective for sulfur removal, as they can function at sizes as small as 150 μm. Work is being done to allow heavy-media separators to work at even finer sizes, thus improving their sulfur-removal abilities.

4.4.1 Heavy Media

The fluid used as the heavy media may be any fluid that has a specific gravity that is between the densities of the coal and of the waste, and the following four types of fluids have been used: (1) organic heavy liquids, (2) high-density solutions of salts in water, (3) beds of solid particles that are fluidized by air, and (4) suspensions of fine, dense particles in water or other fluid.

4.4.1.1 True Heavy Liquids

True heavy liquids are homogeneous liquids that are stable under gravity and will not settle out with time. These include both high-density organic liquids and concentrated salt solutions.

Organic heavy liquids are generally halogenated hydrocarbons, such as perchloroethylene, carbon tetrachloride, bromoform, and tetrabromoethane, although lower-density organic liquids, such as gasoline and benzene, are also sometimes used. The heavy liquids that are most commonly used are given in Table 4.9. These have specific gravities as high as 3.31 and have the advantages that they are stable, immiscible with water, have low viscosities, and the specific gravity of the liquid can be easily regulated by mixing the liquids in the proper proportions. Their disadvantages are that they are for the most part very toxic and quite expensive. As a result, these liquids are used exclusively for laboratory sink–float analyses and other quality-control applications, although some pilot-scale attempts have been made to

TABLE 4.9
Liquids that Have Been Used for Sink/Float Separation of Coal (Keller, 1982; Merck, 1983)

Compound	Formula	Specific Gravity	Normal Boiling Point, °C	Freezing Point, °C	Viscosity, Centipoise at 25°C
Water	H_2O	1.00	100.0	0.0	1.0
Methylene iodide	CH_2I_2	3.32	182.0	6.1	2.6
1,1,2,2-Tetrabromoethane (acetylene tetrabromide)	$CHBr_2$-$CHBr_2$	2.96	151.0	0.1	9.6
Bromoform	$CHBr_3$	2.89	149.0	8.0	1.8
Tribromofluoromethane	CBr_3F	2.75	108.0	−73.9	1.5
Methylene bromide	CH_2Br_2	2.48	97.0	−52.6	1.0
Methylene chlorobromide	CH_2BrCl	1.92	68.1	−88.0	0.6
Pentachloroethane	CCl_3–$CHCl_2$	1.67	161.0	−22.0	2.3
Tetrachloroethylene (perchloroethylene)	CCl_2=CCl_2	1.61	121.0	−22.4	0.9
Carbon tetrachloride	CCl_4	1.59	76.5	−23.0	0.9
Trichlorofluoromethane	CCl_3F	1.49	23.7	−111.0	0.4
Trichloroethylene	CCl_2=$CHCl$	1.46	87.1	−86.8	0.5
1,1,1-Trichloroethane	CCl_3–CH_3	1.33	74.1	−30.4	0.8
Methylene chloride	CH_2=CL_2	1.32	39.7	−95.0	0.4
Ethylene dichloride	CH_2Cl=CH_2Cl	1.25	83.5	−35.7	0.8

Values for water are also included for comparison. The necessary characteristics for such liquids are that they do not react with the coal, have a low viscosity, and are stable. None of these liquids have been used for full-scale industrial applications, although trichlorofluoromethane was used in the Otisca process, and perchloroethylene/methylene chloride mixtures have been used on a pilot scale in some advanced coal-cleaning processes.

develop an industrially practical separator based on true heavy liquids. One of the more successful attempts was the Otisca process, which used trichlorofluoromethane (Freon-11) as the heavy liquid (Keller, 1982). This liquid had the advantage that it was essentially nontoxic and could be easily recovered by evaporation because of its low boiling point. Because of concerns about its effect on atmospheric ozone, use of freon on an industrial scale has been heavily regulated, and it is unlikely that it would be possible to use it for coal cleaning. More recent work has been conducted using methylene chloride/tetrachloroethylene mixtures (Durney et al., 1991, 1992), which are reported to work well in the laboratory, but are toxic. However, at a plant scale, methylene chloride dissolved the pump seals, spilled all over the floor, and was not tried again.

Dissolved salts in water are sometimes also used in the laboratory. These are less volatile than the organic liquids, and density control can be achieved by diluting with water. Clerici solutions (thallium formate-thallium maionate solutions) are used because they can provide specific gravities as high as 4.2 at room temperature. Unfortunately, these salts are extremely toxic and should be used with great caution. A more recent development is sodium or lithium metatungstate solution, which provides specific gravities up to 3.1 and is much less toxic than clerici solutions (Plewinsky and Kamps, 1984).

A number of salts and aqueous solutions have been suggested for use in plant applications, and these are listed in Table 4.10. However, the majority of these are not practical because they have problems such as high chemical reactivity (i.e., sulfuric acid, ferric chloride, and alkali hydroxides), handling difficulties (starches and sugars), or a tendency to add ash to the clean coal (phosphate salts). The most likely candidates for use are potassium carbonate, calcium nitrate, and calcium chloride. The only true heavy liquid that has ever seen industrial success is calcium chloride solution. This was used in the Belknap coal washer, which used calcium chloride solutions with specific gravities between 1.14 and 1.25. Because the calcium chloride density was not high enough in itself to float the coal, the machine had to use a combination of upward-flowing media and suspended slimes to bring the separating density up to between 1.4 and 1.6 (Palowitch et al., 1991). This unit is now considered obsolete and is no longer used.

There is a need for a heavy liquid for an inexpensive, nontoxic heavy liquid which can be used at plant scale.

4.4.1.2 Dense Suspensions

Dense suspensions are used to provide the dense media for all current full-scale heavy-media coal-cleaning plants. The most common material for producing dense media for coal cleaning is finely ground magnetite, which has the advantages of high density (specific gravity of 5.1), reasonably low cost, easy recoverability by magnetic separation, durability, corrosion resistance, and nonreactivity with the coal being cleaned. Sand is also in current use in some plants, and certain separators (such as the water-only cyclone) are designed so that they naturally generate an "autogenous" heavy media from the high-density clays and other minerals in the coal.

The drawback to dense suspensions is that they are not stable. In industrial magnetite suspensions the particle size is typically 90% passing 44 μm, and the particles

TABLE 4.10
Aqueous Solutions That Have Been Considered for Heavy-Media Separation (Durney et al., 1991)

Solute Name	Solute Formula	% wt to Reach 1.35 SG	Viscosity (cp) at 1.35 SG	Cost $/gallon
Potassium carbonate	$K_2CO_3 \cdot 3/2H_2O$	50.9	3–4	2.40
Calcium nitrate	$Ca(NO_3)_2$	43.2	5–7	0.83
Dipotassium phosphate	K_2HPO_4	45.4	—	4.50
Calcium chloride	$CaCl_2$	48.5	6–8	0.60
Ferric sulfate	$Fe_2(SO_4)_3$	39.1	—	0.26
Potassium phosphate	K_2PO_4	42.5	—	4.69
Potassium hydroxide	KOH	50.9	3–4	0.86
Potassium iodide	KI	38.1	1–2	5.80
Sodium iodide	NaI	35.6	—	6.62
Disodium phosphate	$Na_2HPO_4 \cdot 2H_2O$	50.2	—	3.39
Sodium diphosphate	$NaH_2PO_4 \cdot H_2O$	50.9	9–10	3.67
Potassium nitrate	KNO_3	49.3	—	1.61
Sugar (fructose)	$C_{12}H_{22}O_{11}$	69.1	3–400	0.70
Sodium metasilicate	$Na_2O/XSiO_2$	44.5	—	0.85
Ferrous chloride	$FeCl_2 \cdot 4H_2O$	53.8	6–8	2.97
Corn starch	$C_6H_{11}O_6$	77.8	—	1.05
Sodium hydroxide	NaOH	48.9	15–17	1.10
Sulfuric acid	H_2SO_4	56.8	3–4	0.26

Of these, only calcium chloride has ever seen industrial use, although calcium nitrate is being considered for at least one advanced coal-cleaning process. SG, specific gravity.

will rapidly settle out of the suspension if it is left unagitated. As the particles settle, they produce density gradients that significantly affect the separator performance (Cho and Klima, 1994; Maronde et al., 1995). Dense-media separators must therefore be designed to keep the dense-media particles from settling out, while still allowing the coal being cleaned to float properly.

The particle size of the dense media particles obviously places a size limit on the coal that can be processed. If the coal has a size distribution that is close to the media size distribution, it will not be possible to keep the media in suspension while still allowing the coal to float and the ash to sink. Conventional heavy-media separators are therefore limited to approximately 150 μm. Work has been done to clean finer coal using micronized magnetite (90% passing 7 μm or less), which allows coal particles to be cleaned that are as fine as 20 μm (Maronde et al., 1995; Hucko et al., 1994). Using the micronized magnetite is similar to using conventional magnetite media, except that more advanced magnetic separators (such as rare-earth drum and high-gradient magnetic separators) are needed to recover the very fine magnetite particles. The micronized magnetite is expected to be approximately twice as expensive as the finest standard grade of magnetite that is presently produced for heavy-media separators (Miller et al., 1991).

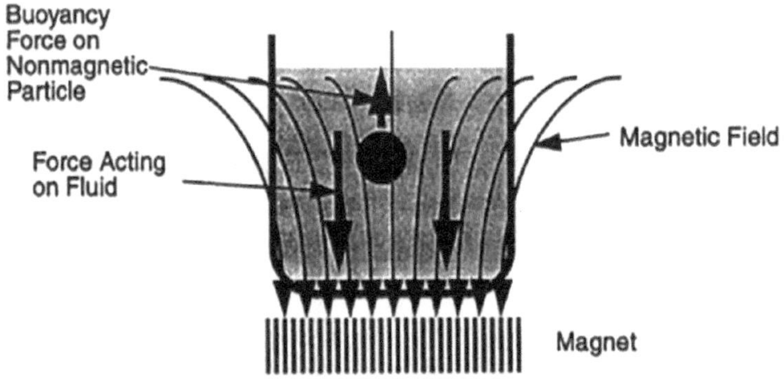

FIGURE 4.4 Principle of magnetically enhanced media. By changing the strength of the magnetic field, the apparent specific gravity of the suspension can be adjusted as desired.

Another recent innovation in dense suspensions is known as "magnetically enhanced media." These fluids are made from a colloidal suspension of magnetic particles (finer than 100 Å) stabilized against flocculation with a lignosulfonate dispersant (Anonymous, 1991; Walker and Devemoe, 1991). Because the particles are so fine, they have no tendency to settle out of suspension. When the fluid is subjected to a magnetic field, the fluid is drawn toward the magnet as shown in Figure 4.4. This has the effect of increasing the apparent weight of the suspension, and so it can be made to appear to have a higher density than it would have without the magnetic field. The magnetic field adjustments can then be used to select any desired density over a wide range (Walker and Devemoe, 1991).

4.5 SULFUR REMOVAL BY HEAVY-MEDIA VESSELS

Heavy-media vessels such as baths, drums, and cones are extremely effective for coarse coal, which will rapidly float or sink under the influence of gravity alone. These units consist of an agitated pool of media, with some means of removing the clean coal and the sinks product, as shown in Figure 4.5 (Hillman, 1994). Cone separators can handle particles as small as 2 mm, which is finer than the approximately 0.635 cm (1/4 in) limit for baths or drums (Palowitch et al., 1991). All heavy-media vessels have similar sulfur-removal ability, with a representative example of the performance that can be expected given in Table 4.11.

4.6 SULFUR REMOVAL BY HEAVY-MEDIA CYCLONES

Heavy-media cyclones were developed during World War II by the Dutch State Mines and were rapidly adopted in Europe, while the first plant using them in the United States was built in 1961 (Sokaski et al., 1991). The basic dense-media cyclone is similar to the classifying cyclones used for sizing and desliming, as can be seen in Figure 4.6. The Dynawhirlpool separator is a variation on this theme, but it is reported not to work as well in coal processing as conventional cyclones and has largely fallen into disuse (Davis, 1995, CRA-ATD, Melbourne. Personal Communication).

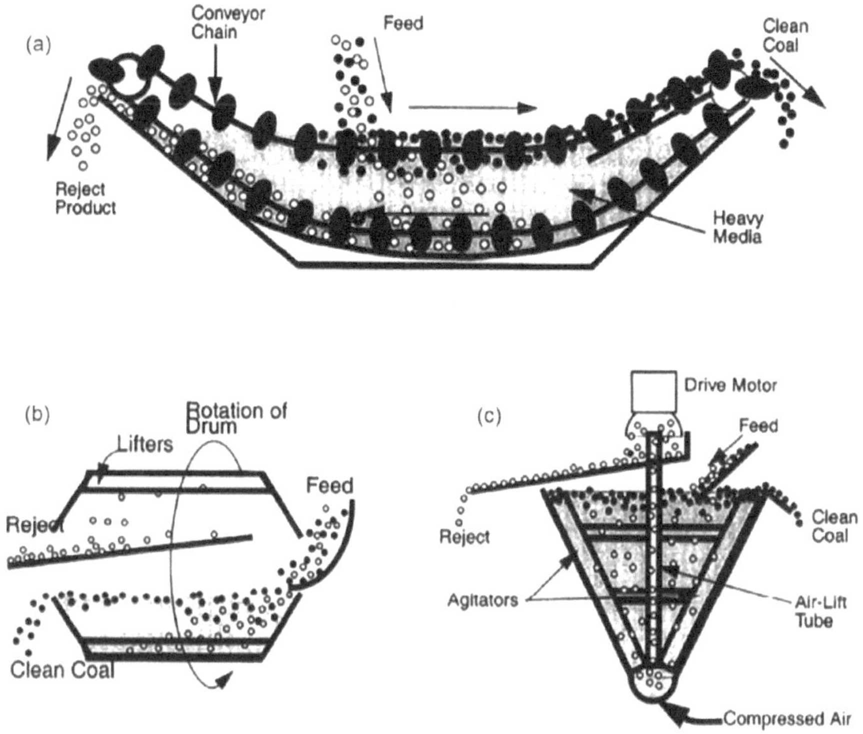

FIGURE 4.5 Schematics of typical heavy-media vessels. In the coal industry, all of these units typically use magnetite suspensions as the heavy media: (a) heavy-media bath, (b) heavy-media drum, and (c) heavy-media cone.

The TriFlow Separator is essentially two Dynawhirlpools in series, and overcomes most of the limitations of the Dynawhirlpool. The Large Coal Dense Medium Separator (LARCODEMS) is much larger in diameter than other heavy-media cyclone separators, and it is used for coarse coal processing. It is equipped with a vortex extractor so that the piping can be made large enough to handle coarse particles without requiring an excessive pumping capacity (Anonymous, 1995; Wills, 1992; Shah, 1988). The first commercial LARCODEMS was installed at the Point of Ayr Colliery, U.K., in 1984, and produced the separation performance shown in Table 4.12 (Hillman, 1994).

Heavy-media cyclones have the greatest potential for desulfurizing coal of any of the heavy-media processes because they work at a finer particle size than baths, cones, or drums. Standard heavy-media cyclones, using standard magnetite suspensions as the heavy media, can separate particles as fine as 150 μm (Baumgartner, 1978). Work is also underway to develop heavy-media cyclone processes that use specialized media such as true heavy liquids, micronized magnetite, and magnetic fluids that allow them to work at even finer particle sizes, to produce ultraclean coals (Maronde et al., 1995; Durney et al., 1992). This is possible because heavy-media cyclone processes are not dependent on gravity to move the particles through the

TABLE 4.11
Performance Summary for a Heavy-Media Bath Separator in a Full-Scale Plant (Kawatra et al., 1995)

Sink/Float Results for Feed				
SG Range	**% wt**	**% S**	**Performance Summary**	
−1.3	15.81	2.70	% Ash in feed	46.27
+1.3/−1.4	17.65	3.99	% Total sulfur in feed	5.63
+1.4/−1.5	4.89	5.60	% Pyritic sulfur in feed	4.58
+1.5/−1.6	2.35	6.72	Feed size range (mm)	152 × 16
+1.6/−1.65	0.78	5.65	% wt recovery (yield)	45.18
+1.65/−1.7	1.08	6.02	Product % ash	14.99
+1.7/−1.75	2.00	8.82	Product % total sulfur	4.28
+1.75/−1.8	5.79	9.13	Product % pyritic sulfur	3.23
+1.8/−1.85	3.03	8.83	% Pyritic sulfur removal	68.1
+1.85/−1.9	1.45	9.09	E_p	0.0351
+1.9	45.16	6.33	ρ_{50}	1.752 ± 0.004
Total	99.99	5.63	A	54.87 ± 5.93

SG, specific gravity.

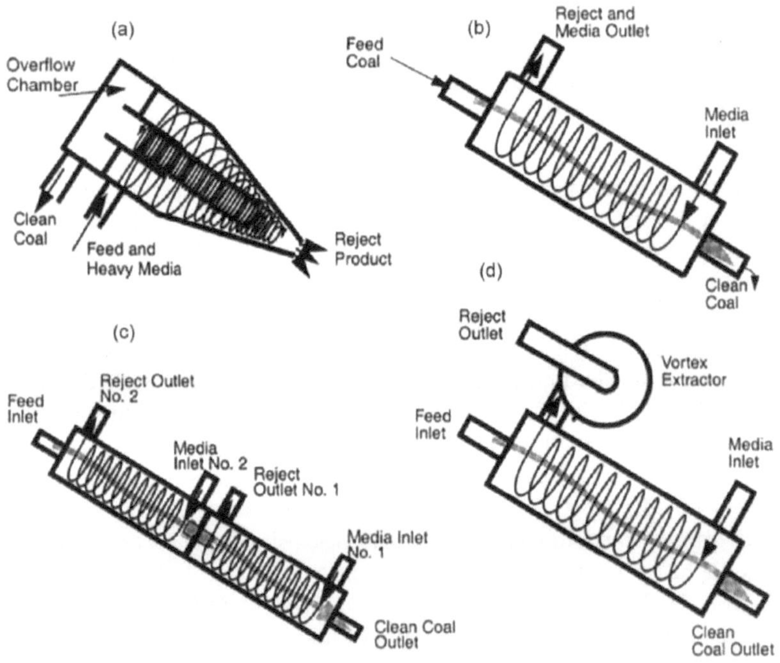

FIGURE 4.6 Schematics of commercial heavy-media cyclone separators: (a) heavy-media cyclone, (b) Dynawhirlpool, (c) TriFlow Separator, and (d) LARCODEMS separator.

TABLE 4.12
Summary of the Performance of LARCODEMS Installed at the Point of Ayr Colliery, U.K.

	Size Fraction (mm)				
	−100/+0.5	**−100/+25**	**−25/+8**	**−8/+2**	**−2/+0.5**
% wt in raw feed	100.00	42.15	35.67	16.78	5.40
% Total sulfur in feed	1.54	1.32	1.46	1.40	1.48
% Ash in feed	26.78	28.45	22.15	29.93	32.88
% wt recovery (yield)	71.81	69.48	77.76	66.90	66.59
Product % total sulfur	1.73	1.57	1.54	1.43	1.39
Product % ash	5.27	4.95	5.09	5.15	9.60
Partition density g/cm^3	1.658	1.655	1.656	1.670	1.818
E_p	0.021	0.020	0.025	0.045	0.114

fluid, and instead use centrifugal force, which can be made considerably stronger than gravitational force. Particles which would take an unreasonable amount of time to float or sink under gravity, therefore, will rapidly move to the float and sink products in a heavy-media cyclone separator.

4.6.1 Heavy-Media Cyclones Using Standard Magnetite Media

A standard heavy-media cyclone uses a magnetite suspension as the dense media, with a particle size no finer than 90% passing 44 μm (Maronde et al., 1995). The heavy-media cyclone is similar to a conventional classification cyclone, but often includes an overflow chamber with a tangential outlet to reduce the backpressure. A cone angle of 20° is typical, and they are generally mounted so that they have just enough of an angle from the vertical so that they can drain completely upon shutdown (approximately 10°). This is done both because the piping is simpler and because inclined cyclones are reported to work better at low inlet pressures than vertical cyclones do. At high inlet pressures, both inclined and vertical cyclones are reported to work equally well (Sokaski et al., 1991). A number of factors are important to the proper operation of the cyclone, including the viscosity and particle size distribution of the media and the ratio of the underflow and overflow flowrates (Collins et al., 1983; Klein et al., 1990; He and Laskowski, 1992, 1993).

A drawback to the use of magnetite suspension in heavy-media cyclones is that the centrifugal force also tends to remove the magnetite media particles from the suspension, which makes the media unstable. As a result, the density of the media exiting with the reject product is higher than the density exiting with the clean coal product. This produces a density gradient in the heavy-media cyclone, which reduces the sharpness of the separation and decreases the ability to remove pyrite. It is also necessary to ensure that the magnetite always has the correct size distribution. In an extensive plant-oriented test work campaign, it was shown that the root cause of inefficient heavy-media cyclone performance was related to incorrect magnetite particle size in the cyclone circuits (Vince and Keast-Jones, 1992). The coarsening of the magnetite was not only due to selective losses of the fine magnetite but also due to a design flaw in the plant that allowed segregation of the media. In extreme cases, coarsening of the magnetite can cause the Ecart Probability value to increase from a normal value of about 0.03 to as much as 0.05 (Sokaski et al., 1991).

Sripriya et al. (2003) investigated an alternative to magnetite for heavy media suspensions: the magnetic fraction of blast furnace flue dust. They claim this was highly successful in laboratory-scale work, but did not succeed in the experimental work at plant scale.

An estimate of the Ecart Probability (E_p) value that would be expected for a particular heavy-media cyclone processing a particular coal can be made before the separation is attempted, using the following formula (Albrecht, 1992):

$$E_p = F_1 \times \left[(0.037)\rho_{50} - 0.015\right] \tag{4.9}$$

where

ρ_{50} = density at which 50% of the particles report to the sinks product,
F_1 = particle-size factor, which can be taken from Table 4.13.

A summary of the sulfur-removal performance of a heavy-media cyclone in an operating coal processing plant is given in Table 4.14.

4.6.6.1 Case Study: Sulfur Removal Using the Dynawhirlpool in Plant Installations

In a study by Maronde et al. (1983), the performance of the Dynawhirlpool separator was evaluated in four different plants. Three of the units were part of the plant process circuits, and the fourth was a 20 tons/h pilot-scale evaluation unit operating in parallel with the main plant. A summary of the feed coal characteristics and separator performances are given in Table 4.15. The value of E_p reported by Maronde et al. was larger than the E_p seen for the heavy-media cyclone described in Table 4.14, showing that the Dynawhirlpool is not as efficient of a separator as a conventional heavy-media cyclone. The lower efficiency of the Dynawhirlpool is even more clearly shown when the values of α are calculated using Equation 4.8. The conventional cyclone showed an α of 64.07, while the Dynawhirlpools all showed values of α of only 22–32. This comparison of the α values is more meaningful than the comparison of the E_p values, because the α value corrects for the effects of differences in the separating density (ρ_{50}).

TABLE 4.13
Particle Size Correction Factors for Prediction of E_p with Heavy-Media Cyclones (Albrecht, 1992)

Ave. Particle Size (mm)	<1	1–2	2–3	3–4	4–5	5–6	6–7	7–8	8–9	>9
Factor F_1	1.45	1.21	0.96	0.86	0.82	0.80	0.82	0.86	0.90	0.95

TABLE 4.14
Performance Summary for a Heavy-Media Cyclone Separator in a Full-Scale Plant, Using Standard Magnetite Media (Kawatra et al., 1995)

Sink/Float Results for Feed			Performance Summary	
SG Range	**% wt**	**% S**		
−1.3	29.86	2.58	% Ash in feed	45.56
+1.3/−1.4	12.04	3.56	% Total sulfur in feed	4.54
+1.4/−1.5	4.62	4.74	% Pyritic sulfur in feed	3.49
+1.5/−1.6	2.45	5.83	Feed size range (mm)	16 × 0.6
+1.6/−1.65	0.73	6.12	% wt recovery (yield)	56.32
+1.65/−1.7	0.79	5.56	Product % ash	12.09
+1.7/−1.75	1.02	7.93	Product % total sulfur	3.80
+1.75/−1.8	2.43	8.96	Product % pyritic sulfur	2.75
+1.8/−1.85	1.51	8.66	% Pyritic sulfur removal	55.6
+1.85/−1.9	1.31	8.25	E_p	0.0319
+1.9	43.24	5.44	ρ_{50}	1.8633 ± 0.0007
Total	100.00	4.54	A	64.07 ± 1.64

SG, specific gravity.

TABLE 4.15
Summary of the Performance of Several Different Dynawhirlpool Installations in Operating Plants (Maronde et al., 1983)

	Unit A	**Unit B**	**Unit C**	**Unit D Test 1**	**Unit D Test 2**
% Ash in feed	18.5	19.5	25.7	9.6	10.1
% Total sulfur in feed	3.99	0.87	0.99	1.98	2.14
% Pyritic sulfur in feed	—	0.49	0.47	—	—
Feed size range (mm)	9.5 × 0.6	38.1 × 0.6	9.5 × 0.6	9.5 × 0.6	9.5 × 0.6
% wt recovery (yield)	53.0	66.5	79.1	87.3	90.5
Product % ash	6.0	7.0	10.3	5.9	6.4
Product % total sulfur	3.62	0.61	0.99	1.38	1.51
% Total sulfur removal	51.8	53.1	21.1	39.1	36.2
Product % pyritic sulfur	—	0.13	0.38	—	—
% Pyritic sulfur removal	—	81.4	36.5	—	—
E_p	0.066	0.050	0.057	0.054	0.060
ρ_{50}	1.375	1.422	1.641	1.423	1.471
A	22.89	31.24	31.63	28.95	26.93

The effects of particle size on the performance are shown by the partition curves in Figure 4.7. As the particles become finer, the quality of the separation made by the Dynawhirlpool degrades. The performance degradation becomes severe when the particle size is reduced to less than 2.36 mm. The partition curves are also seen to be

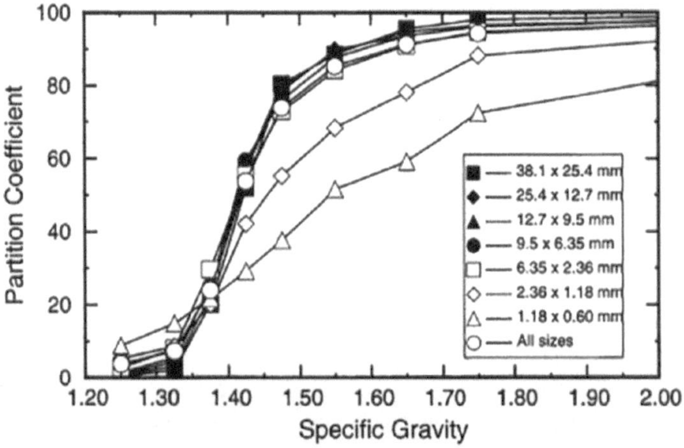

FIGURE 4.7 Partition curves for specific size fractions in a Dynawhirlpool separator in an operating plant (Maronde et al., 1983). As the particle size becomes finer, the sharpness of the separation decreases.

asymmetrical, with more misplaced particles on the high-specific gravity side than on the low specific gravity side. This indicates that there is a significant amount of contamination of the clean coal product by dense particles. This effect is reported to be due to the short residence time of the rejects, which prevents the dense particles from being completely removed from the clean coal (Davis, 1995, CRA-ATD, Melbourne. Personal Communication).

4.6.2 True Heavy-Liquid Cyclones for Super-Clean Coal Production

Conventional heavy-media cyclones are largely limited in their performance by the characteristics of standard magnetite media. Because of the tendency of magnetite to settle out of suspension, the heavy-media cyclones cannot be made too small in diameter (because of the higher forces in small cyclones), and coal that is approaching the particle size of the magnetite cannot be separated.

The suspension settling problem can be eliminated by the use of true heavy liquids, such as perchloroethylene/methylene chloride blends and calcium nitrate solutions. The stability of these liquids allows the use of cyclones of much smaller diameter operating at much higher feed pressures (Thome et al., 1994). This increases the separating forces, allowing separations of coal at finer sizes. Since the liberation of the pyrite is much more complete at the finer particle sizes, these heavy-media cyclones have the potential to produce super-clean coals.

Test work (Durney et al., 1994) has been carried out using a 5.1 cm (2 in) diameter cyclone, with a mixture of methylene chloride and perchloroethylene used as the heavy liquid (specific gravity of 1.45). The feed consisted of 150 μm × 0 μm coal, which is below the normal size limit for conventional magnetite media. Using this cyclone, an E_p value of 0.135 was obtained at a ρ_{50} of 1.53. A heavy solution (calcium nitrate in water) was also tested, but it was found that the viscosity of the calcium nitrate solution was too great for the process to work efficiently.

The methylene chloride/perchloroethylene mixture is low enough in cost to be industrially usable. Unfortunately, both of these compounds are toxic, and a spill in a full-scale plant would have the potential for very large clean-up costs.

4.6.3 Micronized Magnetite Heavy-Media Cyclones for Super-Clean Coal Production

An alternative to the use of true heavy liquids is to increase the stability of the magnetite suspension against settling by grinding the magnetite to a very fine particle size (PETC, 1993). In the micro-mag process, the magnetite is ground to 90% finer than 7 μm, which is considerably finer than the magnetite used in conventional cyclone circuits (Figure 4.8). The increased media stability allows the use of smaller diameter cyclones (25.4 cm diameter, compared to 61.0–91.4 cm diameter for conventional heavy-media cyclones), and the cyclone feed pressures that are 3–10 times higher than conventional circuits can be used (Maronde et al., 1995). The process is currently undergoing continuous bench and pilot-scale studies (Anast, 1994). In the work to date, it has been confirmed that making the magnetite finer has improved the ability to separate finer coal and pyrite particles, and the process appears to be technically feasible for deep-cleaning coal down to 38 μm (Klima and Killmeyer, 1991).

The results of test work with the micro-mag process are given in Table 4.16. The heavy-media cyclone was able to produce a good separation at sizes as fine as 38 μm, with correspondingly good reductions in both ash and pyritic sulfur content. The separating efficiency is comparable to conventional heavy-media cyclones

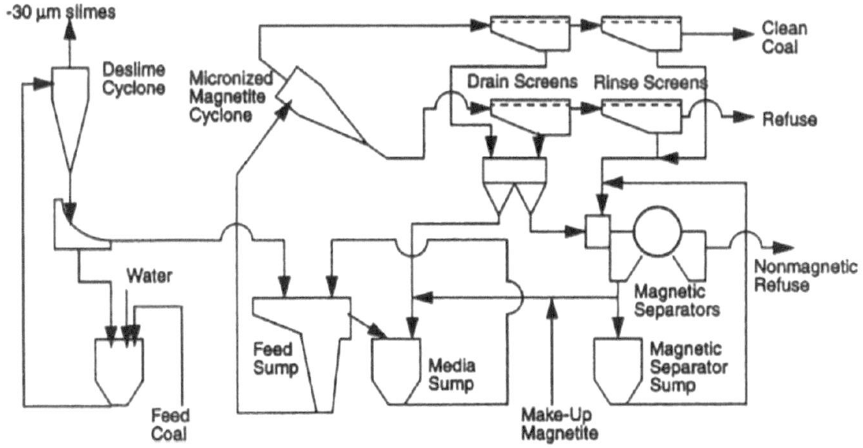

FIGURE 4.8 Flow diagram for the micronized magnetite cyclone test circuit (Maronde et al., 1995). Because of the fine particle size of the magnetite, the magnetic separators are rare-earth drum separators and high-gradient magnetic separators.

TABLE 4.16
Separation Efficiency for a 7.6 cm (3 in) Diameter Cyclone, Using Micronized Magnetite as the Heavy Media (Hucko et al., 1994)

Feed Size (μm)	ρ_{50}	E_p
600 × 150	1.34	0.015
150 × 75	1.40	0.045
75 × 38	1.48	0.100
600 × 38	1.35	0.025

processing coal at a much coarser size (18.9 mm × 600 μm, E_p approximately 0.02–0.04). Since the micro-mag process is capable of high separating efficiency at such a fine size, it has much greater potential for coal desulfurization than conventional heavy-media cyclones.

4.6.4 Magnetic Fluid Cyclones

Units using the magnetic fluid principle in combination with centrifugal forces (Figure 4.9) have been marketed both on a laboratory and a pilot-plant scale, but the equipment needed for a full-scale plant installation is still under development (Durney et al., 1992).

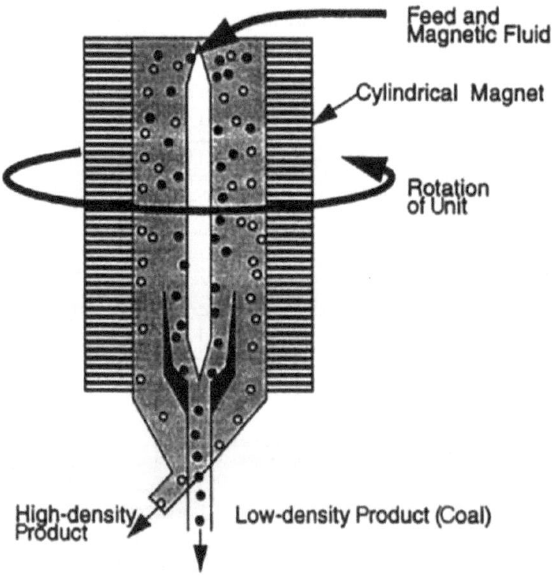

FIGURE 4.9 Schematic of the Magstream separator, which uses a combination of centrifugal forces and magnetic fields to control the apparent density of a magnetic fluid.

Laboratory-scale units are being used for performing sink–float analyses, and are reported to produce results equal to the conventional heavy liquids used for this purpose. Once a full-scale plant unit has been developed, it will be at least equal to the performance of true heavy liquids in past work. The sub-micron particle size and low viscosity of these liquids will make it possible to process even very fine coal (Walker and Devemoe, 1991).

4.6.5 Water-Only Cyclones

Water-only cyclones are widely used in existing plant applications (Figure 4.10). They are often considered to be autogenous heavy-media separators, with the heavy media being the coal slurry feed itself. While they produce a much less accurate separation than other heavy-media separators (Bull et al., 1987), they nevertheless are commonly used because they are simple, compact, have low maintenance, require little ancillary equipment, and have a low capital cost. Also, the high density difference between coal and pyrite makes it possible for these units to make a good separation of pyrite from 0.5 mm × 0 mm coal (Taylor, 1969). They are most suitable as a precleaning stage for other, more precise separators, where they reduce the required capacity of the more expensive high-efficiency separators. Depending on the nature of the coal, they can be flexibly combined with heavy-media cyclones and froth flotation, as well as with ancillary equipment for dewatering and drying (Visman, 1968). The most common application is to use the water-only cyclone to produce a final clean coal product, with the water-only cyclone refuse being re-treated by other means to recover the substantial amount of coal that remains in it (O'Brien and Sharpeta, 1978).

The separation efficiency as a function of particle size is shown in Table 4.17. While the water-only cyclone has a much lower separating efficiency than comparable-size dense-media cyclones, the large density difference between coal and pyrite makes it possible to use these units to remove the bulk of the pyritic sulfur from coal, as shown in Table 4.18.

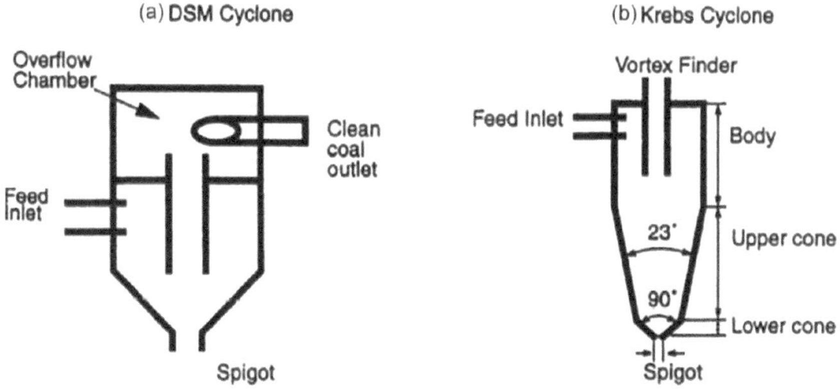

FIGURE 4.10 Schematics of two types of water-only cyclone.

TABLE 4.17
Separating Efficiency for a 38.1cm (15 in) Diameter Water-Only Cyclone in a Plant Application (O'Brien and Sharpeta, 1978)

Feed Particle Size (mm)	E_p
6.35	0.101
6.35 × 4.75	0.108
4.75 × 2.36	0.132
2.36 × 1.18	0.188
1.18 × 0.6	0.264
6.35 × 0.6	0.135

As the particle size becomes smaller, the separation efficiency becomes worse.

TABLE 4.18
Sulfur Removal as a Function of Particle Size by a Typical Industrial-Scale Water-Only Cyclone, Compared to Sulfur Removal by Froth Flotation (Hochscheid, 1981)

Particle Size (μm)	Percent Sulfur Removal by Water-Only Cyclone	Percent Sulfur Removal by Froth Flotation
600 × 150	70–80	50–60
150 × 74	60–70	30–40
74 × 45	35–40	25–30
45 × 0	8–10	20–25

Work has also been done using two-stage water-only cyclone circuits, in attempts to produce a high-quality clean coal at a high BTU recovery. There are three types of two-stage circuits that have been considered, as shown in Figure 4.11. Computer simulations of these three types of circuits show that the two-stage washing circuit provides the highest capacity; the secondary-overflow, middlings recirculation circuit provides higher organic efficiency and clean coal recovery; and the secondary-underflow, middlings recirculation circuit provides better exclusion of high specific gravity fractions from the product (Moorhead and Harrison, 1984).

4.6.6 Decanter Centrifuge

The alfa-laval decanter centrifuge is a device that can achieve higher centrifugal forces than is possible with a cyclone. This device works as shown in Figure 4.12. The entire apparatus spins rapidly, centrifuging the coal and the heavy liquid to the rotor shell. The screw conveyor is designed so that liquid can flow through it,

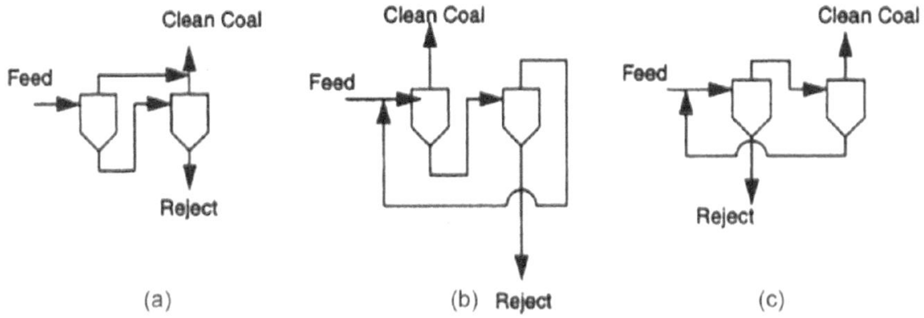

FIGURE 4.11 Three types of two-stage water-only cyclone circuits: (a) two-stage washing; (b) two-stage, secondary-overflow, middlings recirculation; and (c) two-stage, secondary-underflow, middlings recirculation (Moorhead and Harrison, 1984).

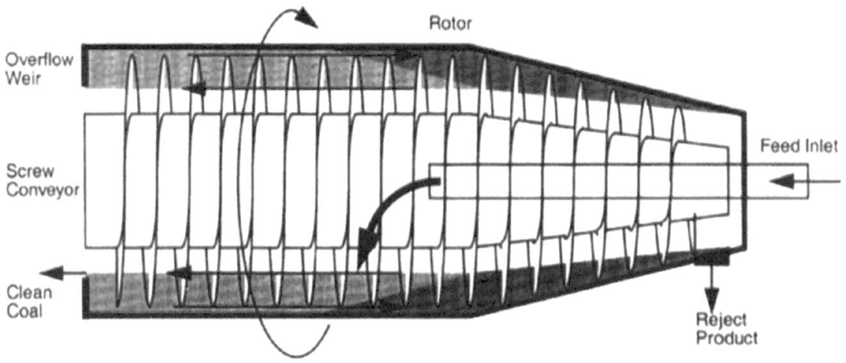

FIGURE 4.12 Schematic diagram of a decanter centrifuge (Durney et al., 1994).

while settled solids are scraped up and conveyed to the sinks product. This is a high-precision separator and is suitable for production of super-clean coals because it can separate ash and pyrite from coal ground finely enough for nearly complete liberation of the mineral particles.

Because of its high centrifugal forces, a decanter centrifuge can use heavy liquids that are more viscous than are practical in heavy-media cyclones. This unit has been tested using a calcium nitrate solution as the heavy liquid (specific gravity of 1.35–1.40) and was used to clean feed coals with size distributions of 150 μm × 0 μm, and 25 μm × 0 μm (Durney et al., 1994). The unit could reach an E_p of 0.035 with the 150 μm × 0 pm feed, and an E_p of 0.195 with the 25 μm × 0 μm. This is a significant improvement over the use of a 5.08 cm (2in) diameter heavy-media cyclone with a methylene chloride/perchloroethylene heavy liquid, which when processing the 150 μm × 0 μm feed could only achieve an E_p of 0.135.

The major drawback to the decanter centrifuge is that it has a high capital and operating cost compared to cyclones. Its advantage is that it can use true heavy liquids that are safer than the toxic organic liquids needed for heavy-liquid cyclones; therefore, its use would avoid high clean-up or medical costs in case of accidents (Durney et al., 1994).

4.7 SUMMARY

The heavy-media processes that are currently used industrially are extremely efficient separators, and are effective for removing pyrite that liberates at sizes coarser than approximately 150 μm. The particle size limitations of these devices are mainly due to their use of magnetite suspensions as the heavy media. By using true heavy liquids or micronized magnetite, their effectiveness can be pushed to finer particle sizes, allowing them to remove finely disseminated pyrite as well. The barriers to using true heavy liquids or micronized magnetite are

1. toxicity and environmental hazards from organic heavy liquids;
2. high viscosity for solutions of heavy salts in water, restricting them to use in high-speed centrifugal machines;
3. difficulty in recovering the heavy media for re-use.

REFERENCES

Albrecht, M.C. (1992), "Selecting Heavy-Media Cyclones," *Coal*, Vol. 97, No. 8, pp. 33–35.

Anast, K. (1994), "Bench Scale Testing of Micronized Magnetite Beneficiation," Quarterly Technical Progress Report 4, October–December 1993, DOE Contract No. DE-AC22-93PC92206.

Anonymous (1991), Product Brochures for Magfluid®, and Magstream® separators, Intermagnetics General Corporation, Guilderland, NY.

Anonymous (1995), "Researchers Fine Tune the Larcodems," *Coal*, Vol. 100, No. 4, p. 66.

ASTM (1994), "Standard Test Methods for Determining the Washability Characteristics of Coal," *Annual Book of ASTM Standards*, Vol. 5.05, Standard No. D4371, American Society for Testing and Materials, Philadelphia.

Baumgartner, F.E. (1978), "Washing Raw Coal to Zero with Heavy Media Cyclones," *Mining Congress Journal*, pp. 45–50.

Browning, J.S. (1961), "Heavy Liquids and Procedures for Laboratory Separation of Minerals," U.S. Bureau of Mines Information Circular IC 8007.

Bull, W.R., Pillai, K.J. and Spottiswood, D.J. (1987), "An Analysis of Water-Only Cyclone Capabilities," *SME Annual Meeting*, Denver, CO, February 24–27, 1987, Preprint No. 87-100.

Burt, R.O. (1984), *Gravity Concentration Technology, Developments in Mineral Processing*, Vol. 5, Elsevier, Amsterdam.

Cho, H. and Klima, M.S. (1994), "Application of a Batch Hindered-Settling Model to Dense-Medium Separations," *Coal Preparation*, Vol. 14, pp. 167–184.

Collins, D.N., Turnbull, T., Wright, R. and Ngan, W. (1983), "Separation Efficiency in Dense Media Cyclones," *Transactions of the Institute of Mining and Metallurgy*, Vol. 92, Section C, pp. C38–C51.

Dennis Jr., J.E., Gay, D.M. and Welsch, R.E. (1981), "An Adaptive Nonlinear Least-Squares Algorithm," *ACM Transactions on Mathematical Software*, Vol. 7, No. 3, pp. 348–368.

Durney, T., Im, C., Suardini, P., Laurila, L., Ferris, D., Harrison, K., Devemoe, A. and Walker, M. (1991), "Evaluation, Engineering, and Development of Advanced Cyclone Processes," *Proceedings of the Seventh Annual Coal Preparation, Utilization, and Environmental Control Contractors Conference*, U.S. DOE/PETC, July 15–18, Pittsburgh, PA, pp. 474–481.

Durney, T., Im, C., Suardini, P., Laurila, L., Ferris, D., Harrison, K., Devemoe, A. and Walker, M. (1992), "Evaluation, Engineering, and Development of Advanced Cyclone Processes," *Proceedings of the Eighth Annual Coal Preparation, Utilization, and Environmental Control Contractors Conference*, U.S. DOE/PETC, July 27–30, Pittsburgh, PA, pp. 471–478.

Durney, T., Cook, C.A., Suardini, P., Laurila, L., Ferris, D. and Devemoe, A. (1994), "Evaluation, Engineering, and Development of Advanced Cyclone Processes," *Proceedings of the Tenth Annual Coal Preparation, Utilization, and Environmental Control Contractors Conference*, U. S. DOE/PETC, July 18–21, Pittsburgh, PA, pp. 75–83.

Fonseca, A.G. (1995), "The Challenge of Coal Preparation," *High Efficiency Coal Preparation* (S.K. Kawatra, ed.), Society for Mining, Metallurgy, and Exploration, Littleton, CO, pp. 2–21.

Gayle, J.B. and Smelley, A.G. (1960), "Selectivities of Laboratory Flotation and Float-Sink Separations of Coal," U.S. Bureau of Mines, Report of Investigations RI5691.

Gottfried, B.S. (1978), "A Generalization of Distribution Data for Characterizing the Performance of Float-Sink Coal Cleaning Devices," *International Journal of Mineral Processing*, Vol. 5, pp. 1–20.

He, Y.B. and Laskowski, J.S. (1992), "The Effect of the Overflow to Underflow Flowrate Ratio on the Performance of a Dense Medium Cyclone," *SME-AIME Annual Meeting*, Phoenix, AZ, February 24–27, Preprint No. 92-58.

He, Y.B. and Laskowski, J.S. (1993), "The Effect of Dense Media Properties on the Performance of a Dense Medium Cyclone," *SME-AIME Annual Meeting*, Reno, NV, February, Preprint No. 93-98.

Hillman, J. (1994), "Dense Medium Separators," *Mine and Quarry*, January/February 1994, pp. 44–47 (Part I), and March 1994, pp. 19–24 (Part II).

Hochscheid, R. (1981), "The Evolving Role of Water-Only Cyclones in Fine Coal Cleaning," *SME-AIME Fall Meeting*, Denver, CO, November 18–20, 1981, Preprint No. 81–395.

Hucko, R.E., Schimmoller, B.K. and Jacobsen, P.S. (1994), "The Prep Plant of Tomorrow," *Coal*, pp. 50–53.

Kawatra, S.K., Eisele, T.C. and Dilley, H.L. (1995), "Procedure for Evaluating Existing Heavy-Media Coal Preparation Circuits to Enhance Sulfur Removal: A Case Study," *High Efficiency Coal Preparation* (S.K. Kawatra, ed.), Society for Mining, Metallurgy, and Exploration, Littleton, CO, pp. 55–68.

Keller Jr., D.V. (1982), "The Otisca Process: An Anhydrous Heavy-Liquid Separation Process for Cleaning Coal," *Physical Cleaning of Coal* (Liu, ed.), Marcel Dekker, New York, pp. 35–86.

Klein, B., Partridge, S J. and Laskowski, J.S. (1990), "Rheology of Unstable Mineral Suspensions," *Coal Preparation*, Vol. 8, pp. 123–134.

Klima, M.S. and Lucide, P.T. (1986), "An Interpolation Methodology for Washability Data," *Coal Preparation*, Vol. 2, pp. 165–177.

Klima, M.S. and Killmeyer, R.P. (1991), "Baseline Performance Evaluation of Micronized-Magnetite Cycloning," U.S. Department of Energy, Pittsburgh Energy Technology Center, DOE/PETC/TR-91/9 (DE91018699).

Leonard, J.W. (ed.) (1991), *Coal Preparation*, 5th edition, Society for Mining, Metallurgy, and Exploration, Littleton, CO.

Lynch, A.J. (1977), *Mineral Crushing and Grinding Circuits, Their Simulation, Optimisation, Design, and Control (Developments in Mineral Processing 1)*, Elsevier, Amsterdam, p. 116.

Maronde, C.P., Killmeyer, R.P. and Deurbrouck, A.W. (1983), "Performance Characteristics of Coal-Washing Equipment: The Dynawhirlpool Separator," U.S. Department of Energy, Pittsburgh Energy Technology Center, Report No. DOE/PETC/TR-83/8 (DE83015049).

Maronde, C.P., Killmeyer, R.P. and Suardini, P.J. (1995), "Integrated Continuous Testing of the Micronized-Magnetite Beneficiation Process in PETC's Coal Preparation Process Research Facility," *High Efficiency Coal Preparation* (S.K. Kawatra, ed.), Society for Mining, Metallurgy, and Exploration, Littleton, CO.

Meloy, T.P. (1982), "Heavy Media Selection Functions—Circuit Analysis," *SME-AIME Annual Meeting*, Dallas, TX, February 14–18, Preprint No. 82-36.

Merck (1983), *The Merck Index*, 10th Edition, Merck & Co., Rahway, NJ.

Miller, K.J., Klima, M.S. and Killmeyer, R.P. (1991), "Selection and Production of Dense-Medium Solids for the Micro-Mag Process," U.S. DOE, Pittsburgh Energy Technology Center, Report No. DOE/PETC/TR-91/5 (DE91011943).

Moorhead, R.G. and Harrison, C.D. (1984), "Evaluation of Two-Stage, Middlings-Recirculation, Water-Only Cyclones," *First Annual Coal Preparation Conference and Exhibition*, March 27–29, Lexington, KY.

O'Brien, E.J. and Sharpeta, K.J. (1978), "Water-Only Cyclones; Their Functions and Performance," *Coal Age Operating Handbook of Preparation* (P.C. Merritt, ed.), McGraw-Hill, New York, pp. 114–118.

Palowitch, E.R., Deurbrouck, A.W., Torak, E. and Akers, D.J. (1991), "Wet Coarse Particle Concentration, Section 1: Dense Media," *Coal Preparation*, 5th edition (Leonard and Hardinge, eds.), Society for Mining Metallurgy, and Exploration, Littleton, CO, pp. 271–300.

Peng, F.F. (1983), "Graphic Routine and Digital Simulation—The Aids in Prediction of Coal Preparation Plant Performance and Yield Optimization," *Proceedings of the First Conference on Use of Computers in the Coal Industry*, AIME, New York, pp. 187–199.

Peng, F.F. and Luckie, P.T. (1991), "Process Control, Part 1: Separation Evaluation," *Coal Preparation*, 5th edition (Leonard and Hardinge, eds.), Society for Mining Metallurgy, and Exploration, Littleton, CO, Chapter 10, pp. 659–716.

PETC (1993), "Micronized Magnetite—Beneficiation and Benefits," *PETC Review*, U.S. Department of Energy, Pittsburgh Energy Technology Center, Issue 9, Fall 1993.

Plewinsky, B. and Kamps, R. (1984), "Sodium Metatungstate, a New Medium for Binary and Tertiary Density Gradient Centrifugation," *Die Makromolekulare Chemie,* Vol. 185, pp. 1429–1439.

Press, W.H., Flannery, B.P., Teukolsky, S.A. and Vetterling, W.T. (1988), *Numerical Recipes in C: The Art of Scientific Computing*, section 14.4, "Nonlinear Models," Cambridge University Press, Cambridge, pp. 540–547.

Rao, K.N. and Rao, T.C. (1975), "Estimating Hydrocyclone Efficiency," *Chemical Engineering*, pp. 121–122.

Shah, C.L. (1988), "The LARCODEMS Separator for 100–0.5 mm Raw Coal," *Mine and Quarry,* Vo. 17, No. 6, pp. 48–52.

Sherrod, P.H. (1992), "Nonlin" (program for PC-compatible computers), available from P.H. Sherrod, 4410 Gerald Place, Nashville, TN 37205–3806.

Sokaski, M., Sands, P.F. and McMorris Ill, W.L. (1991), "Wet Fine Particle Concentration Section 1: Dense Media," *Coal Preparation*, 5th edition (Leonard and Hardinge, eds.), Society for Mining Metallurgy, and Exploration, Littleton, CO, pp. 376–413.

Sripriya, R., Rao, P.V.T., Bapat, J.P., Singh, N.P. and Das, P. (2003), "Development of an Alternative to Magnetite for Use as Heavy Media in Coal Washeries," *International Journal of Mineral Processing*, Vol. 71, pp. 55–71.

Taylor, B.S. (1969), "Removal of Fine Pyrite by Hydrocyclones," *Mining Congress Journal*, Vol. 55, pp. 46–50.

Temple University (1993), "TempleGraph" (program for Unix-based computers), available from Mihalisin Associates, 600 Honey Run Rd., Ambler, PA 19002.

Terra, A. (1954), "Significance of Anamorphosed Partition Curve and the Ecart Probable in Washery Control," *Proceedings of the 2nd International Coal Preparation Congress,* Essen, Germany, paper C–3.

Thome, T.L., Fullerton, K.L. and Lee, S. (1994), "Design and Optimization of a 36 MT/ day pilot plant for the removal of sulfur from coal using the perchloroethylene process," *Proceedings of the Eleventh Annual International Pittsburgh Coal Conferences* (Chiang, ed.), University of Pittsburgh, Pittsburgh, PA, Vol. 2, pp. 1081–1086.

Vince, A. and Keast-Jones, R. (1992), "Appraisal and Optimization of Dense Medium Plant Separations," *Coal Preparation*, Vol. 11, pp. 189–207.

Visman, J. (1968), "Integrated Water Cyclone Plants for Coal Preparation," Preprinted from *The Canadian Mining and Metallurgical Bulletin*, March, 1968.

Walker, M.S. and Devemoe, A.L. (1991), "Mineral Separations using Rotating Magnetic Fluids," *International Journal of Mineral Processing*, Vol. 31, pp. 195–216.

Wills, B.A. (1992), *Mineral Processing Technology*, 5th edition, Pergamon Press, Oxford.

Wizzard, J.T., Killmeyer, R.P. and Gottfried, B.S. (1983), "Computer Program for Evaluating Coal Washer Performance," *Mining Engineering*, pp. 252–257.

5 Gravity Concentration

Other than heavy-media devices, there are a wide range of processes that separate particles based on their densities, including the various types of jigs, spirals, and tables. These gravity separations are the most simple and economical of all of the separation methods. While they generally do not make separations that are as sharp as those made by heavy-media processes, they are less expensive to install and operate. The large density differential between coal and pyrite also makes it possible for even a low-efficiency density separation to remove a considerable fraction of the pyrite, provided that the separator works at a fine enough size for the pyrite to be liberated.

An example of a plant flowsheet that is in current use, and that uses gravity concentration techniques exclusively, is shown in Figure 5.1. The entire feed is first processed using a series of large Baum jigs, which are effective down to approximately 4.75 mm (4 mesh). The material finer than 4.75 mm is then screened out, deslimed using hydrocyclones, and cleaned with spiral concentrators.

Existing plants use gravity concentrators that are not effective for very fine particles. The smaller the particles become, the more important the fluid drag and viscous forces become relative to the force of gravity. As a result, gravity separation effectiveness decreases sharply for fine particle sizes, which are the most important sizes for sulfur removal. To overcome this problem, new methods are being developed where centrifugal forces are used, which are several hundred times stronger than the gravitational forces. For example, the Kelsey centrifugal jig uses centrifugal forces up to 200 times stronger than gravity, and can clean coal down to particle sizes of 10 μm.

5.1 JIGS

5.1.1 Principles of Jigging

Jigging is a stratification process, in which particle sorting results from an alternate expansion and compaction of a bed of particles by pulsating fluid flow (Figure 5.2). The upward movement of the fluid through the particle bed is referred to as the "pulsion stroke," while the downward fluid movement is referred to as the "suction stroke." The pulsion of water through the bed is produced either by using a piston or diaphragm or by applying pulses of pressurized air to a side chamber. Ideally, the flow velocity through the jig bed should be uniform over the entire bed and is often assumed to vary sinusoidally, with the flow switching smoothly between upward and downward flow. The pulsion–suction cycle mechanically impressed on a jig is not the cycle "seen" by a particle in the jig bed, both because of the shape of the jig body and because the particles restrict the flow differently on the pulsion stroke than on the suction stroke (Thomson and Aplan, 1995). The ability of a jig to "see" the

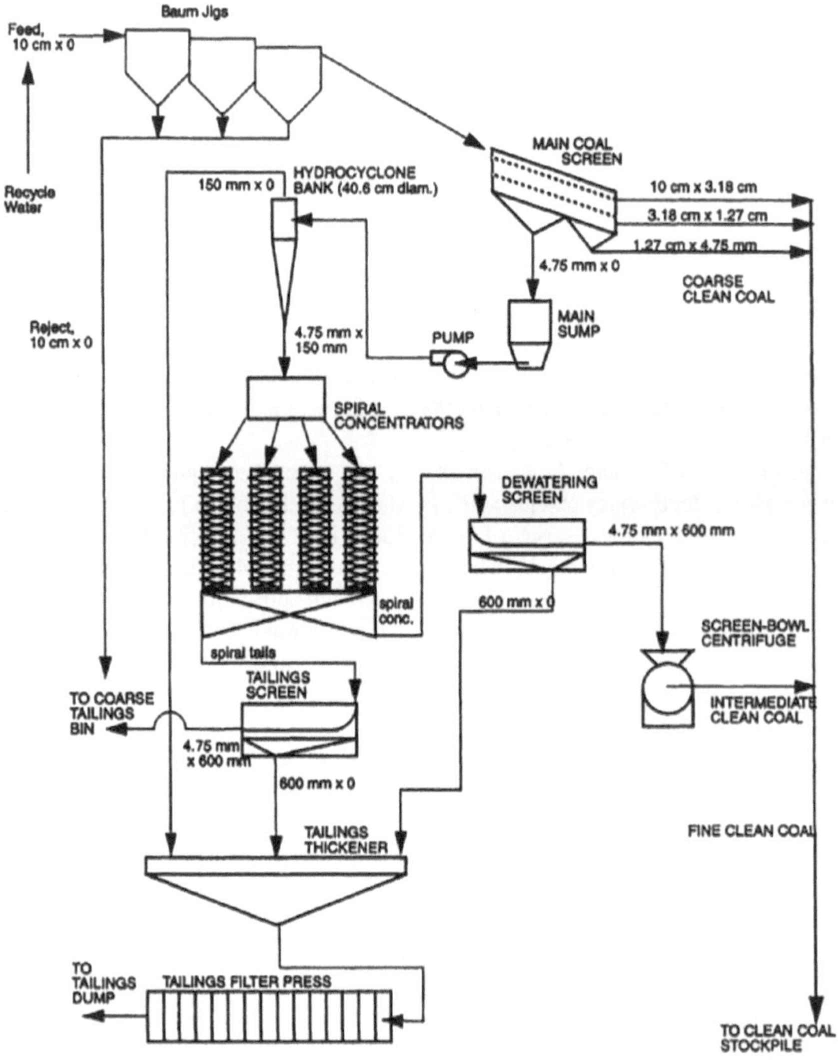

FIGURE 5.1 Flow diagram for a typical gravity-based coal cleaning circuit (Kukura, 1990, Personal Communication).

desired pulsion/suction cycle therefore controls the ability of the jig to discard ash and pyrite as refuse. In the past, it was believed that a sinusoidal flow velocity change would give the best separation. More recent plant tests, however, have shown that by changing the waveform from the conventional quasi-sinusoidal form to a trapezoidal form, the capacity of the jig could be significantly improved (Onodera et al., 1993).

It is still believed that an ideal jig would "see" a sinusoidal force curve as shown in Figure 5.3, but in reality, this is not generally achieved. The actual pulsion and suction stroke on the particle are rarely identical, and so the sine wave is distorted or limited. To develop a perfect jig one would install pressure transducers in the

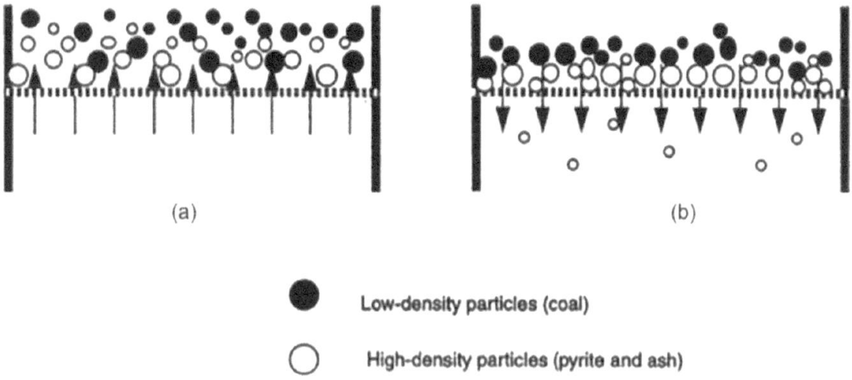

FIGURE 5.2 Diagram of the operating principle of a jig. Water pulses upward through the material on the screen and then is sucked back down, which segregates the particles by density with the low-density particles on top. Dense particles can be either drawn through the screen to be removed, or if too coarse to fit through the screen, they can be scalped from the bottom of the flow as the particles on the bed flow out of the jig. (a) Pulsion stroke; (b) Suction stroke.

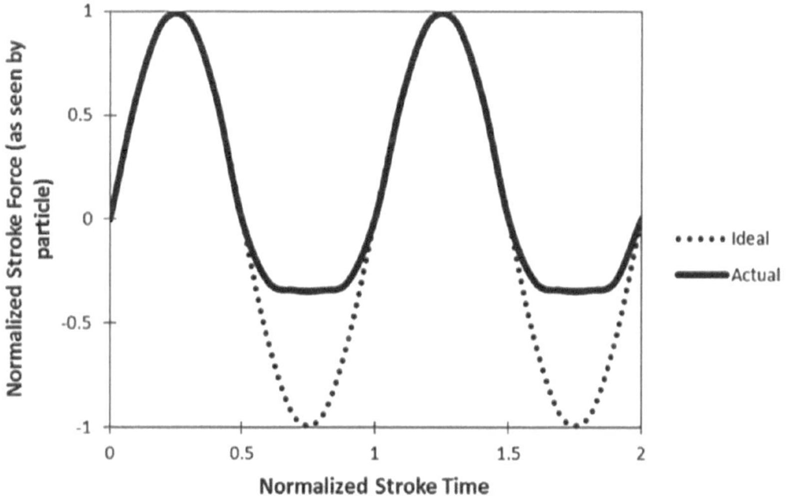

FIGURE 5.3 Diagram of an example force curve as seen by a particle in an operating jig. The ideal curve could be achieved by installing pressure gauges along the bottom of the jig screen (see Figure 5.2) to ensure that the pulsion and suction stroke resulted in the same sinusoidal curve.

screen of the jig to ensure that a full, balanced sine wave develops. However, as in Figures 5.4 and 5.5, this has not been developed in any enhanced gravity jig.

The particle rearrangement in a jig gives rise to a layer of particles that are arranged by density, with the densest particles at the bottom of the bed and the least dense particles at the top. As a result of several hydrodynamic effects, jigs are

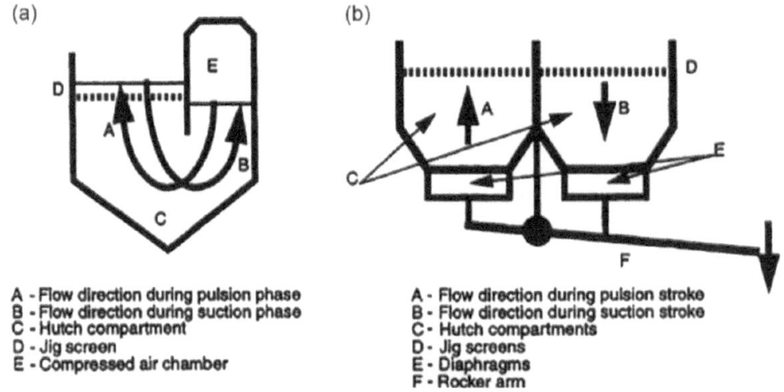

FIGURE 5.4 (a and b) Comparison of a Baum jig and a Pan-Am jig. The main difference between the two is the manner of producing the pulsion stroke, with the Baum jig relying on air pressure while the Pan-Am jig uses a mechanical diaphragm, which makes for a more powerful suction stroke.

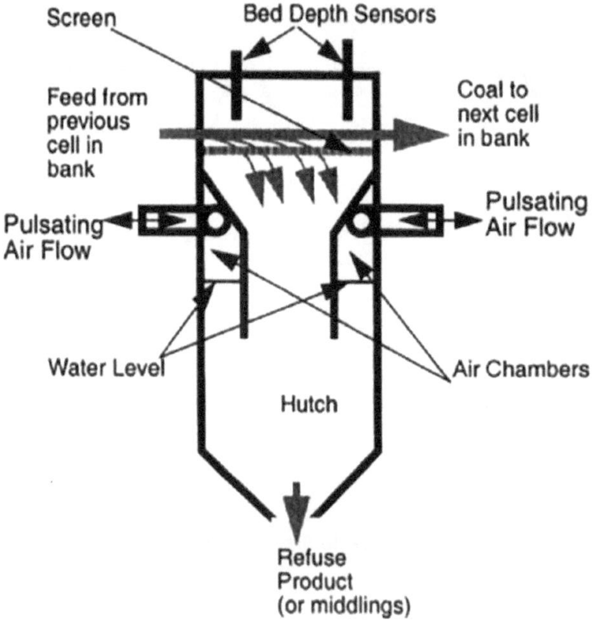

FIGURE 5.5 Schematic of a single cell of a Batac jig. Like the Baum jig, the water pulsation is produced by pulses of air forcing water out of a chamber; however, the air chamber in the Batac jig is in the hutch chamber, directly below the screen. Since the pulse does not have to travel far to reach the screen and the pulse is more symmetrical than in the Baum jig, the Batac jig can work at higher frequencies and clean finer particles. The feldspar ragging on the jig screen also improves the performance with fine particles. Dividing the jig into numerous cells further improves the uniformity of the jigging action. Batac jigs also are typically more heavily instrumented than Baum jigs, with sensors for measuring and controlling both the bed depth and the pulse amplitude.

less sensitive to particle size than many other gravity separators, which makes them quite selective when treating feeds with wide size distributions. The jig feed must be constant and spread evenly, and the jig capacity is determined by the solids-carrying capacity of the feed water and the rate at which the products can be discharged.

Several types of jigs are commercially used, with the major differences being the manner of producing the pulsation and the pulsation rate. In general, the more rapid the pulsation rate, the finer the particles that can be processed. The performance of a jig can also be modified by adding extra water to the jig, which changes the relationship between the suction and pulsion strokes. One of the drawbacks of jigs is that they use a great deal of water, and so a jig plant must rely heavily on recycled water.

5.1.1.1 Concentration Criterion

The driving force of all gravity concentrators is the density difference between the two materials to be separated. The driving force can be correlated with a concentration criterion, C, as per Equation 5.1:

$$C = \frac{\rho_h - \rho_f}{\rho_l - \rho_f} \tag{5.1}$$

where

C = concentration criterion
ρ_h = density of the heavier material
ρ_l = density of the lighter material
ρ_f = density of the fluid media

As the density of the fluid approaches the density of the lighter material, the concentration criterion goes to infinity. Concentration criterions above 2.5 are generally considered relatively easy separations. Concentration criterions below 1.25 are generally impossible to make economically viable.

As an example calculation, the concentration criterion of separating gold (ρ = 19.3 g/cm^3) from arsenopyrite (ρ = 6.0 g/cm^3) in water (ρ = 1.0 g/cm^3) is shown in Equation 5.2. The concentration criterion value suggests that arsenopyrite would be relatively easy to separate from gold via gravity separation.

$$C = \frac{19.3 - 1.0}{6.0 - 1.0} = 3.66 \tag{5.2}$$

On the other end, the concentration criterion of separating apatite (ρ = 3.2 g/cm^3) from dolomite (ρ = 2.84 g/cm^3) in water (ρ = 1.0 g/cm^3) is shown in Equation 5.3. This separation is far more difficult, and would require the use of heavy media if gravity separation is to be made efficient.

$$C = \frac{3.2 - 1.0}{2.84 - 1.0} = 1.2 \tag{5.3}$$

High-density fluids, which were discussed in Chapter 4, can be used to improve the separation efficiency. Most high-density fluids are made by suspending solids in the fluid, which significantly hampers material settling and separation. Similarly, most heavy-media solutions are quite viscous. Thus, rather than maximizing the concentration criterion, many gravity separators instead seek to amplify the settling rate so that low concentration criterions may be better utilized.

As an example, the same apatite from dolomite separation performed in methylene bromide ($\rho = 2.48$ g/cm^3) is shown in Equation 5.4. This gravity separation is far more efficient, but requires handling and recovering of methylene bromide, which is toxic and somewhat volatile at normal operating temperatures. The additional expense involved in using methylene bromide is likely not worth it overall.

$$C = \frac{3.2 - 2.48}{2.84 - 2.48} = 2.0 \tag{5.4}$$

5.1.1.2 Ragging

"Ragging" is a coarse material sometimes added to jigs, usually of an intermediate density between the gangue and valuable materials, which helps prevent light and fine material from passing through the jig screen preemptively. Feldspar is a common ragging for coal jigging, with a specific gravity of 2.56.

5.1.2 Baum Jig

The principle of operation of a Baum jig is illustrated in Figure 5.4. For this jig, compressed air is admitted in pulses through slide or rotary impulse valves. The compressed air acts on the hutch water, producing pulsion that fluidizes the bed; the air pressure is then released, allowing a gravity down the drain of the water through the bed. By reducing the amount of air admitted in each pulse, that is, by reducing the stroke length, and adding feldspar particles to the bed, a Baum jig can be converted to a Baum-feldspar jig, which is better able to deal with feed that contains a relatively small amount of dense material.

A wide range of particles can be processed by this type of jig. The top size is generally 100–200 mm, even though Baum jigs have been used for particles as coarse as 250 mm. The limit for the lower size is generally 5 mm. Generally, in the United States, 30%–40% of coal is cleaned by Baum jigs, which can handle up to 1,000 tons/h of coal. Even though such a large quantity of coal is cleaned by Baum jigs, a limited number of studies have been conducted to understand the mechanism of jigs. In part, this is due to the fact that jig performance is mainly limited by the liberation of pyrite in the feed and not by the actual operation of the jig. Since Baum jigs clean coal at a coarse size, and only the coarsest pyrite is liberated at that size, these jigs have a limited capability to desulfurize coal beyond removing the large "sulfur balls."

5.1.3 Pan-American Placer Jig

The principle of Pan-Am jigs is shown in Figure 5.4. In a Pan-Am type of jig, water pulsion instead of air pulsion is used. It uses a centrally pivoted rocker arm to pulse rubber diaphragms located in the hutch a short distance below the jig bed. In this way, the full effect of the pulsion and suction stroke are immediately transmitted to the particles in the bed (Thomson and Aplan, 1995). The principal difference between a Baum jig and a Pan-Am jig is that in the Baum jig, there is no specific suction to move fine particles into the hutch other than gravity down the drain through the bed. In a Baum jig the pulsion cycle is emphasized, whereas in the Pam-Am jig the suction cycle is equally important. The Pan-Am jig has a better ability than the Baum jig to transfer the pulsion–suction cycle, which is a sine wave, to the jig bed, hence, the Pan-Am jig should have a better ability to desulfurize coal. Thomson and Aplan (1995) and Laros and Aplan (1987) have compared the Pan-Am placer jig to the Baum-feldspar jig and determined the Pan-Am jig to be superior for desulfurizing coal. The results of Thomson and Apian are shown in Table 5.1

The difference in efficiency between the Baum, Baum-feldspar, and Pan-Am Placer jigs is shown by the values of the Ecart probability (E_p) for the various types of jigs, shown in Table 5.2.

5.1.4 Batac Jig

The basic principles of a Batac jig are the same as those of a Baum jig. Both utilize air to produce pulsation, however, in a Baum jig, pulsations are produced in a chamber beside the hutch, as shown in Figure 5.4. This introduces a time delay for the pulse to reach the entire jig bed and limits the machine to a relatively low pulse rate (less than one per second). In a Batac jig, pulsations are produced directly beneath the jig bed, and generally, two chambers are provided for each cell, as shown in Figure 5.5. Thus in a Batac jig more uniform distribution of air is provided, and therefore this type of jig has a better ability to transfer the pulsion–suction cycle to the jig bed. The increased uniformity of the pulse also allows a larger screen area and faster pulse

TABLE 5.1
Comparative Performance of a Baum-Feldspar and a Pam-Am Placer Jig to Remove −1.68 mm Pyrite from −6 mm Coal

Particle Size	Baum-Feldspar Jig	Pam-Am Placer Jig
	% Pyrite Removal	
1,700 × 425 μm (10 × 35 mesh)	91.5	99.5
425 × 74 μm (35 × 200 mesh)	47.1	96.0
−74 μm (−200 mesh)	1.8	34.0
All sizes	76.3	95.4
	Clean Coal Quality	
% Yield of clean coal	92.90	87.91
% Ash	7.60	6.59
% Sulfur	0.99	0.83

Feed rate = 8.9 ton/m²/h.

TABLE 5.2
Efficiency Data for a Baum Jig, a Baum Jig with Feldspar Ragging, and a Pan-Am Placer Jig (Thomson and Aplan, 1995)

Size Fraction	Baum Jig		Baum-Feldspar Jig		Pan-Am Jig	
	ρ_{50}	E_p	ρ_{50}	E_p	ρ_{50}	E_p
6,700 × 600 μm (3 × 28 mesh)	1.61	0.48	1.65	0.22	0.57	0.12
6,700 × 2,360 μm (3 × 8 mesh)	1.50	0.38	1.65	0.08	1.56	0.09
2,360 × 1,180 μm (8 × 14 mesh)	1.68	0.40	1.69	0.27	1.58	0.11
1,180 × 600 μm (14 × 28 mesh)	1.76	0.55	1.68	0.64	1.58	0.16
600 × 74 μm (28 × 200 mesh)			2.30		1.80	0.48
600 × 300 μm (28 × 48 mesh)			1.95		1.62	0.22
300 × 150 μm (48 × 100 mesh)			2.70		1.92	0.50
150 × 74 μm (100 × 200 mesh)					2.55	

rates, which provide a higher total capacity and a higher capacity per unit area than the Baum jigs. A Batac jig, therefore, can be used to efficiently remove pyrite at finer particle size ranges than other types of jigs, as shown in Table 5.3. A Batac jig will therefore provide performance superior to that of a Baum jig for desulfurization purposes, because the pyrite is better liberated at smaller particle sizes.

Batac jigs have been used for coals with particle sizes as small as 0.150 mm (Lovell et al., 1991). While cleaning fine coals, feldspar beds are used in some of the cells of Batac jigs to improve their performance. In spite of the importance of process characterization in Batac jigs, very limited data is available in the literature to determine their desulfurization ability.

TABLE 5.3
Results for Baum and Batac Jigs Cleaning Coals of Various Particle Sizes and Pyritic Sulfur Contents

	Baum No. 1	Baum No. 2	Batac No. 1	Batac No. 2
Size range processed	152.4 × 6.35 mm (6 × 1/4 in)	101.6 × 6.35 mm (4 × 1/4 in)	19.0 × 0.6 mm (3/4 in × 28 mesh)	12.7 × 0.6 mm (1/2 in × 28 mesh)
% Wt Clean coal yield	59.2	45.4	70.5	82.6
BTU/lb	14,449	13,410	14,231	13,204
% BTU recovery	88.2	83.7	95.7	97.0
% Ash	7.1	9.1	8.7	7.6
% Ash removal	88.0	91.0	79.8	67.6
% Pyritic sulfur	0.19	2.09	0.48	0.88
% Pyritic sulfur removal	53.2	68.2	56.6	61.3

The Batac jigs worked at a finer particle size, and could remove the pyrite as effectively as the Baum jig while recovering more of the coal (Hoke, 1976).

In a process reversing the normal use of a jig, Tavares and Rubio (1991) have used a Batac jig to recover pyrite from coal tailings. The tailings were ground to a top size of 12 mm and cleaned in a two-stage Batac jig. They upgraded the concentrate from 3% sulfur to 42%–43% sulfur in the concentrate.

Batac jigs compare favorably with other types of gravity and heavy-media separators that operate over a similar size range, as shown in Table 5.4. Their capacity is high, their separating efficiency is good, and they are capable of processing particles as fine as 150 μm. They are also cheaper to operate than tables and heavy-media cyclones (Adams, 1988).

5.1.5 Implementations of Jigs to Desulfurize Coal

A combination of Baum and Batac jigs can be used to treat coal ranging in size from 4 in down to 100 mesh very effectively, with the Baum jigs treating the 101.6 mm × 12.7 mm (4 in × 1/2 in) fraction and the Batac jigs treating the 12.7 mm × 150 μm (1/2 in × 100 mesh) fraction. One such installation was used at the Old Ben Mine No. 26 plant in Franklin County, Illinois (Hoke, 1976). The estimated pyrite removal in this plant by the Batac jig was approximately 40% at 89.2% wt yield, and the total sulfur content was reduced from 2.46% to only 1.99% after the −100 mesh material (which was mostly clay and pyrite) was discarded. Some of the benefits of such an installation are as follows:

- Batac jigs have a low space requirement, so they can be integrated easily into existing plants.
- Jigs are relatively inexpensive, and are quite easy to operate and control.
- In the +150 μm (100 mesh) size range, the Batac jig gives equal or superior performance to other hydraulic coal-cleaning devices.
- Since Batac jigs use the same sort of support equipment as Baum jigs (blowers, air compressors, feeders, etc.) they are easier to integrate into a plant using Baum jigs than other types of equipment would be.

However, as seen from Figures 5.4 and 5.5 along with the data in Table 5.2, the Pan-Am jig design is superior to the design of the Baum or Batak jigs. Even then, the Pan-Am jig is primarily used in the gold industry, while the Baum and Batak jigs are used in the coal industry.

TABLE 5.4
Comparison of Fine Coal (Batac) Jigs, Tables, Water-Only Cyclones (See Chapter 4), and Heavy-Media Cyclones (after Adams, 1988)

	Fine Coal Jig	Table	Water-Only Cyclone	Heavy-Media Cyclone
Unit capacity	200–800 tons/h	8–12 tons/h	4–90 tons/h	10–120 tons/h
Feed size range	19 mm × 150 μm	12.5 mm × 150 μm	12.5 mm × 106 μm	30 mm × 425 μm
Approximate E_p	0.08	0.10	0.15–0.25	0.04
ρ_{50} range	1.5–2.0	1.6–1.8	1.6–2.0	1.3–1.8

5.2 FLOWING-FILM CONCENTRATORS

Flowing-film concentrators produce a separation from the interaction of gravitational or centrifugal forces with the velocity gradient in a flowing film of liquid. The particles are sorted by both size and density, with the coarsest high-density particles reporting to the same product as the finest low-density particles. As a result, the flowing-film technique works best when the feed has first been screened or classified into separate size fractions. These devices include tables and spirals.

5.2.1 TABLES

Tables consist of an inclined, usually riffled surface that water and feed particles flow across while the table is shaken from side to side. The basic principle is that large and low-density particles are most affected by the flow of water; therefore, they spend very little time on the table and are quickly carried straight across. At the same time, the small and high-density particles spend more time on the table because they are less affected by the water flow, and so the table motion and the action of the riffles carry them sideways. The point where a given particle will leave the table is therefore determined by its size and density, as shown in Figure 5.6. The table performance can be adjusted by altering the water flowrate, the shaking frequency, or the tilt of the table. The riffle pattern on the table surface is designed to suit the density ranges of the particles being separated, and its behavior is shown in Figure 5.7.

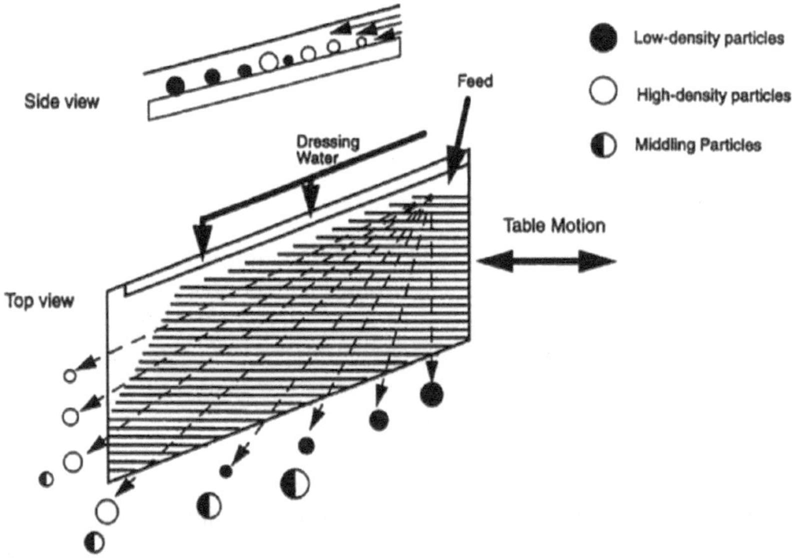

FIGURE 5.6 Schematic of a table concentrator. Since coarser particles project further up from the table surface into the flowing film and the upper part of the film is flowing faster, these particles tend to be carried off the table most rapidly by the flowing water. Small particles and high-density particles spend more time on the table and are caught by the riffles which carry them to the side as the table shakes.

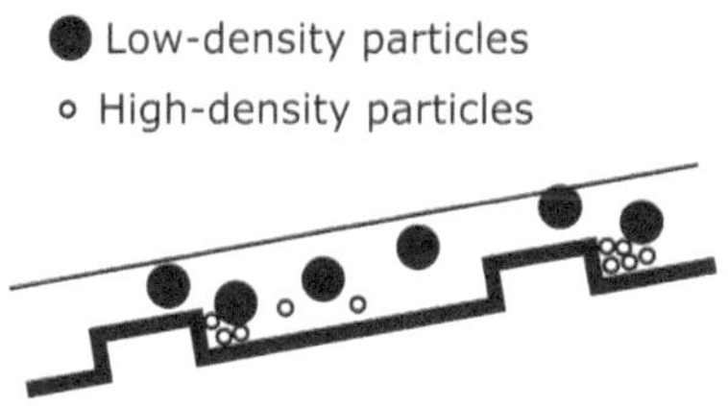

FIGURE 5.7 The riffles in a table concentrator make it more difficult for small, high-density particles, such as pyrite, to make their way down the table. The back-and-forth shaking motion then forces them toward the far end of the table, allowing for a separation from the lower density, typically, coarser particles such as coal.

Tables have been extensively used to clean coal in the intermediate size range, with good results. They have limited capacity, however, because the flowing film cannot be very thick, and so they take up more plant area for a given tonnage than do most other coal-cleaning devices. The shaking action of the table also makes them more prone to breakdown than other devices with fewer moving parts.

If the pyrite liberates at a sufficiently coarse size, then tables are very effective for removing it from the coal. Free pyrite particles down to about 64 μm can be separated quite thoroughly, with some removal down to as fine as 16 μm; however, there is no noticeable separation action for pyrite finer than 7 μm (Saltsman, 1969). Since much of the pyrite in many coals is finer than this, there is still a good deal of room for improvement beyond this performance. Tables are a useful adjunct to water-only cyclones (which were described in Chapter 4), because the waste product from the cyclone is deslimed, and therefore well-suited for tabling. This is illustrated by the results given in Table 5.5, for a table treating a high-sulfur waste

TABLE 5.5
A Coal with Primarily Pyritic Sulfur Was Chosen from the Waste Stream of a Water-Only Cyclone (See Chapter 4) and Further Separated by a Table Concentrator

Size	Feed % Sulfur	% Wt Yield	Clean Coal % Sulfur	% Sulfur Removal
+1,180 μm (+14 mesh)	4.43	62.10	0.94	86.83
1,180 × 600 μm (14 × 28 mesh)	9.12	31.56	1.05	96.37
600 × 425 μm (28 × 35 mesh)	14.25	17.40	0.85	98.96
425 × 300 μm (35 × 48 mesh)	17.73	9.92	0.80	99.55
300 × 212 μm (48 × 65 mesh)	20.81	7.29	0.80	99.72
212 × 150 μm (65 × 100 mesh)	24.77	7.48	0.95	99.71
−150 μm (−100 mesh)	30.73	7.45	2.31	99.44
Total	17.51	16.98	1.03	99.00

The result was a total 99% removal of sulfur content in this reject stream, especially in the smaller size fractions (Miller and Podgursky, 1972).

stream from a water-only cyclone. These results show that the table is very effective at the coarser sizes, but the sulfur removal suffers when the feed is finer than 150 μm (100 mesh).

5.2.2 Spirals

Modern spirals, as shown in Figure 5.8, consist of a spiraling chute. Water and suspended particles flow down this chute, and the action of the flowing water film tends to force particles to the outside of the chute, while gravity tends to pull particles back toward the center. The high-density particles are more affected by gravity and are slowed by friction with the trough, and as a result, they discharge from the inner edge of the spiral. The low-density particles are more affected by fluid drag, and are carried along more rapidly by the flowing water, causing them to move outward and discharge from the outer edge, as shown in Figure 5.9. The spiral is adjusted by changing the position of the divertors which remove the dense products. Important design parameters for spirals include the spiral pitch and the cross-section profile of the trough. Coarse and low-density particles are separated best with shallow pitches, while fine-grained and high-density particles separate better with steeper pitches. For coal processing, it has been found that a wide, flat chute profile is better than the traditional elliptical profile (Gartner, 1994).

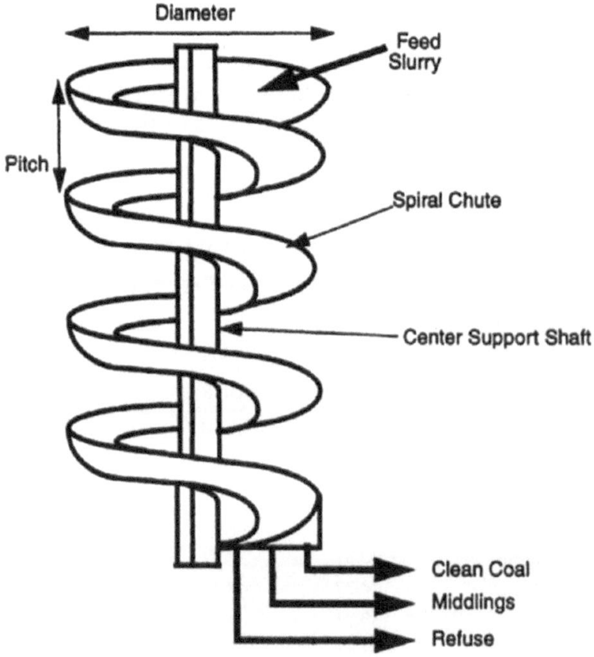

FIGURE 5.8 Schematic of a six-turn, single-start spiral concentrator. Double-start and triple-start spirals have two and three separate helices spiraling around a single center shaft, and therefore have higher capacity for a given amount of floor space than a single-start spiral.

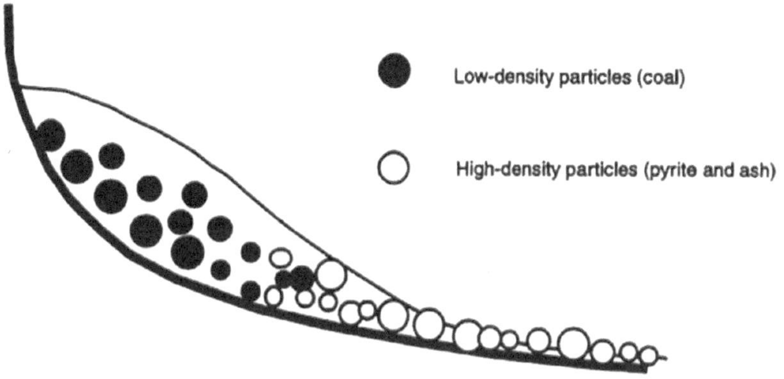

FIGURE 5.9 Cross-section of the chute of a spiral concentrator. As the water flows down the spiral chute, it centrifuges toward the outside edge and tends to carry the particles along with it. The denser particles interact more with the chute walls due to their greater weight, and so they tend to slow down and segregate toward the center, while the less-dense particles are carried to the outside edge by the flow of the water.

Spirals have several positive features, such as having low capital and operating cost, being easy to operate and maintain, and taking up relatively little floor space for a given capacity. For example, a 76 cm (30 in) diameter spiral has a capacity of 2–5 metric tons/h, but only occupies approximately 0.84 m^2 (9 ft^2) of floor space. They are quite well-suited for removing pyrite from coal, as shown in Table 5.6, provided that the pyrite liberates at a sufficiently coarse size. The particle size range over which spirals are effective is similar to that of tables, with a top size of about 1 cm and a lower limit of approximately 50 μm. Spirals are typically used in conjunction with either desliming cyclones or fine screens, to remove the fines that the spiral cannot handle. The desliming operation can be either before or after the spirals.

TABLE 5.6
Results for Physical Pyrite Removal from Middle Kittanning Seam Coals Using a Commercial Size 6 Turn Humphrey's Spiral in a Closed-Loop Test Rig (Zeilinger and Deurbrouck, 1976)

Test No.	Feed % Pyritic Sulfur	Clean Coal % Yield	Clean Coal % Pyritic Sulfur	% Pyritic Sulfur Removal
1 (425 × 74 μm)	1.96	86.0	0.29	87.3
2 (425 × 74 μm)	1.69	91.3	0.31	83.3
3 (425 × 74 μm)	2.10	89.3	0.46	80.4
4 (1,180 × 74 μm)	2.14	89.4	0.57	76.2
5 (1,180 × 74 μm)	2.41	90.6	0.54	79.8
6 (1,180 × 74 μm)	1.96	90.4	0.40	81.6
7 (1,180 × 74 μm)	1.99	91.9	0.40	81.5
8 (1,180 × 74 μm)	2.14	91.2	0.54	77.0

Spirals have been improved markedly since their introduction in 1943 (Otto et al., 1947), with improvements in trough design improving the separation efficiency, increases in capacity with double-start and triple-start units, and reductions in weight and cost by using fiberglass/polyurethane as the construction material. As a result, their small capital cost, sharp separation, and ease of operation are making them more popular than table concentrators (Apodaca, 1988).

5.2.2.1 Single-Stage Spirals

A number of different types of spirals are available, with differing numbers of turns around the central axis and differences in trough profile. The sulfur removal capabilities of some of the most popular units have been compared, as shown in Table 5.7. It was determined that for 1,000 × 150 μm (18 × 100 mesh) material, the number of turns had little effect on the performance, while the spiral with the smallest diameter showed the best sulfur removal performance. It was also found that by using a rougher/cleaner spiral configuration, the sulfur content of the final product could be

TABLE 5.7
Comparison of the Sulfur Removal Capabilities of Five Different Types of Coal-Cleaning Spirals

Spiral Type	Spiral Product	% Yield	% Total Sulfur	% Sulfur Rejection	% Combustibles Recovery
Krebs 3.25 turns	Feed	—	2.38	—	—
	Clean coal	80.7	1.99	32.6	87.5
	Middlings	2.8	2.63	—	—
	Reject	16.5	4.28	—	—
Carpco 4 turns	Feed	—	2.20	—	—
	Clean coal	78.0	1.81	35.1	82.8
	Middlings	16.1	2.57	—	—
	Reject	5.9	5.93	—	—
Carpco 6 turns	Feed	—	2.40	—	—
	Clean coal	79.3	1.98	34.6	85.4
	Middlings	14.6	3.14	—	—
	Reject	6.0	6.35	—	—
Mineral Deposits ltd.	Feed	—	2.32	—	—
LD4 6 turns	Clean coal	77.5	2.01	33.5	83.3
	Middlings	15.4	2.63	—	—
	Reject	7.2	4.95	—	—
Mineral Deposits ltd.	Feed	—	2.27	—	—
LD2 6 turns	Clean coal	79.3	1.79	37.4	84.5
	Middlings	11.2	3.39	—	—
	Reject	9.5	4.93	—	—

The LD2 had a trough diameter of 660 mm and a pitch of 310 mm, while the other four spirals had diameters of 920 mm and pitches of 400 mm (Honaker, 1994).

reduced still further (Honaker, 1994, Personal Communication). Work is also being done to reduce the height of spirals, so that they can be more easily installed in existing coal processing plants (Ivatt, 1993).

When spirals are used in a plant to follow a process such as jigging, as in the flowsheet shown in Figure 5.1, and are operated to maximize the clean coal yield, the sulfur removal by the spirals will not be very large, as can be seen from the plant data given in Table 5.8. Although the reject product was much higher in sulfur than the clean coal product for sizes down to 25 μm (500 mesh), the absolute sulfur removals were low for two reasons: (1) The spiral feed had previously passed through the Baum jigs, which removed the well-liberated, easy-to-separate pyrite from the coal before it ever reached the spirals; and (2) The spirals were mainly intended to improve coal recovery in the circuit, and therefore, pyrite removal and separation efficiency was less important than achieving a high product yield.

5.2.2.2 Two-Stage and Compound Spirals

In most industrial coal-cleaning operations, the need for high throughput at low cost makes multi-stage separation impractical; however, the low cost and simple operation of spirals can make it feasible to use two-stage processing circuits, such as rougher-cleaner or middlings retreatment. It has been found that the best two-stage spiral circuit for pyrite rejection is a rougher-cleaner circuit, as shown in Figure 5.10. When used to process an Illinois No. 5 seam coal, with a feed sulfur content of 2.3%, this circuit was found to improve the sulfur removal by approximately 10% compared to a single-stage spiral (Honaker, 1994, Personal Communication). In work using Reichert LD 9 spirals to process Brazilian Cambui coal, Tavares and Sampaio (1990) found that the sharpness and density of separation increased with increasing particle size, as can be seen in Table 5.9. The spirals were arranged in a rougher-cleaner

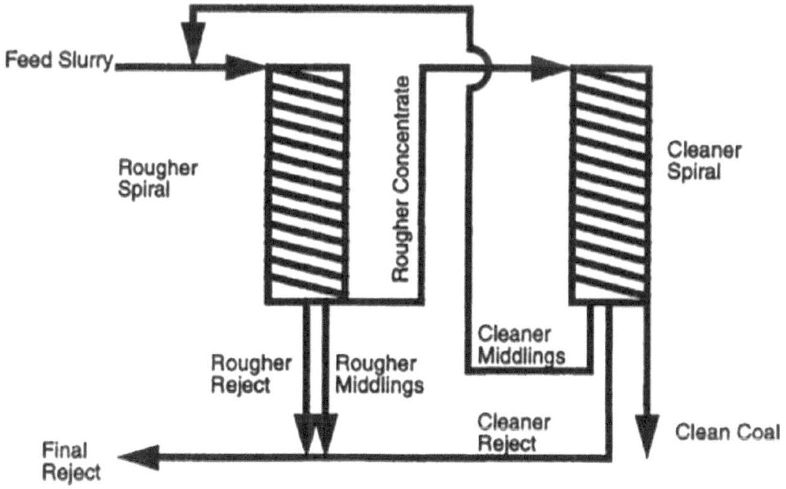

FIGURE 5.10 Two stages of spiral concentration in a rougher-cleaner configuration, to maximize pyrite removal.

TABLE 5.8
Example of the Sulfur-Removal Performance of a Spiral in the Plant Flowsheet Shown in Figure 5.1 (Kukura, 1990, Personal Communication)

Size Fraction	% Yield	% Total Sulfur: Clean Product	% Total Sulfur: Reject Product	ρ_{50}	E_p	% Sulfur Removal
+1,000 μm (+16 mesh)	99.6	2.80	5.16	1.95	0.305	0.4
1,000 × 60 μm (16 × 28 mesh)	99.1	2.95	6.02	1.71	0.250	1.0
600 × 150 μm (28 × 100 mesh)	95.5	3.45	8.18	1.49	0.258	3.2
150 × 74 μm (100 × 200 mesh)	90.6	7.90	18.50	—	—	5.2
74 × 45 μm (200 × 325 mesh)	89.0	10.98	22.73	—	—	5.0
45 × 25 μm (325 × 500 mesh)	93.5	12.38	21.75	—	—	2.7
−25 μm (−500 mesh)	90.7	5.03	4.02	—	—	0.5
Overall	95.8	4.08	10.97	—	—	2.7

These spirals were being operated to maximize the coal yield, not to maximize sulfur removal, and were immediately following a jigging stage that had already removed most of the well-liberated sulfur. As a result, the sulfur removal values are small.

configuration. The feed contained 4.2% pyritic sulfur, which was reduced to 2.2% pyritic sulfur at a clean coal yield of 86.5%.

A compound spiral can be considered to be a two-spiral circuit with middlings retreatment, with both spirals sharing a single central column, as shown in Figure 5.11. The reason for using this arrangement is that in a spiral, much of the separation occurs within the first few turns, and the "middlings" material in the center of the spiral largely consists of misplaced coal and mineral particles and not of true middlings with an intermediate density. It has therefore been argued that if

TABLE 5.9
Separation Performance of Two-Stage Reichert LD9 Spirals Processing Cambui Coal (Tavares and Sampaio, 1990)

Feed Size	Spiral Configuration	ρ_{50}	E_p
2.4 × 0.3 mm	Primary spiral	2.37	0.238
	Secondary spiral	1.84	0.203
	Two-stage spirals	1.81	0.180
0.3 × 0.075 mm	Primary spiral	2.19	0.263
	Secondary spiral	1.70	0.283
	Two -stage spirals	1.68	0.221
2.4 × 0.075 mm	Primary spiral	2.31	0.215
	Secondary spiral	1.81	0.227
	Two -stage spirals	1.79	0.199

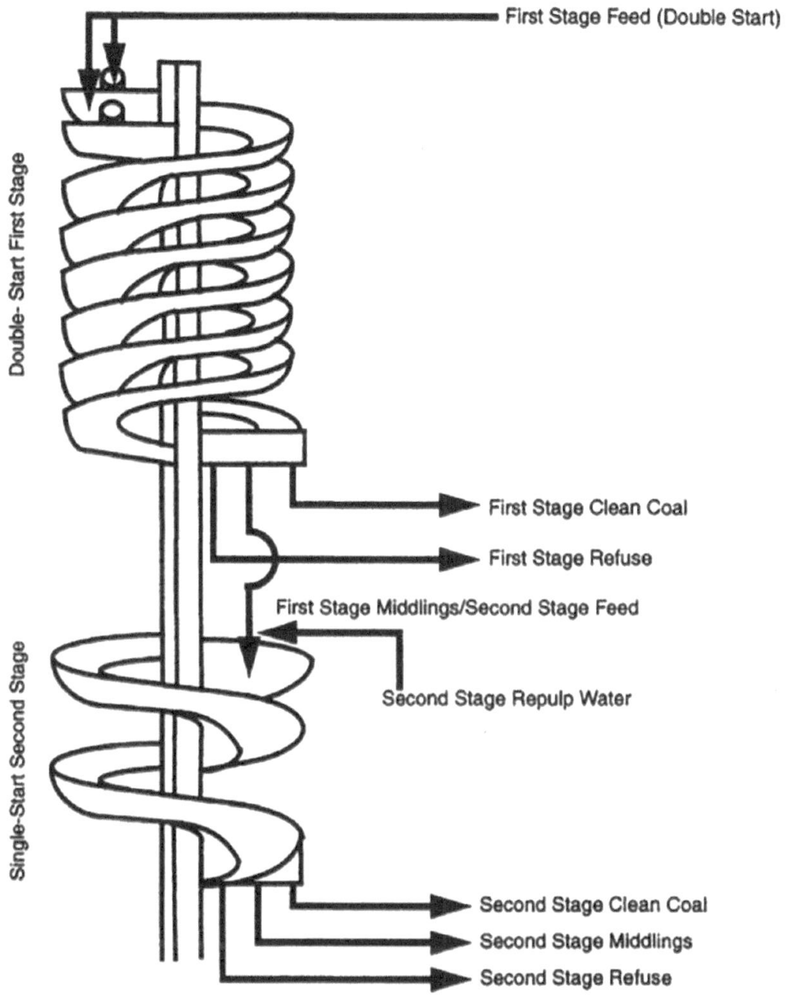

FIGURE 5.11 Schematic of the general arrangement of a compound spiral. The first portion is a high-capacity multiple-start unit, which rapidly separates the easily recoverable coal and the readily rejected mineral matter and removes them to eliminate the danger that they could become re-mixed in later stages of the separation. The second portion is a lower-capacity spiral which treats the more difficult-to-separate middlings.

the particles that are separated early on remain in the spiral, they can interfere with the later separation in the lower portion, and can even become re-mixed, degrading the performance. As a result, a series of short spirals with stages to re-treat the middlings is claimed to be a better approach (MacNamara et al., 1995, 1996). In the compound spiral, the upper high-capacity spiral separates the most easily recovered coal and the most easily rejected ash and removes them from the circuit quickly so that they cannot interfere with the later separation. The middlings can then be distributed across the width of the secondary spiral, which allows a more accurate cut

TABLE 5.10
Comparison of the Performance of Two Individual Spirals to the Performance of Both Spirals Operating in Series

Unit	ρ_{50}	E_p	% Combustibles Recovery	% Yield	% Ash
Primary spirals	1.68	0.150	92.61	76.19	7.34
Secondary spirals	1.66	0.175	92.09	5.68	15.17
Primary + secondary	1.74	0.135	98.01	81.87	7.88

The combination of two spirals results in a sharper separation, allowing a higher combustibles recovery without a great increase in the ash content.

point to be taken and gives a greater tolerance to fluctuations. An added advantage is that the secondary spiral is only a single-start unit, which is much easier to observe and control than the very crowded multi-start first-stage units. It is therefore easier to check the performance of a compound spiral than of a conventional multi-start single-stage spiral of similar capacity.

Compound spirals have been tested both in laboratories and in operating plants. Typical results, obtained at the Pittston Coal Co. McClure River Preparation Plant, are given in Table 5.10. Sulfur removal values are not available, but the compound spirals yield a higher combustibles recovery than single-stage spirals and have a smaller value for the Ecart probability.

5.3 HINDERED-BED SEPARATORS

Hindered-bed separators, or hydrosizers, are very simple separators that are gaining popularity in the U.K. coal industry (Hyde et al., 1988). A hydrosizer is a device in which upward-flowing water classifies a bed of particles according to their density. They have the advantages of low cost and simple automatic operation. Changes in feed rate and composition are automatically compensated for, requiring minimal operator intervention.

Hydrosizers work as shown in Figure 5.12. Coal slurry enters the unit at the top, where it encounters a uniform rising current of water. The low-density particles have a settling rate slower than the upward flowrate of the water, so they are carried into the overflow launder. High-density particles, particularly pyrite, have faster settling rates and therefore discharge through the control valve at the bottom. The optimum solids loading in these units is 40%–60% solids, and the feed slurry is often concentrated to this level by a hydrocyclone. Their separation efficiency is shown in Table 5.11.

It has been reported that a Carpco Floatex hindered-bed separator is most effective for removing pyrite that is coarser than 300 μm (48 mesh) and that using one of these units to pretreat the feed to a spiral concentrator will reduce the load on the spiral by approximately 50% (Mankosa et al., 1995). In plant trials treating a coal with a nominal size of 1,180 × 150 μm, a 45.7 cm (18 in) square Floatex separator was compared with an existing spiral circuit, with the results shown in Table 5.12. This unit had a capacity of 19.5–24.4 metric tons/m^2/h, and was reported to show

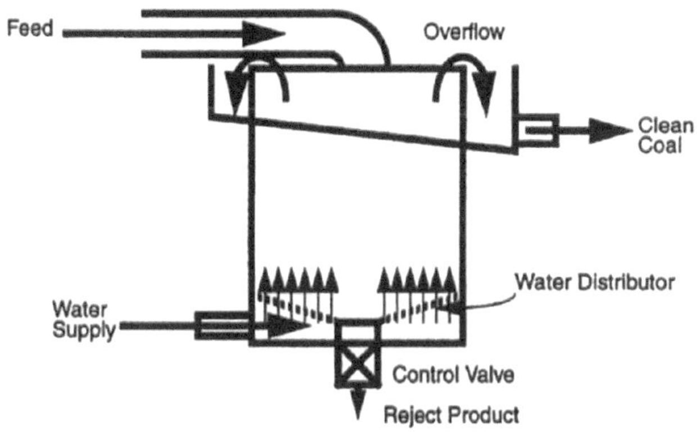

FIGURE 5.12 Schematic of a hydrosizer.

TABLE 5.11
Separating Efficiency of a Hydrosizer as a Function of Particle Size (Hyde et al., 1988)

Particle Size	ρ_{50}	E_p
>1 mm	1.38	0.038
1 mm × 0.5 mm	1.53	0.063
0.5 mm × 0.25 mm	1.89	0.195

TABLE 5.12
In-Plant Comparison of a Carpco Floatex Separator and an Existing Spiral Circuit, with Feed Sized to 1,180 × 150 μm (Reed et al., 1995)

Parameter	Floatex		Spirals
E_p	0.125		0.185
ρ_{50}	1.94		1.82
Product % ash	10.3		9.5
Product % total sulfur	1.67		1.85
% wt Yield	88.2		77.2
Feed % ash		22.19	
Feed % total sulfur		2.43	

performance slightly superior to that of the spirals over this particle size range (Reed et al., 1995).

It has been found that countercurrent hindered settling separators can be used effectively as classifiers as well. When paired with another classification unit, it has been found that the final concentrate quality can sometimes be achieved in a single-stage operation (Das and Sarkar, 2018).

5.4 ENHANCED GRAVITY SEPARATORS

Enhanced gravity separators are recently developed, separating devices that use the same basic principles as jigs, spirals, and tables, but replace the gravitational force with centrifugal force. The centrifugal force can be made considerably stronger than gravity, which greatly increases the particle settling rates, allowing much finer particles to be processed. This makes it possible to perform density separations at particle sizes fine enough that pyrite is well-liberated, making these units well-suited for removing pyrite from coal. It has been reported that some of these units can achieve high pyritic sulfur rejections with feeds containing particles as fine as 10 μm, which is fine enough to be competitive with froth flotation.

Several enhanced gravity separators are commercially available, including the Multi-Gravity Concentrator (MGS), the Kelsey Centrifugal Jig, the Falcon Concentrator, and the Knelson Concentrator.

5.4.1 Kelsey Centrifugal Jig

The Kelsey centrifugal jig is a comparatively recent development, which allows continuous operation with high centrifugal force. A schematic of the unit is shown in Figure 5.13. The rotation of the jig allows it to generate forces over 200 times stronger than gravity, which extends the usefulness of the jigging process to ultrafine sizes and lower separation densities (Riley et al., 1995).

In this unit, a rotating bowl is surrounded by a casing for collecting the high- and low-density fractions. The bowl is a cylindrical screen, with a layer of ragging material that is coarser than the screen perforations. Feed slurry enters the rotating bowl, and the centrifugal force causes it to flow outward to the ragging bed on the screen, very much like the spin-dry action of a washing machine. A mechanical diaphragm is used to apply a high frequency pressure pulse that flows through the screen and ragging bed, producing the jigging action. The low-density coal cannot penetrate

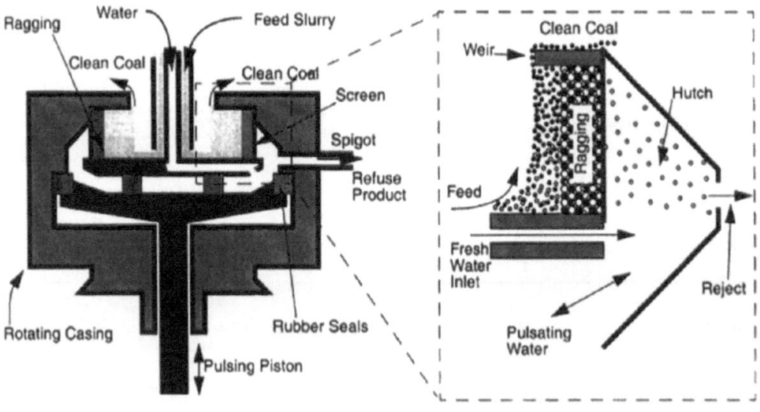

FIGURE 5.13 Schematic of the Kelsey centrifugal jig, with an expanded view of the jigging bed and hutch.

TABLE 5.13
Separation Efficiency as a Function of Particle Size for the Kelsey Centrifugal Jig (Riley et al., 1995)

Size Fraction, μm	ρ_{50}	E_p
500 × 250	2.06	0.12
250 × 106	2.06	0.12
106 × 63	1.92	0.21
63 × 38	1.97	0.28

the ragging and is carried away from inside the screen bowl, while the high-density pyrite and ash go through the screen and exit through the spigot.

The Kelsey centrifugal jig maintains a high separating efficiency down to very fine sizes, as shown by the Ecart probability values in Table 5.13. In laboratory-scale experiments carried out with finely ground coals from the Pittsburgh No. 8 and Illinois No. 6 seams, it was found that this device could remove up to 40% of the pyritic sulfur while recovering 93% of the heating value. Results from these laboratory studies, including jig feed rate, centrifugal force, and pulsation frequency, are given in Table 5.14.

TABLE 5.14
Pyritic Sulfur Removal and BTU Recovery for the Kelsey Jig, under Various Operating Conditions

Coal Type	Feed Rate gallons/min, L/min	Centrifugal Force, Gravities	Frequency, Hz	Feed % Solids	% BTU Recovery	% Pyrite Removal
Illinois	1.75 (6.62)	45	15	7.15	93.59	27.83
No. 6	3.50 (13.25)	20	15	11.44	94.80	18.85
seam	3.50 (13.25)	20	25	11.42	91.31	27.08
	3.50 (13.25)	40	15	9.85	92.26	27.05
	3.50 (13.25)	40	25	8.88	90.11	27.61
	3.50 (13.25)	45	15	7.16	92.33	28.99
	5.00 (18.92)	45	15	9.03	94.79	23.56
Pittsburgh	1.75 (6.62)	45	15	9.28	95.10	33.10
No. 8	3.50 (13.25)	31.8	15	9.56	95.63	—
seam	3.50 (13.25)	31.8	25	9.23	96.80	13.34
	3.50 (13.25)	45	15	9.60	94.26	36.60
	3.50 (13.25)	45	15	9.10	96.80	15.32
	3.50 (13.25)	45	25	9.74	93.23	40.03
	5.00 (18.92)	45	15	8.79	95.03	34.49

The feed coals were approximately 60% passing 25 μm (500 mesh) (Kaiser Engineers, 1994).

5.4.2 Falcon Concentrator

The Falcon concentrator is a centrifugal bowl device that can operate continuously, as shown in Figure 5.14. The unit uses hindered settling and stratification effects for separation of high-density particles from low-density particles.

Results have been reported showing that the Falcon concentrator is capable of functioning at particle sizes as small as 45 μm (325 mesh), while still removing up to 36% of the total sulfur, as shown in Table 5.15. Unfortunately, clay slimes finer than 10 μm tend to be swept along with the clean coal, and a separate desliming device is needed to remove this clay from the coal.

5.4.3 Knelson Concentrator

The Knelson concentrator is a hindered settling device, related to the hydrosizer, with centrifugal force substituting for the force of gravity. It consists of a rotating, ribbed conical basket, with perforations between the ribs, as shown schematically in Figure 5.15. The basket rotates at 40 rpm, generating a centrifugal force of 60 times the force of gravity. Heavy particles are trapped between the ribs, while the low-density coal is carried out by the water flow. The flow of water through the perforations in the bowl keeps the dense particles from packing tightly between the ribs, and also flushes out low-density particles (Knelson, 1992; Knelson and Jones, 1994). One of the disadvantages of the Knelson unit is that it requires two to three times more water as feed to fluidize the particles between the ribs. The water demand is high and the clean coal product is dilute.

Studies of the Knelson concentrator for coal desulfurization have shown that significant ash and sulfur removals are achievable with good calorific value recoveries. With classified feed in the 500 × 300 μm size range, this type of unit could recover 93.7% of the calorific value of the coal, while reducing the total sulfur content of the product by 25.9% and reducing the ash content by 45.4%. With unclassified feed in

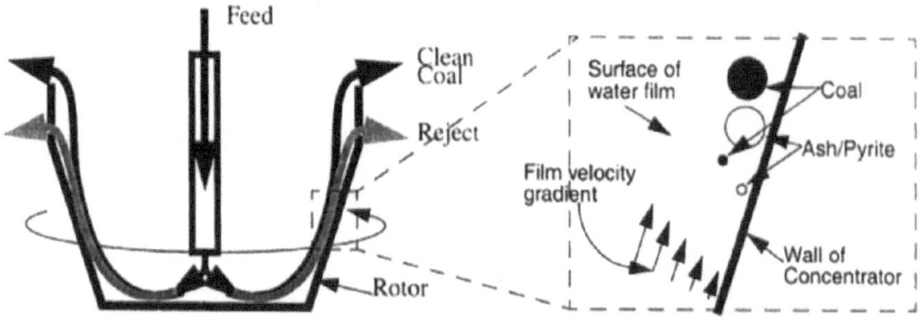

FIGURE 5.14 Schematic of the continuous Falcon concentrator, with a detail of the phenomena occurring at the wall of the rotor. The velocity of the water in the flowing film increases as the distance from the concentrator wall increases. Large particles project into the fast-flowing part of the film, and so large, low-density coal particles tend to be swept to the clean coal product. Hindered settling effects also tend to segregate the dense pyrite particles to the separator wall, and they report to the reject product.

TABLE 5.15
Size-by-Size Sulfur Removal Results Using a Falcon Concentrator to Process a Coal from the Illinois No. 5 Seam (Honaker et al., 1995)

Feed Size Range	Test No.	% Total Sulfur Clean Coal	% Total Sulfur Reject	Clean Coal, BTU/lb (MJ/Kg)	% Coal Recovery	% Sulfur Removal	lb SO_2/Million BTU (g SO_2/MJ)
+150 μm (+100 mesh)	Feed	1.46	—	13,660 (31.69)	—	—	2.14 (0.92)
	1	1.38	7.91	13,470 (31.25)	99.2	6.64	2.05 (0.88)
	2	1.28	5.55	13,610 (31.58)	97.3	16.0	1.88 (0.81)
	3	1.23	4.57	13,610 (31.58)	95.2	21.6	1.80 (0.77)
	4	1.15	2.57	13,610 (31.58)	79.9	38.4	1.69 (0.73)
150 × 45 μm (100 × 325 mesh)	Feed	1.80	—	12,700 (29.46)	—	—	2.84 (1.22)
	1	1.53	8.76	12,525 (29.06)	98.7	18.2	2.44 (1.05)
	2	1.41	5.81	13,365 (31.01)	96.7	28.6	2.11 (0.91)
	3	1.29	4.65	13,430 (31.16)	90.4	39.2	1.92 (0.82)
	4	1.23	3.47	13,635 (31.63)	80.5	49.1	1.80 (0.77)
−45 μm (−325 mesh)	Feed	1.73	—	5,830 (13.52)	—	—	5.93 (2.55)
	1	1.44	7.86	5,855 (13.58)	98.5	20.5	4.92 (2.12)
	2	1.38	6.80	6,175 (14.33)	97.4	25.4	4.47 (1.92)
	3	1.32	4.99	6,220 (14.43)	93.5	36.0	4.25 (1.83)

the 1,000 × 0 μm size range, the calorific value recovery was 94.8%, with a reduction in sulfur content of 21.7% and a reduction in ash of 35.7%.

5.4.4 Multi-Gravity Separator

The Mozley Multi-Gravity Separator (MGS) (Figure 5.16) is a flowing-film device similar to a shaking table, except that the "table" is formed into a cylinder and rotated to provide centrifugal force. The coal particles are swept across the surface

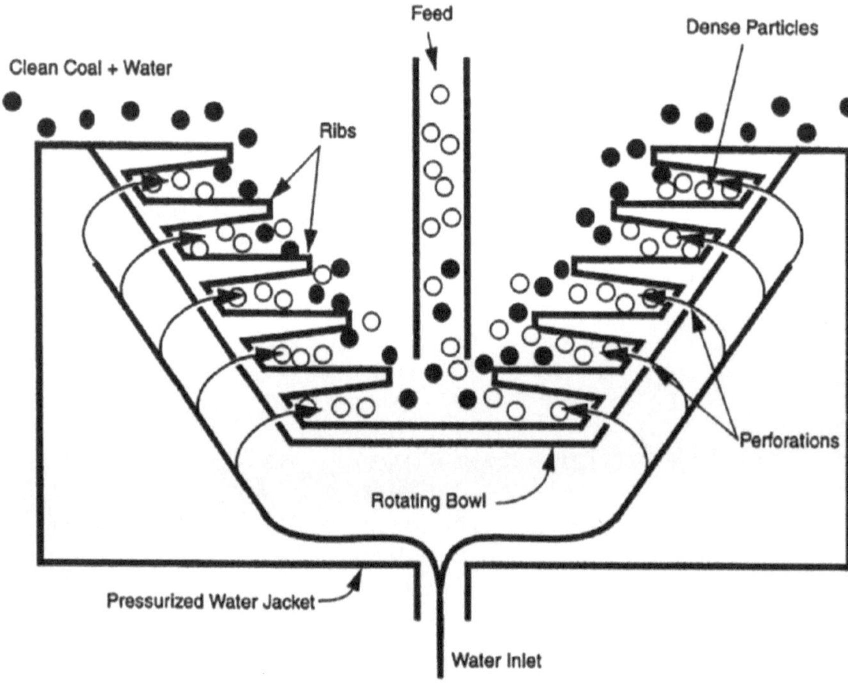

FIGURE 5.15 Schematic of the Knelson concentrator. Water is injected under pressure in the water jacket, and flows through the perforations between the ribs of the rotating bowl. The water flow classifies the particles between the ribs by hindered settling; therefore, the dense particles are trapped between the ribs while the coal is swept out by the water flow.

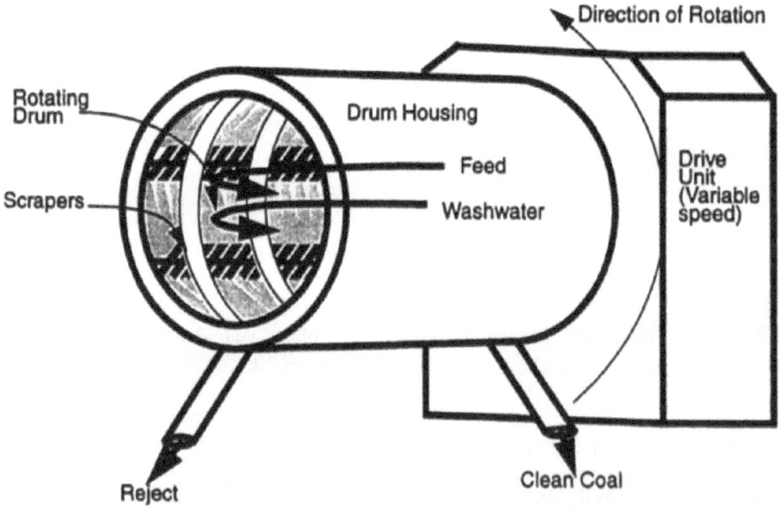

FIGURE 5.16 Drawing of the Mozley pilot-scale MGS.

of the drum and out one end by the flow of water, while the high-density reject (ash and pyrite) interacts more strongly with the drum surface and the scrapers, and are scraped to the other end, as shown in Figure 5.17 (Venkatraman et al., 1995).

The MGS can remove a large proportion of the pyritic sulfur from coal, even at sizes as small as 45 μm (325 mesh), as shown in Table 5.16. It is reported that the performance of the unit is best when the feed is pretreated to remove ultrafine clays, which would otherwise tend to be entrained in the clean coal product. This unit has also been used in combination with column flotation, and the combination is discussed further in Chapter 12. Comparison studies have been performed for the MGS and the Kelsey jig, with results as given in Table 5.14 (Kelsey jig) and Table 5.17 (MGS). From these results, it appears that the pyrite-removal ability of the MGS is inferior to that of the Kelsey jig for finely ground, dilute feeds.

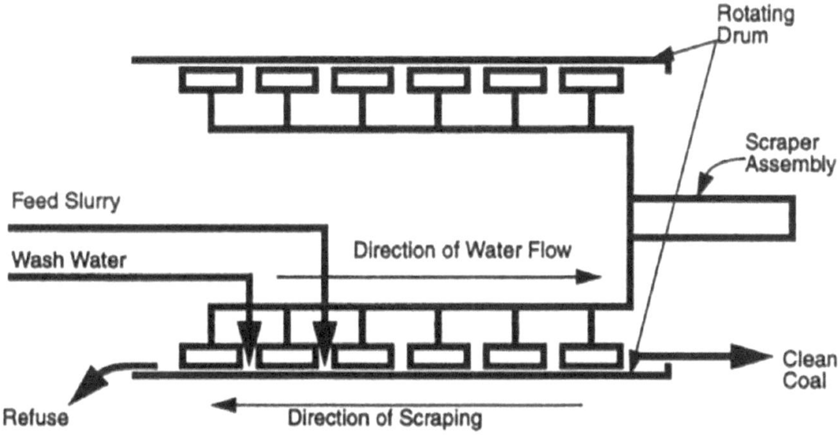

FIGURE 5.17 Schematic of the action of a Mozley MGS. Dense particles (pyrite) sink to the table surface where they are scraped to one end of the drum, while low-density particles (coal) are swept to the other end by the water flow.

TABLE 5.16
Coal Recovery and Removal of Ash and Sulfur by a Mozley MGS, Processing a Run-of-Mine Pittsburgh No. 8 Seam Coal (Venkatraman et al., 1995)

Particle Size	% Yield	% Combustibles Recovery	% Ash Removal	% Pyrite Removal
+300 μm (+48 mesh)	92.12	97.22	46.51	35.04
300 × 150 μm (48 × 100 mesh)	75.10	87.20	69.25	68.10
150 × 74 μm (100 × 200 mesh)	70.42	83.00	78.74	81.04
74 × 45 μm (200 × 325 mesh)	78.69	87.79	71.28	83.52
−45 μm (−325 mesh)	91.29	95.35	13.64	72.61
Total	86.14	93.39	43.53	63.20

TABLE 5.17
Pyritic Sulfur Removal and BTU Recovery for the Mozley MGS, under Various Operating Conditions

Coal Type	Feed Rate, L/min	Rotational Speed, rpm	Wash Water, L/min	Feed % Solids	% BTU Recovery	% Pyrite Removal
Illinois	9.00	380	0.0	10.02	99.85	0.61
No. 6	20.00	320	0.5	9.20	99.70	0.41
seam	19.00	320	1.0	8.37	99.84	0.90
	20.00	380	0.5	9.70	99.83	0.58
	20.00	380	1.0	11.86	99.83	4.26
Pittsburgh	9.00	320	0.5	12.79	99.63	12.17
No.8	16.00	320	0.5	13.25	99.83	0.70
seam	16.00	320	1.0	13.17	99.84	0.24
	16.00	380	0.5	14.28	99.74	1.80
	16.00	380	1.0	13.46	99.78	0.84

The feed coals were approximately 60% passing 500 mesh (25 μm) (Kaiser Engineers, 1994).

5.4.5 Comparison of the Enhanced Gravity Separators

The Falcon concentrator, Kelsey jig, Knelson concentrator, and Mozley MGS have all been tested on a bench scale to determine their coal-cleaning abilities. Luttrell et al. (1995) compiled comparative data for each of these concentrators, with the results given in Table 5.18. It should be noted that the comparison is only qualitative because the various units were processing coals from different sources, and therefore had different feeds. These results suggest that all of the units are capable of rejecting over 40% of the ash and 60% of the sulfur, while recovering more than 80% of the combustible matter.

TABLE 5.18
Typical Results Obtained Using the Various Enhanced Gravity Separators (Luttrell et al., 1995)

Separator	Coal Seam and Size Range	% Combustibles Recovery	% Ash Rejection	% Sulfur Rejection	References
Falcon	Illinois No. 5 212 × 38 μm	88.3	68.1	72.5	Paul and Honaker (1994)
Kelsey jig	Wittingham 500 × 38 μm	96.9	43.6	—	Riley and Firth (1993)
Knelson	Illinois No. 5 212 × 38 μm	84.2	69.1	69.1	Paul and Honaker (1994)
Mozley	Pittsburgh No. 5 212 × 0 μm	86.1	72.9	72.9	Venkatraman et al. (1995)

5.4.5.1 Particle Size Effects in Enhanced Gravity Separators

The important feature of enhanced gravity separators is that they can separate particles at much finer sizes than conventional gravity separators. It is therefore of particular interest to see how they perform as a function of particle size. The graphs shown in Figure 5.18 compare the reported results for all four types of enhanced gravity separators. Again, it should be noted that these were compiled from data generated by different investigators with different coals. The comparison is qualitative and not all data is available for all of the units.

5.4.5.1.1 Falcon Concentrator

Figure 5.18 shows the results for a 25.4 cm (10 in) diameter Falcon concentrator. The poor performance at very fine sizes was due to the presence of ultrafine clay slimes, which were carried along with the water regardless of the densities of the slime particles. The separation of the coarse particles also becomes less effective, both because of poorer liberation of the coarse particles and because they become large enough that the flowing-film effect does not work properly.

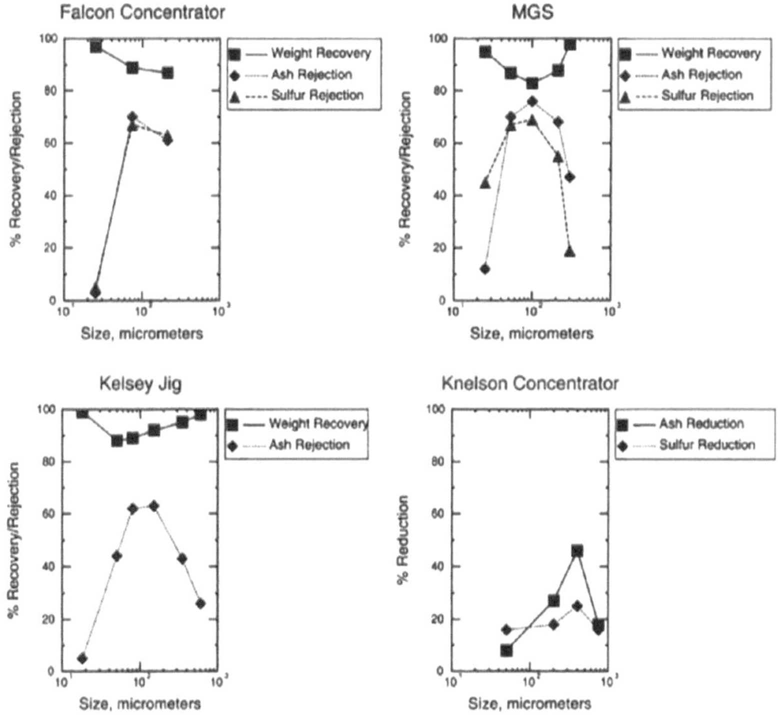

FIGURE 5.18 Performance of each type of enhanced gravity separator as a function of size. It should be noted that the results are presented as percent rejection for the Falcon concentrator (Honaker et al., 1996), MGS (Venkatraman et al., 1995), and Kelsey jig (Riley and Firth, 1993), while they are given as percent reduction for the Knelson concentrator (Butcher and Rowson, 1996).

5.4.5.1.2 Kelsey Jig

The Kelsey jig results in Figure 5.18 also show a loss in performance at coarse and fine sizes (Riley and Firth, 1993). The poorer performance at coarse sizes was due to the passage of large amounts of coarse shale particles through the loosely packed bed of ragging material. Alternative forms of ragging are currently being investigated to reduce this problem. At the fine sizes, the centrifugal force is simply too small to produce a separation of the very fine particles (Merwin, 1995, Personal Communication).

5.4.5.1.3 Mozley Multi-Gravity Separator (MGS)

The MGS has essentially the same particle size limitations as the Kelsey jig, as shown in Figure 5.18. The coarse particles of mineral matter are large enough to project into the high-velocity region of the flowing film, and therefore tend to be carried nonselectively into the clean coal product. Attempts to reject these particles by increasing the drum speed to pin them to the drum wall has the adverse effect of pinning the coarse coal to the wall as well, which reduces coal recovery. The MGS is also incapable of rejecting the ultrafine entrained slimes (Luttrell et al., 1995).

5.4.5.1.4 Knelson Concentrator

The ash and total sulfur reduction for a Knelson concentrator as a function of size are given in Figure 5.18 (Butcher and Rowson, 1996). While the results for the other three separators were reported by the investigators in terms of sulfur and ash rejections, the Knelson results were reported in terms of sulfur and ash reduction, which normally has a lower numerical value than the percent rejection, which is equivalent to the percent removal. As a result, the Knelson results at first glance appear to be inferior to the results from the other units; however, a positive value for the sulfur and ash reduction indicates that a separation is taking place. These results therefore show that the Knelson concentrator is able to make a separation at coarser sizes than is practical with the other units and is also able to continue functioning down to a lower particle size limit similar to that of the Falcon, Kelsey, and Mozley units. Knelson and Falcon separators have been found to complement each other's effective size ranges (Das and Sarkar, 2018).

5.4.6 Capacity of Enhanced Gravity Separators

Capacity for these units can be calculated in two ways. The first is the throughput per unit area of the separating surface (metric tons/meter^2/hour), and the second is the throughput per unit circumference of the separating element (metric tons/meter/hour). The estimated capacities for each of the enhanced gravity separators were determined both ways, and the capacities and estimated capital costs are given in Table 5.19.

The data shown suggests that the Kelsey jig has the highest overall capacity, while the MGS has the lowest. The manufacturer of the Falcon concentrator claims that significant improvements in capacity can be achieved by scaling up their unit. As a result, the Falcon concentrator is claimed to have the highest

TABLE 5.19
Estimated Capital Costs and Capacities for the Falcon, Kelsey, Knelson, and Mozley Enhanced Gravity Separators

Separator	Capacity, metric tons/ m²/h	Capacity, metric tons/m/h	Estimated Capital Cost ($ US)	References
Falcon	4.9–14.6	1.5–3.0	$200,000 for >50 metric tons/h	Paul and Honaker (1994)
Kelsey jig	14.6–19.5	4.5–5.9	$150,000 for 5–15 metric tons/h	Riley and Firth (1993)
Knelson	1.0–3.9	0.7–1.5	$150,000 for 20 metric tons/h	Paul and Honaker (1994)
Mozley MGS	0.1–0.3	0.09–0.18	$200,000 for 5–15 metric tons/h	Venkatraman et al. (1995)

Costs were estimated from verbal quotes and/or advertised materials, and may be subject to considerable variations (Luttrell et al., 1995).

capacity per dollar of capital invested when very large units are used. The low capacity of the MGS concentrator is a significant setback, and implies that it would be more useful near the final stages of separation where there is less material to work with (Das and Sarkar, 2018).

5.5 SUMMARY

The large density difference between pyrite and coal makes this system an excellent candidate for treatment by processes that separate particles using differences in their densities. Unfortunately, such processes work best with coarser particles and do not give good results for very fine particles. While the conventional gravity-based devices that exist are extremely effective for removing pyrite, which liberates at sizes larger than about 50 μm, these machines are not suitable for removing the last traces of finely dispersed pyrite, or for separating very fine coal from clay slimes.

In an effort to extend gravity separators to finer particle sizes, several new separators have been developed, which spin rapidly to produce "enhanced gravity" centrifugal forces, several hundred times stronger than gravity. These separators can process particles that are as fine as 10 μm, which is fine enough for the bulk of the pyrite to be liberated. They therefore are promising new technology for desulfurization of very fine coal. Their primary limitation is that the need for precisely balancing the rotating elements tends to increase the capital costs and limit the equipment size.

These advanced gravity separators have been employed very successfully in many areas of mineral processing, particularly including the processing of coal and coal preparation tailings. This technology is well developed, and optimal operating conditions have been identified for many important use cases (Das and Sarkar, 2018).

Despite that, there are open routes of advancement for gravity separation technology. As such, due to the tested longevity of the technology in actual plant work, overall low capital and operating costs for separation, alongside the potential for yet further improvement, it is likely that gravity separation will remain a staple of coal and mineral processing for many years to come.

REFERENCES

Adams, R.J. (1988), "Applications of Fine Coal Jigs," *Industrial Practice of Fine Coal Processing* (Klimpel and Lucide, eds.), Society of Mining Engineers, Littleton, CO, pp. 65–70.

Apodaca, L.E. (1988), "Applications of Spiral Concentrators in Fine Coal Processing," *Industrial Practice of Fine Coal Processing* (Klimpel and Luckie, eds.), Society of Mining Engineers, Littleton, CO, pp. 87–96.

Butcher, D.A. and Rowson, N.A. (1996), "Desulfurization of Coal in Intensified Magnetic, Electrostatic, and Gravitational Fields," *New Trends in Coal Preparation Technologies and Equipment* (Blaschke, ed.), Gordon and Breach, Amsterdam, pp. 223–234.

Das, A. and Sarkar, B. (2018), "Advanced Gravity Concentration of Fine Particles: A Review," *Mineral Processing and Extractive Metallurgy Review*, Vol. 39, No. 6, pp. 359–394.

Gartner, H.G. (1994), "Spiral Classifiers for Preparation of Smalls and Fines," *Glueckauf: Zeitschrift fuer Technik und Wirtschaft des Bergbaus*, Vol. 130, pp. 332–336.

Hoke, W.D. (1976), "Application of the Batac Jig for Processing Fine Coal," *Mining Congress Journal*, Vol. 62, No. 9, pp. 52–55.

Honaker, R.Q., Paul, B.C., Wang, D. and Ho, K. (1995), "Enhanced Gravity Separation: An Alternative to Flotation," *High Efficiency Coal Preparation* (Kawatra, ed.), Society for Mining, Metallurgy, and Exploration, Littleton, CO, pp. 69–78.

Honaker, R.Q., Wang, D. and Ho, K. (1996), "Application of the Falcon Concentrator for Fine Coal Cleaning," *Minerals Engineering*, Vol. 9, No. 11, pp. 1143–1156.

Hyde, D.A., Williams, K.P., Morris, A.N. and Yexley, P.M. (1988), "The Beneficiation of Fine Coal Using the Hydrosizer," *Mine and Quarry*, Vol. 17, No. 3, pp. 50–54.

Ivatt, S. (1993), "Spiral Concentrators for UK Coal," *Mine and Quarry*, Vol. 22, No. 6, pp. 38–41.

Kaiser Engineers (1994), "Engineering Development of Advanced Physical Fine Coal Cleaning Technologies—Froth Flotation," Quarterly Technical Progress Report No. 22, January 1, 1994–March 31, 1994, DOE Contract No. DE-AC22-88PC88881 (DE95006661).

Knelson, B. (1992), "The Knelson Concentrator: Metamorphosis from Crude Beginning to Sophisticated World Wide Acceptance," *Minerals Engineering*, Vol. 5, No. 10–12, pp. 1091–1097.

Knelson, B. and Jones, R. (1994), "A New Generation of Knelson Concentrators: A Totally Secure System Goes On-Line," *Minerals Engineering*, Vol. 7, No. 2–3, pp. 201–207.

Laros, T.J. and Aplan, F.F. (1987), "Comparative Performance of a Placer Ore Jig and a Baum-Feldspar Jig for Pyrite and Ash Removal form Coal," *Proceedings of the 2nd International Conference on Processing and Utilization of High-Sulfur Coals* (Chugh and Caudle, eds.), Elsevier, New York, pp. 130–139.

Lovell, H.L., Moorehead, R.G., Lucide, P.T. and Kindig, J.K. (1991), "Hydraulic Wet Coarse Particle Concentration," *Coal Preparation*, 5th Edition (Leonard, ed.), Society for Mining, Metallurgy, and Exploration, Littleton, CO, pp. 301–377.

Luttrell, G.H., Honaker, R.Q. and Phillips, D.I. (1995), "Enhanced Gravity Separators: New Alternatives for Fine Coal Cleaning," *Proceedings of the 12th International Coal Preparation Exhibition and Conference*, Lexington, KY, pp. 282–292.

MacNamara, L., Addison, F., Miles, N.J., Bethell, P. and Davis, P. (1995), "The Application of New Configurations of Coal Spirals," *Proceedings of the 12th International Coal Preparation Exhibition and Conference*, Lexington, KY, pp. 119–144.

MacNamara, L., Moorhead, R.G., Davis, P., Miles, N.J., Bethell, P. and Everitt, B. (1996), "On-Site Testing of the Compound Spiral," *Proceedings of the 13th International Coal Preparation Exhibition and Conference*, Lexington, KY, pp. 253–276.

Mankosa, M.J., Stanley, F.L. and Honaker, R.Q. (1995), "Combining Hydraulic Classification and Spiral Concentration for Improved Efficiency in Fine Coal Recovery Circuits," *High Efficiency Coal Preparation* (Kawatra, ed.), Society for Mining, Metallurgy, and Exploration, Littleton, CO, pp. 99–107.

Miller, F.G. and Podgursky, J.M. (1972), "Tables in Combination with Hydrocyclones for Fine Coal Processing," *SME Transactions*, Vol. 252, pp. 7–11.

Onodera, J., Koyanagi, N. and Kubo, Y. (1993), "Effect of Reducing the Width of the Screen Chamber of a 'Variwave' Air-Pulsed Jig," *Shigen to Sozai*, Vol. 109, No. 9, pp. 33–36.

Otto, H.H., Wilson, V.W. and Denner, W.L. (1947), "Preparation of Anthracite Silt for Boiler Fuel in Humphreys Spiral Test Plant," *Transactions of the 5th Anthracite Conference*, Lehigh University, Bethlehem, PA, pp. 269–291.

Paul, B.C. and Honaker, R.Q. (1994), "Production of Illinois Base Compliance Coal Using Enhanced Gravity Separation," Final Technical Report, Illinois Clean Coal Institute Project No. 93-1/5.1B-1P (DOE Grant No. DE-FC22-92PC92521), September 1, 1993–August 31, 1994, 27 pp.

Reed, S., Riffey, R., Honaker, R. and Mankosa, M. (1995), "In-Plant Testing of the Floatex Density Separator for Fine Coal Cleaning," *Proceedings of the 12th International Coal Preparation Exhibit and Conference*, Lexington, KY, May, 1995, pp. 164–173.

Riley, D.M. and Firth, B.A. (1993), "Applications of Enhanced Gravity Separator for Cleaning Fine Coal," *Proceedings of Coal Prep '93, 10th International Coal Preparation Exhibition and Conference*, May 4–6, Lexington, KY, pp. 45–72.

Riley, D.M., Firth, B.A. and Lockhart, N.C. (1995), "Enhanced Gravity Separation," *High Efficiency Coal Preparation* (Kawatra, ed.), Society for Mining, Metallurgy, and Exploration, Littleton, CO, pp. 79–88.

Saltsman, R.D. (1969), "Facing Up to Sulfur Content of Coal," *American Society of Mechanical Engineers Meeting*, February 11–13, 1969, St. Louis, MO, Paper No. 69-FU-5, 8 pp.

Tavares, L.M. and Rubio, J. (1991), "Performance Evaluation and Simulation of a Batac Jig Cleaning Pyrite from Coal Washery Tailings," Coal Science and Technology, *Processing and Utilization of High-Sulfur Coals IV*, Elsevier, New York City, NY, Vol. 18, pp. 597–607.

Tavares, L.M. and Sampaio, C.H. (1990), "Evaluation of Spirals in Cleaning Fines of a High-Sulfur Brazilian Coal," *Processing and Utilization of High-Sulfur Coals III* (Markuszewski and Wheelock, eds.), Elsevier, Amsterdam, pp. 301–320.

Thomson, R.S. and Aplan, F.F. (1995), "Fine Pyrite Removal from Coal Using a Placer Ore Jig," *High Efficiency Coal Preparation* (Kawatra, ed.), Society for Mining, Metallurgy, and Exploration, Littleton, CO, pp. 89–98.

Venkatraman, P., Luttrell, G.H., Yoon, R.-H., Knoll, F.S., Kow, W.S. and Mankosa, M.J. (1995), "Fine Coal Cleaning Using the Multi-Gravity Separator," *High Efficiency Coal Preparation* (Kawatra, ed.), Society for Mining, Metallurgy, and Exploration, Littleton, CO, pp. 109–117.

Zeilinger, J.E. and Deurbrouck, A.W. (1976), "Physical Desulfurization of Fine-Size Coals on a Spiral Concentrator," U.S. Bureau of Mines Report of Investigations # 8152.

6 Dry Separation

Most coal separation is done at the mine, but it is equally possible to perform separation steps at the power plant. For plants processing pulverized coal, adding an additional separation step to take advantage of the fact that coal is already being grinded finely is an attractive option for removing additional sulfur from the coal. The primary requirements for this separation are that the process must be inexpensive, and that it must not add water.

There are other reasons to shy away from adding water in separation as well. In particular, if drying becomes necessary it is often a very considerable expense in mineral processing, involving pumping, screening, filtering, and perhaps centrifuging (Das and Sarkar, 2018). If water is not added, the likelihood that water will need to be removed decreases. In the case of highly porous coals, wet separation may result in too much water being trapped to be economical, requiring dry separation (Sampaio et al., 2008).

Even if expensive drying steps end up not being required, the water addition alone might become too expensive in some arid climates. Such conditions are common over large portions of China, and the water required for wet operations in coal processing represents an untenable drain on the water infrastructure in such regions (Fan et al., 2001; Luo et al., 2008).

In colder climates, wet processing may also mandate significant increases in capital and operating costs during the winter – these costs arise from the potential for the wet processes to simply freeze entirely if the temperature drops enough. Thus, these processes may need to be housed inside of heated structures, raising the overall cost of facility significantly. On the other hand dry processing can usually continue in cold weather, making such processes more suitable for use in operations in colder climates.

Dry separation efficiently avoids the issues presented by adding water, providing an inexpensive separation operation suitable for use in many conditions.

6.1 PRINCIPLES OF DRY SEPARATION

The dry separation processes used in coal processing are typically density separations, size separations, electromagnetic separations, or mass separations. These properties can be influenced and measured without the addition of water, allowing for the separations to be performed without a solvent. These separations typically take place either on a fluidized bed of particles which can more or less move past each other, or on falling particle streams which do not directly interact with each other at all.

Beds of coal are typically fluidized by passing air through them to impart a semblance of mobility on the particles. Some fluidized beds are magnetically stabilized instead (Fan et al., 2001; Luo et al., 2002), which may be practical if magnetic material is readily available, such as intermingled magnetite ores.

In either case, to achieve dry separation the solid particles need to be able to move past each other or otherwise act independently. After that, as with all separations, the goal is to induce a positional difference between one category of particles and another. Many dry separations are density separators, which rely on the migration of denser particles toward the bottom of a fluidized bed. Mass separations that rely on the comparative inertia of more massive particles are also possible. For the sake of completeness, size separations can be achieved with sieves as a dry process.

Electromagnetic separations can be divided into electrical separations (e.g. triboelectrostatic separators) or magnetic separations, which may be particularly useful in separating coal from entirely different minerals. These techniques will not be covered here, as they are discussed in detail in Chapters 11 and 12. However, it is worth noting that both electrical and magnetic separations can be adapted to (or must be run as) dry processes.

6.2 DRY SEPARATION EQUIPMENT

Some simple wet gravity separation processes can be re-designed directly to run without water, such as the dry spiral. The dry spiral is essentially the same as the wet spiral, except that water is not used. The main difference between the dry spiral and the wet spiral is that the resistive force in a dry spiral is friction rather than buoyancy, and thus the final product is ultimately a friction separation process rather than a gravity separation process (Gartner, 1994). Dry spirals were originally developed in the late nineteenth century for cleaning anthracite, but are no longer used industrially.

The dry pinched sluice, shown in Figure 6.1, is likely the purest example of a dry density separation. The dry pinched sluice is an inclined, porous, tapering chute with air blown through the deck to fluidize the particle bed. The feed enters at the widest end and flows toward the narrow end. As the chute becomes narrower, the particle bed becomes thicker and the higher density particles migrate toward the bottom. A simple mechanical splitter at the end divides the sluice into heavy and light streams.

Bignell and Gooriah (1990) carried out pilot-scale tests using a dry pinched sluice to treat a pulverized coal classifier return stream and found that they could remove 48% of the pyrite while only losing 8% of the heating value. This sort of treatment is most suitable for coals that contain fairly coarse pyrite, as the dry pinched sluice cannot generate as sharp of a separation as wet separators do.

A third option is the aerodynamic separator. An aerodynamic separator consists of a chamber where a horizontal air flow is used to deflect a falling stream. Low-density and fine particles are deflected more readily, so the heavier and coarser particles tend to drop straight down. To achieve the most meaningful density separation, the input to an aerodynamic separator should be uniformly sized. Figure 6.2 presents a typical aerodynamic separator which carries the fine coal out the far end of the separator. In the case of a setup like in Figure 6.2, the air flow needs to be laminar.

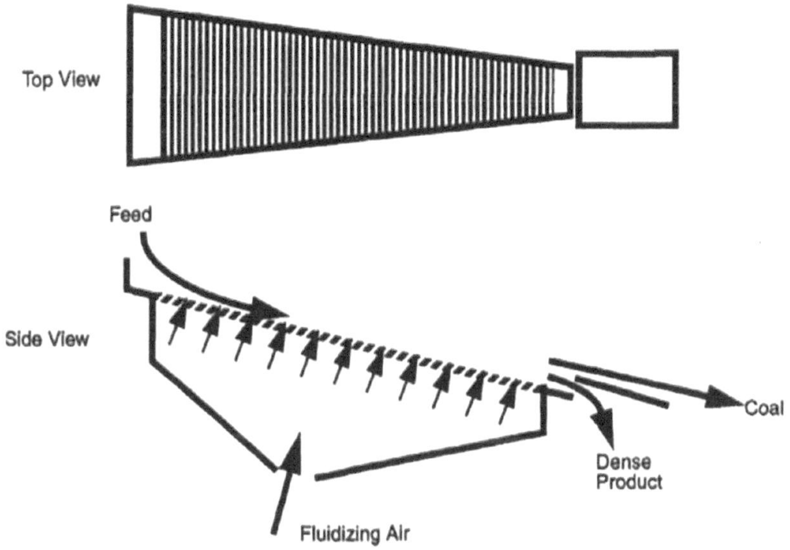

FIGURE 6.1 Schematic of a dry pinched sluice. Particles are segregated by the flow of fluidizing air as they flow down the chute. The point of feed entry is wider than the point where the product is removed, so the products are crowded together and increase in depth, simplifying the removal of the products.

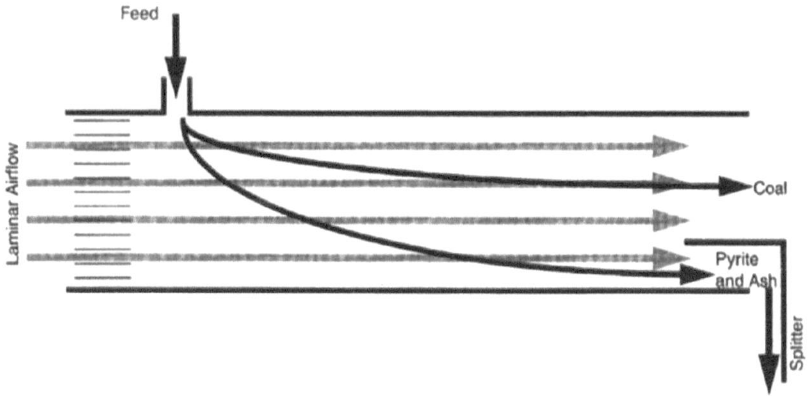

FIGURE 6.2 Schematic of an aerodynamic separator which uses a horizontal, laminar air flow to carry coal to the product stream. Such units have not yet been applied to coal cleaning.

Rather than applying a continuous air stream through the entirety of a separator, a planar air jet (or air knife) can be used to impart a comparatively instantaneous force on a falling particle stream. This instantaneous force deflects the falling particles for an instant, which then return to parabolic motion as shown in Figure 6.3 (Yang et al., 2018). The use of this technology for coal processing is a new development, and it has not to date been deployed to plant-scale work.

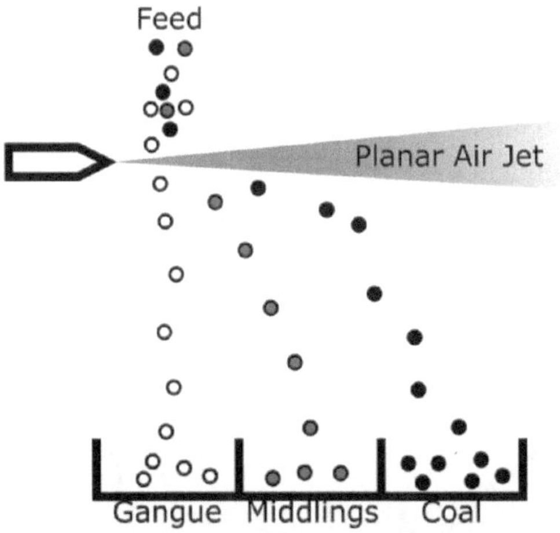

FIGURE 6.3 Schematic of an aerodynamic separator operating with an instantaneous impulse rather than a continuous driving force (Yang et al., 2018).

A fourth common variety of dry separation is dry jigging. A dry jig, or air jig, works by the same basic principles as a wet jig as described in Chapter 5, except that the fluid medium is air instead of water.

Another style of dry separation unit is the vibrational fluidized bed separator. The vibration stabilizes the fluidized bed even for fine particle distributions, allowing the particles to readily move past each other. Using this technique and a magnetite medium, coal particles as fine as 0.5 mm have been separated in the laboratory (Luo et al., 2008). In this case as with water-only cyclones, the solid material itself provides a dense medium for density separation.

Other separators include air tables and FGX separators, but the primary mechanism of each is typically a combination of the dry sluice, air jig, and vibrational fluidized bed separators.

These separation units are most effective for coarser particle sizes between 6 and 75 mm (Das and Sarkar, 2018), but modest separation efficiencies have been achieved in the finer ranges of 1 to 50 mm (Zhenfu and Qingru, 2001) and down to 0.5 mm (Luo et al., 2008). For coal finer than 1 mm, electrical separation techniques tend to provide better results (Fan et al., 2001; Chen and Honaker, 2015).

6.3 SUMMARY

Dry separation can be performed using a variety of separators. The dry separations in this chapter are performed against density, mass, and particle size. It is also possible to perform dry separations against magnetic and electrical properties, described in Chapters 11 and 12. Many dry separations are based heavily on wet separations where the water is not strictly necessary and can be omitted.

The dry separation techniques described here can be used to separate fine coals down to 0.5 mm in diameter, though they are most efficient between 6 and 75 mm.

Sizes between 1 and 6 mm are a comparatively difficult regime, but below 1 mm electrical separation techniques are advised.

The use of dry separation allows for a final removal of sulfur from coal at the power facility, for the processing of fine, porous coals which would otherwise acquire an excess of water, or for the processing of coals in climates where water is scarce.

REFERENCES

Bignell, J.D. and Gooriah, B. (1990), "The Physical Removal of Pyrite at the Power Plant," *Processing and Utilization of High-Sulfur Coals III* (Markuszewski and Whee-lock, eds.), Elsevier, Amsterdam, pp. 321–330.

Chen, J. and Honaker, R. (2015), "Dry Separation of Coal-Silica Mixture Using Rotary Triboelectrostatic Separator," *Fuel Processing Technology*, Vol. 131, pp. 317–324.

Das, A. and Sarkar, B. (2018), "Advanced Gravity Concentration of Fine Particles: A Review," *Mineral Processing and Extractive Metallurgy Review*, Vol. 39, No. 6, pp. 359–394.

Fan, M., Chen, Q., Zhao, Y. and Luo, Z. (2001), "Fine Coal (6-1mm) Separation in Magnetically Stabilized Fluidized Beds," *International Journal of Mineral Processing*, Vol. 63, pp. 225–232.

Gartner, H.G. (1994), "Spiral Classifiers for Preparation of Smalls and Fines," *Glueckauf: Zeitschrift fuer Technik und Wirtschaft des Bergbaus*, Vol. 130, pp. 332–336.

Luo, Z., Zhao, Y., Chen, Q., Fan, M. and Tao, X. (2002), "Separation Characteristics for Fine Coal of the Magnetically Fluidized Bed," *Fuel Processing Technology*, Vol. 79, pp. 63–69.

Luo, Z., Fan, M., Zhao, Y., Tao, X., Chen, Q. and Chen, Z. (2008), "Density-Dependent Separation of Dry Fine Coal in a Vibrated Fluidized Bed," *Powder Technology*, Vol. 187, pp. 119–123.

Sampaio, C.H., Aliaga, W., Pacheco, E.T., Petter, E. and Wotruba, H. (2008), "Coal Beneficiation of Candiota Mine by Dry Jigging," *Fuel Processing Technology*, Vol. 89, pp. 198–202.

Yang, X., Wang, S., Zhang, Y., Zhao, Y. and Luo, Z. (2018), "Dry Beneficiation of Fine Coal Using Planar Air Jets," *Powder Technology*, Vol. 323, pp. 518–524.

Zhenfu, L. and Qingru, C. (2001), "Dry Beneficiation Technology of Coal with an Air Dense-Medium Fluidized Bed," *International Journal of Mineral Processing*, Vol. 63, pp. 167–175.

7 Froth Flotation

Froth flotation is an effective method for cleaning many coals at nominal particle sizes finer than 600 μm (28 mesh). This fine coal can be dried and re-blended with the coarse coal for sale, or it can be used to produce coal-water fuels or as feedstock for slurry-fed gasifiers (Ehrlinger et al., 1995). Unlike the processes discussed in previous chapters which separate coal from mineral matter based on density, froth flotation separates particles based on their surface chemistry. This allows it to be very effective for particle sizes coarser than about 10–25 μm but finer than approximately 1–2 mm (Klassen and Mokrousov, 1963; Laskowski and Walters, 1992).

The basis of froth flotation of coal is the difference in the wettabilities of coal and mineral matter. Since coal is composed of organic compounds, it tends to be water-repellent, or *hydrophobic*. In contrast, most of the associated mineral matter is easily wetted by water, or *hydrophilic*. So, if particles of hydrophobic coal are suspended in water and air is bubbled through the suspension, then the coal particles will tend to attach to the air bubbles and float to the surface. The froth layer that forms on the surface will then be heavily loaded with coal and can be removed as a clean coal product. The mineral matter will have much less tendency to attach to air bubbles; therefore, it will remain in suspension and be flushed away (Whelan and Brown, 1956). A basic schematic of the flotation process is shown in Figure 7.1. Conventional froth flotation circuits all use basically the same type of machine, and performance differences are mainly due to changes in operating parameters and reagents. The theory of froth flotation is described in standard textbooks (Leja, 1982; Wills, 1992).

Froth flotation is sensitive to factors that change the surface chemistry, such as coal rank and degree of oxidation. As a result, it will not be suitable for use with all coals (Leja, 1982; Taggart, 1945). In particular, low-rank coals (such as lignite and sub-bituminous coals) are difficult to clean by froth flotation. Very high-rank coals such as anthracite are also poorly floatable. In addition, coals that have been stockpiled for an extended time will oxidize upon contact with oxygen and water, causing them to accumulate hydrophilic functional groups on their surface, which makes them difficult to float.

Coal flotation is a good example of an engineering "system," in that the various important parameters are highly interrelated, as shown in Figure 7.2. Changes in the settings of one factor (such as feed rate) will automatically cause or demand changes in other parts of the system (such as flotation rate, particle size recovery, air flow, and pulp density) As a result, it is difficult to study the effects of any single factor in isolation, and compensation effects within the system can keep process changes from producing the expected effects (Klimpel, 1995). This makes it difficult to develop predictive models for froth flotation, although work is being done to develop simple models that can predict the performance of the circuit from

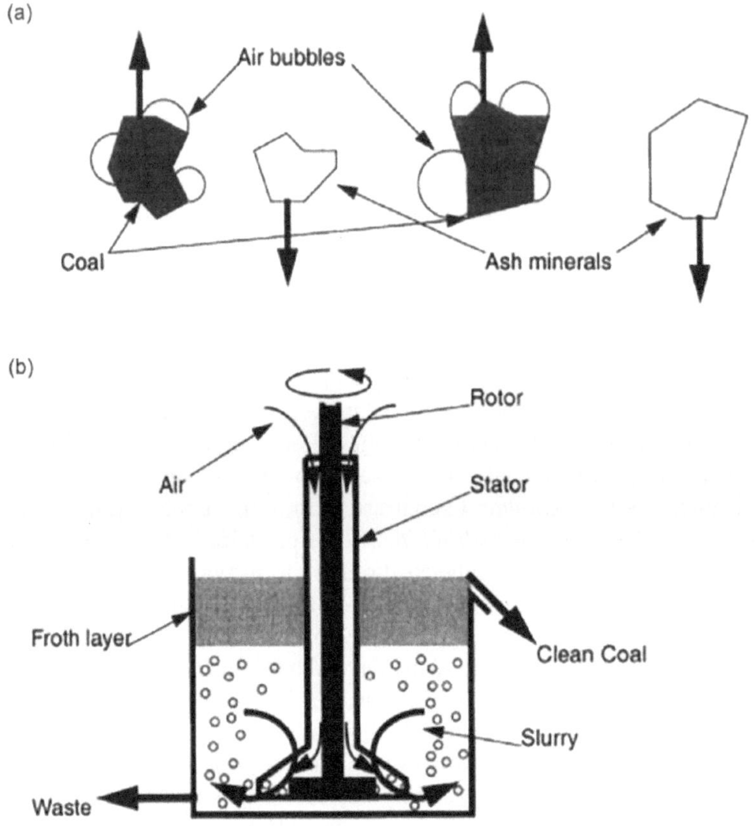

FIGURE 7.1 (a) Selective attachment of air bubbles to coal. The buoyancy of the bubbles then carries the coal particles to the surface. (b) Simplified schematic of a conventional froth flotation cell. Air bubbles are generated by a rotating impeller (rotor) spinning inside of a stationary sheath (stator). The circulation of the slurry pulls air down the stator, where the rotor breaks the air up into bubbles, which are dispersed through the slurry. The bubbles then rise to the froth layer, carrying coal particles with them, and are removed from the cell.

easily measurable parameters such as solids recovery and tailings solid content (Rao et al., 1995).

One major mechanical choice is whether to use one large flotation machine or several smaller flotation machines. The main consideration here is whether the additional separation efficiency gained from the larger cell is sufficient to offset the greater residence time of the coal particles within it. After all, the coal must mechanically be transported from the inlet to the froth layer, so the larger cells, and especially with taller cells, this can begin to take a considerably longer time. When refining copper, for example, these large cells are necessary to effectively concentrate copper ore starting at 2% copper grades. However, the coal particles which are worth recovering are already on the order of 80% coal. As the surface chemistries of coal are quite different from its typical impurities, a flotation cell capable

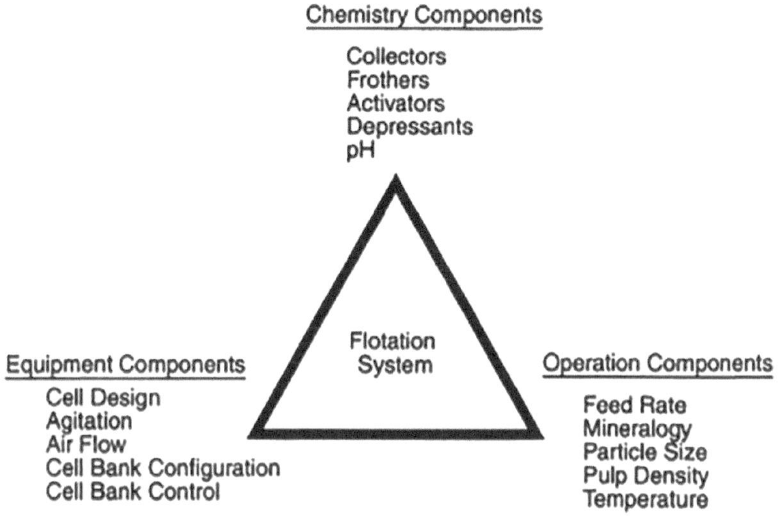

FIGURE 7.2 The flotation system includes many interrelated components, and changes in one area will produce compensating effects in other areas (Klimpel, 1995).

of achieving good concentration from a 2% grade is largely unnecessary – rather, it has been this author's experience that lower residence times are more desirable than a fractionally higher separation efficiency. Even with small cells, very good separations of coal are typically possible.

However, the broadest overview of recommendations for coal flotation would be

- Use as little frother as possible to minimize entrainment. In most cases, ash and pyrite are recovered by entrainment, which is the mechanical transportation of materials into the froth layer. Because these particles are recovered with no concern for surface chemistry, pyrite and ash depressants generally will not improve the separation efficiency.
- The use of smaller flotation cells allows for lower residence times largely, without sacrificing product quality. Flotation cells have traditionally been designed for valuable minerals that are present in comparatively low concentrations. Coal particles have quite a high concentration compared to, for example, copper, and do not benefit as much from the increased residence time.

7.1 CHEMICAL CONSIDERATIONS

The hydrophobicity of the particle surfaces and the stability of the froth phase are both determined by surface chemistry. As a result, the flotation process is strongly influenced by chemical as well as physical effects. It is therefore important to consider the chemical influences in froth flotation.

7.1.1 Basic Reagents Used in Coal Flotation

For froth flotation to work properly, certain chemical reagents must be added to the slurry. These reagents can be classified as follows.

- *Frothers.* These compounds act to stabilize the air bubbles so that they will remain well-dispersed in the slurry, and will form a stable froth layer that can be removed before the bubbles burst. The most commonly used frothers are alcohols, particularly MIBC (methyl isobutyl carbinol, or 4-methyl-2-pentanol, a branched-chain aliphatic alcohol) or any of a number of polyglycols. The polyglycols in particular are very versatile and can be tailored to give a wide range of froth properties. Many other frothers are available, such as cresols and pine oils, but most of these are considered obsolete and are not as widely used as they once were. Some work has also been done using saltwater (particularly seawater) as the frothing agent, and the process has been used industrially in Russia. In addition to stabilizing the froth, high ion concentrations are reported to increase the hydrophobicity of coals, due to thinning of the hydrated layer as the ionic strength increases (Klassen and Mokrousov, 1963). High ionic strength, however, is reported to decrease the selectivity of flotation of coal from mineral matter (Tyunikova and Naumov, 1981).
- *Collectors.* Hydrocarbon oils, and similar compounds, have an affinity for hydrophobic surfaces such as coal surfaces. They selectively adsorb on the coal and increase its hydrophobicity. This improves the recovery of the coal and increases the selectivity between coal particles and mineral matter. The most common collectors in coal flotation are No. 2 fuel oil and kerosene, which have the following advantages: (1) they have low enough viscosity to disperse in the slurry and spread over the coal particles easily and (2) they are very low cost compared to other compounds that can be used as coal collectors.
- *Modifiers.* These chemicals include acids and alkalis for pH control; flocculants and dispersants to control the way that particles interact with each other; and oxidizing, reducing, and coating agents to depress flotation of undesired particles, particularly pyrite. Additives are also available for the kerosene and fuel oil, which improve their ability to coat oxidized coals, which would otherwise be very difficult to float. An example of the type of compounds that are useful for this purpose is the alkanolamides, which have an affinity for the oxidized groups on the coal surfaces. Modifiers for purposes such as pH control, ash dispersion, and recovery of oxidized coal have been used commercially. Pyrite depressants for coal flotation have been studied in the laboratory, but have seen no application in full-scale plants.

The quantities of reagents used will vary a great deal, depending on the coal being processed and on the specific type of reagent. Frother dosages are typically 0.12–0.25 kg/metric ton of coal. Collector dosages are much more variable, depending on the characteristics of the coal, and range from less than 0.5 kg/metric ton for strongly hydrophobic bituminous coals, up to as high as 50 kg/metric ton for some lignites (Aplan and Arnold, 1991). The experience of the author is that the ratio of collector to frother should vary between 3:1 and 5:1. The amount of frother added should be as little as possible, as increasing the amount of froth results in more entrainment and more ash reporting to the concentrate by the entrainment process. Increasing oxidation of the coal tends to increase the amount of collector needed. Higher collector dosages in turn can require more frother to be added to keep the froth layer stable because of reagent interaction. There has been some concern about the possibility of pollution due to flotation reagents leaching out of coal tailings disposal areas, but Hoberg et al. (1995) determined that the bulk of the reagents remain adsorbed onto the clean coal, and only traces remain in the flotation tailings. As a result, the risk of pollution from this source is small.

7.1.1.1 Coal Frothers

All coal flotation circuits use a frothing reagent. The frother is necessary for producing fine air bubble dispersions and for producing a froth layer that lasts long enough to remove the clean coal product. Differences in the chemical structures of frothers lead to different bubble and froth characteristics, and so it is important to select the correct frother for the application. In extensive studies over several years, Klimpel (1992, 1995) has evaluated three major chemical families of frothers:

- Aliphatic alcohols (R-OH), of which MIBC, with the structure $(CH_3)_2CHCH_2CH(OH)CH_3$, was selected as the representative;
- Water-soluble polymers based on propylene oxide (PO), such as polypropylene glycols, with the following three selected as representatives:
  - Dowfroth 200, with the structure $CH_3\text{-}(O\text{-}C_3H_6)_n\text{-}OH$, and molecular weight = 200
  - Dowfroth 1012, with the structure $CH_3\text{-}(O\text{-}C_3H_6)_n\text{-}OH$, and molecular weight = 400
  - Dowfroth 400, with the structure $H\text{-}(O\text{-}C_3H_6)_n\text{-}OH$, and molecular weight = 400
- PO and/or butylene oxide adducts of aliphatic alcohols, with H-200 (Hexanol · 2PO) chosen as the representative.

Equivalent frothers, such as certain members of Cytec's Aerofloat product line, are also available from other manufacturers.

To compare the effects of these frothers, the flotation performance was modeled using the following relationship, which includes a term for both flotation rate and ultimate flotation recovery:

$$r = R\left\{1 - \frac{1 - \exp(-Kt)}{Kt}\right\} \tag{7.1}$$

where: r = total weight of component recovered at time t,
t = time,
K = rate constant,
R = ultimate theoretical weight recovery at "infinite" time.

This model is particularly useful for correlation of laboratory results with plant results. In conventional laboratory test work, it is common for the parameter *R* to be the most important in determining the flotation performance because laboratory tests are often run until all floatable coal is recovered. In the plant, it is common for the parameter *K* to be most important because it is too expensive to provide enough cell volume to recover all coal that does not float in a short time. Because of this difference in operation, the results of laboratory studies can be poor predictors of plant performance. To correct this, it is best to run timed flotation laboratory tests, which can produce kinetic data, so that the *R* and *K* performance can both be determined. Then, based on the residence time of the plant-scale units, it can be determined whether the plant performance is being dominated by kinetics (*K*) or ultimate recovery (*R*).

Klimpel (1995) found that the use of different frothers produced changes in the flotation rate (*K*) and recovery (*R*) values in coal flotation and reached the following conclusions:

- When frother dosage was held constant while collector dosage was increased, it was found that the flotation rate went through a maximum and then decreased. This was observed for all frother types and all particle size fractions. The difference between the frother families studied was that the collector dosage that produced the maximum value of *K* was different.
- For all of the frother types, the finest (–88 µm) and coarsest (+500 µm) particles tend to float more slowly than the intermediate-size particles.
- Changes in flotation rate were due to both changes in the coal particle size and to frother/collector dosage. While the contribution of particle size was generally more significant, the reagent dosage effect provides a useful means for adjusting *K* in the plant.
- With aliphatic alcohol frothers, the flotation rate maximum was much more pronounced than for the PO and PO-alcohol adduct frothers.
- Regardless of frother type, increasing the frother dosage to increase recovery always leads to less selective flotation.
- The PO and PO-alcohol adduct frothers are more powerful recovery agents than alcohol frothers and should therefore be used at lower dosages.
- Overdosing with alcohol frothers leads to a slower flotation rate, because excess of these frothers tends to destabilize the froth. This effect does not occur with the PO and PO-alcohol frothers, and so overdosing with these frothers leads to high recovery with poor selectivity.
- PO frothers with molecular weights of 300–500 are optimal for coal recovery.
- Alcohol frothers tend to be more effective for fine-particle recovery than for coarse-particle recovery. To recover coarse particles, the alcohol frother and the hydrocarbon collector dosages should both be high. The alcohol will still provide reasonable selectivity at these high dosages.
- The high-molecular-weight PO-based frothers are more effective for coarse-particle flotation than for alcohol or low-molecular-weight PO frothers, but

also have a lower selectivity. For both good coarse-particle recovery and selectivity, PO frothers should be used at low dosage, with low collector dosage as well. The PO-alcohol adduct frothers are even more effective for coarse-particle recovery and need to be used at even lower dosages.

- The optimal frother for high recovery with good selectivity will often be a blend of members of the various frother classes examined. It is reported that such frother blending will give enough benefit to be worth the effort in approximately half of all coal flotation operations.
- None of the frothers in the three categories studied will change the shape of the grade/recovery curve. Changes in frother type and dosage simply move the flotation results along the curve. Similarly, changes in hydrocarbon collector dosage also mainly move the performance along the grade/recovery curve.
- For medium and coarse coal size fractions, the total gangue recovered is linearly related to the total coal recovered. It is only for the finest particles that the gangue recovery increases nonlinearly with increasing coal recovery.
- When floating coals with a broad particle size range, the majority of the gangue reaching the froth is from the finer particle size fractions.
- As the rate of coal flotation increases, the rate of gangue flotation increases proportionately. This is typical of a froth entrainment process acting on the gangue.

The effect of frother addition on the bubble size depends on the nature of the frother, but some representative frothers are shown in Figure 7.3.

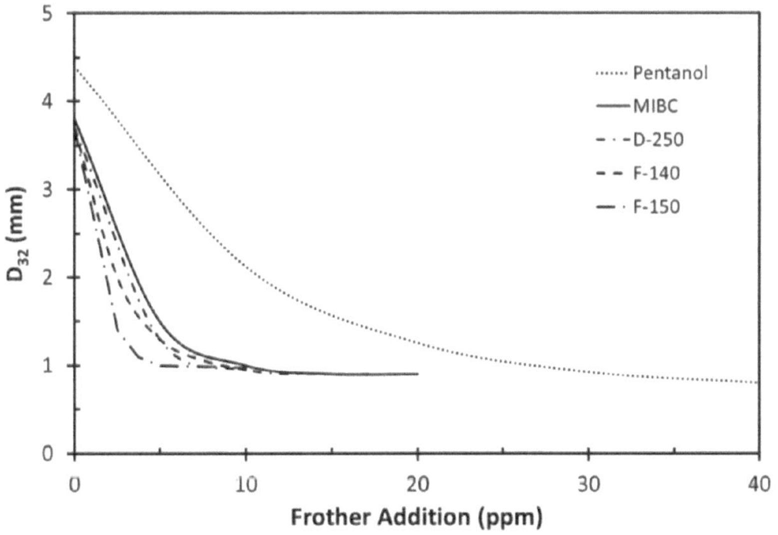

FIGURE 7.3 Bubble size versus amount of frother added for some common frothers used in mineral processing. Based on data from Finch et al. (2008).

7.1.1.2 Coal Collectors

Some coals can be floated without a collector, particularly freshly mined, unoxidized bituminous coals; however, a collector will generally improve the flotation performance for any coal. Many coals that have become oxidized, or are either of very high or very low rank, will not float at all without a collector. The surface of unoxidized coal is naturally hydrophobic because coal is mainly composed of nonpolar hydrocarbons. Since water is a polar liquid, it has little tendency to wet nonpolar materials such as coal. Oils, on the other hand, have a strong affinity for nonpolar surfaces and wet them easily. The various ash minerals, such as silicates and clays, are composed of strongly polar compounds, and therefore water wets them but oils do not. This basic difference in structure is then responsible for the ease of cleaning coal by froth flotation. The most floatable coals will therefore be those which are most nonpolar, which are those that are high in carbon and hydrogen and low in oxygen and moisture. The surface chemistry is also of importance in other areas, such as production of coal/water fuels. Flowable slurries can be produced using strongly hydrophobic coals that are more than 70% solids, but hydrophilic coals require more water to be fluid, and so are much less attractive for this application.

Coalification is the process of conversion of organic material into coal, and coal rank is the measure of the extent of coalification. These terms are therefore synonymous with progressive enrichment of coal in organically bound carbon. The chemical composition of coal therefore varies with changing rank, with the low-rank coals containing less carbon in aromatic or aliphatic forms and more carbon in etheric forms, as summarized in Figure 7.4 (Laskowski, 1987, 1991, 1992, 1995; Laskowski and Miller, 1984; Laskowski and Romero, 1995). Of most importance to coal surface chemistry is the oxygen-containing groups in the coal, which tend to be hydrophilic.

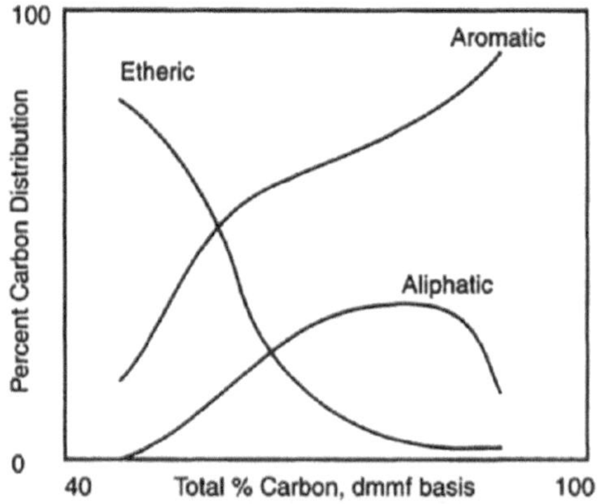

FIGURE 7.4 The distribution of carbon atom phases in coal as a function of the coal rank. Low-rank coals primarily consist of etheric phases, while high-rank coals primarily consist of aromatic phases. These phases determine the coal's overall hydrophobicity (after Whitehurst et al., 1980). Percent carbon, expressed on a dry, mineral matter free (dmmf) basis, is one of the several measures of coal rank. Etheric carbon is bonded to oxygen atoms and is more hydrophilic than aromatic or aliphatic carbon.

A coal surface that is free of oxygen-containing functional groups would therefore be expected to be more hydrophobic than coal surfaces that contain such groups. This has been confirmed by the determination of "contact angles," which are a measure of hydrophobicity. The contact angle is the angle made between a solid surface immersed in a liquid and the surface of an attached bubble. A strongly hydrophobic surface will have a large contact angle, while a hydrophilic surface will have a contact angle of 0°. An ideal coal's surface, with no oxygen-containing groups, would be expected to have a water contact angle of 100°, but real coals generally do not exceed 70° and are always much lower than 100° (Laskowski, 1995; He and Laskowski, 1992). While this lower-than-theoretical contact angle is partly due to porosity effects, the presence of oxygen functional groups is also a major factor (Xiao et al., 1989). As a result of variations in the content of oxygen-rich functional groups, the wettability of coal varies with changing rank, as shown in Figure 7.5. The wettability (and therefore floatability) of coal is lowest for low-rank coal and increases as the rank increases, reaching a maximum for bituminous coals. As the rank increases further, the hydrophobicity of coal decreases because of the increasing proportion of carbon that is in the less-hydrophobic aromatic form.

The more difficult-to-float coals (anthracites, sub-bituminous coals, etc.) require an oily collector such as fuel oil or kerosene to increase their contact angle and render them floatable. Since hydrocarbon oils are not soluble in water, it is best to emulsify them before adding them to the pulp. This emulsification improves their dispersion throughout the pulp and results in more uniform coating of coal particles and better flotation performance. Emulsification can be accomplished either mechanically or by the addition of surfactants (Tyurnikova and Naumov, 1981; Laskowski, 1986, 1995).

Mechanical emulsification of collectors is beneficial because it increases the available surface area of the oily collector and improves the uniformity of reagent

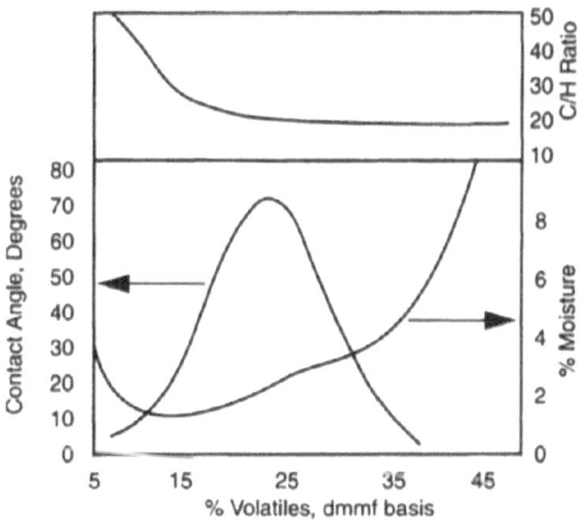

FIGURE 7.5 Effect of coal rank on wettability by water. Higher contact angles indicate that the coal is more difficult to wet, and therefore, more hydrophobic and more amenable to flotation (after Klassen, 1963).

distribution over the coal surface. If this is not done, then relatively large droplets will collect a small amount of coal and be less effective than they would be if better distributed. If only mechanical emulsification is used, then the oil droplets can re-coalesce, so emulsifying agents can be used to prevent coalescence of the particles. Many frothers are effective emulsifying agents for oily collectors; therefore, it is not necessary to add a separate emulsifying agent in addition to the collector and frother addition. The emulsifying agents can also be used to bring the Zeta potentials of the coal and collector droplet surfaces closer to zero, so that they will not repel each other electrostatically. This reduction in Zeta potential tends to increase the rate of collector/coal interaction and improves the collector performance.

An interesting alternative method for addition of oily collectors is as an aerosol in the air used to produce bubbles in the flotation cell (Laskowski and Miller, 1984). If the collector is sprayed into the flotation air supply, then the bubbles produced in the cell will be coated by the oily collector. This is reported to decrease the bubble attachment time from greater than 100 ms to less than 5 ms and also increases the contact angle, which indicates that the bubble is more firmly attached to the coal surface. This effect increases the flotation yield and also results in an increased flotation rate and a more consistent, uniform froth. This can be seen in the results shown in Figure 7.6.

The effectiveness of a collector can sometimes also be improved by stage addition of the reagent in combination with collector emulsification. Laskowski and Poling (1995) modified a six-cell plant circuit that had originally been using a single addition of unemulsified kerosene to the coal before it entered the flotation bank. This addition scheme produced the performance shown in Figure 7.7. From these results, it can be seen that most of the flotation was occurring in the first three cells, with very little flotation in the subsequent cells. The total concentrate yield was 83%. The circuit was then modified so that reagents were added in two stages and were

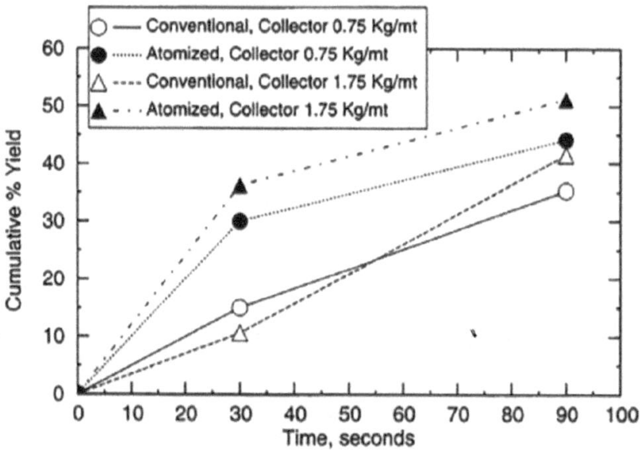

FIGURE 7.6 Effect of adding atomized collector to the flotation air during froth flotation of a Beaver Creek Seam coal. Frother was MIBC, added at a constant dosage of 0.5 kg/metric ton. Atomized oily collector produced a significant increase in the flotation rate.

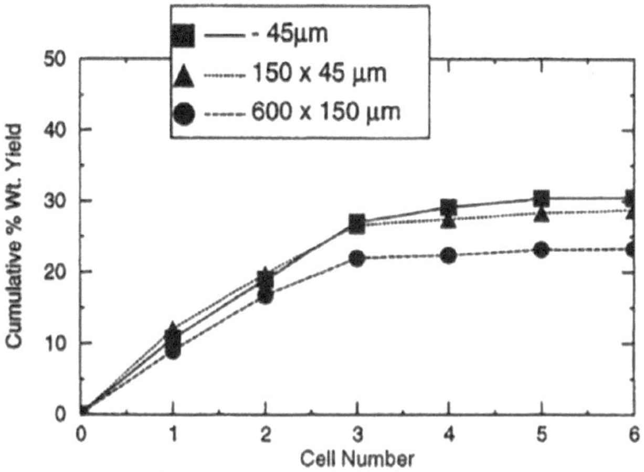

FIGURE 7.7 Cumulative yield of various size fractions of coal in the original plant circuit, using a single addition of unemulsified collector. Total concentrate yield was 83% (Laskowski and Poling, 1995).

emulsified before addition. The first addition was 71% of the kerosene and 66% of the MIBC before entering cell number 1, and the second addition was 29% of the kerosene and 34% of the MIBC at cell number 3. With the modified reagent addition scheme, the results shown in Figure 7.8 were produced. In this configuration, the coal was floating more uniformly from all of the cells, which improved the circuit utilization and increased the concentrate yield to 88.2%.

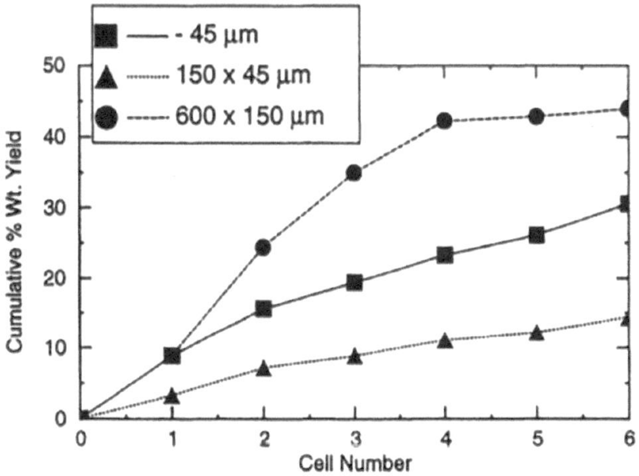

FIGURE 7.8 Cumulative yield of various size fractions of coal in the plant circuit after modification for two-stage addition of emulsified reagents. Total concentrate yield was 88.2% (Laskowski and Poling, 1995).

Coal has an unfortunate tendency to oxidize slowly when it is exposed to air, even at ambient temperatures. This oxidation is most pronounced for lower-rank coals (Samylin et al., 1993). One common sequence of the oxidation takes place as follows (Tyurnikova and Naumov, 1981):

$$\text{Free radicals}\left(O_2^{\,-}\right) \Rightarrow \text{Peroxide group}\,(-OO-) \Rightarrow \text{Hydroxyl Group}\,(-OH)$$

$$\Rightarrow \text{Carbonyl Groups}\,(-CO) \Rightarrow \text{Carboxyl group}\,(-COOH)$$

As coal oxidizes, it accumulates phenolic, carboxylic, and carbonic groups on its surface (Fuerstenau et al., 1983, 1987, 1990; Xiao et al., 1989; Aplan and Arnold, 1991; Wen and Sun, 1981). Since these groups are strongly polar, they make it easier for water to wet the coal and so reduce the coal's hydrophobicity. Adding oil counters this effect by wetting the less-oxidized regions and bridging over the oxidized areas.

Another problem with coal oxidation is that it produces soluble humic acids, which are highly reactive, amorphous, acidic, predominantly aromatic, chemically complex polyelectrolytes. Humic acids form complexes with metal ions, clays, and aluminosilicate minerals (Wong and Laskowski, 1984). At acidic pH, these acids can act as collectors for the mineral matter, while simultaneously depressing the coal, which can seriously reduce the flotation selectivity (Firth and Nicol, 1980; Laskowski et al., 1986; Liu and Laskowski, 1988). This effect is most pronounced when the ionic strength of the solution is high (Wong and Laskowski, 1984); however, under neutral or alkaline conditions, humic acids tend to lose their affinity for the mineral surfaces (Firth and Nicol, 1980).

The surface chemistry of coal can also be affected by inorganic ions in solution. These ions are released by dissolution of particular minerals in the coal and can adsorb on the coal surface, making it less hydrophobic. The effect of ions increases as the valency of the ions increases. The presence of finely dispersed ash-forming minerals in the coal also tends to make the coal less hydrophobic (Laskowski and Parfitt, 1989; Laskowski et al., 1992).

Oxidation of pyrite, according to the reaction,

$$FeS_2\ (s) + \frac{7}{2}O_2 + H_2O \rightarrow Fe^{2+} + 2SO_4^{\,2-} + 2H^+ \tag{7.2}$$

can also affect coal flotation. In raw coals that contain significant amounts of clay minerals, coal recovery drops as the oxidation of pyrite releases acid that reduces the pH. This decrease in coal recovery is believed to be due to increased adsorption of humic acids by the clay particles at acidic pH. These humic acids cause the clay to adsorb a larger proportion of neutral oil collectors, which results in the coal being starved of collector and floating less readily. This effect is not observed in cleaned coals that do not contain clay minerals (Firth and Nicol, 1984).

Pure simple hydrocarbons, composed of a single type of molecule, have been shown to be rather inefficient collectors. Hydrocarbons containing more than eight carbons, and particularly cyclic hydrocarbons such as decalin, are preferable to shorter-chain hydrocarbons. The compounds that show the greatest flotation activity are those that boil in the temperature range of 200°C–250°C, which have an average molecular weight of 185 and have between 13 and 15 carbon atoms per molecule. In general, unsaturated hydrocarbons are reported to be the most effective, with the effectiveness of hydrocarbon collectors decreasing in the following order (Laskowski, 1995; Tyumikova and Naumov, 1981):

$$(\text{Most effective})\ \text{Unsaturated} \Rightarrow \text{Aromatic} \Rightarrow \text{Naphthalenic}$$

$$\Rightarrow \text{Paraffinic}\ (\text{Least effective})$$

The most popular collector is No. 2 fuel oil, which is both low cost and appears to be generally applicable to a wide range of coals. This oil works well as a general-purpose coal collector because it is naturally composed of a mixture of many different hydrocarbon chain lengths and structures and has a low enough viscosity to spread easily over the coal surfaces. Many other oils, however, either alone or in combination with No. 2 fuel oil, give superior results for particular coals. By adding No. 6 fuel oil to No. 2 fuel oil, the flotation of lower rank and/or oxidized coals can be improved. Also, selected carboxylates and amines improve the ability of fuel oil to collect oxidized coals, as do aromatic reagents such as cresols and phenols.

The advantage of adding No. 6 fuel oil for oxidized coals likely stems from a greater concentration of unsaturated, aromatic, or partially oxidized carbon chains. The similarity in the chemical structures is expected the extent to which the collector can coat the coal.

Alkanolamides are also useful reagents for improving flotation of oxidized coals, as they have been demonstrated to produce sometimes dramatic increases in clean coal recoveries, while simultaneously reducing reagent usage (Nimerick and Scott, 1980). A wide range of "promoters" have been sold commercially, which include these and related proprietary compounds, as well as emulsifying agents, to improve the flotation of oxidized or poorly floatable coals. Rubinshtein and Yutifenko (1993) have noted that emulsification of the oils can boost coal yield while improving mineral matter rejection and still reduce reagent consumption. It has also been reported that at low dosage levels, anionic polymers such as carboxymethyl cellulose can promote coal flotation while reducing flotation of mineral matter, and that sodium sulfide can increase the coal hydrophobicity (Tyurnikova and Naumov, 1981).

Xanthates, which are popular collectors for sulfide minerals, are not suitable collectors for coal. Any collection action attributed to xanthates is a result of their oxidation to form dixanthogen, which is a powerful collector for all coals of higher rank than lignite (Horsley et al., 1952). Klassen (1953) has also pointed out that xanthates do not interact with the carbonaceous portion of the coal, but that they do interact with any sulfur that is present.

In addition to more conventional reagents, laboratory studies have shown that coal particles have a higher affinity for carbon dioxide bubbles than for air bubbles. When carbon dioxide is used as the flotation gas, the coal yield and flotation rate are improved, and the rejection of pyrite and ash minerals is also increased (Jin, 1989). The expense of adding carbon dioxide to a flotation cell has kept this process from being used industrially.

Although many improvements to flotation collectors have been developed in the laboratory, these are only rarely translated into plant applications. This is largely due to the great differences between laboratory- and plant-scale flotations. For example, in a study of various collectors and promoters on a plant scale, it was determined that once oil dispersion and adsorption were adequate, the coal recovery was largely dependent on froth stability (Franzidis and Anderson, 1990).

7.1.2 Surface Chemistry of Coal Pyrite

It is often reported that flotation is ineffective for removing pyritic sulfur from coal, and it has been widely claimed in the literature that this is due to hydrophobicity of the pyrite. Pyrite has been found to interact with the compounds that are commonly used as coal collectors (Janczuk et al., 1989), which could conceivably result in the pyrite being recovered along with the coal. Based on this assumption, many investigators have developed depressants that are supposed to prevent pyrite flotation. None of these depressants have ever been successfully used on an industrial scale (Kawatra et al., 1991). The natural question that then arises is whether the pyrite is truly hydrophobic enough to float in the first place? To determine this, a series of experiments were carried out at Michigan Technological University (MTU) (Kawatra et al., 1991; Kawatra and Eisele, 1992a, b; Kawatra and Eisele, 1997).

Timed flotation studies were carried out at MTU with a Pittsburgh seam coal, containing 2.9% total sulfur, 2.2% pyritic sulfur, and 37.2% ash. The coal was ground to 80% passing 36 μm, and was floated at a pH of 7.6 and 5.8% solids. The frother was Dowfroth 200, added at a rate of 0.03 g/L (0.5 kg/metric ton of solids), and the collector was No. 2 fuel oil added at dosages ranging from 0.0 to 0.7 kg/metric ton. Timed froth increments were collected for a total of 9 min. When analyzed, the products from these tests showed no significant hydrophobic flotation of pyrite. The pyrite reaching the froth could all be adequately accounted for by other recovery mechanisms (flotation of locked particles and entrainment of fine pyrite). These experiments are described in more detail later in this chapter (see Section 7.2.2.2). This indicates that for fresh unoxidized coal, pyrite is not naturally hydrophobic, and so pyrite depressants will be of no benefit for such coals.

Like coal, pyrite is slowly oxidized by air. In the case of pyrite, however, the oxidation does not necessarily cause the surface to become hydrophilic. In pyrite oxidation, it is possible to form elemental sulfur on the pyrite surface (Chou, 1990; Stock and Wolny, 1990; Shi et al., 1997). Smith et al. (1968) and Stokes (1901) found that elemental sulfur was produced in the initial stages of oxidation of freshly prepared pyrite, but that in highly oxidized sulfide mineral deposits, little or no elemental sulfur was found (Kinkel et al., 1956). It was documented by Bergholm (1955) that elemental sulfur is formed from pyrite at low pH in the presence of ferric iron, and that once a layer of sulfur had formed it could retard further oxidation of the pyrite (Nordstrom, 1982).

There are two basic reactions that lead to elemental sulfur formation from pyrite. The simplest is

$$2Fe^{3+}(aq) + FeS_2(s) \rightarrow {}_3Fe^{2+}(aq) + {}_2S(s) \tag{7.3}$$

This reaction occurs during microbial oxidation of pyrite, and elemental sulfur will be formed when pyrite oxidizes in moist acidic conditions that are needed for pyrite-oxidizing bacteria to grow (Kawatra and Eisele, 1992a, b; Kawatra et al., 1991).

The second reaction that can produce elemental sulfur from pyrite is a reaction with water (Ahlberg et al., 1990; Wadsworth et al., 1992) of the form:

$$FeS_2 + 3H_2O \rightarrow Fe(OH)_3 + 2S^0 + 3H^+ + 3e^- \tag{7.4}$$

In both cases, the elemental sulfur is normally an intermediate product in a series of reactions that ultimately produce sulfate (SO_4^{-2}); however, if conditions are suitable, detectable amounts of elemental sulfur can form on the pyrite surfaces. Cyclic voltammetry and Raman spectroscopy have been used to determine that elemental sulfur forms at oxidation potentials in the range of +0.4 to +1.1 V relative to a standard calomel electrode (SCE). The optimum oxidation potential to form elemental sulfur on mineral pyrite was found to be +0.8 to +0.9 V, while pyrite from an Illinois coal most readily formed elemental sulfur at oxidation potentials of +0.7 to +0.8 V (Wadsworth et al., 1992).

Elemental sulfur, like coal, is a nonpolar material, and it is also hydrophobic. Even before oxidation, pyrite is very close to the borderline between hydrophobic and hydrophilic; therefore, a small accumulation of hydrophobic elemental sulfur can act as a collector for pyrite (Kawatra and Eisele, 1992a). This can be seen in Table 7.1, which shows the change in the flotation recovery for a pyrite that has been oxidized under various conditions. The pyrite is not hydrophobic when it is fresh

TABLE 7.1
Flotation Results for a Mineral Pyrite (from Custer, South Dakota) at Three Levels of Oxidation: Fresh; Aged at –15°C for 1 year; and Oxidized at 100°C for 44 Days (Kawatra and Eisele, 1997)

Treatment	Flotation pH	% Wt Floating
Fresh	8.3	5
Fresh	7.5	4
Fresh	7.5	3
Fresh	2.3	98
Fresh	2.2	98
Fresh	2.0	99
Aged at –15°C	9.0	46
Aged at –15°C	6.8	37
Aged at –15°C	6.0	35
Aged at –15°C	1.9	92
Aged at 100°C	6.2	7
Aged at 100°C	2.0	82

A high weight % floating indicates that the pyrite has become hydrophobic. The reagents used were MIBC frother (0.38 kg/metric ton) and No. 2 fuel oil collector (3.0 kg/metric ton).

or when its surface is thoroughly oxidized, but it is hydrophobic when it is only partially oxidized.

The observed increase in floatability at near-neutral pH, upon oxidation at −15°C, is believed to be due to elemental sulfur formation on the surface, according to the reactions described above. Fresh pyrite apparently has little or no elemental sulfur on its surface, and so it is not naturally hydrophobic. Partial oxidation under the proper condition leads to formation of elemental sulfur, and the pyrite becomes hydrophobic. The hydrophobicity of the pyrite will be greatest when conditions encourage the dissolution of any iron hydroxides from the pyrite surface (Ahlberg et al., 1990; Kocabag et al., 1990). More complete oxidation of the surface converts the elemental sulfur to sulfates, which are hydrophilic, and the hydrophobicity of the pyrite is lost when the surface is more completely oxidized. This hypothesis also accounts for the results reported by Gayle et al. (1965), who noted that rejection of pyrite from coal by flotation became poorer when coal was briefly oxidized, but that the rejection then improved upon further oxidation.

Table 7.1 also shows that the hydrophobicity of mineral pyrite is very strongly pH dependent. This is due to the dissolution and reprecipitation of iron salts from the pyrite surface. When the pH is less than 4, iron salts are very soluble. Oxidized pyrite has on its surface both iron sulfate (which is hydrophilic) and elemental sulfur (which is hydrophobic). Under acid conditions, the iron sulfate dissolves, leaving only the hydrophobic elemental sulfur on the surface. As a result, pyrite becomes floatable when the pH is lowered below 4.

It should be noted that the work described in Table 7.1 was carried out using mineral pyrite. It has been observed that oxidation rates of mineral pyrite can be quite different from that of coal pyrite. For example, Lai et al. (1990) noted that pyrite collected from an Ohio coal oxidized more rapidly than mineral pyrite collected from Colorado.

Coal flotation is normally carried out at a pH between 6 and 8 because this is the range where coal is the most floatable (Zimmerman, 1948). This is also the range where pyrite is the least likely to be hydrophobic because the iron sulfates that form on its surface cannot dissolve easily under these conditions (Liu et al., 1994). In the majority of cases, therefore, froth flotation will be capable of removing pyrite from coal, and in fact, a large fraction of pyrite is frequently removed from the coal in the froth flotation stage (Monostory and Kubitza, 1989). While it should be kept in mind that there are some conditions where the pyrite will float along with the coal at neutral pH, this is more of an occasional problem to watch for than a universal occurrence (Kawatra et al., 1991; Kawatra and Eisele, 1992a, b).

Even though the hydrophobicity of pyrite particles is only a problem for a few coals in special circumstances, a great deal of work has been devoted to finding chemicals that will depress pyrite. These chemicals are intended to either coat or chemically attack the pyrite surface and make these particles hydrophilic without affecting the floatability of the coal particles (Luo et al., 1993; Ye et al., 1990; Purcell and Aplan, 1991). Some representative pyrite depressants that have been reported in the coal flotation literature are listed in Table 7.2. These chemicals are intended to work by a wide range of processes, including both physical and chemical adsorption, and both oxidation and reduction of the pyrite surface. In laboratory testing for particular coals, however, these depressants frequently do not perform well enough

TABLE 7.2
Some Representative Depressants for Coal Pyrite That Have Been Reported in the Literature

Reagent	Action	Reference
Alkali	pH	Liu et al. (1993)
Lime, CaO	pH, hydrolyzed ion adsorption	Chander and Aplan (1989)
Potassium dichromate, $K_2Cr_2O_7$	Oxidizing agent	Chander and Aplan (1989)
Sodium hypochlorite, NaClO	Oxidizing agent	Chander and Aplan (1989)
Sodium sulfide, Na_2S	Reducing agent	Chander and Aplan (1989)
Sodium sulfite, Na_2SO_3	Reducing agent	Chander and Aplan (1989)
Bisulfites, HSO_3^-	Reducing agent	Mu et al. (2016)
Metabisulfites, $S_2O_5^{2-}$	Reducing agent	Mu et al. (2016)
Sulfur dioxide, SO_2	Reducing agent	Mu et al. (2016)
Corn starch	Physically adsorbed colloid	Chander and Aplan (1989)
High-amylose starch	Physically adsorbed colloid	Chander and Aplan (1989)
Cationic starch	Physically adsorbed colloid	Chander and Aplan (1989)
Xanthated cationic starch	Physically adsorbed colloid	Chander and Aplan (1989)
Dextrin	Physically adsorbed colloid	Mu et al. (2016)
Guar gum	Physically adsorbed colloid	Mu et al. (2016)
Carboxymethyl cellulose	Physically adsorbed colloid	Mu et al. (2016)
Chitosan	Chelating agent	Mu et al. (2016)
Diethylenetriamine	Chelating agent	Mu et al. (2016)
Modified lignosulfonates	Chelating and dispersing agent	Mu et al. (2016)
Polyacrylamides	Flocculant and adsorbed colloid	Mu et al. (2016)
Congo Red	Dye	Chander and Aplan (1989)
Nigrosine	Dye, chelating agent	Raleigh and Aplan (1990,1993)
Sodium silicate	Dispersing agent	Raleigh and Aplan (1990,1993)
Aerosol MA-80	Dispersing agent	Raleigh and Aplan (1990,1993)
Citric acid	pH, complexing agent	Raleigh and Aplan (1990,1993)
Marasperse CB	Dispersing agent	Raleigh and Aplan (1990,1993)
Potassium hypochlorite, KClO	Oxidation	Raleigh and Aplan (1990,1993)
Ferric chloride, $FeCl_3$	Hydrolyzed ion adsorption	Baker and Miller (1971)
Aluminum chloride, $AlCl_3$	Hydrolyzed ion adsorption	Baker and Miller (1971)
Chromium chloride, $CrCl_3$	Hydrolyzed ion adsorption	Baker and Miller (1971)
Copper sulfate, $CuSO_4$	Hydrolyzed ion adsorption	Baker and Miller (1971)
Calcium chloride, $CaCl_2$	Hydrolyzed ion adsorption	Baker and Miller (1971)
Sodium cyanide, NaCN	Complexing agent	Yancey and Taylor (1935)
Potassium cyanide, KCN	Complexing agent	Moslemi and Gharabaghi (2017)
Ferrous sulfate, $FeSO_4$	Hydrolyzed ion adsorption	Yancey and Taylor (1935)
Ferric sulfate, $Fe_2(SO_4)_3$	Hydrolyzed ion adsorption	Yancey and Taylor (1935)
Para-aminophenol	Complexing agent	Yancey and Taylor (1935)
Carboxymethyl cellulose	Physically adsorbed colloid	Laskowski et al. (1985)

(Continued)

TABLE 7.2 (*Continued*)
Some Representative Depressants for Coal Pyrite That Have Been Reported in the Literature

Reagent	Action	Reference
Thiobacillus ferrooxidans	Oxidation, surface attachment	Elzeky and Attia (1987)
Other microorganisms	Surface attachment	Townsley and Atkins (1986)
Potassium monopersulfate, KSO_5	Oxidation	Miller et al. (1989), Ye et al. (1990)
Cato 2	Dispersing agent	Arnold and Aplan (1990)
Hylon VII	Dispersing agent	Arnold and Aplan (1990)
Aerosol OT	Dispersing agent	Arnold and Aplan (1990)
Quebracho	Dispersing agent	Arnold and Aplan (1990)
Daxad II KLS	Dispersing agent	Arnold and Aplan (1990)
Tergitol NPX	Dispersing agent	Arnold and Aplan (1990)
Sodium thiosulfate, $Na_2S_2O_4$	Reducing agent	Arnold and Aplan (1990)
Thioglycolic acid, $HSCH_2COOH$	Complexing agent	Chmielewski and Wheelock (1991)
Methanol	Surface tension modifier	Feeley (1991)
Ethanol	Surface tension modifier	Feeley (1991)
Butylbenzaldehyde	Surface tension modifier	Feeley (1991)
Fe^{+2} + Hylon VII	Surface activator + Colloid coating	Xu and Aplan (1994)

None of these are widely used as pyrite depressants in industrial coal flotation.

to warrant larger-scale testing (Kawatra et al., 1991). In practice, there are no coal flotation plants that actually use pyrite depressants because they have never been demonstrated to be beneficial enough to warrant the cost of their use.

In addition to chemical depressants, a number of bacteria and other biological products have also been claimed to be effective depressants for mineral pyrite (Atkins, 1990; Eisele and Kawatra, 1993a, 1994; Kawatra et al., 1989). Further experiments showed that these depressants were much less effective for coal pyrite than for mineral pyrite (Eisele and Kawatra, 1993a, 1994). The effects of bacteria in froth flotation are further discussed in Chapter 14.

7.1.2.1 Reasons for the Lack of Industrial Use of Pyrite Depressants

It should be noted that while many depressants are available, only lime is actually used in industrial coal flotation circuits, and it is used mainly for plant water pH control rather than for pyrite depression. There are several reasons for this lack of depressant use:

- Most of the time, the pyrite in coal will not be hydrophobic in the first place, and adding pyrite depressants will have no effect. This is demonstrated in a case study later in this chapter (see Section 7.2.2.2). While depressants

can conceivably help when the conditions are right for pyrite flotation, adding them will normally only be a precautionary measure. Since adding depressants can more than double the total reagent consumption of coal flotation, the occasional possible benefits are not enough to justify the continuous cost.
- Coal flotation is not the primary cleaning operation in any existing coal plants and typically accounts for only about 15% of the total clean coal produced. Thus, even when the flotation product is high in sulfur, the effect on the sulfur content of the entire plant product is usually tolerable. Plant managers are therefore unwilling to spend a great deal of time, effort, and money experimenting with chemicals that the process does not actually need, especially when the benefit will be minor.
- Even when the pyrite in coal is hydrophobic, the fraction of liberated pyrite being recovered by bubble attachment is small compared to the amount that is recovered by other mechanisms, such as locked particles and entrainment. These other mechanisms are described in Section 7.2. The potential benefits of pyrite depression are, even in theory, quite small in comparison.

7.2 PHYSICAL CONSIDERATIONS

In addition to the chemical effects acting in flotation, there are also physical effects that can cause particles to be misplaced into the wrong product. The importance of these effects can easily be as great as many of the chemical effects, and they should not be ignored. In fact, Zimmerman (1948) pointed out as early as 1948 that the greatest single factor interfering with removing pyritic material was that of mechanical entrapment in the froth itself due to the large bulk of coal being floated off, and that the best and the most effective method was to re-clean the froth by one or more passes through the machine rather than using depressant reagents. Miller (1963) also concluded that three stages of flotation was the best removal method in a series of tests to remove pyritic sulfur by froth flotation. He also pointed out that froth sprinkling has the same effect on flotation selectivity as the addition of another cleaning stage. To a large extent, it is these physical factors that place the limits on the ability of a given froth flotation circuit to produce clean coal.

7.2.1 Entrainment

In a conventional flotation machine, there are two ways that a particle can reach the froth layer. It can be carried into the froth by attachment to an air bubble (true flotation), or it can be suspended in the water trapped between the bubbles (entrainment). This is shown schematically in Figure 7.9. While true flotation is selective between hydrophobic and hydrophilic particles, entrainment is nonselective, and so entrained particles are just as likely to be ash minerals or pyrite as they are to be coal. In order for flotation to be as selective as possible, the entrainment must be kept to a minimum.

If particles are sufficiently coarse, then they settle rapidly enough that they are not carried into the froth by entrainment. As they become finer, particles settle more slowly and have more time to be entrapped in the froth and have less tendency to

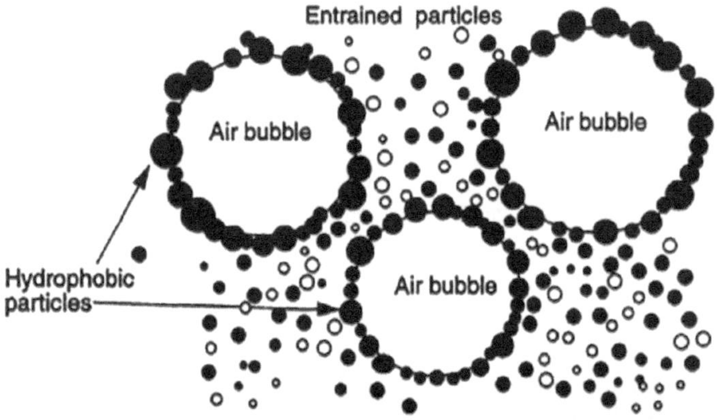

FIGURE 7.9 Entrainment versus bubble attachment. Hydrophobic particles selectively attach to air bubbles, while entrained particles are unselectively carried into the froth. Fine particles are entrained more than coarse particles because they settle more slowly and are therefore more easily carried upward.

drain away. Clay particles in particular, which are only a few micrometers in size, are very easily entrained. For particles that are less than a few micrometers in size, their rate of recovery into the froth by entrainment is equal to the rate of recovery of water into the froth. For example, if 20% of the water entering a flotation cell is carried into the froth, then up to 20% of the fine particles entering the cell will be entrained. The entrainment of coarser particles will be less than 20%, due to their greater ability to drain from the froth.

In addition to clay slimes, a large proportion of pyrite also reaches the clean coal product by entrainment. This is a particular problem for the finest fraction of the coal, where particles are smaller than about 25 μm. The very fine pyrite particles, even if well-liberated from the coal, are still heavily entrained into the froth. As a result, grinding coal to a fine particle size to ensure liberation of the pyrite inclusions is not as beneficial as one would think because the fine grinding also increases the proportion of entrained pyrite.

7.2.1.1 Determining Entrainment Levels

Determining the precise amount of entrainment in froth flotation can be difficult, because it can vary greatly depending on the following factors (Ross, 1990a, b):

- froth depth and flowrate;
- particle size, density, and shape;
- reagent types and dosages;
- pulp viscosity.

Whenever the settling rate of the particles decreases, or the amount of water in the froth increases, or the residence time of material in the froth decreases, then the level of entrainment will increase. If the particles of a particular type are completely hydrophilic and all of the above factors are held constant, then the recovery of these particles into the froth will be entirely by entrainment and will be linearly related to the water recovery. When this is true, the particle recovery by entrainment can be determined precisely. If the particles are even partially hydrophobic or locked to hydrophobic particles, however, or any of the listed factors are changing in the course of the flotation experiments, then the best that can be done is a rough approximation of the entrainment level.

There are a number of methods that have been proposed for measuring entrainment, which vary a good deal in both difficulty and accuracy. Samples for determining entrainment can be either collected in the plant or with a laboratory arrangement as follows:

- *In-plant sampling.* To determine entrainment in a plant, it is best if the flotation circuit consists of banks of five or more discrete cells that can be sampled individually. If possible, each cell should also have the same froth depth, as changes in froth depth can alter the entrainment behavior. Froth samples should be collected by carefully cutting the entire froth overflow of each cell in such a way that the mass balance for the flotation bank can be calculated, and the total mass flowrate of the froth from each cell can be determined. The parameters that need to be measured or known are (1) froth flowrate, (2) quantity of water in the froth from each cell, (3) quantity of each solid species of interest (coal, clay minerals, pyrite) in the froth from each cell, (4) feed slurry flowrate, (5) feed slurry composition, (6) volume of each cell in the bank (for calculating residence times), and (7) composition and flowrate of the final tailings slurry.
- *Laboratory experiments.* Laboratory experiments are most easily run as batch tests, and water is continually removed from the cell in the course of the experiment. This means that addition of makeup water is needed to ensure that the froth depth does not change in the course of the test. The volume of froth produced in coal flotation is very high, because most of the solid particles report to the froth. Because of this, a great deal of water can be carried out of the cell in a short time; therefore, the cell must be designed to very rapidly replace this water. For entrainment studies, a cell similar to the one shown in Figure 7.10 should be constructed so that water can be replaced as fast as it is removed. The test should be run with the froth collected over at least five specific time intervals, with the test continuing until all of the hydrophobic material has been collected (generally 5–10 min). Products should be analyzed to determine water recovery and composition of the solid products. A problem with this type of laboratory setup is that the continuous addition of freshwater constantly dilutes the solids remaining in the cell. Since calculation of the degree of entrainment depends on correlating the recovery of solids with the recovery of water, this dilution must be allowed for. If it is assumed that the flotation cell is well-mixed, then a correction for the dilution water can be made using the following formula (Dankwerts, 1953):

$$F(t) = 1 - \exp\left(\frac{-R}{V}\right) \tag{7.5}$$

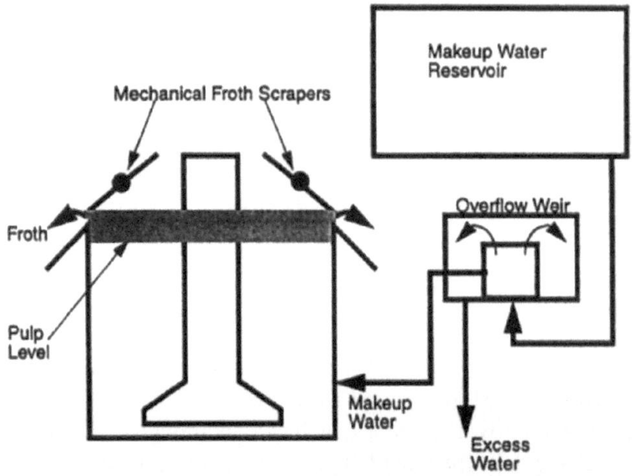

FIGURE 7.10 Laboratory flotation cell for measuring entrainment and flotation kinetics. As water is removed in the froth, makeup water flows into the cell from the overflow weir, until the pulp level is restored to its original value. To change the froth depth between tests, the height of the overflow weir is changed to adjust the pulp level setpoint. The makeup water should contain frother at the same concentration as the pulp in the cell at the beginning of the test, as otherwise some peculiar effects can occur.

where $F(t)$ = the volume fraction of the water originally in the cell at the beginning of the test, which is recovered into the froth by time t,

R = total volume of water that is removed from the cell and replaced by makeup water, by time t,

V = working volume of the flotation cell.

Once this is calculated, then the corrected water recovery, $F(t)$, can be used for calculations of entrainment.

If for some reason the flotation cell cannot be treated as a well-mixed system, then similar results can be obtained by using a water tracer such as potassium chloride or sodium fluorescein dye. The tracer is added to the cell, but not to the makeup water, at a known concentration. The decreasing concentration of tracer in the froth water in the course of the test is then used to correct for the makeup water addition.

Once experiments have been run to collect the necessary data, there are a number of ways that the data can be analyzed to determine the entrainment level. The various analysis methods sometimes give different answers, and so the results should be used with caution. The available methods are as follows:

7.2.1.1.1 Sample Correlation (Lynch et al. 1981)

This simply consists of measuring the percentage of the feed water that is recovered into the froth and plotting it against the percentage of each solid component that is recovered into the froth. Purely entrained particles will follow a straight line in this

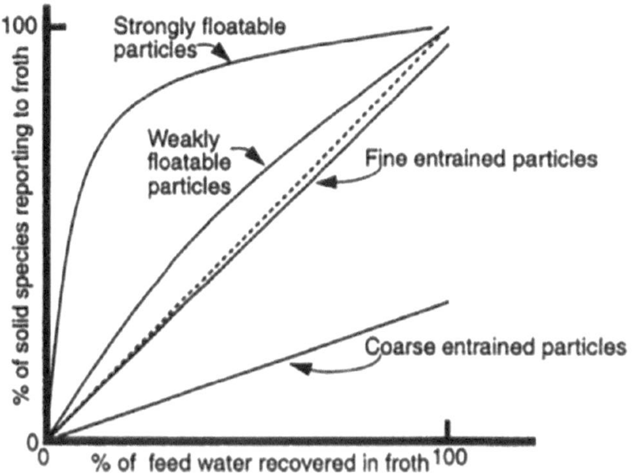

FIGURE 7.11 Results from simple correlation analysis of entrainment. The percent recovery of very fine entrained particles into the froth will be almost identical to the percent recovery of water. Recoveries higher than that of the feed water indicate true, hydrophobic flotation, while recoveries lower than that of the feed water indicate particles coarse enough to sediment out of the froth. A perfectly linear plot indicates pure entrainment, while a curved plot indicates a contribution from true flotation.

plot, and deviations from linearity show that part of the material is being recovered by true flotation. For very fine particles, their percent recovery by entrainment will equal the percent water recovery. For rapid, simple, qualitative determination of entrainment levels, this method will generally be sufficient. Some representative results from this type of analysis are shown in Figure 7.11. This technique is recommended for rapid evaluation and comparison of tests because it gives a clear visual indication of the degree of entrainment with a minimum of calculation and can be used when the data is of somewhat variable quality.

7.2.1.1.2 Trahar's Method (Trahar, 1981)

In this method, duplicate samples are floated with and without collector. It is then assumed that the material that floats without collector is the amount being entrained, and that the difference in flotation between the two tests is the amount recovered by true flotation, as shown in Figure 7.12. This will obviously not work with naturally hydrophobic particles such as coal, because the coal will float in the absence of collector. It will usually be suitable for pyrite particles and ash-forming minerals. This method is generally limited to laboratory flotation experiments because applying it to an operating plant requires that the collector addition be turned off for a period, which will disrupt operations. Another problem with Trahar's method is that the presence of collector affects the characteristics of the froth, and so the entrainment with collector is not necessarily the same as the entrainment without collector. This method is therefore not recommended.

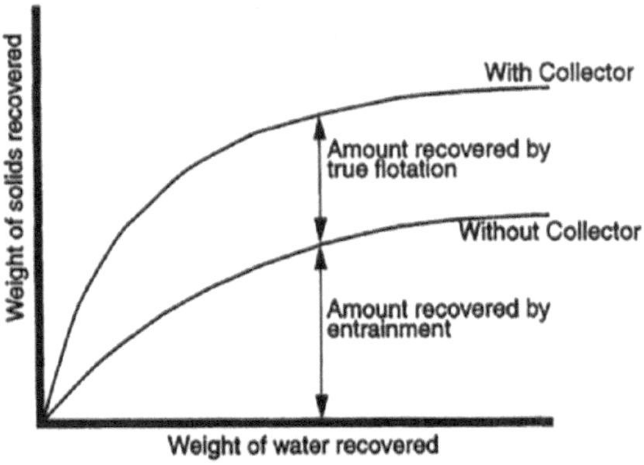

FIGURE 7.12 Determination of true flotation and entrainment by Trahar's method. This method assumes that in the absence of collector, the only particles reaching the froth are entrained. It is therefore not suitable for the hydrophobic species, such as coal, but will be suitable for the ash and pyrite components.

7.2.1.1.3 Warren's Method (Warren, 1985)

This method is more accurate than the previous ones because it allows for changes in froth depth, but requires a much greater amount of data. Numerous timed tests are run, with the froth depth varied from test to test to change the relative amounts of solids and water in the froth. For each data point, the solids recovery is plotted against water recovery, and linear regression is used to calculate the correlation, as shown in Figure 7.13. Extrapolating the regression line to 0% water recovery gives the amount of material that floats by true flotation. This method gives the most reliable results, but it is not recommended for routine work because of the large number of experiments required.

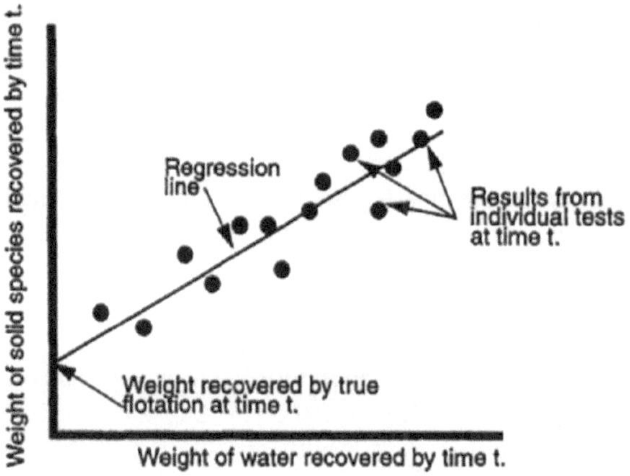

FIGURE 7.13 Determination of true flotation and entrainment by Warren's method. The slope of the regression line is the measure of the degree of entrainment, while the intercept of the regression line with zero water recovery is the mass recovered by true flotation.

7.2.1.1.4 Ross's Method (Ross, 1990a, b)

This technique is quite a bit more mathematically involved than the other techniques, but it only requires a single timed flotation test to be run. It also does not require a correction for added dilution water. It is therefore well-suited for studying how entrainment changes as various froth flotation parameters are varied. This method requires the results of a timed flotation test, which is run past the point where all floatable material has been recovered. From the test results, two dimensionless "transfer functions" are calculated for each time interval as follows:

$$X(t)=\frac{E(t)C_w(t)}{W(t)C_m(t)} \tag{7.6}$$

$$Y(t)=\frac{R(t)C_w(t)}{W(t)C_m(t)} \tag{7.7}$$

where $E(t)$ = Recovery rate of entrained particles of the species of interest into froth at time t in g/min;
$W(t)$ = Recovery rate of water into froth at time t in g/min;
$R(t)$ = Recovery rate of all particles of the species of interest (entrained and floated) into froth at time t in g/min;
$C_w(t)$ = Concentration of water in the cell pulp at time t, in g/mL;
$C_m(t)$ = Concentration of particles of the species of interest in the cell pulp at time t, in g/mL.

The parameter $X(t)$ is a measure of the amount of entrainment, and $Y(t)$ is a measure of the total material floating. To calculate these values, it is first necessary to determine the total mass flowrate of the froth for each time interval and to analyze the products to determine the composition. If makeup water is being added to the cell to maintain a constant pulp level, then the concentration of the solid species of interest in the pulp, $C_m(t)$, can be calculated at any time t from the initial quantities of material, and from the amount of material removed from the cell, provided that the working volume of the cell is known.

The values of $Y(t)$ are then calculated and plotted against time. To calculate $X(t)$, it is necessary to assume that at the end of the test, no hydrophobic material remains in the cell, and particles are being recovered only by entrainment. So, since $X(t)$ is the entrained material floating and $Y(t)$ is the total material floating, at the end of the test $X(t) = Y(t)$. The values of $X(t)$ at earlier times in the test are then determined by extrapolating back from the tail of the $Y(t)$ curve, as shown in Figure 7.14. This obviously requires that the test be run long enough to ensure that no hydrophobic material remains in the cell for at least the last two or three sample intervals.

Ross's method is useful for comparison of multiple tests, particularly when numerical values are needed, but it can only be used reliably when the data is of excellent quality. The primary drawback is that the numerous mathematical operations involved in this method tend to magnify any measurement errors. This method is therefore only recommended when very high-quality data is available.

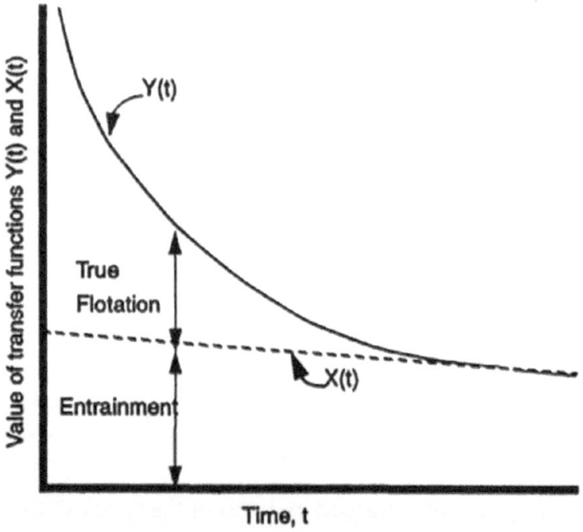

FIGURE 7.14 Determination of true flotation and entrainment by Ross's method. For explanation of symbols, see text. The value of $Y(t)$ at any given time is a measure of the total material floating, and the value of $X(t)$ is a measure of the degree of entrainment at that point in the test.

7.2.1.2 Example Entrainment Calculations Using Ross's Method

Problem Statement: Consider a timed coal flotation test that was carried out at MTU with a Pittsburgh seam coal. The coal characteristics and the flotation conditions were as given in Table 7.3. The flotation cell was equipped with mechanical paddles for froth removal, and makeup water was automatically supplied to maintain a constant level, as was illustrated in Figure 7.10. Analysis of the products from this test

TABLE 7.3
Test Parameters for Timed Flotation of Pittsburgh Seam Coal

Parameter	Value
Percent total sulfur	2.9%
Percent pyritic sulfur	2.3%
Percent ash	37.2%
80% passing size	36 μm
Average specific gravity of solid particles	1.6
Initial pulp percent solids	5.73%
Pulp pH	7.6
Frother type and dosage	Dowfroth 200, 0.03 g/L (0.5 kg/metric ton solids)
Collector type and dosage	No. 2 fuel oil, 0.7 kg/metric ton solids
Froth collection times	0.5, 1,2, 3, 5, and 9 min
Working volume of cell	4,250 mL

TABLE 7.4
Results of Analysis of the Froth Products from a Timed Flotation Test with Pittsburgh Seam Coal

Time Interval	Slurry Weight, g	Dry Weight, g	Water, g	% Py. S.	Pyrite, g
0–0.5 min	958	139.8	818.2	1.4	3.6616
0.5–1 min	369	22.0	347.0	2.8	1.1524
1–2 min	964	17.1	946.9	3.7	1.1837
2–3 min	1,058	13.3	1,044.7	3.3	0.8211
3–5 min	1,914	13.9	1,900.1	3.4	0.8841
5–9 min	3,171	13.7	3,157.3	3.2	0.8202
Final tails	4,268	29.0	4,239.0	4.0	2.1701
Total (calculated)	12,702	248.8	12,453.2	2.297	10.6932

The weight of pyrite (in grams) was calculated from the % pyritic sulfur, by assuming pyrite to be 53.45% sulfur. Water was continually added to the cell to replace water removed in the froth and maintain a constant froth level, and so the total amount of water used was considerably more than the 4,250 mL capacity of the flotation cell.

produced the results given in Table 7.4. Using this data, determine the flowrates into the froth of entrained particles and of the entrained pyrite.

Solution: The following calculations were performed using a spreadsheet, and the values shown were rounded off from the values calculated by the spreadsheet. Using the data in Table 7.4, and given that the average specific gravity of the solid particles was 1.6 and the total cell volume was 4,250 mL, the concentrations of water, total solids, and pyrite at the beginning of the test ($t = 0$) were calculated as follows:

Solids Weight = 248.8 g (total from Table 7.4),
Solids Volume = weight/density = 248.8/1.6= 155.5 mL,
Total Pulp Volume = 4,250 mL (working volume of cell, from Table 7.3),
Water Volume = (total pulp volume) − (solids volume) = 4,250 − 155.5 = 4,094.5 mL, Water Weight = 4,094.5 g,
Pyrite Weight = 10.6932 g (total from Table 7.4),
$C_w(0)$ = (water weight)/(total pulp volume) = 4,094.5/4,250 = 0.9634 g/mL,
(Total Solids) $C_m(0)$ = 248.8g/4250 mL = 0.05854 g/mL,
(Pyrite) $C_m(0)$ = 10.6932g/4250 mL = 0.002515 g/mL.

After 0.5 min had passed, 139.8 g of solids and 3.6616 g of pyrite have been removed from the pulp by being carried into the froth. Since makeup water was added to the cell to maintain a constant volume, the total pulp volume was still 4,250 mL. Assuming that the density of the solids in the cell had not changed significantly, the material remaining in the cell was now

Solids Weight = 248.8 g −139.8 g = 109.0 g,
Solids Volume = weight/density = 109.0/1.6 = 68.12 mL,

Total Pulp Volume = 4,250 mL,
Water Volume = (total pulp volume) – (solids volume) = 4,250 – 68.12 = 4,181.88 mL, Water Weight = 4,181.88 g,
Pyrite Weight = 10.6932 g – 3.6616 g = 7.0316 g,
$C_w(0.5)$ = (water weight)/(total pulp volume) = 4,181.88/4,250 = 0.9840 g/mL,
(Total Solids) $C_m(0.5)$ = 109.0 g/4,250 mL = 0.02565 g/mL,
(Pyrite) $C_w(0.5)$ = 7.0316 g/4,250 mL = 0.001655 g/mL.

The values for $C_w(t)$ and $C_m(t)$ at the end of each of the remaining time intervals were then calculated similarly, with the results given in Table 7.5.

The values for $R(t)$ and $W(t)$ were then estimated as follows. First, the time at the middle of the time interval was determined (for the 0–0.5 min interval, this is 0.25 min). Then, the average flowrates during the interval were calculated by dividing the amount of material recovered in a time interval by the length of the interval, as illustrated below for the 0–0.5 min time interval:

$W(0.25)$ = (grams of water recovered to froth)/(elapsed time) = 818.2 g/0.5 min = 1,636.4 g/min,
(Total Solids) $R(0.25)$ = 139.8 g/0.5 min = 279.6 g/min,
(Pyrite) $R(0.25)$ = 3.6616 g/0.5 min = 7.3231 g/min.

The solids and water concentrations in Table 7.5 are for the beginning and end of time intervals, while the solids and water flowrates are calculated at the center of the time intervals. To determine the concentrations at the center of the time intervals, the beginning and ending concentrations were averaged, as shown:

TABLE 7.5
Calculation of the Concentrations of Water ($C_w(t)$), Total Solids ($C_m(t)$ Total Solids), and Pyrite ($C_m(t)$ Pyrite) at the Beginning and End of Each Time Interval, Using the Data from Table 7.4

Time, min	Solids in Cell, g	Solids in Cell, mL	$C_w(t)$ Total Water	$C_w(t)$ Total Solids	Pyrite in Cell, g	$C_m(t)$ Pyrite
0	248.8	155.50	0.9634	0.05854	10.6932	0.002515
0.5	109.0	68.12	0.9840	0.02565	7.0316	0.001655
1	87.0	54.38	0.9872	0.02047	5.8792	0.001383
2	69.9	43.69	0.9897	0.01645	4.6956	0.001105
3	56.6	35.38	0.9917	0.01332	3.8744	0.000912
5	42.7	26.69	0.9937	0.01005	2.9903	0.000704
9	29.0	18.12	0.9957	0.00682	2.1701	0.000511

All concentrations are in g/mL and are calculated from the amount of material removed from the cell, assuming an average solids specific gravity of 1.6 and a constant slurry volume of 4,250 mL.

$C_w(0.25) = (C_w(0) + C_w(0.5))/2 = (0.9634 + 0.9840)/2 = 0.9737$ g/mL,
(Total Solids) $C_m(0.25) = (0.05854 + 0.02565)/2 = 0.04209$ g/mL,
(Pyrite) $C_m(0.25) = (0.002515 + 0.001655)/2 = 0.002085$ g/mL.

This provides all of the necessary information for calculation of $Y(0.25)$, using Equation 7.7:

$$\text{(Total Solids) } Y(0.25) = \left[R(0.25)C_w(0.25)\right]/\left[W(0.25)C_m(0.25)\right]$$

$$= [(279.6)(0.9737)]/[(1636.4)(0.04209)]$$

$$= 3.9523(\text{dimensionless})$$

$$(\textit{Pyrite})\ Y(0.25) = \left[R(0.25)\ C_w(0.25)\right]/\left[W(0.25)\ C_m(0.25)\right]$$

$$= [(7.32)(0.9737)]/[(1636.4)(0.002085)]$$

$$= 2.0896(\text{dimensionless})$$

These calculations were repeated for each of the time intervals, producing the results shown in Table 7.6. Once the values for $Y(t)$ were determined, they were plotted as shown in Figure 7.15.

The values for $X(t)$ were estimated from Figure 7.15 by assuming that by the end of the test, all of the hydrophobic material had been recovered, and so all particles reaching the froth were entrained particles, and $X(t) = Y(t)$. By extrapolating the line

TABLE 7.6
Values for Calculation of $Y(t)$ for Each Flotation Time Interval, Based on the Data Given in Table 7.4

Time,			Total Solids			Pyrite		
min	$C_w(t)$	$W(t)$	$C_m(t)$	$R(t)$	$Y(t)$	$C_m(t)$	$R(t)$	$Y(t)$
0.25	0.9737	1636.4	0.04209	279.60	3.9520	0.002085	7.3231	2.0900
0.75	0.9856	694.0	0.02306	44.00	2.7100	0.001519	2.3048	2.1550
1.5	0.9885	946.9	0.01846	17.10	0.9670	0.001244	1.1837	0.9932
2.5	0.9907	1044.7	0.01488	13.30	0.8475	0.001008	0.8211	0.7723
4.0	0.9927	950.1	0.01168	6.95	0.6216	0.000808	0.4421	0.5719
7.0	0.9947	789.3	0.00844	3.43	0.5117	0.000607	0.2050	0.4256

The times given are the middle of each of the sampling intervals, and all values are the mean recovery rates during each interval. Concentrations at the interval midpoints are calculated by taking the values at the beginning and end of each interval (Table 7.5) and averaging them. Values of $Y(t)$ are determined using Equation 7.7.

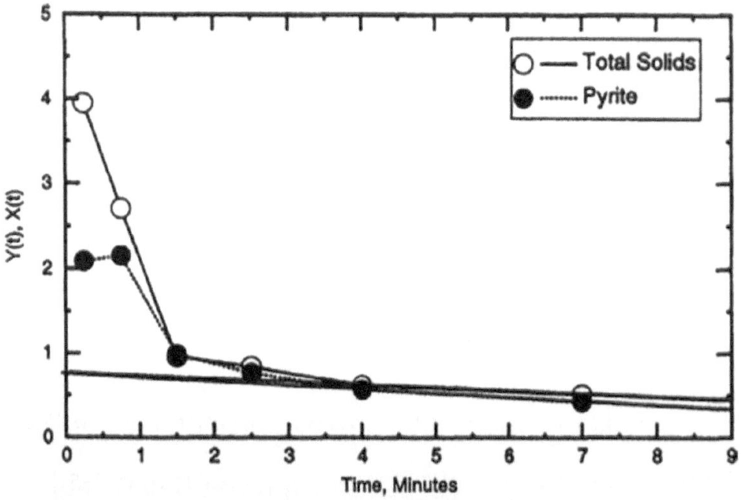

FIGURE 7.15 Example of determination of $X(t)$, for calculation of entrainment by Ross's method. The equations of the extrapolated lines are $X(t) = 0.7682 - 0.0366(t)$ for total solids, and $X(t) = 0.7671 - 0.0488(t)$ for the pyrite.

that passes through the last two data points in the test, the equation of $X(t)$ for the entire test could be determined by linear regression, as illustrated.

Given the equation of the line for $X(t)$, the rate of recovery of entrained material, $E(t)$, could then be calculated from the $X(t)$ values using Equation 7.6, as follows:

A. Total solids at $t = 0.25$ min,
$X(0.25) = 0.7682 - 0.0366(0.25) = 0.7590$,
$E(0.25) = (X(0.25))(W(0.25))(C_m(0.25))/(C_w(0.25))$,
$= (0.7590)(1636.4)(0.04209)/(0.9737) = 53.70$ g/min

B. Pyrite at $t = 0.25$ min,
$X(0.25) = 0.7671 - 0.0488(0.25) = 0.7549$,
$E(0.25) = (0.7549)(1636.4)(0.002085)/(0.9737) = 2.645$ g/min

Repeating these calculations for each time interval produced the results shown in Table 7.7.

The proportion of entrained material in the froth was then determined by comparing the rate of recovery of the entrained material ($E(t)$ from Table 7.7) with the rate of recovery of total floated material ($R(t)$ from Table 7.6). In the 0–0.5 min time interval, total solids were floating at a rate of 279.6 g/min, of which 53.70 g/min was entrained solids; therefore, 19.2% of the solids were reaching the froth by entrainment during this time interval. Similarly, pyrite was floating at a rate of 7.32 g/min, with 2.645 g/min being entrained, and so 36% of the pyrite that floated in the first 0.5 min was entrained. During the 0.5–1 mi interval, 34% of the pyrite that floated was recovered by entrainment. Reducing or eliminating entrainment could therefore have reduced the pyrite content of the froth quite markedly.

TABLE 7.7
Calculation of Rates of Recovery of Entrained Material ($E(t)$) in g/min, as Calculated Using Ross's Method

Time,			Total Solids			Pyrite		
min	$C_w(t)$	$W(t)$	$X(t)$	$C_m(t)$	$E(t)$	$X(t)$	$C_m(t)$	$E(t)$
0.25	0.9737	1,636.4	0.7590	0.04209	53.700	0.7549	0.002085	2.6450
0.75	0.9856	694.0	0.7407	0.02306	12.030	0.7305	0.001519	0.7813
1.5	0.9884	946.9	0.7132	0.01846	12.610	0.6939	0.001244	0.8270
2.5	0.9907	1,044.7	0.6766	0.01488	10.620	0.6451	0.001008	0.6859
4.0	0.9927	950.1	0.6216	0.01168	6.950	0.5720	0.000808	0.4421
7.0	0.9947	789.3	0.5117	0.00844	3.425	0.4256	0.000607	0.2050

The values for $X(t)$ were determined from Figure 7.15, and $E(t)$ was calculated using Equation 7.6.

It should be emphasized that in the above example calculation, no attempt was made to distinguish between pyrite, which floated because it was mechanically locked to hydrophobic coal particles, and pyrite, which was truly hydrophobic. Methods for distinguishing between locked pyrite and hydrophobic pyrite will be discussed in the following section.

7.2.2 Locked Particles

In addition to true flotation and entrainment, pyrite can also reach the froth product as composite coal/pyrite particles. These particles float because of the hydrophobicity of the coal portion. Unlike various gravity or inertial methods for separating pyrite from coal, froth flotation has a particular difficulty in rejecting locked particles, because a very small amount of coal on the pyrite surface can cause it to behave as if the entire particle were coal. In particular, pyrite particles are "rimmed" with coal, so that while they are predominantly pyrite, most or all of their surfaces are coated with coal. It has been demonstrated that virtually all of the pyrite particles that report to the froth product have at least some coal on their surfaces (Wang et al., 1993). Such locked coal/pyrite particles are often mistaken for hydrophobic pyrite, which leads to fruitless attempts to prevent them from floating by using pyrite depressants. Dealing with these particles is one of the most persistent problems in desulfurizing coal by froth flotation. The most obvious solution that has been proposed is to grind the coal to a small enough size that the pyrite is completely liberated. Unfortunately, this requires grinding to finer than about 10 μm, which is very expensive, reduces the effectiveness of froth flotation, and tends to increase entrainment problems.

7.2.2.1 Determination of Locked Particles

Locked particles can be detected and analyzed by microscopic examination, as was described in Chapter 2. Point counting and image analysis can be used to measure the size distribution of pyrite grains and to determine the proportion of pyrite which is locked to coal. It is also sometimes useful to determine the quantity of sulfur (both pyritic and organic), which is so intimately combined with coal that flotation cannot remove it. This "unliberated" sulfur can be determined based on the following assumptions:

- The liberated pyrite is always less floatable than the most floatable coal particles when no collector is added; therefore, the fastest floating coal in the absence of collector will be free of liberated pyrite if entrainment is minimized.
- The locked (unliberated) pyrite and organic sulfur are uniformly distributed throughout the combustible fraction of the coal.

Given these assumptions, a value for organic sulfur plus unliberated pyritic sulfur in the combustible fraction of the coal sample can be determined by any of the three different methods:

- The coal can be successively refloated several times by conventional flotation to recover only the fastest floating particles;
- The coal can be floated using column flotation, using care to eliminate entrainment, with the initial, fastest floating, lowest sulfur product being considered free of liberated pyritic sulfur;
- A heavy-liquid sink/float test can be carried out at the lowest density that will provide enough coal for analysis, with the float fraction being considered free of liberated pyrite.

Once the value for locked and organic sulfur has been determined in the combustible fraction for a sample by one of these three methods, it is possible to use Equation 7.8 (derived from Hirt, 1973) to calculate the liberated pyritic sulfur content:

$$\%\,S_{LP} = \%\,S_T - \%\,S_{LO}(W_C) \tag{7.8}$$

where $\%\,S_{LP}$ = liberated pyritic sulfur content of the sample,
$\%\,S_T$ = total sulfur content of the sample,
W_C = weight fraction of combustible matter in the sample,
$\%\,S_{LO}$ = locked and organic sulfur content of the combustible fraction.

7.2.2.2 Case Study: Recovery of Pyrite – Entrainment or Hydrophobicity?

A series of laboratory experiments were carried out at MTU to separate the effects of entrainment, locked particles, and true hydrophobic entrainment on the recovery of pyrite during flotation of an unoxidized Pittsburgh seam coal (Kawatra et al., 1991; Kawatra and Eisele, 1992a, b). This was done to determine whether or not hydrophobic flotation was really responsible for a significant amount of pyrite recovery. Two sets of experiments were conducted. The first set used conventional flotation, where entrainment would be significant. The second set of experiments used column flotation, which largely eliminates entrainment (Kawatra and Eisele, 1987). Column flotation is described further in Chapter 8.

7.2.2.2.1 *Pittsburgh Coal Conventional Flotation*

These experiments were carried out to determine the quantity of pyrite reporting to a coal froth by entrainment. These tests used Dowfroth 200 (a polypropylene glycol methyl ether, molecular weight of 200) as the frother. Entrainment was determined by correlating the water recovery with the recoveries of coal, ash, water, and the various forms of sulfur as a function of time. The dosage of collector (No. 2 fuel oil) was varied to change the coal flotation rate. The dosage of frother was held constant for these tests.

Material: These experiments used 250 g charges of Pittsburgh seam coal. The coal contained 37.2% ash, 2.3% pyritic sulfur, and 2.9% total sulfur and had previously been stage-crushed to pass 20 mesh. The coal was prepared for flotation by grinding 900 g of coal in 1.5 L of distilled water (37.5% solids) at its natural pH in a steel rod mill to 80% passing 36 μm. The freshly ground coal was then filtered, and the cake was divided into three 250 g charges for flotation, with the remaining material used as a head sample. The coal was then floated immediately.

Procedure: Timed flotation experiments were carried out in a 4.25-L flotation cell and agitated at 1,200 rpm with a Wemco agitator. The cell was equipped with gravity-feed pulp level control and mechanical froth scrapers (Figure 7.10, on page 164). The frother was Dowfroth 200 and was added to all of the water used in the flotation tests at a rate of 0.03 g/L (corresponding to a frother dosage of 0.5 kg/metric ton). Distilled water was used throughout, and the tests were carried out at a pH of 7.6 (which was the natural pH for this coal). Frother was added to all of the water before adding the coal. For tests using collector, the fuel oil was added to the pulp in the cell and conditioned at 1,200 rpm for 5 min before beginning flotation. The air flow was then started, and froth was collected over the time intervals 0–30 s, 30 s^{-1} min, 1–2 min, 2–3 min, 3–5 min, and 5–9 min. The froth products were weighed, filtered, dried, and re-weighed to measure the water content of each. All products were analyzed for ash content, using ASTM method D3174 (ASTM, 1988a), and total sulfur content was determined using a LECO SC-132 sulfur determinator. Pyritic sulfur was measured using ASTM method D2492 (ASTM, 1988b), with the nitric-acid dissolution step carried out by leaching at room temperature overnight (16 h), followed by boiling for 30 min to ensure complete dissolution.

7.2.2.2.2 *Pittsburgh Coal Column Flotation*

These experiments were intended to produce a coal that was as free as possible of entrained particles, so that the quantity of locked pyritic and organic sulfur in the coal could be determined. Flotation experiments were carried out in a laboratory-scale horizontally baffled flotation column, using coal prepared in the same manner as that used in the conventional coal flotation tests. The column had been modified by the introduction of horizontal baffles, which had been found to improve the performance of the column in earlier work. Tracer experiments previously conducted using sodium fluorescein dye showed that less than 4% of the feed water was entrained in the froth product using this column (Eisele and Kawatra, 1993b; Kawatra and Eisele, 1995a, b).

Procedure: A 1.62 cm (3-in) diameter horizontally baffled column, 1.83 m (6 ft) tall, was used for these experiments. A froth depth of 45.7 cm (18 in) was maintained, with 1 L/min of washwater. Aeration was provided by an aspirator, with a water pressure of 103 kPa (15 psi) and a flowrate of 5 L/min. The water injected into the column through the bubble generator and the washwater ring was tap water, with a nominal pH of 7 and a nominal hardness of 150 ppm. The frother was Dowfroth 200, introduced with the aspirated air as a 1 % solution to give a concentration of 0.03 g/L (0.5 kg/metric ton coal), and also added to the washwater at the same concentration. No collector was added in the column flotation tests.

The column flotation experiments were batch tests. Feed coal was prepared as for the conventional flotation experiments, except that each 900 g of ground coal was divided into two flotation feed charges of 400 g each, with the remaining 100 g reserved as a head sample. Timed flotation experiments were carried out by suspending the coal in distilled water at 25% solids and adding the coal slurry as a single increment over a 30-s time interval while the column was being filled at the beginning of the test. After adding the feed, 30 s was required before coal began flowing into the overflow launder. From this time, froth increments were removed over the 0–30 s, 30 s–1 min, 1–2 min, 2–3 min, 3–5 min, and 5–9 min time intervals. These samples were analyzed in the same manner as the samples from the conventional flotation tests.

The initial products from the flotation column contained only 1.7% total sulfur. Since the column minimized entrainment, and any hydrophobic pyrite particles would be expected to float more slowly than coal particles, the value of 1.7% sulfur was taken to be the locked/organic sulfur content of this coal.

7.2.2.2.3 Results

Using Equation 7.5 to correct for makeup water dilution, the cumulative percent recovery of the water originally in the cell during conventional flotation can be calculated. This corrected water recovery can now be used to distinguish between entrainment and true flotation, which is most conveniently done using the "simple correlation" method (described on p. 163).

Figure 7.16 shows the correlations of combustible matter, ash, total pyritic sulfur, and liberated pyritic sulfur with the corrected water recovery for all six conventional tests. It is immediately evident that the ash shows behavior strongly characteristic of entrainment, while the combustible matter behaves as a strongly floatable species. The pyritic sulfur behaves in an intermediate fashion, with recovery being initially slightly faster than that of water, and then becoming slightly slower. When the liberated pyritic sulfur, calculated using Equation 7.8, is plotted, the curve is similar to that of the total pyritic sulfur (as determined using the ASTM nitric-acid leach method) and between the ash recovery and total pyritic sulfur curves. When the liberated pyritic sulfur recovery is plotted against the ash recovery (Figure 7.16), it is seen that there is a good correlation between the two recoveries, and that they follow very close to the 45° line. This indicates that the recovery mechanisms for the liberated pyrite and the ash are similar.

Because the pyrite in Figure 7.16 is primarily being recovered through entrainment, the use of pyrite depressants is very unlikely to improve product grade. Instead, measures should be taken to reduce entrainment. The same reasoning applies to ash recovery in this example as well.

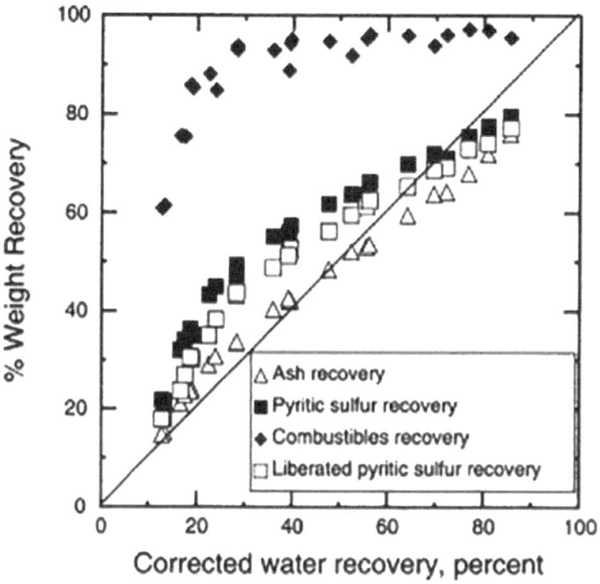

FIGURE 7.16 Conventional timed flotation cell results, with a fixed frother dosage (0.5 kg/metric ton) and a varying collector dosage (0.0, 0.35, and 0.7 kg/metric ton). The feed coal contained 37.2% ash, 2.3% pyritic sulfur, 2.9% total sulfur, and 1.7% locked/organic sulfur. Correlations of combustible matter, ash, total pyritic sulfur, and liberated pyritic sulfur recoveries with water recovery are shown (corrected for makeup water addition using Equation 7.8). Perfect entrainment would follow the 45° line.

The ash is primarily composed of fine clay particles, which are expected to be strongly hydrophilic and are well-liberated due to the high feed ash content (37.2% ash); therefore, the ash is mainly recovered by entrainment. Since the behavior of the liberated pyrite is similar to the behavior of the ash, this strongly implies that entrainment is the most important recovery mechanism for the liberated pyrite as well (Figure 7.17).

Based on the data given here, it is seen that for this coal, the coal pyrite is floating due to locking to floatable coal (incomplete liberation) and to entrainment, and not due to natural hydrophobicity of the pyrite. It appears that the pyrite in this coal has a very low inherent floatability and is therefore floated primarily as either locked or entrained particles.

7.2.2.3 Means for Reducing Entrainment

There are numerous ways to reduce the amount of entrainment, which range from small modifications of normal practice, to completely rebuilding the plant's fine coal cleaning circuit. Of course, the smallest changes are also the ones that give the smallest benefit. These changes are as follows:

- *Change quantities of reagents.* The level of entrainment in coal flotation can be reduced by adding the frother at a starvation level and increasing the quantity of oily collector that is added. The effects of these changes

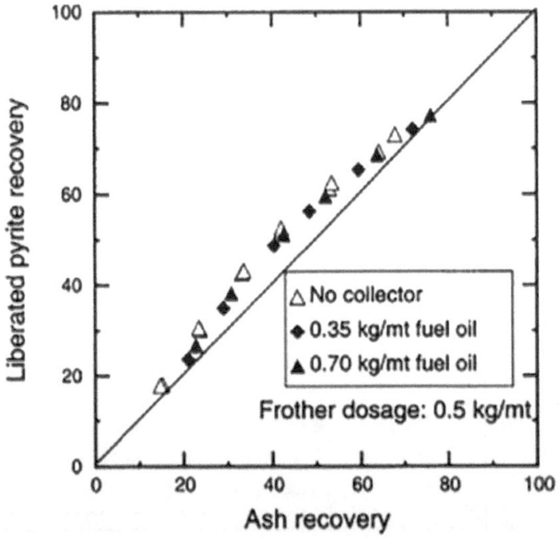

FIGURE 7.17 Conventional timed flotation results, showing the correlation of liberated pyritic sulfur recovery with ash recovery, at a fixed frother dosage and a varying collector dosage. The close approach to the 45° line demonstrates that the liberated pyrite recovery correlates very closely to the ash recovery. This indicates that both are being recovered by the same mechanism (entrainment).

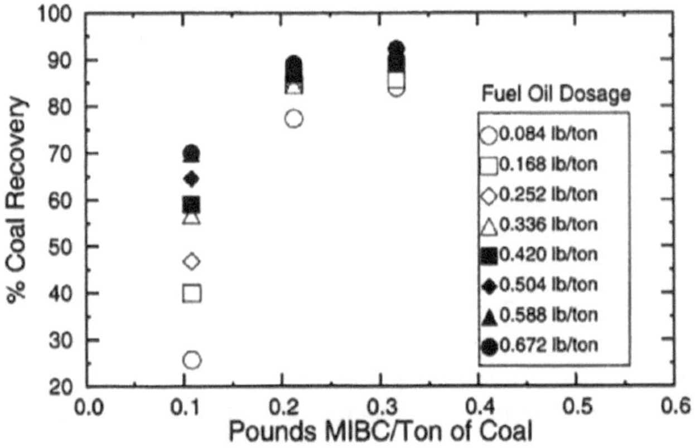

FIGURE 7.18 Effect of changes in MIBC addition level on the percent coal recovery at different fuel oil addition levels. Increasing either MIBC dosage or fuel oil dosage leads to an increase in recovery.

are illustrated in Figures 7.18 and 7.19. It is possible to achieve the same coal recovery with a high frother and low oil dosage as with a low frother and high oil dosage. The use of minimal frother with a high oil dosage minimizes the percent solids of the froth, and so keeps the entrainment low. Similar effects can be achieved by the use of a deeper froth layer (Kawatra and Waters, 1982; Kawatra et al., 1984; Seitz and Kawatra, 1985, 1987).

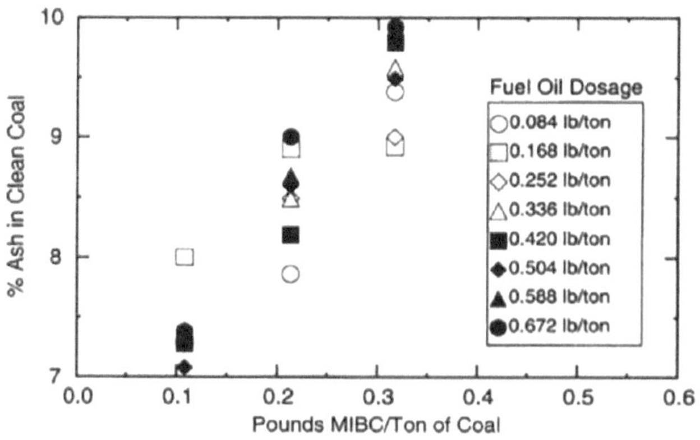

FIGURE 7.19 Effect of changes in MIBC addition on the percent ash in the clean coal product at different fuel oil addition levels. Increasing frother dosage produces a large increase in the clean coal ash content, as a result of increased entrainment.

- *Use a deeper froth layer.* As the froth layer becomes deeper, more water drains out of the froth before it is removed from the cell. Since the draining water carries out many of the entrained particles with it, entrainment is reduced. This can only be taken so far, however, because if too much water drains out of the froth, the bubbles will burst and the froth will collapse before it can be removed. It should also be noted that coal flotation cells are often operated in an "overloaded" condition. This means that the coal is floating at such a rate that it already contains the minimum amount of water needed for stability, and it will not be practical to reduce water content any more (Suardini and Kawatra, 1995). Even under "overloaded" conditions, entrainment will always be present. This is illustrated by the case study of the Panther Valley plant, given in Section 7.2.2.4.
- *Switch to a multi-stage flotation circuit.* If the froth from the primary flotation cells is resuspended in a second bank of flotation cells and refloated, as shown in Figure 7.20, the amount of ash minerals and pyrite entrained in the second stage will be drastically lower than that in the first stage. The tailings from the second stage can then be recycled to the first stage or simply disposed of. A large number of variations on this circuit, including various conditioning methods and circuit layouts, are discussed in the literature (Seitz and Kawatra, 1993; Nicol and Bensley, 1988).
- *Deslime the feed.* In many coals, the finest particles are mainly clay slimes and pyrite fines, with relatively few coal particles. In these cases, it may be best to simply remove the finest particles and dispose them of, and only feed the coarser particles to flotation, as shown in Figure 7.21. Since the fines are the particles that present the biggest entrainment problem, this can reduce entrainment to a great deal without greatly increasing the cost or complexity of the flotation circuit. The difficulty is, of course, that any very fine coal present will be lost to the waste product. High levels of fines in the coal tend to raise the moisture content and therefore reduce its value

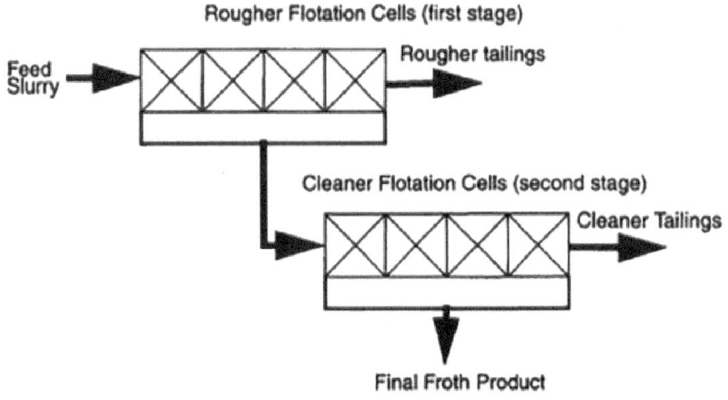

FIGURE 7.20 Schematic of a multi-stage flotation circuit for coal. If the tailings product from the cleaner cells contains a large enough amount of recoverable coal, it can be recirculated to the rougher feed slurry to improve recovery. Otherwise, the cleaner tailings are discarded along with the rougher tailings.

to many customers, so this loss of fines is not necessarily bad. It should be kept in mind that the fine coal in the waste will be more prone to spontaneous combustion than coarser coal would be, and so it should be disposed of carefully.

- *Use a frother that produces thinner films between bubbles.* The characteristics of the froth layer in flotation depend to a large extent on the particular chemicals being used as frothers. "Strong" frothers, such as longer-chain polypropylene glycols, produce a very fine-grained froth with thick water films around the bubbles that drain slowly. While this is excellent for maximizing recovery, it also tends to produce a great deal of entrainment. "Weaker" frothers, such as short-chain glycols or MIBC, produce coarser froths with thinner water films. Since they drain more rapidly, they carry less water into the froth product, and therefore less ash and pyrite is entrained. This results in some sacrifice of recovery.
- *Add a dispersant to reduce occlusion of particles.* In addition to being suspended in the water carried into the froth, some ash and pyrite is flocculated with coal particles and carried into the froth that way. These floccules can be broken up with dispersants such as sodium silicate.
- *Add a flocculant to increase the settling rate of the fines.* In some cases, it may be possible to add a small amount of a flocculant, such as a polyacrylamide, to more-or-less selectively flocculate the ash and pyrite fines into larger flocs. Since these flocs will have a faster settling rate, they will be entrained less readily than the original fines. This can be counterproductive, however, if the floccules also incorporate a significant amount of coal, as then the floccules will be hydrophobic and will be carried into the froth by bubble attachment.
- *Reduce the % solids of the feed.* The concentration of suspended particles in the froth water can obviously be no higher than their concentration in the

slurry in the cell. If the percent solids of the feed slurry is reduced, say from 20% to 10% solids, then the amount of entrained material per unit of water in the froth will be cut in half. This unfortunately also cuts the capacity of the cell bank in half, and the cell volume will have to be doubled with the accompanying increase in capital and operating costs.

- *Sprinkle clean water on the froth.* Some plants have taken to gently spraying clean water on the froth surface. This flushes the solids-laden water back into the pulp and replaces it with solids-free water. While this is helpful, this type of froth washing is done much more effectively by flotation columns, which are discussed in Chapter 8.

7.2.2.4 Case Study: Panther Valley Plant, PA, USA

A study of the Panther Valley plant was conducted, which illustrates the benefits both of desliming the flotation feed and of controlling the reagent additions (Kawatra and Seitz, 1993). This plant processed an anthracite coal, primarily mined from the Mammoth seam. The feed to the plant contained 15% material finer than 297 μm and 8% material finer than 74 μm. The –297 μm fines were too fine for the heavy-media circuit, so they were treated by froth flotation. The –74 μm fraction was very high in ash; therefore, the most effective means for dealing with this fine material was to remove and discard it.

The flotation circuit was as shown in Figure 7.21, with the cyclone desliming the feed at 74 μm. The feed to the desliming cyclone was 90 tons/h, of which 40 tons/h reported to the underflow was used as flotation feed. Removal of the –74 μm fraction by the cyclone was of course not perfect, and so there was still material in this size range in the flotation feed. The flotation circuit consisted of three 150 ft^3 (4.25 m^3) Wemco cells. The reagents used were fuel oil as collector and polypropylene glycol (molecular weight of 200) as frother. Samples were collected from each of the three cells in the flotation bank, so that the grade/recovery performance could be determined. Samples were collected at collector levels of 0.7 and 1.4 lb/ton (0.35 and 0.7 kg/metric ton) and frother levels of 0.2, 0.4, and 0.6 lb/ton (0.1, 0.2, and 0.3 kg/metric ton). The samples were then dried and screened into size fractions.

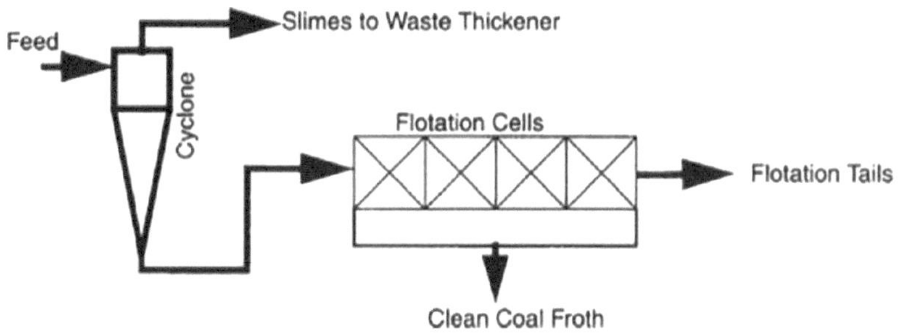

FIGURE 7.21 Coal flotation with desliming. A classifying cyclone provides a dependable, low-cost method for removing fines from coal and simultaneously increases the percent solids of the flotation feed.

The following size fractions were then selected for further analysis: 1,190 × 595 μm (10.0% ash), 297 × 212 μm (18.4% ash), and −74 μm (44.9% ash).

Flotation results for the 1,190 × 595 pm and 297 × 212 μm size fractions are shown in Figures 7.22 and 7.23, respectively. For both of these size fractions, changing the reagent dosages simply moved the plant performance along a single grade-recovery curve. This happened because these particles were coarse enough that they were entrained at very low rates. As a result, changes in water recovery had little effect on the recovery of ash in these size ranges, and flotation was very selective.

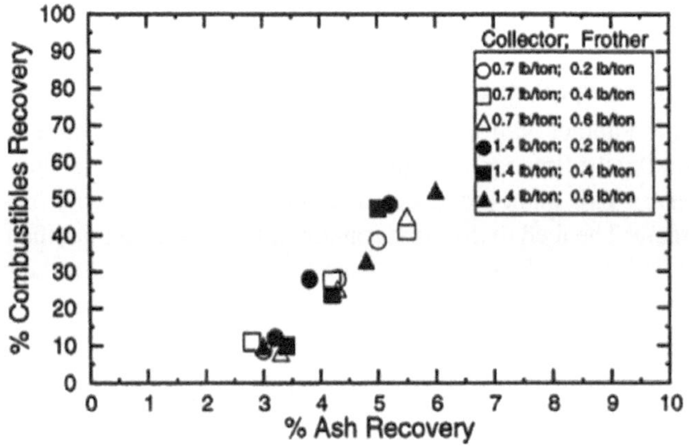

FIGURE 7.22 Combustibles recovery versus ash recovery for the 1190 × 595 μm size fraction at the Panther Valley plant. All results fall along a single grade/recovery curve, regardless of whether recovery is increased by increasing collector dosage or frother dosage. Recovery of combustibles is low because this size fraction is too coarse to float readily.

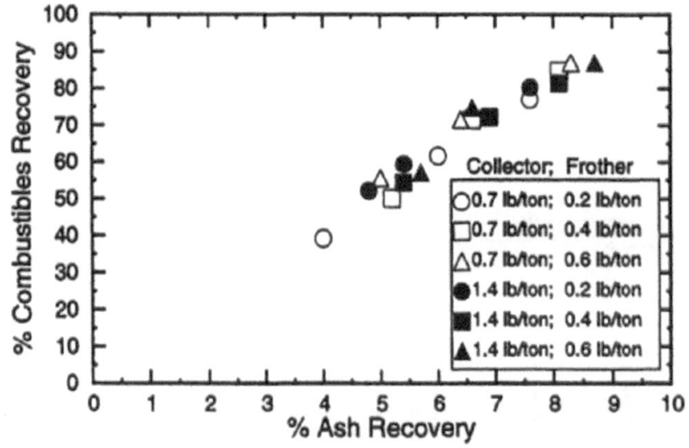

FIGURE 7.23 Combustibles recovery versus ash recovery for the 297 × 210 μm size fraction at the Panther Valley plant. All results fall along a single grade/recovery curve, regardless of whether recovery is increased by increasing collector dosage or frother dosage. The grade/recovery curve for this particle size is similar to the one seen in Figure 7.22, but the recovery is higher at this finer particle size.

In contrast, flotation of the −74 μm fraction was highly sensitive to frother dosage, with the grade-recovery curves shifting to higher ash contents as the frother dosage was increased (Figure 7.24). This was a direct result of the increased amount of water carried into the froth by high frother dosages. Since particles in this size range are easily entrained, the increased water recovery resulted in a great deal of clay slimes being carried into the froth, raising the ash content to high levels.

Based on this data, it is seen that the best selectivity was achieved in this plant by minimizing the frother dosage to prevent entrainment and by desliming the feed with as high efficiency as possible to reduce the levels of easily entrained clay slimes. It can be seen from Figure 7.24 that increasing the collector dosage did not result in an increase in entrainment, and so high collector/low frother was the best reagent scheme for maximum recovery with minimum entrainment.

7.2.2.5 Grab-and-Run Flotation

It has been reported that locked particles of coal/pyrite float somewhat more slowly than pure coal particles. This observation has led to the "grab and run" process, where the coal is initially floated using a starvation level of frother and gentle agitation. Approximately half of the floatable coal is then recovered as a comparatively low-sulfur product. The remaining coal can be either discarded or recovered in a scavenger circuit and sold separately as higher-sulfur coal (Aplan et al., 1976). The very nature of this process results in low recovery rates because it is sacrificing recovery to improve the grade. As a result, it has not seen a great deal of industrial application due to the lack of a significant price premium when selling low-sulfur coal.

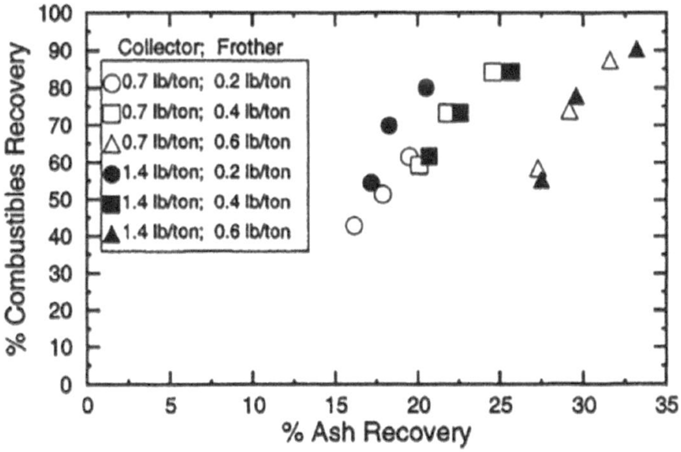

FIGURE 7.24 Combustibles recovery versus ash recovery for the −74 μm size fraction at the Panther Valley plant. Increasing frother dosage causes a pronounced shift in the grade/recovery performance, which is a result of increased entrainment at high frother dosages. Increasing collector dosage does not cause an increase in entrainment.

7.2.2.6 Reverse Flotation

One solution to the problem of locked coal-pyrite particles is to devise a means to specifically remove particles that contain pyrite. A technique that has been developed to do this is the reverse flotation process (Miller, 1973, 1978, 1988; Miller et al., 1984). This is a two-step flotation procedure as follows:

- Conventionally float the coal to recover as much of the coal as possible. This step rejects most of the ash and the liberated pyrite.
- Resuspend the froth from the first stage and add a reagent to prevent the coal particles from floating. Suitable depressants include dextrin and Aero Depressant 633. Then add a collector for pyrite (such as ethyl or amyl xanthate), which will selectively make the pyrite surfaces hydrophobic. Froth flotation will then recover the particles that contain pyrite into the froth, with the free coal particles remaining in suspension.

The reverse flotation process works because xanthates do not function as collectors for coal (Horsley et al., 1952), and dextrin has a strong affinity for coal but relatively little affinity for pyrite. Dextrin is most strongly adsorbed by coals that are fresh, unoxidized, and highly hydrophobic, and its adsorption by the coal surface decreases as the coal becomes more oxidized, which shows that the dextrin is attaching to the coal surface due to hydrophobic bonding (Laskowski and Miller, 1984). The dextrin adsorption is also largely independent of the solution pH, and so the pH can be adjusted for the optimum flotation of pyrite by xanthate.

Results obtained with the reverse flotation process on two different coals are given in Table 7.8. The advantage of this technique is that in the pyrite flotation step, the particles that float are the locked coal-pyrite particles, which will be hydrophobic regardless of whether coal is being floated or pyrite is being floated (Miller and Lin, 1986). The locked pyrite can therefore be selectively removed without the need for grinding to the pyrite liberation size. There are, however, several disadvantages that keep this process from being used industrially:

- Processing costs are high, because this method needs twice as much flotation capacity as a single-stage circuit, and also consumes a considerable amount of coal depressant and pyrite collector. For example, it has been estimated for a particular Illinois No. 6 seam coal that the necessary depressant dosage was 0.68 kg/metric ton, and the required xanthate dosage was 0.41 kg/metric ton (Laskowski and Miller, 1984). When xanthate is used as the pyrite collector, the consumption of the reagent by coal pyrite is much higher than would be predicted from flotation of mineral pyrite because of surface chemistry differences between mineral pyrite and coal pyrite (Chemosky and Lyon, 1972).
- If the coal is finer than about 80% passing 425 μm (35 mesh), then complete depression of the coal is impossible with any reasonable levels of depressant. The fines are then floated along with the pyrite, and the selectivity is lost.
- If the pyrite level becomes too low, then the reverse flotation froth does not contain enough solids to be stable. The froth layer then collapses, and no pyrite is removed.

TABLE 7.8
Results from Two-Stage Reverse Flotation of Coal for Two Coals

Coal Source	Product	% Yield	% Ash	% Sulfur	% Py. S.	Cumulative % Py. S Removal
Poland	Feed	100	18.45	0.55	0.45	
	Stage 1 froth (coal + pyrite)	61.6	3.88	0.33	0.24	67.1
	Stage 1 sinks (ash minerals)	38.3	41.85	0.90	0.79	
	Stage 2 froth (pyrite)	3.5	4.19	0.58	0.58	71.7
	Stage 2 sinks (clean coal)	58.1	3.86	0.32	0.22	
Eastern	Feed	100	8.37	0.56	0.20	
U.S.	Stage 1 froth (coal + pyrite)	92.4	6.84	0.51	0.17	21.2
	Stage 1 sinks (ash minerals)	7.6	26.9	1.15	0.56	
	Stage 2 froth (pyrite)	4.7	5.83	0.73	0.73	24.5
	Stage 2 sinks (clean coal)	87.5	6.92	0.50	0.14	

In the first stage, coal was floated using MIBC frother and no collector, at a pH of 7.8. In the second stage, the pH was lowered to 5.9 with sulfuric acid, and coal was depressed with Aerodepressant 633 while collecting pyrite with potassium amyl xanthate (Forssberg and Ryk, 1995).

It should be noted that pyrite content of any given coal seam is highly variable, with variations from 0% to several percentage points for any given plant feed. In a commercial reverse flotation process plant, these rapid shifts in pyrite content will regularly cause the pyrite levels to become so low that nothing floats in the second stage. This sort of fluctuation causes the whole circuit to become unstable, and therefore, the usefulness of the reverse pyrite flotation is limited. The action of the coal depressant also presents a limitation on the use of reverse flotation to remove pyrite from oxidized coal. If the pyrite is naturally hydrophobic due to mild oxidation, then the coal is also likely to be oxidized. The dextrin will then have an increased tendency to depress the flotation of the hydrophobic pyrite, while it will be less able to depress the oxidized coal. This could tend to reduce the selectivity of the reverse flotation process for oxidized coals.

7.3 SUMMARY

Conventional froth flotation is useful for recovering fine coal from material that would otherwise be discarded as waste. While it is very effective for separating coal from ash-forming minerals, it is somewhat less effective for coal desulfurization, for the following reasons:

- Pyrite in mildly oxidized coals can, in some circumstances, become naturally hydrophobic. When this occurs, pyrite particles can be floated along with the coal by true hydrophobic flotation. This effect can be reduced by the use of depressants.

- Small particles of pyrite are carried into the froth by entrainment in the water or by flocculation to floating coal particles. This effect can be reduced by minimizing the amount of water carried over into the froth product.
- Pyrite grains are often locked to coal particles. These locked particles behave very similarly to pure coal particles, and so are carried into the froth. This effect can be reduced by grinding the coal to a finer size to more completely liberate the pyrite.

While all three of these effects can be reduced by various means, many researchers have concentrated on finding and developing depressants that can prevent true hydrophobic flotation of pyrite. Unfortunately, this approach can only succeed if a substantial portion of the pyrite in the coal is naturally hydrophobic in the first place. Studies of coal-pyrite flotation have shown that the bulk of the pyrite entering the froth product is recovered either by entrainment or as part of locked particles, and that pyrite recovery by true flotation is negligible for most coals.

Pyrite depressants have not been commercially successful, because in most cases, the pyrite recovery that is blamed on true hydrophobic flotation is actually due to entrainment and locked particles, neither of which is affected by pyrite depressants. There is also a limit to how much entrainment can be reduced in a conventional flotation cell, because some water must always be carried into the froth product. Ultrafine grinding of the coal to eliminate locked coal/pyrite particles is also not commercially viable because very fine coal is difficult to float and tends to aggravate the entrainment problem. It is therefore not practical to greatly improve the desulfurization capability of conventional flotation cells. To desulfurize coals by froth flotation, it is best to use more advanced processes such as the column flotation method described in Chapter 8.

REFERENCES

Ahlberg, E., Forssberg, K.S.E. and Wang, X. (1990), "The Surface Oxidation of Pyrite in Alkaline Solution," *Journal of Applied Electrochemistry*, Vol. 20, pp. 1033–1039.

Aplan, F.F. and Arnold, B.J. (1991), "Wet Fine Particle Concentration, Section 3: Flotation," *Coal Preparation*, 5th edition (Leonard, ed.), Society for Mining, Metallurgy, and Exploration, Littleton, CO, pp. 450–485.

Aplan, F.F. Bonner, C.M., Hirt, W.C. and Rastogi, R.C. (1976), "Recent Advances in Coal Flotation," *Papers Presented before the Symposium on Coal Preparation*, Vol. 2, pp. 39–49.

Arnold, B.J. and Aplan, F.F. (1990), "The Use of Pyrite Depressants to Reduce the Sulfur Content of Upper Freeport Seam Coal," *Processing and Utilization of High-Sulfur Coals III* (Markuszewski and Wheelock, eds.), Elsevier, Amsterdam, pp. 171–186.

ASTM (1988a), "Standard Test Method for Ash in the Analysis Sample of Coal and Coke from Coal," *1988 Annual Book of ASTM Standards*, Vol. 5.05 (gaseous fuels; coal and coke), American Society for Testing and Materials, Standard No. D3174–82.

ASTM (1988b), "Standard Test Method for Forms of Sulfur in Coal," *1988 Annual Book of ASTM Standards*, Vol. 5.05 (gaseous fuels; coal and coke), American Society for Testing and Materials, Standard No. D2492–84.

Atkins, A.S. (1990), "Developments in the Biological Suppression of Pyritic Sulfur in Coal Flotation," *Bioprocessing and Biotreatment of Coal* (Wise, ed.), Marcel Dekker, New York, pp. 507–548.

Baker, A.F. and Miller, K.J. (1971), "Hydrolyzed Metal Ions as Pyrite Depressants in Coal Flotation: A Laboratory Study," U.S. Bureau of Mines, Report of Investigations RI 7518.

Bergholm, A. (1955), "Oxidation av Pyrit," *Jemkontorets Annaler*, Vol. 139, No. 8, pp. 531–549.

Chander, S. and Aplan, F.F. (1989), "Surface and Electrochemical Studies in Coal Cleaning," Final Report to the U.S. Department of Energy, DOE/PC/80523-T11 (DE 90007603).

Chemosky, F.J. and Lyon, F.M. (1972), "Comparison of the Flotation and Adsorption Characteristics of Ore and Coal-Pyrite with Ethyl Xanthate," *AIME Transactions*, Vol. 252, pp. 11–14.

Chmielewski, T. and Wheelock, T.D. (1991), "Thioglycolic Acid as a Flotation Depressant for Pyrite," *Processing and Utilization of High-Sulfur Coals IV* (Dugan, Quigley, and Attia, eds.), Elsevier, Amsterdam, pp. 295–308.

Chou, C.-L. (1990), "Geochemistry of Sulfur in Coal," *Geochemistry of Sulfur in Fossil Fuels* (Orr and White, eds.), ACS Symposium Series, Vol. 429, American Chemical Society, Washington, DC, pp. 30–52.

Dankwerts, P.V. (1953), "Continuous Flow Systems: Distribution of Residence Times," *Chemical Engineering Science*, Vol. 2, pp. 11–14.

Ehrlinger III, H.P., Lytle, J.M. and Khan, L. (1995), "Use of Illinois Coal Preparation Plant Fines in Slurry-Fed Gasifiers," *High-Efficiency Coal Preparation* (Kawatra, ed.), Society for Mining, Metallurgy, and Exploration, Littleton, CO, pp. 261–269.

Eisele, T.C. and Kawatra, S.K. (1994), "Use of Fungi and Bacteria for the Depression of Mineral and Coal Pyrite," Chapter 11, *Reagents for Better Metallurgy* (Mulukutla, ed.), Society for Mining, Metallurgy, and Exploration, Littleton, CO, pp. 91–100.

Eisele, T.C. and Kawatra, S.K. (1993a), "Depression of Pyrite Flotation by Microorganisms as a Function of pH," *Proceedings of the International Conference on Processing and Utilization of High Sulfur Coals*, Vol. V, Elsevier, Amsterdam, pp. 139–148.

Eisele, T.C. and Kawatra, S.K. (1993b), "The Use of Horizontal Baffles to Improve the Effectiveness of Column Flotation of Coal," *Proceedings of the XVIII International Mineral Processing Congress*, Sydney, Australia, 23–28 May, pp. 771–778.

Elzeky, M. and Attia, Y.A. (1987), "Coal Slurry Desulfurization by Flotation Using Thiophilic Bacteria for Pyrite Depression," *Coal Preparation*, Vol. 5, pp. 15–37.

Feeley, T J. (1991), "Coal Surface Control for Advanced Physical Beneficiation," *Processing and Utilization of High-Sulfur Coals IV* (Dugan, Quigley and Attia, eds.), Elsevier, Amsterdam, pp. 195–204.

Finch, J.A., Nesset, J.E. and Acuña, C. (2008), "Role of Frother on Bubble Production and Behavior in Flotation," *Minerals Engineering*, Vol. 21, pp. 949–957.

Firth, B.A. and Nicol, S.A. (1980), "The Influence of Pyritic Sulphur on the Recovery of Fine Coal by Froth Flotation," End of Grant Report, Department of National Development and Energy, Australia, Report No. NERDDP/EG/80/13.

Firth, B.A. and Nicol, S.A. (1984), "The Effect of Oxidized Pyritic Sulphur on Coal Flotation," *Coal Preparation*, Vol. 1, No. 1, pp. 53–70.

Forssberg, K.S.E. and Ryk, L. (1995), "Cleaning and Utilization of Coals for Low Emission at Swedish Power Plants," *High-Efficiency Coal Preparation* (Kawatra, ed.), Society for Mining, Metallurgy, and Exploration, Littleton, CO, pp. 247–259.

Franzidis, J.-P. and Anderson, G.V.M. (1990), "Influence of Various Collectors in Coal Flotation," *Proceedings of the 14th Congress of the Council of Mining and Metallurgical Institutions*, IMM, London, pp. 295–309.

Fuerstenau, D.W., Diao, J. and Hanson, J.S. (1990), "Estimation of the Distribution of Surface Sites and Contact Angles on Coal Particles from Film Flotation Data," *Energy and Fuels*, Vol. 4, pp. 34–37.

Fuerstenau, D.W., Yang, G.C.C. and Laskowski, J.S. (1987), "Oxidation Phenomena in Coal Flotation, Part 1: Correlation Between Oxygen Functional Group Concentration, Immersion Wettability, and Salt Flotation Response," *Coal Preparation*, Vol. 4, pp. 161–182.

Fuerstenau, D.W., Rosenbaum, J.M. and Laskowski, J.S. (1983), "Effect of Surface Functional Groups on the Flotation of Coal," *Colloids and Surfaces*, Vol. 8, pp. 153–174.

Gayle, J.B., Eddy, W.H. and Shotts, R.Q. (1965), "Laboratory Investigation of the Effect of Oxidation on Coal Flotation," U.S. Bureau of Mines Report of Investigations RI 6620.

He, Y.B. and Laskowski, J.S. (1992), "Contact Angle Measurements on Discs Compressed from Fine Coal," *Coal Preparation*, Vol. 10, pp. 19–36.

Hirt, W.C. (1973), *Separation of Pyrite from Coal by Froth Flotation*, unpublished Master's thesis, Pennsylvania State University.

Hoberg, H., Diegel, R. and Schneider, F.U. (1995), "Adsorption and Mobilization of Coal Flotation Reagents with Regard to Detrimental Effects on the Environment," *High-Efficiency Coal Preparation* (Kawatra, ed.), Society for Mining, Metallurgy, and Exploration, Littleton, CO, pp. 153–162.

Horsley, R.M., El-Sinbawy, H. and Smith, H.G. (1952), "Xanthates in Coal Flotation," *Fuel,* Vol. 31, pp. 302–311.

Janczuk, B., Wojcik, W., Staszczuk, P., Bialopiotrowicz, T. and Chibowski, E. (1989), "Detachment Forces of Air Bubble from Pyrite Surface Covered with Apolar and Polar Liquids," *Journal of Mines, Metals & Fuels*, Vol. 37, No. 9, pp. 380–384.

Jin, R. (1989), "Coal Flotation Chemistry," Ph.D. Thesis, University of Utah.

Kawatra, S.K. and Eisele, T.C. (1987), "Column flotation of coal," *Fine Coal Processing* (Mishra and Klimpel, eds.), Noyes Publications, Park Ridge, NJ, pp. 414–426.

Kawatra, S.K. and Eisele, T.C. (1992a), "Removal of Pyrite in Coal Flotation," *Mineral Processing and Extractive Metallurgy Review*, Vol. 8, pp. 205–218.

Kawatra, S.K. and Eisele, T.C. (1992b), "Pyrite Recovery Mechanisms in Coal Flotation," *Presented at the SME Annual Meeting*, Phoenix, AZ, February 24–27, Preprint No. 92–64.

Kawatra, S.K. and Eisele, T.C. (1995a), "Laboratory Baffled-Column Flotation of Mixed Lower/Middle Kittanning Seam Bituminous Coal," *Minerals and Metallurgical Processing*, Vol. 12, No. 2, pp. 103–107.

Kawatra, S.K. and Eisele, T.C. (1995b), "Baffled-Column Flotation of a Coal Plant Fine-Waste Stream," *Minerals and Metallurgical Processing*, Vol. 12, No. 3, pp. 138–142.

Kawatra, S.K. and Eisele, T.C. (1997), "Pyrite Recovery Mechanisms in Coal Flotation," *International Journal of Mineral Processing*, Vol. 50, No. 3, pp. 187–201.

Kawatra, S.K., Eisele, T.C. and Bagley, S.T. (1989), "Studies of Pyrite Dissolution in Pachuca Tanks and Depression of Pyrite Flotation by Bacteria," *Biotechnology in Minerals and Metal Processing* (Scheiner, Doyle and Kawatra, eds.), Society of Mining Engineers, Littleton, CO, pp. 55–62.

Kawatra, S.K., Eisele, T.C. and Johnson, H. (1991), "Recovery of Liberated Pyrite in Coal Flotation: Entrainment or Hydrophobicity?" *Processing and Utilization of High-Sulfur Coals IV* (Dugan, Quigley and Attia, eds.), Elsevier, Amsterdam, pp. 255–277.

Kawatra, S.K., Seitz, R.A. and Suardini, P.J. (1984), "The Control of Coal Flotation Circuits," Chapter 23, *Control 84* (Herbst, George, and Sastry, eds.), Society of Mining Engineers, New York, pp. 225–233.

Kawatra, S.K. and Seitz, R.A. (1993), "Conditioning Methods and Design Philosophy of Various Coal Flotation Circuits," *Flotation Plants: Are They Optimized?* (Malhotra, ed.), Society for Mining, Metallurgy, and Exploration, Littleton, CO, pp. 137–146.

Kawatra, S.K. and Waters, J.W. (1982), "An Investigation on the Effect of Reagent Addition on the Response of a Fine Coal Flotation Circuit," *Proceedings of the 14th International Mineral Processing Congress*, Toronto, Ontario, Canada, pp. VII.3.1–VII.3.14.

Kinkel, A.R., Hall, W.E. and Albers, J.P. (1956), "Geology and Base-Metal Deposits of West Shasta Copper-Zinc District, Shasta County, California," U.S. Geological Survey Professional Paper No. 285.

Klassen, V.I. (1953), *Elements of the Theory of Coal Flotation*, translated, summarized, and edited by the National Coal Board, U.K.

Klassen, V.I. (1963), *Coal Flotation*, 2nd edition, Gosgortekhisdat, Moscow.

Klassen, V.I. and Mokrousov, V.A. (1963), *"An Introduction to the Theory of Flotation"* (translated by J. Leja and G.W. Poling), Butterworths, London.

Klimpel, R.R. (1992), "Some Results of Various New Chemical Reagents for Modifying Coal Flotation Performance," *Coal Preparation*, Vol. 10, pp. 159–175.

Klimpel, R.R. (1995), "The Influence of Frother Structure on Industrial Coal Flotation," *High-Efficiency Coal Preparation* (Kawatra, ed.), Society for Mining, Metallurgy, and Exploration, Littleton, CO, pp. 141–151.

Kocabag, D., Shergold, H.L. and Kelsall, G.H. (1990), "Natural Oleophilicity/Hydrophobicity of Sulphide Minerals, II: Pyrite," *International Journal of Mineral Processing*, Vol. 29, pp. 211–219.

Lai, R.W., Diehl, J.R., Hammack, R.W. and Khan, S.U.M. (1990), "Comparative Study of the Surface Properties and the Reactivity of Coal Pyrite and Mineral Pyrite," *Minerals and Metallurgical Processing*, Vol. 7, pp. 43–48.

Laskowski, J.S. (1995), "Coal Surface Chemistry and Its Effect on Fine Coal Processing," *High-Efficiency Coal Preparation* (Kawatra, ed.), Society for Mining, Metallurgy, and Exploration, Littleton, CO, pp. 163–176.

Laskowski, J.S. (1992), "Oil Assisted Fine Particle Processing," *Colloid Chemistry in Mineral Processing* (Laskowski and Ralston, eds.), Elsevier, Amsterdam, pp. 361–394.

Laskowski, J.S. (1991), "Fizycochemiczne Problemy Mechnicznej Przeróbki Wegla. Czesc П. Zawiesiny Weglowo-Wodne," *Fizykochemiczne Problemy Mineralurgii*, Vol. 24, pp. 11–31.

Laskowski, J.S. (1987), "Coal Electrokinetics: The Origin of Charge at Coal/Water Interface," American Chemical Society, Division of Fuel Chemistry, Preprint of Paper Presented at Denver, CO, Vol. 32, No. 1, pp. 367–377.

Laskowski, J.S. (1986), "Flotation of Difficult-to-Float Coals," *Proceedings of the 10th International Coal Preparation Congress*, Edmonton, Canada.

Laskowski, J.S., Bustin, M., Moon, K.S. and Sirois, L.L. (1985), "Desulfurizing Flotation of Eastern Canadian High-Sulfur Coal," *Proceedings of the First International Conference on Processing and Utilization of High Sulfur Coals*, Elsevier, Amsterdam, pp. 247–265.

Laskowski, J.S., He, Y.B. and Zhan, Y. (1992), "Coal Wettability and Its Correlation with Floatability," *SME Annual Meeting*, Phoenix, AZ, February 24–27, Preprint No. 92–125.

Laskowski, J.S. and Miller, J.D. (1984), "New Reagents in Coal Flotation," *Reagents in the Minerals Industry* (Jones and Oblatt, eds.), IMM, London, pp. 145–154.

Laskowski, J.S. and Parfitt, G.D. (1989), "Electrokinetics of Coal-Water Suspensions," Chapter 7, *Interfacial Phenomena in Coal Technology* (Botsaris and Glazman, eds.), Marcel Dekker, New York, pp. 279–327.

Laskowski, J.S. and Poling, G.W. (1995), "Processing of Hydrophobic Minerals and Fine Coal," *Proceedings of the 1st UBC-McGill Bi-Annual International Symposium on Fundamentals of Mineral Processing*, Vancouver, BC, pp. 191–197.

Laskowski, J.S. and Romero, D. (1995), "The Use of Reagents in Coal Flotation," *Processing of Hydrophobic Minerals and Fine Coal* (Laskowski and Poling, eds.), Metallurgical Society of CIM, Vancouver, pp. 191–200.

Laskowski, J.S. and Walters, A.D. (1992), "Coal Preparation," *Encyclopedia of Physical Science and Technology*, Vol. 3, pp. 409–435.

Laskowski, J.S., Sirois, L.L. and Moon, K.S. (1986), "Effect of Humic Acids on Coal Flotation, Part 1: Coal Flotation Selectivity in the Presence of Humic Acids," *Coal Preparation*, Vol. 3, pp. 133–154.

Leja, J. (1982), *Surface Chemistry of Froth Flotation*, Plenum Press, New York.

Liu, D., Somasundaran, P., Vasudevan, T.V. and Harris, C.C. (1994), "Role of pH and Dissolved Mineral Species in Pittsburgh No. 8 Coal Flotation System—II. Separation of Pyrite and Non-pyritic Minerals from Coal," *International Journal of Mineral Processing*, Vol. 41, Nos. 3–4, pp. 215–225.

Liu, D., Somasundaran, P. and Duby, P.F. (1993), "Electrochemical Equilibria in Coal Flotation Systems and Their Role in Determining the Interfacial Properties of Pyrite and Its Separation from Coal," *Processing and Utilization of High-Sulfur Coals V* (Parekh and Groppo, eds.), Elsevier, Amsterdam, pp. 47–54.

Liu, Q. and Laskowski, J.S. (1988), "Effect of Humic Acids on Coal Flotation, Part II: The Role of pH," *SME Annual Meeting*, Phoenix, AZ, Preprint No. 88–31.

Luo, S., McClelland, J.F. and Wheelock, T.D. (1993), "The Interaction of Thioglycolic Acid and Pyrite," *Processing and Utilization of High-Sulfur Coals V* (Parekh and Groppo, eds.), Elsevier, Amsterdam, pp. 55–70.

Lynch, A.J., Johnson, N.W., Manlapig, E.V. and Thome, C.G. (1981), *Mineral and Coal Flotation Circuits: Their Simulation and Control*, Elsevier, Amsterdam.

Miller, J.D. and Lin, C.L. (1986), "Characterization of Pyrite in Products from the Reverse Flotation of Coal," *Coal Preparation*, Vol. 2, pp. 243–261.

Miller, J.D., Lin, C.L. and Chang, S.S. (1984), "Coadsorption Phenomena in the Separation of Pyrite from Coal by Reverse Flotation," *Coal Preparation*, Vol. 1, No. 1, pp. 21–38.

Miller, J.D., Ye, Y. and Jin, R. (1989), "Improved Pyrite Rejection by Chemically-Modified Fine Coal Flotation," *Coal Preparation*, Vol. 6, pp. 151–166.

Miller, K.J. (1978), "Desulfurization of Various Midwestern Coals by Flotation," U.S. Bureau of Mines, Report of Investigations RI 8262.

Miller, K.J. (1973), "Flotation of Pyrite from Coal: Pilot Plant Study," U.S. Bureau of Mines, Report of Investigations RI 7822.

Miller, K.J. (1988), "Novel Flotation Technology: A Survey of Equipment and Processes," Chapter 33, *Industrial Practice of Fine Coal Processing* (Klimpel and Luckie, eds.), Society of Mining Engineers, Littleton, CO, pp. 347–363.

Miller, F.G. (1963), "Sulfur Reduction on Minus 28 Mesh Bituminous Coal," *Rocky Mountain Minerals Conference, Society of Mining Engineers Fall Meeting*, Salt Lake City, UT, Preprint No. 63F322.

Monostory, F.P. and Kubitza, K.H. (1989), "Trennergebnisse von Betriebsanlagen fuer die Schaumflotation der Schlamme in Steinkohlen-Aufbereitungsanlagen [Separation results of operational plants for froth flotation of slurries in coal preparation plants]" *Aufbereitungs-Technik*, Vol. 30, No. 5, pp. 277–286.

Moslemi, H. and Gharabaghi, M. (2017), "A Review on Electrochemical Behavior of Pyrite in the Froth Flotation Process," *Journal of Industrial and Engineering Chemistry*, Vol. 47, pp. 1–18.

Mu, Y., Peng, Y. and Lauten, R.A. (2016), "The Depression of Pyrite in Selective Flotation by Different Reagent Systems – A Literature Review," *Minerals Engineering*, Vol. 96–97, pp. 143–156.

Nicol, S.K. and Bensley, C.N. (1988), "Recent Developments in Fine Coal Preparation in Australia," *Industrial Practice of Fine Coal Processing* (Klimpel and Luckie, eds.), Society of Mining Engineers, Littleton, CO, pp. 147–158.

Nimerick, K.H. and Scott, B.E. (1980), "New Method of Oxidized Coal Flotation," *Presented at the American Mining Congress Coal Show*, Chicago, IL, May 5–8, 1980.

Nordstrom, D.K. (1982), "Aqueous Pyrite Oxidation and the Consequent Formation of Secondary Iron Minerals," Chapter 3, *Acid Sulfate Weathering* (Kittrick, Fanning, Hossner, Kral, and Hawkins, eds.), Soil Science Society of America, Madison, WI.

Purcell Jr., R.J. and Aplan, F.F. (1991), "Pyrite and Ash Depression During Coal Flotation by Flotation Rate, Recovery, and Reagent Control," *Processing and Utilization of HighSulfur Coals IV* (Dugan, Quigley and Attia, eds.), Elsevier, Amsterdam, pp. 279–293.

Raleigh, C.E. and Aplan, F.F. (1990), "The Effect of Feed Particle Size and Reagents on Coal-Mineral Matter Selectivity During the Flotation of Bituminous Coals," *SME Annual Meeting*, Salt Lake City, UT, Preprint No. 90–118.

Raleigh, C.E. and Aplan, F.F. (1993), "The Use of Mineral Matter Dispersants and Depressants during the Flotation of Bituminous Coals," *Processing and Utilization of High-Sulfur Coals V* (Parekh and Groppo, eds), Elsevier, Amsterdam, pp. 71–90.

Rao, T.C., Govindarajan, B. and Barnwal, J.P. (1995), "A Simple Model for Industrial Coal Flotation Operation," *High-Efficiency Coal Preparation* (Kawatra, ed.), Society for Mining, Metallurgy, and Exploration, Littleton, CO, pp. 177–185.

Ross, V.E. (1990a), "Comparison of Methods for Evaluation of True Flotation and Entrainment," *Transactions of the Institute of Mining and Metallurgy (Section C: Mineral Processing)*, Vol. 100, pp. C121–C126.

Ross, V.E. (1990b), "Flotation and Entrainment of Particles During Batch Flotation Tests," *Minerals Engineering*, Vol. 3, No. 3/4, pp. 245–256.

Rubinshtein, Yu. B. and Yutifenko, S.N. (1993), "Investigating Aerosol Technology for Coal Flotation," *Koks i Khimiya*, Vol. 4, pp. 10–14.

Samylin, V.N., Beletskii, V.S. and Sergeev, P.V. (1993), "Characteristics of Flotation of Oxidized Coal," *Izvestiya Vysshikh Uchebnykh Zavedenii, Gornyi Zhtirnal*, Vol. 1, pp. 118–120.

Seitz, R.A. and Kawatra, S.K. (1985), "Control Strategies for Coal Flotation Circuits," *Proceedings of the Symposium on Automation for Mineral Resource Development*, Brisbane, Australia, pp. 241–246.

Seitz, R.A. and Kawatra, S.K. (1987), "The Role of Nonpolar Oils as Flotation Reagents," Chapter 19, *Chemical Reagents in the Mineral Processing Industry* (Malhotra and Riggs, eds.), Society of Mining Engineers, Littleton, CO, pp. 171–180.

Seitz, R.A. and Kawatra, S.K. (1993), "Conditioning Methods and Design Philosophy of Various Coal Flotation Circuits," Chapter 20, *Flotation Plants: Are They Optimized?* (Malhotra, ed.) Society for Mining, Metallurgy, and Exploration, Inc., Littleton, CO, pp. 137–146.

Shi, X., Zhu, H. and Ou, Z. (1997), "Effect of Coal Pyrite Surface Oxidation on Coal Desulfurization by Flotation and Methods of Pyrite Depression," *Journal Wuhan University of Technology, Materials Science Edition*, Vol. 12, No. 1–2, pp. 37–41.

Smith, E.E., Svanks, K. and Shumate, K.S. (1968), "Sulfide to Sulfate Reaction Studies," *2nd Symposium on Coal Mine Drainage Research*, Pittsburgh, PA, pp. 1–11.

Stock, L.M. and Wolny, R. (1990), "Elemental Sulfur in Bituminous Coals," *Geochemistry of Sulfur in Fossil Fuels* (W.L. Orr and C.M. White, eds.), ACS Symposium Series, Vol. 429, American Chemical Society, Washington, D.C., pp. 241–248.

Stokes, H.N. (1901), "On Pyrite and Marcasite," U.S. Geological Survey Bulletin No. 186.

Suardini, P.J. and Kawatra, S.K. (1995), "Response of an Industrial Coal Flotation Circuit to Changing Reagent Dosages," *High-Efficiency Coal Preparation* (Kawatra, ed.), Society for Mining, Metallurgy, and Exploration, Littleton, CO, pp. 307–323.

Taggart, A.F. (1945), *Handbook of Mineral Dressing*, John Wiley & Sons, New York.

Townsley, C.C. and Atkins, A.S. (1986), "Comparative Coal Fines Desulphurization Using the Iron Oxidizing Bacterium Thiobacillus ferrooxidans and the Yeast Saccharomyces cerevisiae During Simulated Froth Flotation," *Process Biochemistry*, Vol. 21, No. 6, pp. 188–191.

Trahar, W.J. (1981), "A Rational Interpretation of the Role of Particle Size in Flotation," *International Journal of Mineral Processing*, Vol. 8, pp. 289–327.

Tyumikova, V.I. and Naumov, M.E. (1981), *Improving the Effectiveness of Flotation*, English edition translated by C.D. Zundorf, Technicopy Ltd., Stonehouse, England (originally published by Izd. Nedra, Moscow).

Wadsworth, M.E., Zhu, X., Li, J., Zhong, T., Hu, W. and Bodily, D.M. (1992), "Surface Electrochemical Control for Fine Coal and Pyrite Separation," *Proceedings of the Eighth Annual Pittsburgh Coal Conference*, Pittsburgh, PA, July 1992.

Wang, D., Adel, G.T. and Yoon, R.-H. (1993), "Image Analysis Characterization of Pyrite in Fine Coal Flotation," *Minerals and Metallurgical Processing*, Vol. 10, No. 3, pp. 154–159.

Warren, L.J. (1985), "Determination of the Contributions of True Flotation and Entrainment in Batch Flotation Tests," *International Journal of Mineral Processing*, Vol. 14, pp. 33–44.

Wen, W.W. and Sun, S.C. (1981), "An Electrokinetic Study on the Oil Flotation of Oxidized Coal," *Separation Science*, Vol. 16, pp. 1491–1521.

Whelan, P.F. and Brown, D.J. (1956), "Particle-Bubble Attachment in Froth Flotation," *Bulletin of the Institute of Mining and Metallurgy*, Vol. 591, pp. 181–192.

Whitehurst, D.D., Mitchell, T.O. and Farcasiu, M. (1980), *Coal Liquefaction*, Academic Press, New York.

Wills, B.A. (1992), *Mineral Processing Technology*, 5th edition, Pergamon Press, Oxford.

Wong, K. and Laskowski, J.S. (1984), "Effect of Humic Acids on the Properties of Graphite Aqueous Suspensions," *Colloids and Surfaces*, Vol. 12, pp. 319–332.

Xiao, L., Somasundaran, P. and Vasudevan, T.V. (1989), "Air Oxidation of Bituminous Coals and its Effect on Floatability," *Proceedings of the 6th International Pittsburgh Coal Conference*, University of Pittsburgh, pp. 1173–1183.

Yancey, H.F. and Taylor, J.A. (1935), "Froth Flotation of Coal: Sulphur and Ash Reduction," U.S. Bureau of Mines, Report of Investigations RI 3263.

Ye, Y., Liu, Y., Miller, J.D. and Fuerstenau, M.C. (1990), "Preliminary Analysis of Pyrite Rejection During Chemically Modified Fine Coal Flotation by Conditioning with Monopersulfate," *Processing and Utilization of High-Sulfur Coals III* (Markuszewski and Wheelock, eds.), Elsevier, Amsterdam, pp. 159–170.

Xu, D.D. and Aplan, F.F. (1994), "Joint Use of Metal Ion Hydroxy Complexes and Organic Polymers to Depress Pyrite and Ash during Coal Flotation," *Minerals and Metallurgical Processing*, Vol. 11, No. 4, pp. 223–230.

Zimmerman, R.E. (1948), "Flotation of Bituminous Coal," Technical Paper 2397, AIME.

8 Column Flotation

It is well-known that one of the most effective techniques currently available for cleaning fine coal is froth flotation. Conventional froth flotation, however, is generally considered to be insufficiently selective to be used for desulfurizing coal, as conventional machines are incapable of removing all of the liberated mineral matter in a single pass because of entrainment effects (Sutherland and Wark, 1955; Lynch et al., 1981). Since froth flotation is only a secondary means of coal recovery in most plants, the investment in floor space and equipment is kept to a minimum, which means that in general only a single stage of flotation is used. It is therefore desirable to redesign the flotation process to produce the effect of multi-stage cleaning, while maintaining low capital and operating costs of a single stage.

Flotation columns are a promising method for improving the effectiveness of froth flotation (Eberts, 1986). A great deal of excellent work has been carried out with flotation columns, showing that they are highly effective for fine-coal cleaning (Sorokin et al., 1978; Davis et al., 1995; Peng and Li, 1995; Von Holt and Franzidis, 1994). Much of the column flotation work has been carried out with coals that are low in sulfur, and so only a limited amount of sulfur removal data is available.

A column essentially performs as if it were a multi-stage flotation circuit arranged vertically (Dell, 1985), with slurry flowing downward while the air bubbles travel upward, producing a countercurrent flow. The first flotation machine design to use a countercurrent flow of slurry and air was developed by Town and Flynn in 1919. It was not until the work of Boutin and Tremblay in the early 1960s that a new generation of countercurrent columns was developed that ultimately became industrially successful (Rubinstein, 1995).

8.1 THEORY OF OPERATION

The basic principle of column flotation is the use of countercurrent flow of air bubbles and solid particles. This is achieved by injecting air at the base of the column and feed near the midpoint. The particles then sink through a rising swarm of air bubbles. Countercurrent flow is accentuated in most columns by the addition of washwater at the top of the column, which forces all of the water that entered with the feed downward, to the tailings outlet. This flow pattern is in direct contrast to that found in conventional cells, where both air and solid particles are driven in the same direction. The result is that columns provide improved hydrodynamic conditions for flotation and thus produce a cleaner product while maintaining high recovery and low power consumption. The performance differences between columns and conventional cells may best be described in terms of the following factors: collection zone size, particle/bubble contact efficiency, and fines entrainment (Kawatra and Eisele, 1987).

8.1.1 Collection Zone

The collection zone is the volume where particle/bubble contact occurs, and it differs greatly in size between column and conventional flotation. In conventional cells, contact occurs primarily in the region surrounding the mechanical impeller. The remainder of the cell acts mainly as a storage volume for material that has not yet been through the collection zone. This creates a bottleneck, which keeps the flotation rate down. In contrast, flotation columns have a collection zone that fills the entire volume of the machine, so that there are more opportunities for particle/bubble collisions. The reduced level of turbulence needed to achieve a good rate of recovery in columns also reduces the tendency of coarse particles to be torn away from the bubbles that they attach to, and therefore columns are more effective for floating coarser coal particles (Kawatra and Eisele, 1987).

8.1.2 Particle–Bubble Contact Efficiency

Columns exhibit a higher particle/bubble contact efficiency than conventional machines due to the particles colliding head-on with the bubbles. As a result, the energy intensity needed to promote contact is less, and so the power consumption is reduced.

A second beneficial effect in certain types of flotation columns is the reduction of bubble diameter (Yoon and Luttrell, 1986). As the bubble diameter is reduced, the flotation rate of both coarser and finer particles is improved. The effects producing the improvement are illustrated in Figure 8.1. Coarse particles can attach to more than one bubble if the bubbles are small, and therefore the chance of the particle being tom loose and sinking again is reduced. For fine particles, the probability of collision with the bubble is improved if the bubble is small, as then the hydrodynamic forces tending to sweep the particle away from a collision are reduced. The reduction of bubble diameter has the added benefit of increasing the available bubble surface area for the same amount of injected air. It is therefore desirable to produce bubbles as fine as possible.

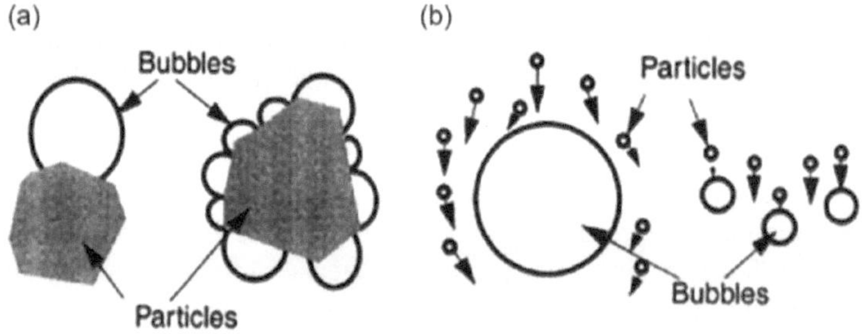

FIGURE 8.1 Effect of bubble size on flotation. Smaller bubbles are more effective due to their greater surface area, which provides more attachment surface for large particles and reduces the probability that hydrodynamic effects will prevent contact with small particles. (a) Large Particles (b) Small Particles.

8.1.3 Fines Entrainment

The entrainment of fine waste material in the froth product is a serious failing of conventional flotation machines. It results from the need for a certain amount of water to be carried into the froth as the film surrounds the air bubbles. As a result, fine suspended particles are swept into the froth with this water, even though they are not physically attached to the air bubbles. In most column flotation machines, the entrainment problem is addressed through the use of washwater, as shown in Figure 8.2. Where the conventional cell must allow a certain amount of feed water to enter the froth, the washwater in the column cell displaces this feed water to the tailings, thus preventing entrained contaminants from reaching the froth. The only drawback to the use of froth washing is that the demand for clean water is increased, which may cause water problems in some situations.

The net effect of the relatively gentle mixing, countercurrent flow, and use of washwater in columns is that there is a distance of several meters between the clean coal discharging in the froth and the concentrated gangue discharging at the tailings, with a gradual gradient of concentration between the two extremes, as shown in Figure 8.3. There is therefore a reduced possibility of coal being misplaced into the tails or of gangue short-circuiting to the froth. The result is that a column is typically equivalent to between three and five stages of conventional flotation, depending on the column design.

8.1.4 Washwater Distribution Methods

In small-diameter columns, it is relatively easy to distribute the washwater across the froth surface without introducing problems. As column diameter increases, however, it takes progressively more time for the froth at the center of the column to make its way to the edge, and so it has more time to drain and collapse, as shown in Figure 8.4 (Luttrell, 1990). If the froth collapses in the center, the material will tend to be drawn back from the overflow lip, which will slow the froth recovery. One solution is simply to install a

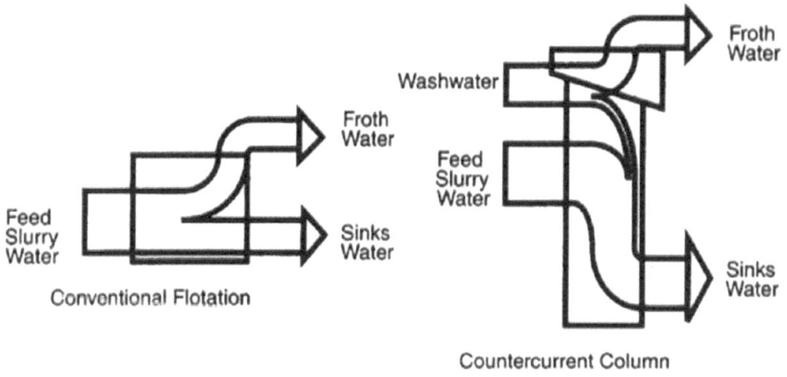

FIGURE 8.2 Effect of washwater on the entrainment of feed water into the froth. By displacing feed water with clean water, mechanical entrainment of hydrophilic particles into column froth products is largely eliminated.

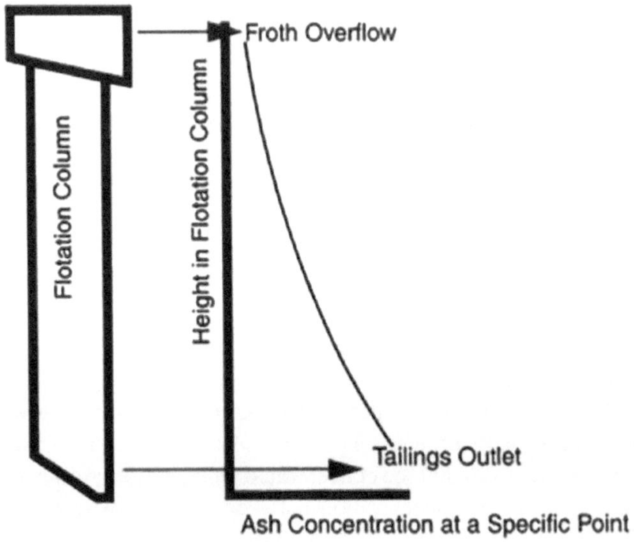

FIGURE 8.3 Concentration gradient as a function of the depth below the froth overflow in a flotation column. The ash concentration in the pulp is lowest near the froth overflow and highest near the tailings outlet. The necessary column height to achieve the desired froth and tailings grades is a function of the steepness of this gradient.

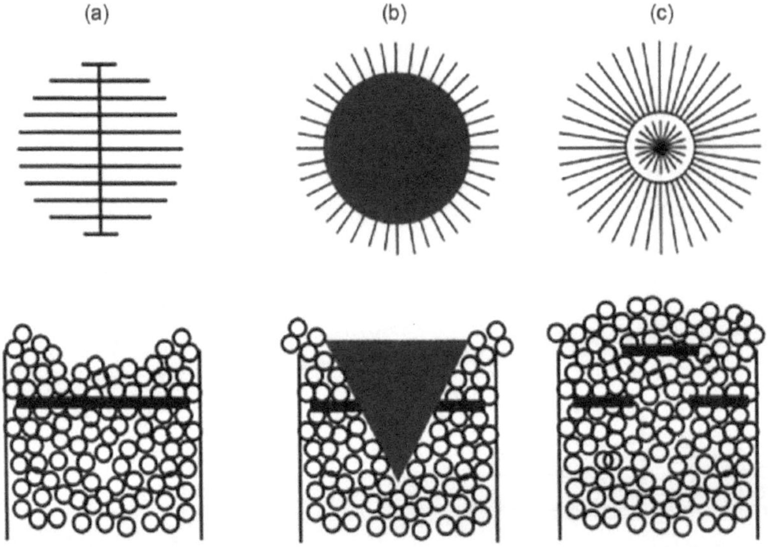

FIGURE 8.4 Systems for distributing washwater to a column froth. A simple distribution system (a) can be used for small-diameter columns, but as the diameter increases this will develop a tendency for collapse of the froth in the center of the column. A froth divertor (b) will prevent this problem, but only at the cost of reducing the available area for froth removal. Multi-level addition (c) can be used to keep the center of the froth layer fluidized and stable, while still keeping the washwater at the rim below the overflow lip, which allows adequate drainage.

divertor in the center of the column to divert froth away from the center before it can collapse. However, this tends to restrict the froth and can reduce capacity. A better solution is to use multi-level washwater addition. By raising the addition point of the washwater in the center, the middle part of the froth can be kept fluid and stable, while still allowing adequate drainage of excess water before the froth reaches the overflow lip.

8.1.5 Column Capacity

The capacity of a coal flotation column is commonly measured as the tons per hour of raw coal fed to the column. A better measure of the capacity is the quantity of clean coal that it can deliver to the froth product. This is because columns have very little restriction on material flowing toward the tailings; therefore, the quantity of ash exiting through the tailings outlet will not be the limiting factor in the column flotation of coal. The column capacity is much more strongly controlled by the rate at which solids can be transported into the froth overflow launder. It has been found that for columns <1 m in diameter, the important factor for determining capacity is the cross-sectional area of the column, with the maximum mass flowrate increasing linearly with increasing cross-sectional area. It has been suggested that the length of the overflow lip available for removing the froth is also important to the froth flowrate, but the addition of washwater makes the froth very fluid, and it can easily flow sideways over the froth lip. As a result, the length of the froth overflow lip does not become important to the capacity until the column diameter becomes very large.

If too much floatable material is fed into a column, the carrying capacity of the froth will be exceeded. When this happens, the coarser particles will be preferentially displaced from the froth and will be lost to the tailings. This results in a loss of recovery and causes the particle size distribution of the froth product to become finer. For flotation of coal, where most of the solids fed to the column report to the froth, the capacity is controlled by the amount of clean coal that can be transported by the froth layer. The main body of the column is not obstructed by air bubbles to the degree that the froth layer is, and as a result the solids can move more freely when they are not in the froth layer. Since there is less volume of material in the sinks product than in the froth product, the capacity of the column to handle tailings is unlikely to control the quantity of coal that can be processed. As a result, modifications to the column that change or moderately restrict the flow in the pulp phase, but which do not affect the froth layer, are unlikely to have a significant effect on the column capacity (Finch and Dobby, 1990).

8.1.6 Flow Regimes in Column Flotation

A column gives the best performance when it is operating under "plug-flow" conditions. This condition is met when there is little turbulent mixing along the vertical axis of the column. The desirability of plug flow places a limit on the rate at which air can be injected into the column, as illustrated in Figure 8.5 (Luttrell, 1990). Excessive amounts of air lead to large air bubbles, which cause vertical mixing and destroy the desired plug-flow condition. It is therefore important to avoid the use of excess air in a flotation column and to prevent the formation of large bubbles.

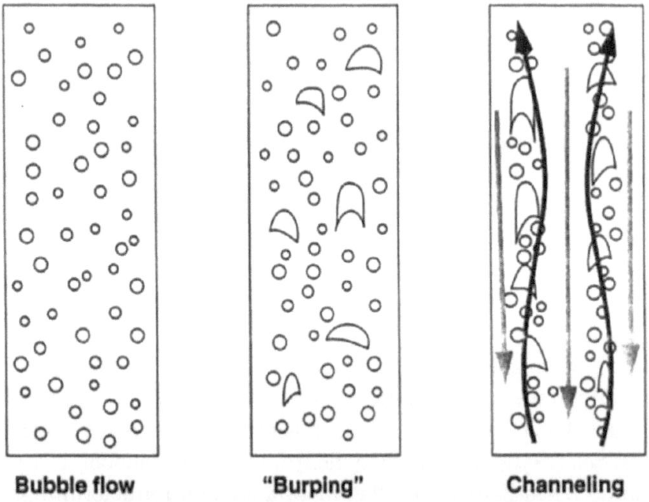

FIGURE 8.5 Effects of bubbles on the flow in a column. When bubbles are relatively small, they rise slowly through the slurry, and the flow in the column is uniform, with little turbulence. When bubbles coalesce and become large, they rise very rapidly, giving rise to "burping," which introduces a great deal of turbulence and can disrupt the froth. When bubbles become still larger, the upward flow of water that they cause begins to drag the other bubbles upward as well, producing fast "channels," which cause vigorous mixing along the vertical axis of the column. Channeling reduces the effectiveness of the column and must be avoided.

8.1.7 Other Advantages of Flotation Columns

Most flotation columns do not contain rotating impellers and have a minimum number of moving parts. As a result, they have lower operating and maintenance costs than conventional cells. Their mechanical simplicity, and their ability to perform a separation with a single machine that would otherwise require an entire bank of conventional cells to perform, also translates to lower capital costs. It should also be noted that the vertical orientation of a flotation column results in a smaller floor space requirement than conventional cells.

8.1.8 Scale-Up of Flotation Columns

Scale-up models are available for predicting the behavior of a full-scale column from laboratory- and pilot-scale experiments. A useful approach is to make use of the collection zone rate constant in combination with the froth zone recovery, which is more comprehensive than models based only on the overall rate constant. Overall recovery, R_{fc}, is given by

$$R_{fc} = \frac{R_c R_f}{R_c R_f + 1 - R_c} \tag{8.1}$$

where R_c is collection zone recovery and R_f is froth zone recovery.

For calculation of the collection zone recovery, a suitable expression is (Finch and Dobby, 1990)

$$R_c = 1 - \frac{4\alpha \exp\left(\frac{1}{2N_d}\right)}{(1+a)^2 \exp\left(\frac{\alpha}{2N_d}\right) - (1-a)^2 \exp\left(\frac{-\alpha}{2N_d}\right)} \tag{8.2}$$

where

$$\alpha = \left(1 + 4k_c\tau_p N_d\right)^{1/2} \tag{8.3}$$

$$N_d = \frac{0.063\, d_c \left(J_g/1.6\right)^{0.3}}{\left[\left(\frac{J_{sl}}{1-\varepsilon_g}\right) + U_{sp}\right] H_c} = \text{Vessel dispersion number} \tag{8.4}$$

$$U_{sp} = \left[\frac{g\, d_p^2 \left(\rho_p - \rho_{sl}\right)\left(1-\varphi_s\right)^{2.7}}{18\mu_f\left(1+0.15Re_p^{0.687}\right)}\right] = \text{Slip velocity between particles and water} \tag{8.5}$$

$$Re_p = \frac{d_p U_{sp} \rho_l \left(1-\varphi_s\right)}{\mu_f} = \text{Reynolds number of a single particle} \tag{8.6}$$

$$\tau_p = \tau_l \left[\frac{J_{sl}/\left(1-\varepsilon_g\right)}{\left(J_{sl}/\left(1-\varepsilon_g\right)\right) + U_s p}\right] = \text{Mean residence time of particles} \tag{8.7}$$

$$\tau_l = \frac{H_c\left(1-\varepsilon_g\right)}{J_{sl}} = \text{Mean residence time of liquid} \tag{8.8}$$

k_c = Flotation rate constant of a particle type in the collection zone,
N = Number of flotation columns needed to reach the target recovery,
J_{sl}, J_g = Superficial velocities of slurry and gas, respectively,
ε_g = Gas holdup fraction,
H_c = Column height,
d_c = Column diameter,
g = Acceleration due to gravity,
d_p = Particle diameter,
ρ_p, ρ_l, ρ_{sl} = Densities of particles, liquid phase, and bulk slurry, respectively,
φ_s = Volume fraction solids in the slurry,
μ_f = Viscosity of the fluid.

Froth zone recovery, R_f, is less amenable to analysis. In general, the froth zone recovery is dependent on "drop back," which is the return of particles from the froth phase back to the pulp phase. Drop back is known to be size dependent, increasing as the particle size increases. The overall drop back is typically near 50%, or even higher in larger diameter columns, with a greater degree of mixing in the froth (Finch and Dobby, 1990).

8.2 TYPES OF FLOTATION COLUMNS

Obviously, a description of all of these column types is far beyond the scope of this book. What follows is a discussion of a limited number of column types, concentrating on those designs that are most commonly used for coal desulfurization work. Detailed descriptions of flotation columns, including mathematical treatments of their operation, are given by Finch and Dobby (1990) and Rubinstein (1995).

Flotation columns can be most easily divided into two general categories: unobstructed designs, and partially or fully packed designs. While the first flotation columns were predominantly of the unobstructed type, there are certain situations where the somewhat more complex packed or partially packed columns have advantages.

8.2.1 Unobstructed (Conventional) Columns

The unobstructed column types are nearly identical to each other, consisting of open chambers ~10–14 m tall with washwater added at the top and air bubbles introduced at the bottom. The major differences between unobstructed columns from various manufacturers are mainly in the nature of bubble generators and in the degree of instrumentation and control with which they are equipped. A number of different columns of this type are commercially available, including the Microcel, Deister Flotaire, and Cominco column. The mechanical simplicity of columns has also made it possible for many mineral processing plants and research organizations to build their own units. While columns of this type have become common in metallic ore concentrators, they are not yet widely introduced in coal-cleaning plants.

8.2.1.1 Conventional Column Designs

The basic configuration of unobstructed columns is illustrated in Figure 8.6, which is based on the columns that were made by the Column Flotation Corporation of Canada (CFCC) (Dobby and Finch, 1986; Dobby et al., 1985; Moon and Sirois, 1983; Kawatra and Eisele, 1987). Air is injected through a diffuser at the base, and the level is controlled using a level sensor and an automatic control valve on the tailings line. A froth depth of 0.6–1.2 m is maintained, and feed is introduced a short distance below the base of the froth layer. In the original design, the air diffuser was sintered ceramic, which produced very fine air bubbles. This was found to suffer from plugging problems, however, particularly in hard water, and so cloth and perforated rubber sheeting were adopted instead (Dobby et al., 1985; Boutin and Wheeler, 1967). The flexible diffuser materials resist plugging due to the flexing that they undergo during operation, which breaks loose the minerals that crystallize on the diffuser. Internal spargers will still eventually plug over a period of weeks, requiring that the column be drained so that they can be replaced.

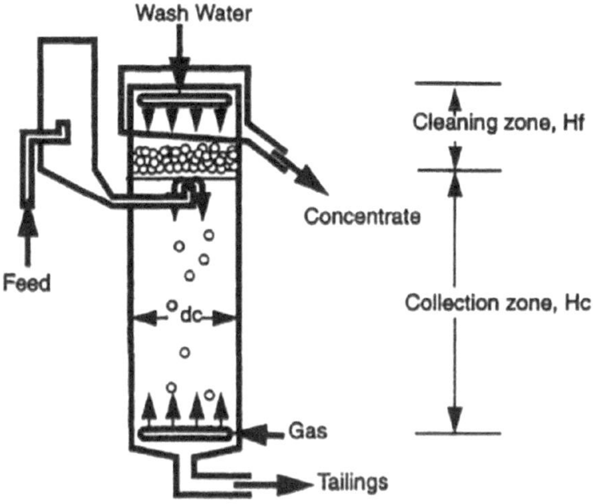

FIGURE 8.6 Schematic showing the critical features of a flotation column.

The Deister Flotaire column, which was originally developed for phosphate flotation, initially had a number of important differences from the CFCC column, although the two types of machines have become steadily more similar over the years. The early Flotaire machines were <5 m tall, did not use washwater, and introduced the use of an external bubble generator rather than internal spargers (Hollingsworth, 1981; Im and Wolfe, 1986; Zipperian, 1984; Zipperian and Christophersen, 1985). Current versions have added washwater sprays similar to the CFCC design and have increased in height. The external bubble generator has evolved to a type that uses less water than the simple aspirators that were used originally. The external bubble generators can be serviced without draining the column, thus greatly simplifying column maintenance. A schematic diagram of the Flotaire bubble generator is shown in Figure 8.7. Air is diffused through the porous plastic sleeve, and the bubbles are sheared off by the water flowing through the center of the sleeve. The air/water mixture is then injected directly into the column immediately after leaving the generator, to minimize bubble coalescence.

The volume ratio of air to water for the Deister system can be as high as 8:1, which is a considerable improvement over aspirators which have an air-to-water ratio on the order of 0.15:1. It is important to maintain a high ratio of air to water in these bubble generators, because water that is provided to the bubble generators must be clean and also tends to dilute the column slurry, which can cause difficulties with water supplies and recycling in a plant situation.

Similar systems are used by the Bureau of Mines external bubble generator shown in Figure 8.8 (McKay et al., 1988), and by the microbubble generator developed for the Microcel column, illustrated in Figure 8.9 (Yoon et al., 1990). The U.S. Bureau of Mines (USBM) Turbo-Air bubble generator is similar in design to the system used by the Cominco column (Rubinstein, 1995). The Cominco aerators have been installed on more than 100 columns manufactured by other companies, and over 75 Cominco columns have been installed around the world. The Cominco column has been used for applications similar to coal flotation, such as the de-inking of recycled

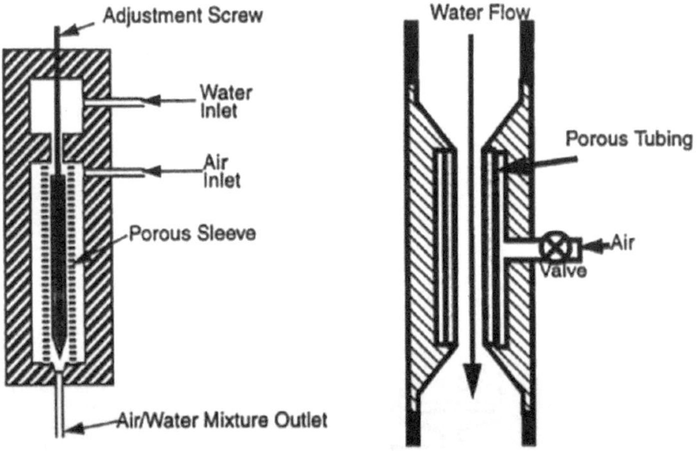

FIGURE 8.7 Schematics of two designs used for the Deister Flotaire external bubble generators. These generators have a much higher air/water ratio than the original aspirator types and are much easier to service than generators inside the column.

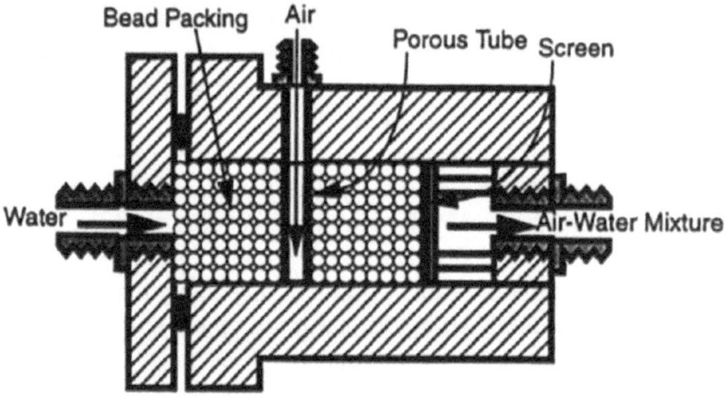

FIGURE 8.8 Turbo-Air bubble generator developed by the Bureau of Mines.

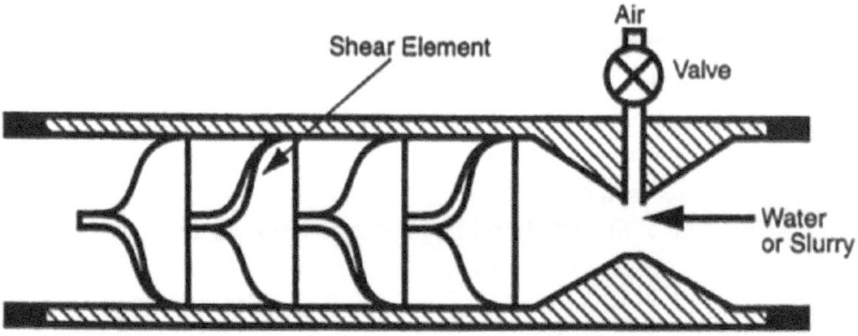

FIGURE 8.9 Bubble generator used with the Microcel column. This generator is designed to allow recycled slurry to be used as the liquid instead of clean water. The static shear elements introduce a vigorous mixing action as the slurry flows past them.

newsprint (Murdock, 1995, Personal Communication), but they have not yet been used by the coal industry.

Most research work with unobstructed columns is done using some type of external bubble generator due to the greater flexibility, ease of servicing, and finer bubble size distribution produced by these units. The primary drawback of the external bubble generators is that they require the addition of clean water to form the bubbles and convey them to the column. While recycled water can be used for aspirator-type bubble generators, the recycled water in coal plants is often too dirty to be used in the more sophisticated bubble generators, which have a lower water demand. The use of these devices can therefore result in high plant water consumption, and also may produce excessively dilute column tailings products which can be difficult to thicken. The Microcel bubble generator manages to avoid many of these problems by using a design that does not require porous frits to produce the bubbles, and therefore can be used with dirty water or even slurries. A schematic representation of a typical Microcel unit is shown in Figure 8.9. In this device, air bubbles ranging from 0.1 to 0.4 mm in diameter are generated by passing air and a portion of the flotation pulp through an in-line static mixer. The intense high-shear agitation provided by the in-line mixers generates smaller air bubbles than other commercially available air sparging systems. The Microcel bubble generators are resistant to plugging and can be serviced without column shutdown. A diagram of the complete Microcel column is shown in Figure 8.10 (Davis et al., 1995).

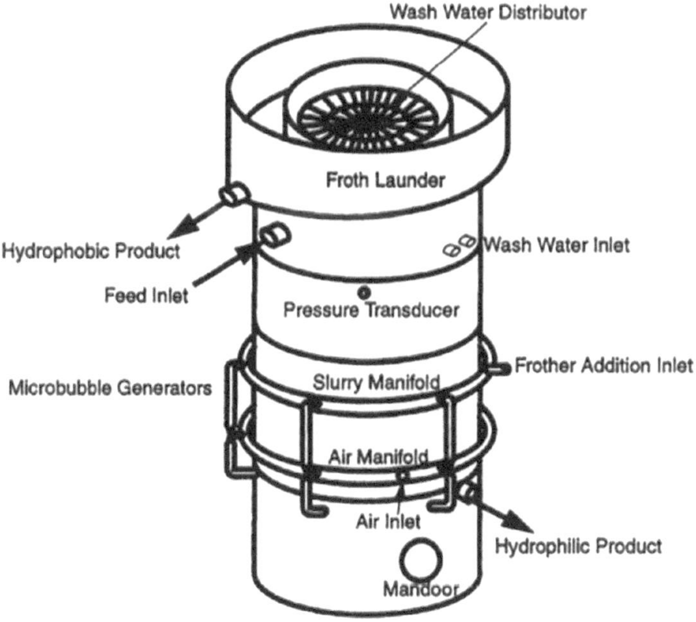

FIGURE 8.10 Schematic of the Microcel column. The main difference between this unit and other conventional columns is that the bubble generator (Figure 8.9) allows slurry to be used to generate bubbles, which reduces the column's demand for clean water.

A Microcel was installed at the Lady of Dunn coal operations, USA. It did not achieve the performance guarantees. This resulted in a significant fall in Microcel coal installations (Phillips et al., 1997).

Li et al. (2003) have applied a cyclone-microbubble column flotation approach to fine coal. While initial experimental results appear promising, it is worth remembering that the main advantage of a column is it does not have moving parts. By introducing a cyclone in a flotation column, this advantage is lost.

8.2.1.2 Performance of Conventional Columns

8.2.1.2.1 Release Analysis

To evaluate the performance of a flotation column, some method is needed to determine what a perfect flotation separation would be with a particular coal. To do this, a procedure is needed that will do for froth flotation what float–sink analysis does for density separations. The "release analysis" test procedure has been developed for this purpose by Dell et al. (1972) and was intended to produce the best grade/recovery curve that can be accomplished by froth flotation of a particular material using a particular reagent. The procedure is as follows:

- Using a conventional laboratory flotation machine, run a flotation test until no more floatable material is present, which will represent the maximum recovery.
- Re-float the froth product from the first stage, again taking care to maximize the recovery. Add additional reagents if necessary. Repeat this for several re-cleaning stages (as many as seven or eight), until the quantity of nonfloatable material is negligible.
- Take the final cleaned froth product and fractionally float it without adding any reagents. For each froth fraction, increase the air flow and the mixing intensity. Continue collecting froth products until the cell is barren of floatable material. Analyze each fraction of the froth separately and plot the results to produce the release analysis grade/recovery curve.

The release analysis procedure is illustrated in Figure 8.11.

In recent work (Honaker and Paul, 1994), it has been found that the release analysis procedure does not actually give the theoretically optimum flotation performance because some advanced column flotation technologies can produce results that are superior to the release analysis results.

Release analysis is very important. If the results of flotation are well below the values indicated by the release analysis, it is necessary to recheck the procedures and analytical methods.

8.2.1.2.2 Processing Results

Groppo (1986) studied column flotation of coal with a 5.08 cm (2 in) diameter column using the CFCC design and compared its results with conventional flotation. Both the conventional flotation and the column flotation produced concentrates with 5%–8% ash at 80%–90% energy recovery; however, the column produced these results while operating at 15%–17% solids, while the conventional cell had to be

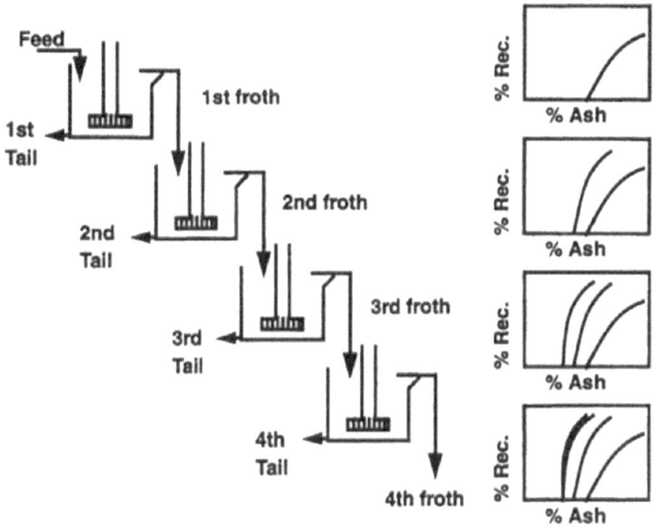

FIGURE 8.11 Schematic of the release analysis procedure. Material is re-floated numerous times, until all of the nonhydrophobic material is removed. Timed flotation of this final product then gives the release analysis curve.

diluted to only 4%–6% solids. The effect of washwater flowrate on the column was also determined, while the column was operating at 11.5% solids, with 0.5 kg/metric ton of fuel oil added as collector, and methyl isobutyl carbinol (MIBC) added to the water at a dosage of 20–30 µL/L. When the washwater flowrate was increased from 200 to 600 mL/min, the concentrate grade improved from 9% ash to only 6.5% ash, while the recovery of coal appeared to be unaffected.

Based on pilot test results (Groppo and Parekh, 1990; Parekh et al., 1986), four columns were installed at the Powell Mountain Coal Co., each of which were 2.4 m in diameter and 6.7 m tall (Groppo et al., 1993). These columns were used to process the classifying cyclone overflow, which previously had been diverted to the tailings dump. The feed slurry was fed to the columns at a rate of 800 m^3/h, at <20% solids, and the coal contained 24%–48% ash. The ash content of the clean product was 5.98%–8.63%, and a comparison between the pilot-scale tests and the plant tests is given in Figure 8.12. Groppo et al. (1993) also evaluated various bubble-generating devices and found that there was no measurable difference in column performance between an external bubble generator and static spargers.

Davis et al. (1994) carried out an extensive study of column flotation at the Middle Fork Preparation Facility of the Pittston Coal Co., using laboratory-, pilot-, and full-scale columns. The performance of the full-scale column was compared with the existing conventional flotation banks at the plant, with the results shown in Figure 8.13. The conventional flotation circuit consisted of a five-cell rougher bank with a total volume of 8.5 m^3, while the column flotation circuit consisted of five 3-m diameter Microcel flotation columns. The flotation feed was 100 tons/h of raw coal finer than 150 µm (Davis, 1993).

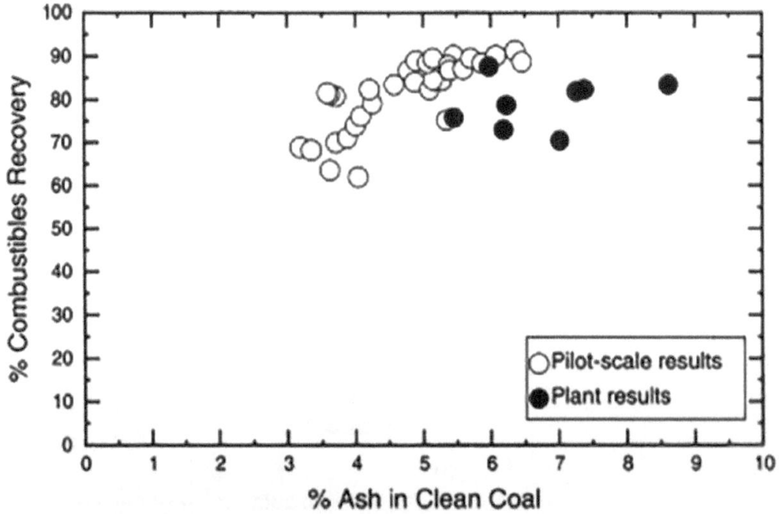

FIGURE 8.12 Comparison of pilot- and full-scale flotation performance at the Powell Mountain coal preparation facility.

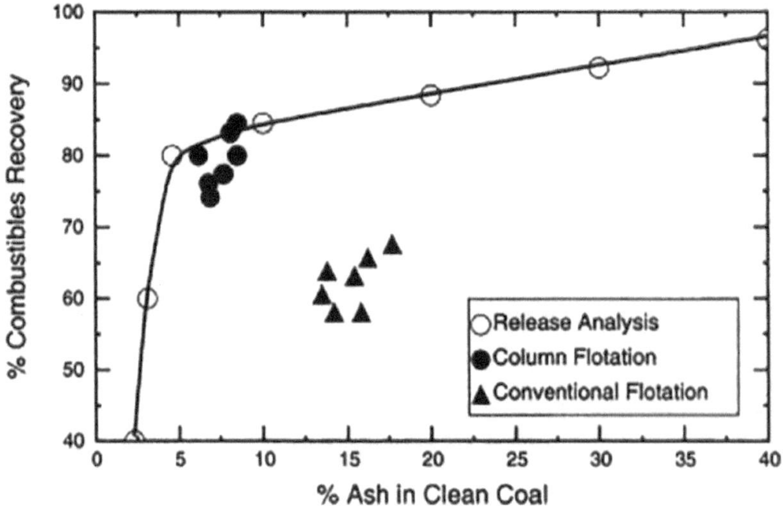

FIGURE 8.13 Comparison of full-scale column flotation and conventional flotation at the Middle Fork coal preparation plant (Davis, 1993).

The Microcel was also tested at the Pittsburgh Energy Technology Center (VPI, 1995), using a 50.8 cm (20 in) diameter column 9.1 m (30 ft) tall. Results of this work are given in Figure 8.14, which shows that the column could remove approximately 80% of the ash and 60% of the pyritic sulfur, while maintaining a combustibles recovery of 80%–90%.

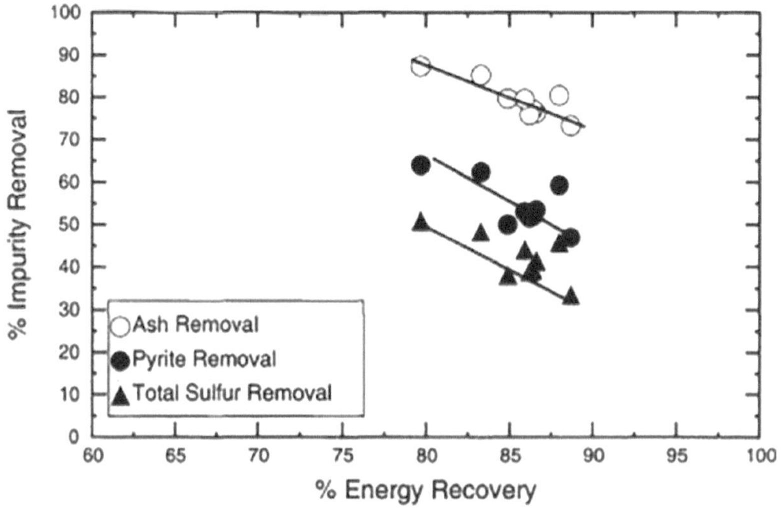

FIGURE 8.14 Results of pilot-scale coal desulfurization tests using the Microcel flotation machine (VPI, 1995).

8.2.1.3 The Jameson Cell

The Jameson cell is an Australian development, and it is a significant departure from conventional column designs (Kennedy, 1990). The bubble-particle contact and attachment are carried out rapidly, outside of the main body of the column. This makes it possible for the Jameson cell to be much more compact than other columns while still maintaining high recovery.

8.2.1.3.1 Principles of Operation

The Jameson cell consists of two main parts, as shown in Figure 8.15. In the downcomer, the particles and air come in contact with each other and an intimate mixture of air and slurry is produced. This mixture is accomplished by injecting the feed at high pressure through an orifice plate, producing a slurry jet that pulls air into the downcomer by aspiration. The air/slurry mixture that is produced has up to 60% void space (Manlapig et al., 1993; Atkinson et al., 1995). Due to the high mixing velocity and extensive bubble surface area in the downcomer, there is rapid contact and capture of the hydrophobic particles by the bubbles during their 5–10 s residence time in this zone.

When the slurry from the downcomer enters the flotation tank, the bubbles are free to rise to the surface along with the attached hydrophobic particles, while the hydrophilic particles and water are free to discharge to the tailings. In some cases, washwater is added to the top of the froth to aid in washing out the entrained fine hydrophilic particles.

In coal flotation with the Jameson cell, it has been found that a single stage can recover 70%–80% of the total combustible material. To achieve higher recoveries, a second stage can be added to recover up to 95% of the combustibles. A full

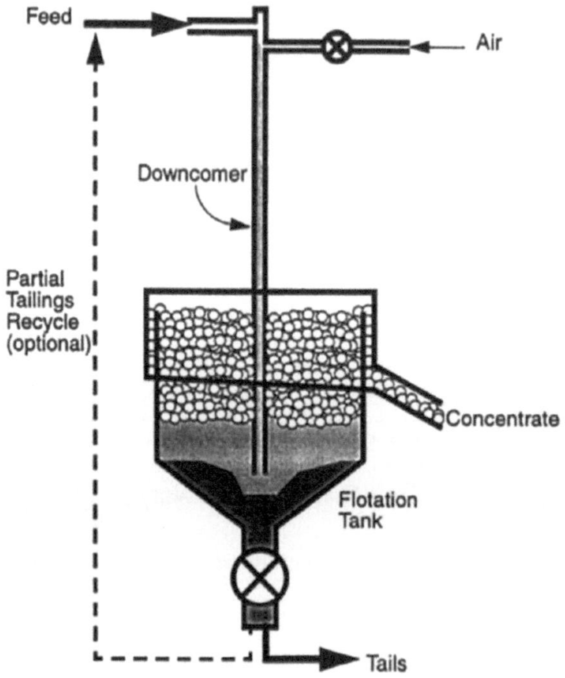

FIGURE 8.15 Schematic of the Jameson flotation cell. A portion of the tailings can be split off and returned to the feed line to increase recovery, if desired.

second stage, however, is generally not necessary, as it is more economical to use partial recycle of the tailings back to the feed. With partial tailings recycle, there is a significant probability that a hydrophobic particle that missed being floated in the first pass will run through the column a second time and be collected. When a feed is relatively low in ash, the performance of a single stage is limited by the amount of available bubble surface area, and partial recycle is effectively a method for increasing the amount of bubble surface area that is exposed to the feed.

8.2.1.3.2 Jameson Cell Performance

The Jameson cell has been installed in several Australian coal preparation plants (Atkinson et al., 1995). The results from one such plant installation are shown in Table 8.1. In parallel tests with conventional cells, the Jameson cell has been found to give superior performance for both fine and coarse coal flotation.

8.2.2 Packed and Baffled Columns

While the standard unobstructed column designs are quite well-accepted, their great height is somewhat inconvenient, both due to headspace requirements and to the large amount of water that is held up in them. This height is necessary to produce good froth grades and product recoveries, as otherwise backmixing of gangue into

TABLE 8.1
Results of Flotation of Various Process Streams in an Australian Coal Preparation Plant, Using a Jameson Cell (Atkinson et al., 1995)

Coal Source	Cyclone Overflow		Cyclone Feed		Thickener Underflow	
Froth washwater used?	Yes	No	Yes	No	Yes	No
Feed % solids	4.7	5.2	11.8	13.2	10.0	10.0
Feed % ash	33.3	30.0	22.9	23.5	57.2	55.9
Concentrate % solids	17.6	16.3	26.1	26.0	12.0	12.0
Concentrate % ash	4.8	9.0	6.0	9.7	7.5	8.3
Concentrate % yield	57.2	66.2	70.2	74.7	30.7	34.0
% Combustibles recovery	81.6	86.0	85.6	88.2	66.2	70.7
Carrying capacity (g/min/cm^2)	1.3	3.6	3.6	4.5	1.25	1.25

the froth would degrade product quality, while short-circuiting of coal to the tailings would reduce recovery. If the hydrodynamics of the column are changed by introducing flow obstructions of various types, backmixing and short-circuiting can be reduced considerably. Flow obstructions can also improve particle-bubble contact efficiency and thus increase the flotation rate of very fine particles, which is of great importance for producing superclean coals. Unfortunately, such obstructions also tend to increase the likelihood that the column will plug. To date, only two types of obstructed column have been used industrially; the Wemco-Leeds column (Dell, 1985) and the Static Tube (Yang, 1984).

8.2.2.1 The Wemco-Leeds Column

The Wemco-Leeds column was originally designed with the specific goal of approximating the operation of a multi-stage flotation circuit with less space and lower mechanical overhead (Dell, 1976, 1978, 1985; Dell and Jenkins, 1976). It resembles a very deep conventional flotation cell bank, as shown in Figure 8.16. The upper portion of the cell is divided into a series of chambers by several horizontal barriers. Each barrier consists of a rack of fixed rods supporting a set of movable plastic rods, as shown in Figure 8.17. The intent of this barrier design is to trap a thin layer of froth underneath each, with the movable rods lifting to allow froth into the upper chambers when the froth depth underneath them is sufficiently great. Washwater is added to the uppermost chamber of this machine, which descends to wash each froth layer in succession. Since bubbles only remain in the froth layers beneath the barriers for a short time, the bubbles do not coalesce significantly. Also, the action of the rods will tend to break up any bubbles that do become too large.

Aside from the horizontal barriers and the washwater, the Wemco-Leeds column is operated in essentially the same fashion as a conventional bank of flotation cells. The large available surface area for removing froth is also beneficial, as this reduces the chances of the machine becoming choked with froth, which can be a problem with flotation of coal using other types of columns.

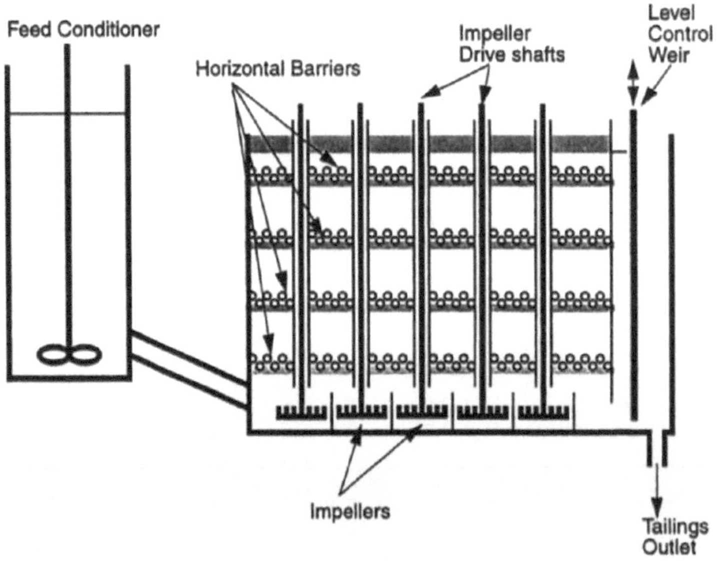

FIGURE 8.16 Schematic of the Wemco-Leeds flotation machine.

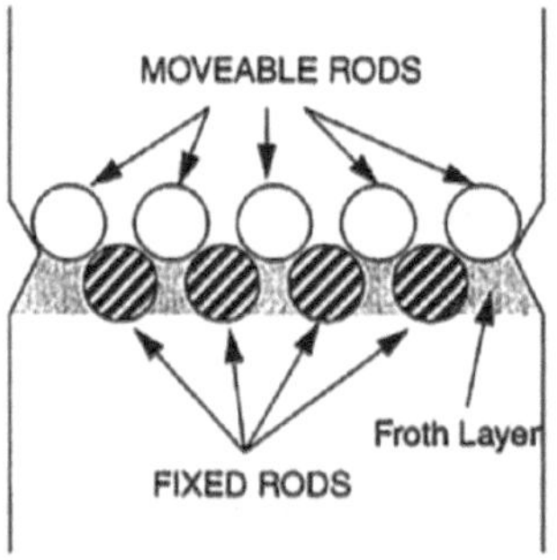

FIGURE 8.17 Detail of the Wemco-Leeds froth barrier design. The movable rods act as valves to restrict the flow of froth and thus improve the cleaning action.

The Leeds column was mainly developed as a coal-cleaning device and was designed to be able to deal with the large froth volumes characteristic of coal flotation. In bench-scale studies, the performance of the unit was compared with the performance of conventional mechanical flotation machines of the type described in Chapter 6, and also compared with the results produced by "release analysis" (see Section 7.2.1.2.1). The results, see Figure 8.18, show that the Wemco-Leeds column can more closely approach the ideal performance than can be done using either a single stage of conventional flotation or a two-stage rougher/cleaner circuit (Degner and Person, 1991).

A pilot-scale Wemco-Leeds column was also tested, using a row of four cells. Each cell had independent tankage, rotors, froth skimmers, and rod barriers, and had a square cross-section. The minimum volume of individual cells was 0.028 m^3. In the

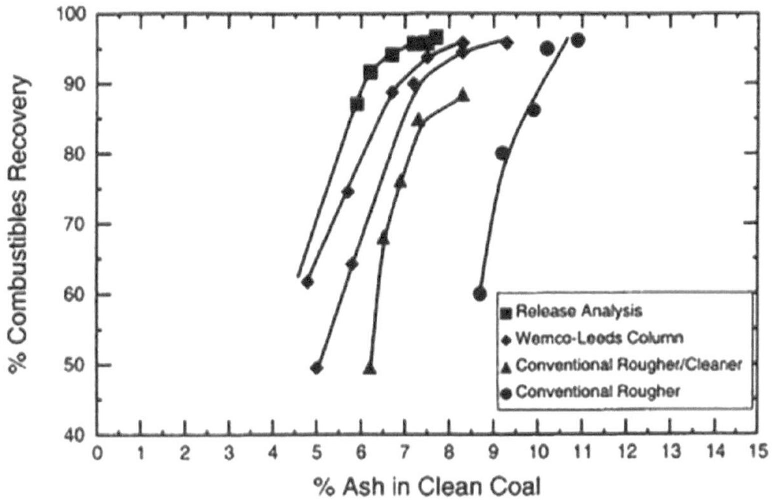

FIGURE 8.18 Bench-scale comparison of the Wemco-Leeds column with release analysis results, single-stage conventional flotation, and two-stage rougher/cleaner flotation.

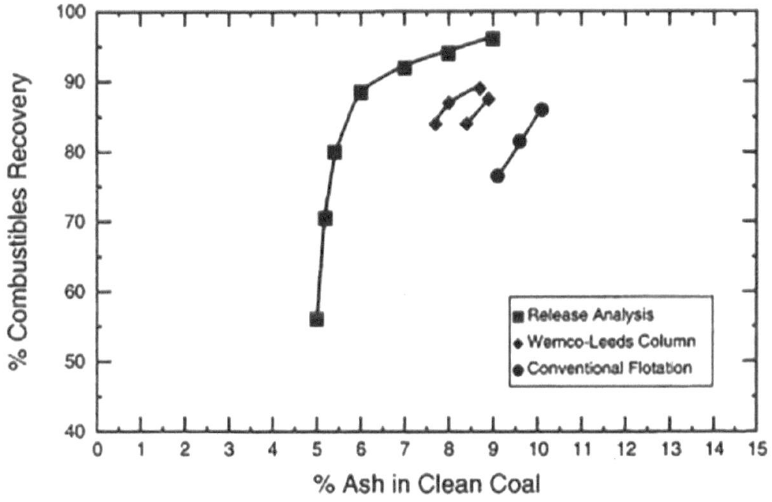

FIGURE 8.19 Comparative pilot-scale results using a Wemco-Leeds column and a conventional flotation cell. The Wemco-Leeds column is significantly closer to the theoretically ideal performance shown by release analysis, showing that it provides a more selective separation than conventional flotation.

course of pilot plant testing, the number of rod barrier sets was varied from three to six. Conventional pilot-scale flotation machines were tested in parallel with the Wemco-Leeds column. The results, shown in Figure 8.19, confirm that the Wemco-Leeds column can more closely approach the ideal performance shown by release analysis than is possible with a conventional flotation cell.

Analysis of the pilot-scale results showed that for most applications, four sets of rod barriers are sufficient. A progressively cleaner product can still be produced by adding additional barrier sets, but the improvement per additional barrier stage becomes small when there are more than four stages (Degner and Person, 1991).

8.2.2.2 The Static Tube (or Packed Column)

The Static Tube, shown in Figure 8.20, carries the introduction of flow restriction to an extreme case, in order to provide high collection efficiency and selectivity for micron-sized particles. The collection zone is completely packed with corrugated plates, which produce very tortuous flow channels only a few millimeters across. Due to the small diameter of the channels, no diffuser or bubble generator is needed, as the mere act of traveling through the packing breaks up the air bubbles (Yang, 1984, 1988). The presence of the packing prevents hydrodynamic forces from sweeping bubbles and particles apart, and so forces bubble–particle contact regardless of the size of the particle.

An additional benefit that is claimed for this unit is that it is reportedly easier to scale-up than conventional columns, because the presence of the packing makes the height-to-diameter ratio less important. It can therefore maintain plug flow even for large diameters. The packing is also claimed to provide mechanical support for the froth layer, and so a deeper froth can be maintained than is practical with other columns.

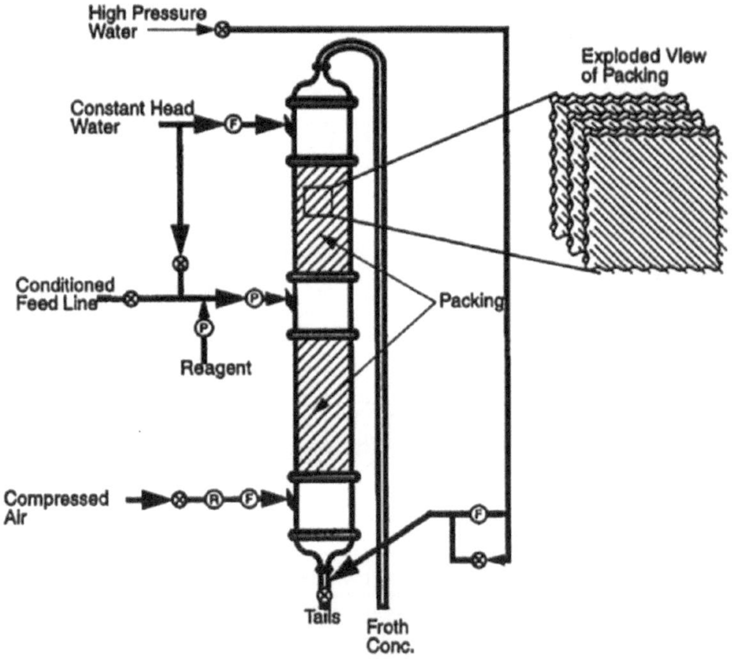

FIGURE 8.20 The Static Tube flotation machine. The packing material is corrugated sheets, which force the air and slurry to follow tortuous paths through the column. This is designed to allow flotation of particles that are too fine for efficient froth flotation using other types of flotation machine.

While performance is good for coal that is ground to sizes finer than 20 μm, coarser particles tend to abrade and plug the column. Also, if the froth at the top of the column becomes too dry, it can solidify and plug the machine, making it necessary to backwash the packing periodically. A static tube was installed at the Copper Range Co. processing plant in White Pine, MI, and while it was found that the results were superior to conventional flotation machines, it was not ultimately used in the plant because of the plugging problems. It was therefore replaced by a Deister Flotaire column (Rusesky, 1989, Personal Communication).

While some investigators have claimed good performance from the packed column, others have had poor results. Shiao (1994) carried out a series of CO_2 flotation tests with the packed column and found that the packing in the packed column seemed to hinder the separation. The rationale of inserting the packing in the packed column was that the packing can break large bubbles into smaller bubbles, prevent vertical mixing of the froth, increase bubble–particle interaction, and stabilize the froth on the top of the column. In experiments using CO_2 as the flotation gas, however, packing retarded the coal from floating toward the top of the column. As a result the separation efficiency was reduced. When the same tests were carried out with a standard microbubble column, the microbubble column gave superior results. The packed bed column has no future in coal processing, however, as the solids settle in the packing.

8.3 COMPARISONS OF COMMERCIALLY AVAILABLE COLUMNS

A comparative survey of column performance, carried out by the Pittsburgh Energy Technology Center, obtained interesting results by sending identical coals to various laboratories that were conducting column flotation research. The tests were therefore performed by investigators who were experienced in operating each type of column (Jacobsen et al., 1989). The flotation machines that were compared were a conventional Denver machine, and the Kentucky, Flotaire, Microcel, and Packed columns. The results of this survey are given in Table 8.2, where the "Ash Efficiency Index" and the "Pyritic Sulfur Efficiency Index" were calculated from the formulas:

$$E_a = Y\left(\frac{A_t}{A_c}\right) \tag{8.9}$$

$$E_{py} = Y\left(\frac{Py_t}{Py_c}\right) \tag{8.10}$$

where E_a = Ash efficiency index;
E_{py} = Pyrite efficiency index;
A_t = Ash content of the tailings;
Py_t = Pyrite content of the tailings;
A_c = Ash content of the clean-coal product;
Py_c = Pyrite content of the clean-coal product; and Y = Concentrate yield.

TABLE 8.2
Results of the PETC Interlaboratory Comparison of Conventional Flotation with Several Types of Flotation Column (Jacobsen et al., 1989)

	Conventional Flotation Cell	CAER[a] Column	Packed Column	Deister Column	Microcel Column
Clean-coal % weight recovery (yield)	87.3 ± 6.1	84.5 ± 5.2	77.5 ± 3.6	79.9 ± 4.4	77.7 ± 7.7
% Energy recovery	91.9 ± 5.5	98.3 ± 1.0	90.5 ± 4.9	88.2 ± 4.2	89.4 ± 7.1
Clean-coal % ash	7.4 ± 0.6	4.1 ± 1.5	2.9 ± 0.4	4.8 ± 1.0	3.3 ± 1.2
Ash efficiency index	500 ± 166	1,662 ± 378	1,537 ± 195	810 ± 203	1323 ± 65
Clean-coal % pyritic Sulfur	1.02 ± 0.54	0.39 ± 0.28	0.29 ± 0.17	0.44 ± 0.18	0.56 ± 0.24
Pyritic sulfur efficiency Index	489 ± 202	2,813 ± 2,399	2,124 ± 1,212	1,062 ± 625	890 ± 359

[a] CAER, Countercurrent column at the Center for Advanced Energy Research, University of Kentucky.

A large value for the efficiency index indicates an efficient separation, because an efficient separation will have a low ash or pyrite content in the concentrate, a high ash/pyrite content in the tailings, and a high yield.

The operating parameters were not given in this study, so the factors such as reagent dosages and aeration rates are not known. The lowest ash and pyritic sulfur contents were produced by the packed column, although its comparatively low recovery resulted in efficiency indexes that were comparable to those of the Countercurrent column at the Center for Advanced Energy Research (CAER) column.

An additional comparison of the various column types was carried out by Honaker and Paul (1994), including the Jameson Cell, Flotaire Column, Turbo-Air Column,

TABLE 8.3
Cell Geometries and Critical Operating Parameters of the Columns Used in the Comparative Study Reported by Honaker and Paul (1994)

	Canadian Column	Flotaire Column	Jameson Cell	Microcel Column	Packed Column	Turbo-Air Column
Cell height (m)	4.8	4.8	1.5	4.8	4.8	4.8
Diameter (cm)	5	5	16.5	5	10	5
Feed inlet depth (m)	1.5	1.5	1.2	1.5	2	1.5
Washwater depth (cm)	13	13	7.5	13	0	13
Superficial gas rate (cm/s)	2.4–5.2	3.0–4.0	1.0–1.5	2.0–2.5	6.0–8.0	2.8–3.5
Aeration pressure (kPa)	276	242	N/A	249	152	415
Frother concentration (ppm)	20–30	20–30	15–25	20–30	30–45	20–30
Washwater rate (cm/s)	0.40–0.60	0.40–0.60	0.15–0.23	0.40–0.60	0.30–0.50	0.40–0.60
Froth height (cm)	30–40	30–40	20–46	30–40	250–390	30–40

Packed Column, Microcel, and Canadian column. The geometries of the columns examined, along with the critical operating parameters, are given in Table 8.3. Results of these comparative tests are given in Figures 8.21 and 8.22. From these removal/recovery curves, it is seen that all of the columns have comparable levels of performance as they all follow the same general curves. The Jameson cell achieves lower energy recovery at a given level of impurity removal than the other columns do. Feed dilution or partial tailings recycle can be used to boost the recovery of the Jameson cell.

China dominates the flotation column installation market. More than 1,000 flotation columns have been installed in China (Harbort and Clarke, 2017).

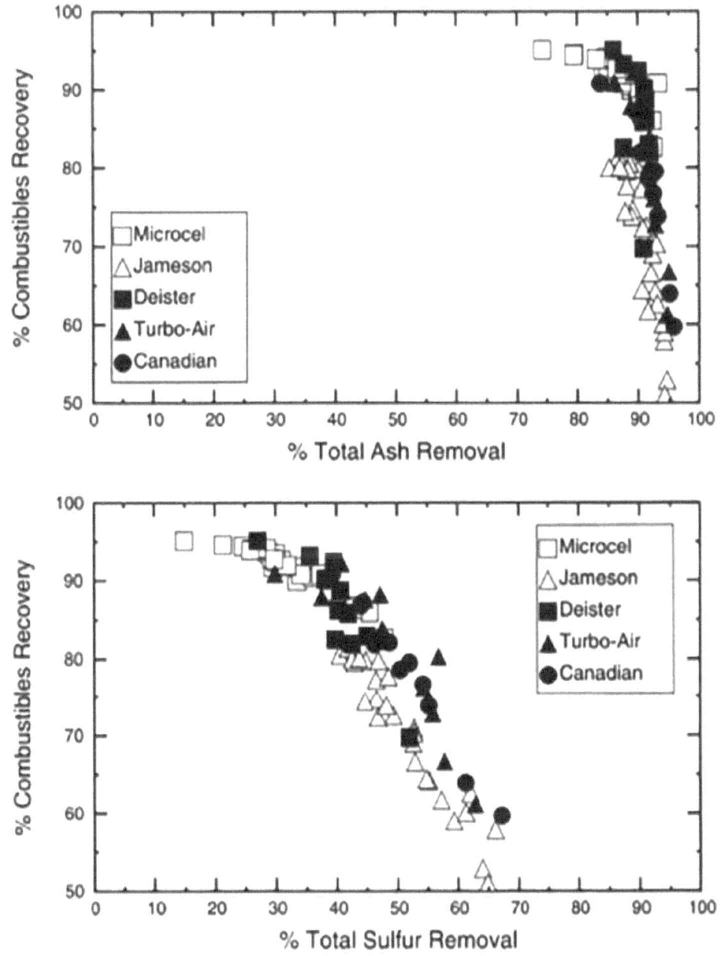

FIGURE 8.21 Comparison of the performance of five types of columns, processing an Illinois No. 5 seam coal containing approximately 30% ash and 1.3% total sulfur. The Jameson cell tended to show a lower removal of impurities at a given combustibles recovery.

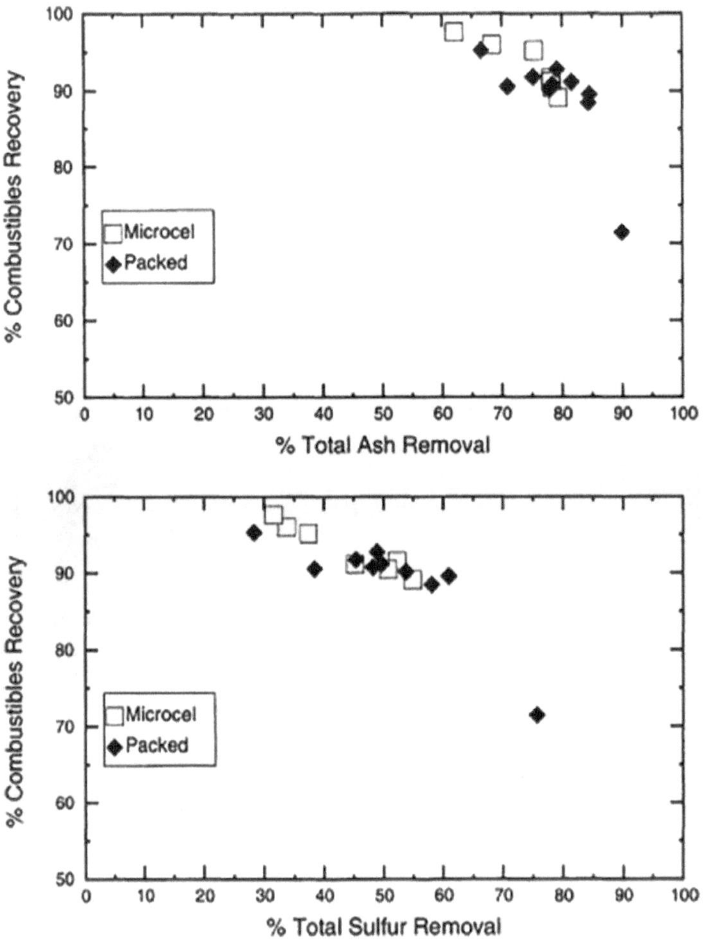

FIGURE 8.22 Comparison of a 10.2 cm. (4 in) diameter Microcel column with a 4-in diameter packed column. The coal being processed was from the Illinois No. 5 seam and contained approximately 23% ash and 1.5% total sulfur. The removal/recovery curve for both types of column was similar, with slightly better performance by the packed column.

A comparison of the minimum sulfur levels obtained for each type of column is given in Table 8.4. From these results, it is seen that while certain columns, such as the packed column, could produce a lower sulfur content in the product, the accompanying losses in combustibles recovery were quite substantial. Since these tests were carried out using small-diameter laboratory-scale units, any problems that might be found with larger-scale units were not identified. From these small-scale comparative studies, the following conclusions were reached (Honaker and Paul, 1994):

- Each of the column flotation techniques could achieve a near-optimum ash removal performance, as determined by the release analysis curve. The packed column was claimed to provide better performance than the

TABLE 8.4
Comparison of the Lowest Sulfur Products Produced by Column Flotation, Using Six Different Column Types on a Laboratory Scale

Feed Coal	Column Type	% Total Sulfur in Clean Coal	% Combustibles Recovery
Illinois No. 5 seam,	Microcel	1.22	86.0
30% ash, 1.3% sulfur	Jameson	1.17	70.3
	Deister	1.19	92.4
	Turbo-Air	1.10	61.2
	Canadian	1.18	73.8
Illinois No. 5 seam,	Microcel	0.77	89.1
23% ash, 1.5% sulfur	Packed	0.64	71.5

While some columns could produce a lower sulfur product than others, the lower sulfur values were accompanied by a loss of recovery.

other columns, and it was reported to be superior to release analysis as well. This superior performance is believed to be due to the very deep froth, the vigorous refluxing action, and the close approach to plug flow.

- Each of the columns could equal or exceed the sulfur-removal levels obtained in release analysis. As was seen for the ash, the packed column also gave superior total sulfur removal. Of the other columns, the Microcel achieved the best total sulfur removal, probably as a result of the closer approach to plug flow due to the comparatively low aeration rates of this column.
- The fact that the packed column could outperform release analysis and that some of the other columns could outperform it as well, indicates that release analysis does not adequately predict the theoretically optimum separation by froth flotation. This is apparently due to the lack of a refluxing action in the release analysis procedure.
- At high recoveries, the packed column and the Microcel had higher capacities than the other columns; however, when the recovery was <84%, the Jameson cell had a higher capacity than any of the other columns.
- The amount of air required to achieve the maximum throughout was reported to decrease in the order: Packed column > Canadian column > Flotaire > Turbo-Air > Microcel > Jameson Cell. Since the flotation of coal is limited by the amount of bubble surface area available, the differences in air requirements are believed to be due to differences in the air bubble size distribution. Therefore, the Jameson Cell had the lowest air demand because it generated and maintained the finest bubbles.
- The bubble size in column flotation is a function of the frother dosage and the amount of mechanical shear provided by the machine. The Jameson cell required the least amount of water, while also apparently producing the finest bubbles, indicating that it has a higher level of shear than the other columns.
- A positive bias (that is, complete displacement of entrained water from the froth by washwater) is necessary for producing good removal of hydrophilic

gangue such as ash-forming minerals. In order to displace weakly hydrophobic material such as locked ash/coal particles and coal pyrite, it is most important to provide a deep froth so that the refluxing action will return them to the pulp.

- Coal maceral separation by column flotation is discussed by Honaker et al. (1996).

8.4 HORIZONTALLY BAFFLED COLUMNS

The conventional columns described previously suffer from the following two serious problems, which degrade their performance, particularly in large-scale machines:

- *Axial mixing.* In the most common type of flotation columns, there is no restriction to flow along the vertical axis of the column. Rising air bubbles are therefore free to carry slurry up along the axis, with the slurry then descending again along the column walls, producing a strong axial mixing as shown in Figure 8.23. The axial mixing causes the slurry flow in the column to deviate from plug flow. This is harmful to the column performance, as it tends to reduce the product recovery and make the separation less effective. As columns are increased in size from laboratory or pilot scale to full scale, the degree of axial mixing increases, as larger columns generally have a smaller height/diameter ratio than smaller columns. The axial mixing not only harms column performance but also makes scale-up calculations more difficult. Some investigators have attempted to reduce axial mixing by the introduction of vertical baffles, which were used to subdivide the column into smaller-diameter sections so that the apparent diameter of the column would be smaller. This has been found, however, to introduce bubble distribution problems (Finch and Dobby, 1990), and, in any case, was found to have only a small effect on performance (Alford, 1992).
- *Bubble coalescence.* As bubbles rise in the column, they collide and coalesce. Even if very fine bubbles are produced using a microbubble generator, they will begin to coalesce and enlarge as soon as they enter the column. As the bubbles enlarge, they increase the degree of entrainment, and so it is desirable to keep the bubble diameter small. In packed columns, the bubbles are continuously broken down, so the bubbles never become enlarged, which accounts for the improved performance of the packed column compared to other column types. Packed columns are prone to plugging, however, and so they are not a fully satisfactory solution to bubble coalescence.

From the foregoing discussion, it is obvious that there is tremendous scope for improving the performance of flotation column by reducing axial mixing and preventing bubble coalescence. A horizontally baffled column was therefore designed at the Michigan Technological University to accomplish this (Kawatra and Eisele, 1993, 1994, 1995a,b; Kawatra et al., 1988; Eisele and Kawatra, 1995).

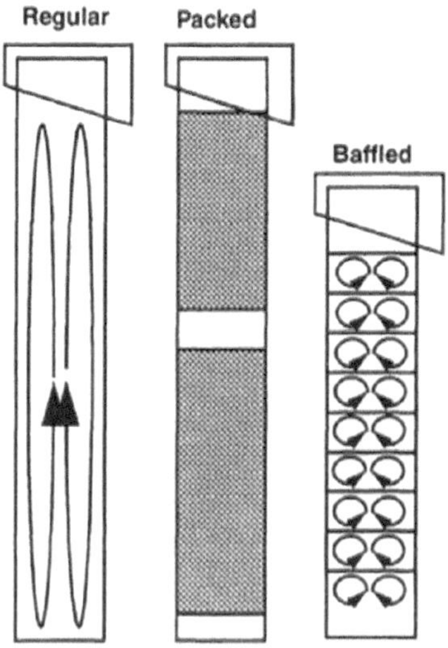

FIGURE 8.23 Comparison of flow patterns in regular, packed, and horizontally baffled columns.

8.4.1 Horizontal Baffle Designs

The horizontally baffled column provides many of the advantages of both the regular and packed columns while minimizing the drawbacks. The horizontal baffles consist of simple perforated plates, with openings large enough to keep them from being plugged by solid particles but small enough to break up vertical mixing currents, as shown in Figure 8.23. This prevents slurry from being rapidly swept along the axis of the column and provides a much closer approximation to plug flow, which improves the separation. The perforated-plate baffles are much simpler and take up less volume than the rod-racks used in the Wemco-Leeds column (Dell, 1976) and are more resistant to plugging and wear than the closely spaced packing material used in packed columns. The baffles are also suitable for retrofitting existing columns, as they are simple to make and install.

The action of the baffles in preventing bubble coalescence is shown in Figure 8.24. As the bubbles rise, they strike the baffles and are sheared back into smaller bubbles. The baffles are designed to intercept all of the bubbles, while leaving an area for coarse trash to pass through, so that it will not become trapped in the column and lead to a plugging problem.

8.4.2 Effects of Baffle Location

Baffles can be installed either above the feed inlet ("upper baffles") or below the feed inlet ("lower baffles"). The function of the baffles is different for each location, as follows:

- *Upper baffles*. These baffles are installed above the feed inlet point. Their primary function is to prevent churning and mixing of the froth layer by eliminating the coarse bubbles that cause these problems in conventional

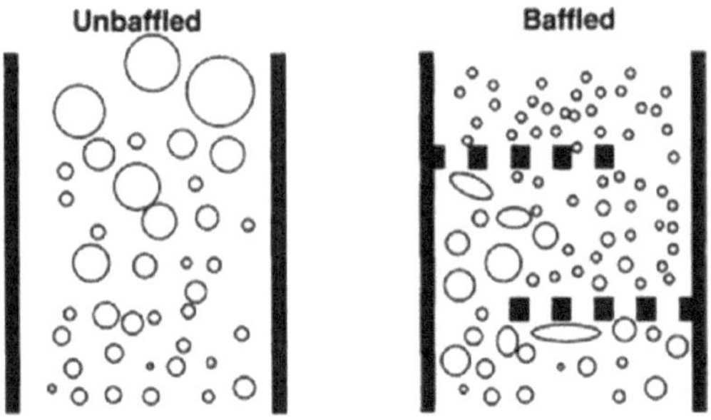

FIGURE 8.24 Effect of horizontal baffles on the maximum bubble size. Bubbles enlarge with time in unbaffled columns, due to both coalescence and decreasing hydrostatic pressure as they rise. Horizontal baffles break up the large bubbles, which keeps them from disrupting the froth layer.

columns. Froth washing is more effective when the froth layer is not being mixed, and so the upper baffles improve the product quality.

- *Lower baffles.* Baffles installed below the feed inlet point are effective for increasing the minimum residence time of particles in the column. They accomplish this by preventing particles and slurry from rapidly short-circuiting to the tailings outlet. The increased residence time gives more opportunity for hydrophobic particles to contact air bubbles, and the recovery is improved.

8.4.3 Baffled Column Bench-Scale Results

8.4.3.1 Initial Studies: One Upper Baffle, Three Lower Baffles

A 7.62 cm (3 in) diameter Flotaire column, 1.83 m (6 ft) tall, was initially fitted with four horizontal baffles spaced 30.5 cm (1 ft) apart. The feed inlet was at a depth of 91 cm (3 ft), and the froth depth was 46 cm (18 in). One baffle was installed above the feed inlet, and the remaining three baffles were below the feed inlet. Washwater was injected into the froth at a flowrate of 1.0 L/min, and the aspirator-type bubble generator used 5 L/min of water.

This column was used for froth flotation of three different coals, whose washability data is shown in Table 8.5. All three of these coals were from the coal mines in Ohio (midwestern US coals). A two-stage column flotation process was used, with intermediate grinding between the stages, as shown in Figure 8.25. The two-stage procedure was used so that the clay minerals in the coal would not interfere with the fine-coal flotation in the second stage.

For these experiments, the frother used was a strong polyglycol frother (Dowfroth 1012, molecular weight of 400) added to the bubble generator water at a rate of 0.04 g/L. The Pittsburgh Seam coal was floated without any collector. The Middle Kittanning Seam and Meigs Creek Seam coals both used No. 2 fuel oil as collector, added at a rate of 0.22 kg/metric ton in the first stage, with an additional 0.054 kg/metric ton added in the second stage.

TABLE 8.5
Washability Data for the Coals Used in the Initial Baffled Column Flotation Tests

	Specific Gravity	% Wt.	% Ash	% Pyritic Sulfur	% Sulfur	MJ/kg (BTU/lb)	Grams SO_2/MJ (lb. SO_2 million BTU)
Meigs	Float-1.3	21.1	3.9	0.5	4.1	32.17(13,838)	2.54 (5.9)
Creek	Float-1.4	61.7	7.7	0.9	4.2	30.77 (13,236)	2.71 (6.3)
Seam	Float-1.6	81.0	10.8	1.3	4.3	29.73 (12,789)	2.88 (6.7)
	Total	100.0	18.0	2.0	4.8	26.52 (11,406)	3.60 (8.4)
Pittsburgh	Float-1.3	14.6	5.8	0.5	2.4	34.77 (14,957)	1.38 (3.2)
Seam	Float-1.4	43.3	8.3	0.9	2.6	34.35 (14,775)	1.50 (3.5)
	Float-1.6	58.7	12.4	1.5	3.0	32.75 (14,086)	1.85 (4.3)
	Total	100.0	29.1	2.8	4.3	23.38 (10,059)	3.66 (8.5)
Middle	Float-1.3	71.2	3.0	0.2	1.9	30.81 (13,253)	1.25 (2.9)
Kittaning	Float-1.4	92.4	3.9	0.3	2.0	30.50 (13,120)	1.29 (3.0)
Seam	Float-1.6	95.8	4.4	0.4	2.1	30.32 (13,040)	1.38 (3.2)
	Total	100.0	7.9	1.1	2.2	29.60 (12,734)	1.50 (3.5)

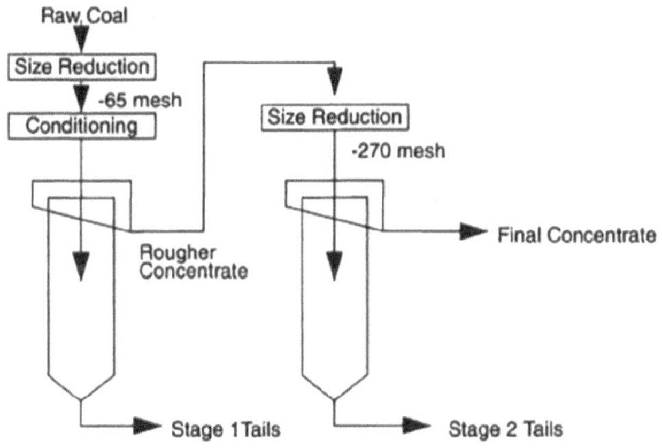

FIGURE 8.25 Flow diagram of the column flotation process used in initial bench-scale studies of the baffled column. The first stage was carried out at a relatively coarse size to remove the bulk of the clay while recovering most of the calorific value of the coal, while the second stage was carried out at a finer size to remove the ash and pyrite that was locked to coal in the first stage.

Using this procedure, the results presented in Table 8.6 were obtained. Even with only a single stage, the column could produce a high-quality product at a high energy recovery. With the second stage, the removal of pyritic sulfur was also over 76%.

TABLE 8.6
Results of Two-Stage Column Flotation of Three Coals, in Bench-Scale Tests of the Baffled Column with One Upper Baffle and Three Lower Baffles

	Product	% Wt.	% Ash	% Sulfur	MJ/kg (BTU/lb)	% Pyritic Sulfur	G. SO_2/MJ (Lb. SO_2 MM BTU)
Pittsburgh	Feed	100	29.1	4.3	23.38 (10,059)	2.8	3.66 (8.5)
Seam	Stage 1 Froth	71	6.6	3.4	31.92 (13,730)	1.4	2.11 (4.9)
	Stage 1 Tails	29	84.7	6.5	2.34 (1,009)	6.3	—
	Stage 2 Froth	59.8	3.6	2.7	33.12(14,246)	0.6	1.63 (3.8)
	Stage 2 Tails	11.2	22.3	7.2	25.99 (11,180)	5.5	—
	Stage 1: Yield = 71.0%, energy recovery = 96.9%, pyritic sulfur removal = 65.2%						
	Stage 2: Yield = 59.8%, energy recovery = 84.7%, pyritic sulfur removal = 87.2%						
Middle	Feed	100	7.9	2.2	29.60 (12,734)	1.1	1.51 (3.5)
Kittanning	Stage 1 Froth	92.1	4.7	1.8	30.94 (13,309)	0.6	1.16(2.7)
Seam	Stage 1 Tails	7.9	49.7	6.9	13.91 (5,982)	6.3	—
	Stage 2 Froth	81	2.7	1.5	31.89 (13,717)	0.3	0.94 (2.2)
	Stage 2 Tails	11.1	23.5	4	24.07 (10,355)	3.1	—
	Stage 1: Yield = 92.1%, energy recovery = 96.2%, pyritic sulfur removal = 45.2%						
	Stage 2: Yield = 81.0%, energy recovery = 87.2%, pyritic sulfur removal = 76.5%						
Meigs	Feed	100	18	4.8	26.52 (11,406)	2	3.61 (8.4)
Creek	Stage 1 Froth	85	9.6	4.4	29.98 (12,896)	1.2	2.92 (6.8)
Seam	Stage 1 Tails	15	67.4	7.1	7.64 (3,287)	6.4	—
	Stage 2 Froth	56	4.9	3.7	31.78 (13,670)	0.4	2.32 (5.4)
	Stage 2 Tails	29	18.9	5.7	26.58 (11,434)	2.9	—
	Stage 1: Yield = 85.0%, energy recovery = 96.1%, pyritic sulfur removal = 48.0%						
	Stage 2: Yield = 56.0%, energy recovery = 67.1%, pyritic sulfur removal = 90.0%						

Column diameter was 7.6 cm (3 in). A polyglycol frother (Dowfroth 1012) was used. The Middle Kittanning and Meigs Creek seam coals used No. 2 fuel oil as a collector.

8.4.3.2 Effects of Increasing Baffle Numbers

Kawatra and Eisele (1995a,b) have studied the effects of horizontal baffles on column performance in the laboratories of the Michigan Technological University. The column used was 7.6cm in diameter and 1.83m tall. A schematic of the column is shown in Figure 8.26. This column was significantly shorter than the columns of similar diameter used by other investigators, which are commonly 4–5m tall. The short column was used so that its height/diameter ratio would be similar to that of a plant-scale column.

The column was fitted with an aspirator-type bubble generator and a washwater spray ring that supplied 1 L/min of water. An overflow weir was used to control the pulp level in the column, and frother was metered directly into the aspirator. Frother was also added to the washwater reservoir, so that all of the water entering the column would have the same frother concentration. The froth depth was maintained

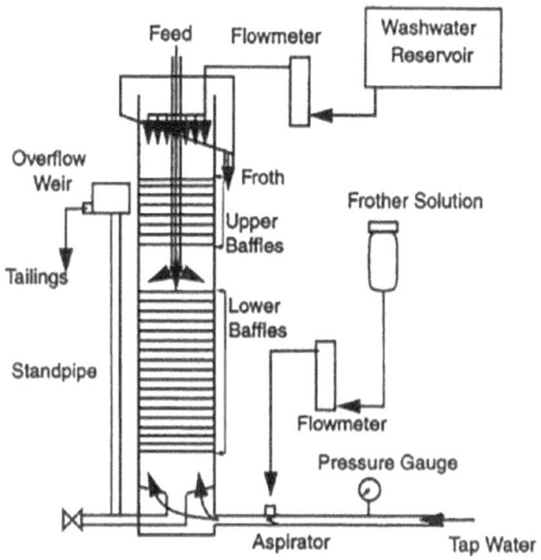

FIGURE 8.26 Schematic of a laboratory-scale, horizontally baffled flotation column.

at 0.5 m, and the feed was introduced at a depth of 1 m. The aspirator required 7 L/min of water at 138 kPa in order to operate.

The column was constructed of 30 cm sections of plexiglas tubing, with flanges between the sections. The baffles installed were of two types: The first type consisted of plates that could be placed between the tubing flanges and bolted into place. The second type of baffles consisted of disks that were threaded onto support rods at 3.5 cm intervals, with the support rods held in place by the bolted baffles. All of the baffles had an open area of 29%, and all could be removed when desired.

The baffles were installed as two sets, as shown in Figure 8.26. Up to eight baffles could be installed above the feed inlet (upper baffles), and 18 baffles could be installed below the feed inlet (lower baffles). The reason for having more lower baffles than upper baffles was that the upper baffles tended to restrict the flow of solids to the froth and forced more material to the tailings. To compensate for this effect, it was considered necessary to use more lower baffles than upper baffles.

8.4.3.2.1 Entrainment Studies

Entrainment in flotation depends upon the rate of recovery of feed water into the froth product. To study the effects of baffles on entrainment, a series of experiments were carried out without added solids, using a dye tracer to track the water flows.

The tracer used was a solution of sodium fluorescein, which was added to the feed water. A strong frother, Dowfroth 1012, was used at a concentration of 0.03 g/L to ensure that there would be a stable froth overflow in the absence of solid particles. The experiments used tap water, at a pH of 7 and a hardness of 150 ppm. The froth and tailings products were collected at selected time intervals, and the concentration of sodium fluorescein dye in each was determined by comparison to standards using a spectrophotometer, measuring the absorbance at a wavelength of 492 nm.

In these experiments, the column was first operated for 5 min, so that the froth overflow would reach a steady state. A pulse of fluorescein solution was then added, and the timer was started. The entire froth and tailings output was then collected over the following time intervals: 0.5–2.5, 2.5–4.5, 4.5–6.5, 6.5–8.5, and 8.5–10.5 min. Each product was then weighed, and the fluorescein concentration in each was measured so that the fraction of the total tracer reporting to each product could be determined.

Results of the tracer studies are shown in Figure 8.27. It was found in these tests that if the column was operated with no baffles at all, it was prone to an operating instability where the froth level would gradually surge over a period of several minutes, carrying a considerable amount of water and nearly half of the tracer into the froth. When even a single baffle was present in the column, this instability did not occur, which shows that the baffles had a stabilizing effect on the column operation.

In Figure 8.27a, it can be seen that the upper baffles are most effective for reducing the amount of tracer carried into the froth. They are therefore useful for reducing

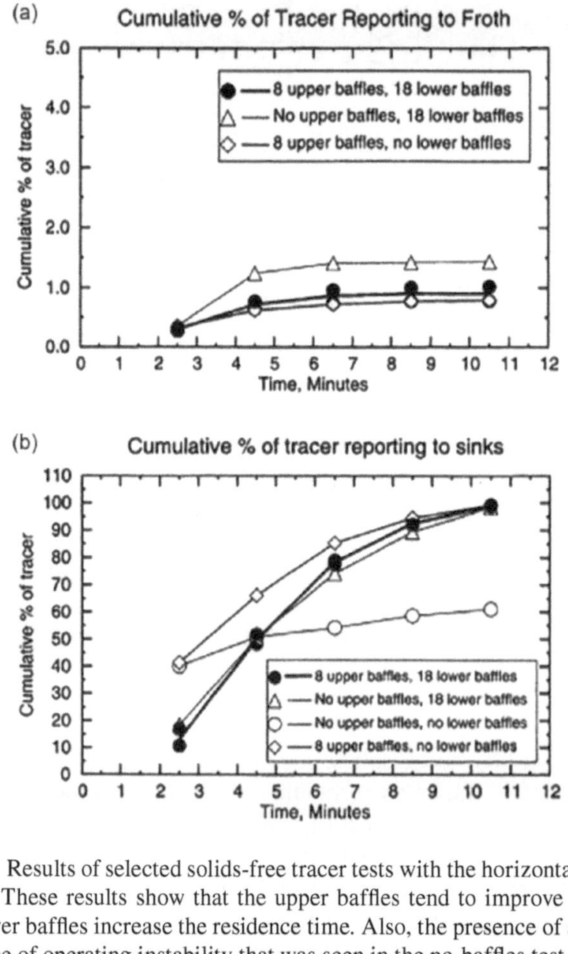

FIGURE 8.27 Results of selected solids-free tracer tests with the horizontally baffled laboratory column. These results show that the upper baffles tend to improve the froth grade, and that the lower baffles increase the residence time. Also, the presence of any baffles at all prevents the type of operating instability that was seen in the no-baffles test, where over half of the tracer was carried off in the froth product.

entrainment of gangue into the clean product. Figure 8.27b shows that the lower baffles have a significant effect on the time needed for tracer to reach the sinks product. When all 18 lower baffles are present, the time needed for 50% of the tracer to reach the tailings outlet is 4.5 min, compared to only 3 min when no lower baffles are present. This illustrates that the baffles increase the minimum residence time in the column, which will improve the recovery by giving particles more opportunity to attach to air bubbles and be carried to the froth.

It should be noted that the capacity of the column is not affected by the presence of lower baffles for two reasons:

1. The capacity of a coal flotation column is limited by the carrying capacity of the froth, not by the flow of tailings in the body of the column.
2. The baffles do not change the volume flowrate of slurry through the tailings outlet. Their effect is to cause the flow to more closely approximate plug flow, so that all of the tailings spend similar amounts of time in the column. This is in contrast to an unbaffled column, where some material can reach the tailings outlet very rapidly, while other material remains in the column for extended periods.

8.4.3.2.2 Effects of Baffles on Recovery and Grade in Coal Flotation

The coal used in these studies was collected from the Empire coal processing plant, Gnadenhutten, OH, and consisted of a blend of bituminous coals from the Lower Kittanning (No. 5) and Middle Kittanning (No. 6) seams. The coal sample was prepared by stage-crushing to pass 850 μm, and it was then stored at –20°C to prevent oxidation. Feed samples for the column tests were prepared by grinding 900 g lots of the coal for 45 min in 1,500 mL distilled water, using a 20.3 cm diameter rod mill. The rod-milled coal was 13.5% ash, 3.41% total sulfur, 28.13 MJ/kg or 12,100 British Thermal Units per pound (BTU/lb), 80% passing 40 μm, and 10% passing 4.5 μm.

The effects of variations in the number and position of the baffles on the grade and recovery are shown in Figure 8.28. Increasing the number of baffles in the column markedly improved the grade-recovery performance. The best results (i.e., higher grade with little loss of recovery) were reproducibly obtained when the complete set of eight upper baffles and 18 lower baffles was installed. It was also determined that the effect of the baffles on the recovery rate was small, as shown in Figure 8.29. Since the baffles did not greatly slow down the rate of coal recovery, this indicates that baffles of this design will not reduce the column capacity.

Fluorescein dye was also added to the feed coal for these tests, to use as a tracer for measuring entrainment. The purpose of these tracer studies was to confirm that the results from the solids-free tests were still valid for flotation with solid particles present. The amount of tracer reporting to the froth was greatly reduced by the baffles, as shown in Figure 8.30. This indicates that the baffles are very effective for reducing entrainment, and this accounts for the improvement in the clean-coal grade with baffles present. Again, the best results were obtained with the complete set of baffles present. The quantity of tracer reaching the tailings as a function of time was also determined. The results, given in Figure 8.31, confirm that the lower baffles act to increase the minimum residence time, as was seen in the solids-free tracer tests.

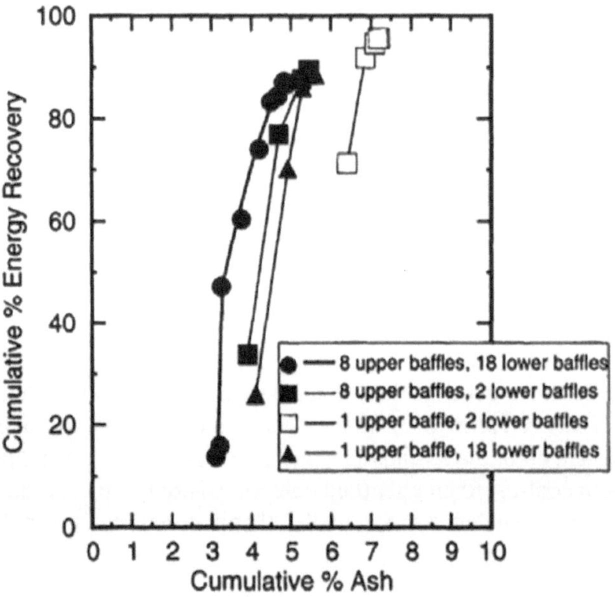

FIGURE 8.28 Grade-recovery performance of the horizontally baffled laboratory column with varying numbers of baffles. The column showed best results when the full set of baffles was installed. The feed was 13.5% ash and 28.13 MJ/kg (12,100 BTU/lb).

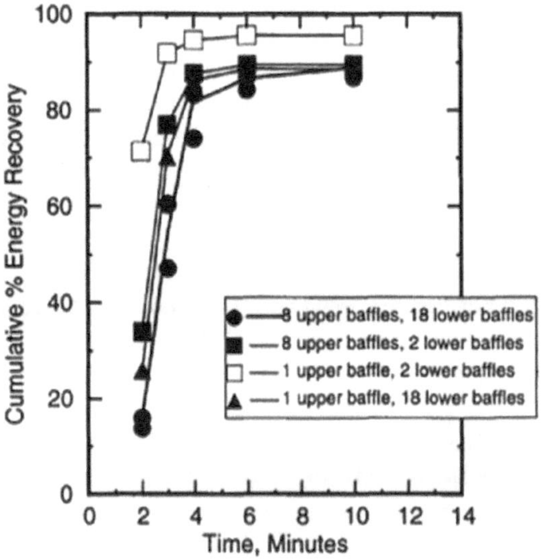

FIGURE 8.29 Energy recovery as a function of time for the horizontally baffled laboratory column with varying numbers of baffles.

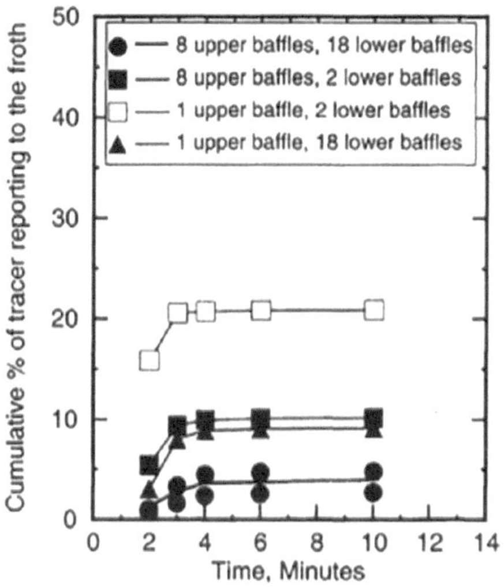

FIGURE 8.30 Recovery of tracer dye into the froth as a function of time during laboratory coal flotation tests with the horizontally baffled column. Increasing the number of baffles, and installing baffles both above and below the feed inlet, greatly reduces the amount of tracer reaching the froth, showing that entrainment is reduced.

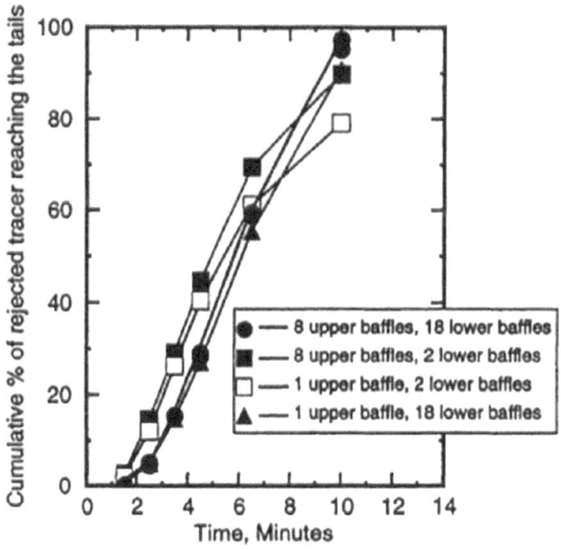

FIGURE 8.31 Quantity of tracer dye reaching the tailings product as a function of time in the horizontally baffled column. Greater numbers of lower baffles consistently cause the tracer to reach the tailings more slowly, regardless of the number of upper baffles. The addition of baffles below the feed inlet reduces short-circuiting to the tailings.

8.4.4 Baffled Column In-Plant Studies

Based on the very promising results found in the bench-scale studies, further work was carried out using a larger column, installed on-line in an operating coal-processing plant. These studies were intended to determine whether there would be any problems in scaling up the baffles and to confirm that a baffled column would work satisfactorily in a plant situation.

8.4.4.1 Pilot-Scale Baffled Column Design

The column used in the plant was derived from a Deister Flotaire unit, 20.3 cm in diameter and 9.1 m tall. A schematic of the column, with baffles installed, is given in Figure 8.32. The column contained 9 upper baffles and 17 lower baffles, and each baffle had 34% open area. Deister bubble generators were used, which injected an air/water mixture at a volume ratio of 7.5:1. Each of the two bubble generators had

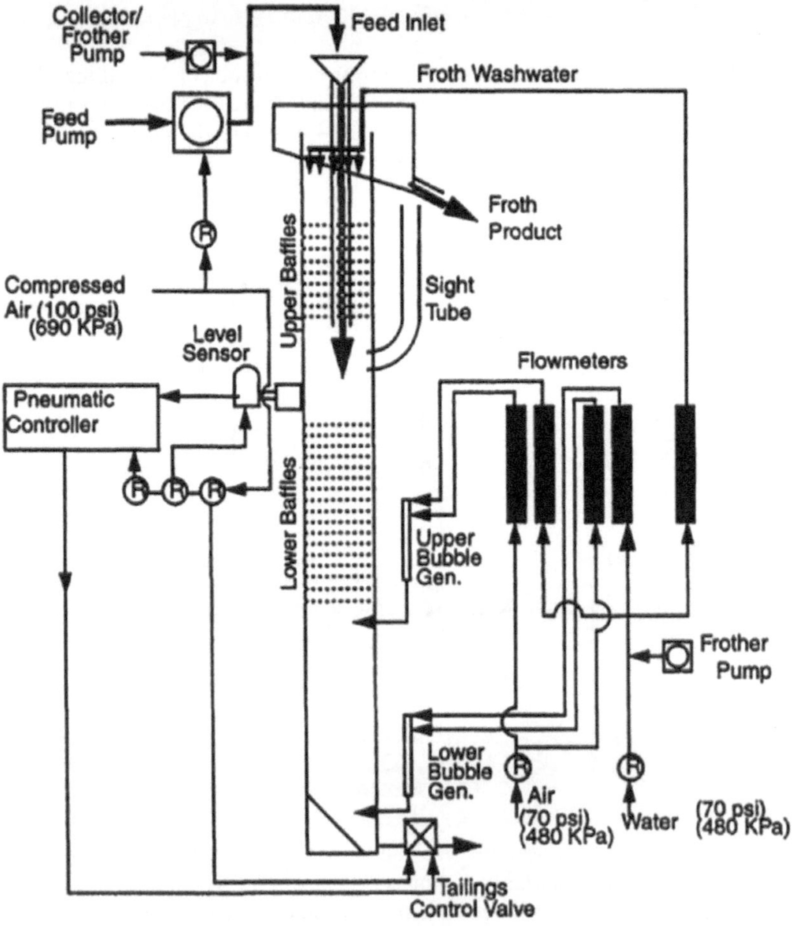

FIGURE 8.32 Schematic of the pilot-scale baffled flotation column.

maximum flowrates of 28.3 standard L/min of air and 3.78 L/min of water. The air/water mixture was injected at depths of 4.5 and 9.1 m below the froth overflow lip.

The column was operated with a froth depth of 61 cm. Measuring from the froth overflow lip, the end of the feed tube was at a depth of 122 cm. The feed tube was designed to be long enough to provide room to install the upper baffles between the base of the froth and the feed entry point. The upper baffles extended from a depth of 71 to 117 cm, and the lower baffles extended from 147 to 234 cm. The washwater spray ring was immersed 5 cm below the froth surface, and the washwater flowrate was maintained at 7.57 L/min.

In initial tests, it was found that the froth layer would collapse unless frother was added to the washwater. This was due to the upper baffles increasing the effectiveness of the washwater. As a result, frother rising from the feed or from the bubble generator water would be flushed back down before it could reach the froth, and there would be insufficient frother to maintain the froth layer.

The baffles caused the froth to be much more stable, because they broke up the fast-rising large bubbles before they could disrupt the froth. This was generally beneficial, as it reduced mixing of the froth and made the cleaning action of the washwater more effective. The more quiescent froth, however, then tended to dry slightly on its surface, forming a semisolid mass that would, over time, gradually form a "cap" on the froth and plug the top of the column. This was easily corrected by installing additional spray nozzles above the froth, spraying a mist of water at 1 L/min to keep the top of the froth moist and fluid. The mist sprayers were in addition to the main washwater ring.

8.4.4.2 Plant Flowsheet and Column Installation

The column was installed in the fine-coal processing section of the Empire coal processing plant, as shown in Figure 8.33. This plant processed a mixture of bituminous coals from the Lower Kittanning (No. 5) and Middle Kittanning (No. 6) seams, with the mineral matter being primarily fine clay.

In laboratory tests, it had been found that the most effective collector for the coals processed at Empire was a mixture of 80% No. 2 fuel oil with 20% Dow M210 froth conditioner. It was also determined that polyglycol frothers, such as Dow DF1012, were suitable for use in flotation columns.

8.4.4.3 Baffled Column Flotation of Fine Tailings

The feed used for the tests described here was taken from the plant filter press, which dewatered the solids from the fine tailings thickener (feed stream A in Figure 8.33). This material contained 39.8% ash, 2.83% total sulfur, 2.04% pyritic sulfur, and had a heating value of 19.53 MJ/kg (8,401 BTU/lb). The size distribution was 80% passing 176 μm and 10% passing 3.7 μm. The plant flowsheet had originally included conventional flotation to treat the fine coal, but the flotation circuit had to be abandoned because it could not produce a sufficiently high quality product from the coal being processed. Plant personnel felt that the fine coal was too heavily oxidized and contaminated by flocculants, and dewatering aids for conventional flotation to be practical. The feed was a 10% solids slurry and was added to the column at a rate of 0.75 kg/min of solids.

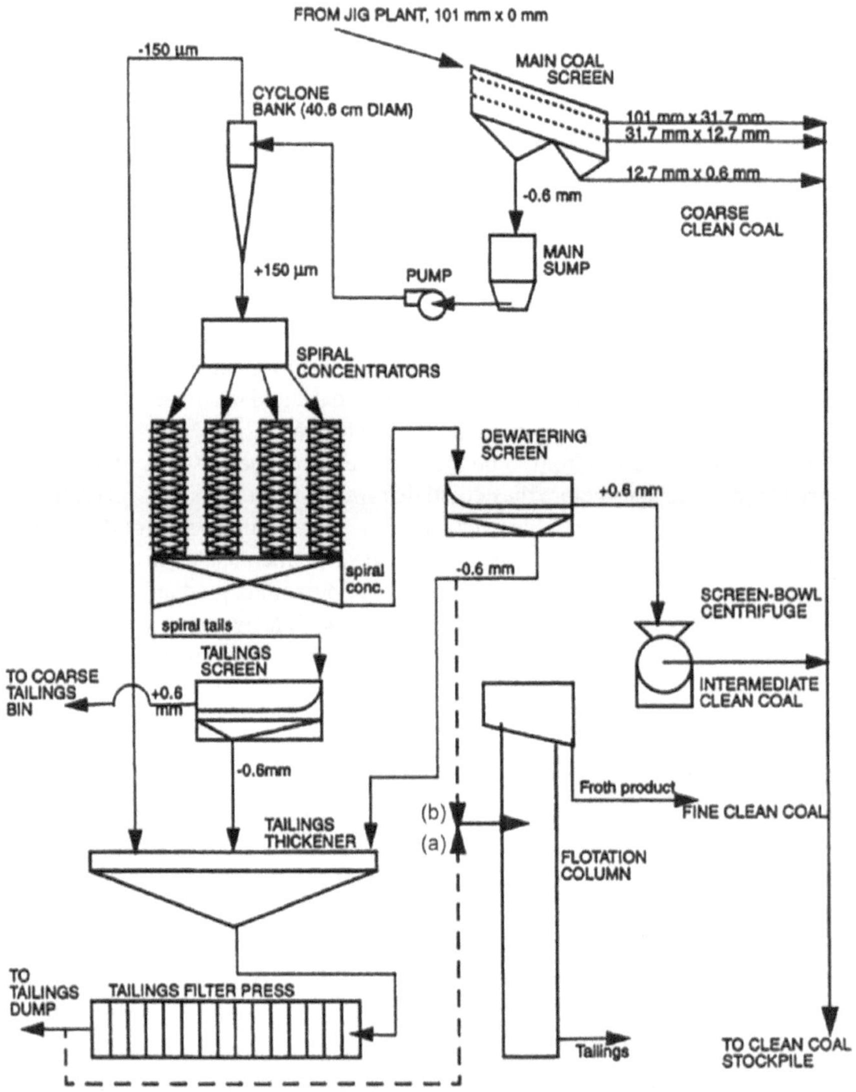

FIGURE 8.33 Flowsheet for the fine-coal processing circuit at the Empire coal processing plant. Column feed could be taken from either (a) the filtered thickener product or (b) the clean-coal dewatering screen underflow.

8.4.4.3.1 Baffle Effects on Grade/Recovery Performance

A series of experiments were carried out at a fixed reagent dosage while varying the number of baffles in the column, to determine the effect of baffles on the grade/recovery performance. The reagent dosage used was 1.1 kg/metric ton collector (80% No. 2 fuel oil, 20% M210) and 0.36 kg/metric ton frother (DF1012). When no baffles were present, the energy recovery at this reagent dosage was only 15%, but when the baffles were installed, the energy recovery increased markedly, to 54%, as shown in Figure 8.34.

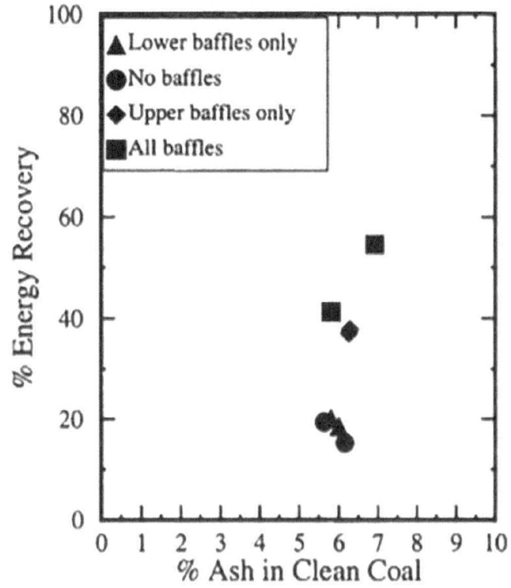

FIGURE 8.34 Effect of baffles above and below the feed inlet on the grade-recovery performance of a pilot-scale column with respect to ash. (Feed—39.8% ash; Collector—80% No. 2 fuel oil, 20% Dow M210, 1.1 kg/metric ton; Frother-Dowfroth DF1012, 0.36 kg/metric ton.)

This marked improvement upon adding baffles is due both to the improved residence time and the decreased disturbance of the froth layer when the baffles are present. These results clearly show that the horizontal baffles can greatly improve column performance in situations where the coal being processed is marginal for flotation.

A second series of tests was conducted which varied the dosages of reagents, and which showed that the column could produce good results in a single stage of flotation, even when processing a very high-ash and difficult-to-float coal. The reagent dosages and results are given in Table 8.7, and it can be seen that at the best reagent dosage, a product containing only 10.5% ash could be produced from the 39.8% as feed stream, at 85% energy recovery. This feed was not considered to be suitable flotation feed by plant personnel and was being discarded, but the column flotation product produced from it was sufficiently clean to be salable. It was also found that upon reflotation of the froth product in a second stage, there was no significant improvement in the grade of the clean coal at a given recovery, as shown in Figure 8.35. This indicates that the column flotation was producing a highly cleaned coal in a single stage, with relatively little misplaced material that could be removed by a second stage of flotation. The column performance was therefore very close to the results that would be obtained by release analysis.

8.4.4.3.2 Pyritic Sulfur Removal from Fine Tailings

Using the baffled column to treat the plant fine tailings, pyritic sulfur removals as high as 70% could be achieved at approximately 60% energy recovery, as shown in Figure 8.36. This illustrates that the baffled column performs well even with a low-quality feed stream.

TABLE 8.7
Reagent Dosages Used in Column Flotation Tests of the Plant Fine Tailings Stream with a Complete Set of Baffles and the Corresponding Results

Collector (Kg/mt)	Frother (kg/mt)	% Energy Recovery	% Pyrite Removal	% Ash Removal
0.55	0.18	18.19	94.6	98.1
0.55	0.36	45.55	81.4	94.5
0.55	1.26	70.55	65.8	88.2
1.10	0.18	11.33	98.2	98.9
1.10	0.36	55.90	80.3	93.9
1.10	1.26	65.26	70.6	91.0
1.30	0.36	30.31	92.4	97.3
2.25	0.54	24.65	94.9	98.0
2.25	0.72	51.12	84.1	95.6
2.25	1.26	83.57	53.3	85.3
2.25	2.52	85.26	51.7	84.0

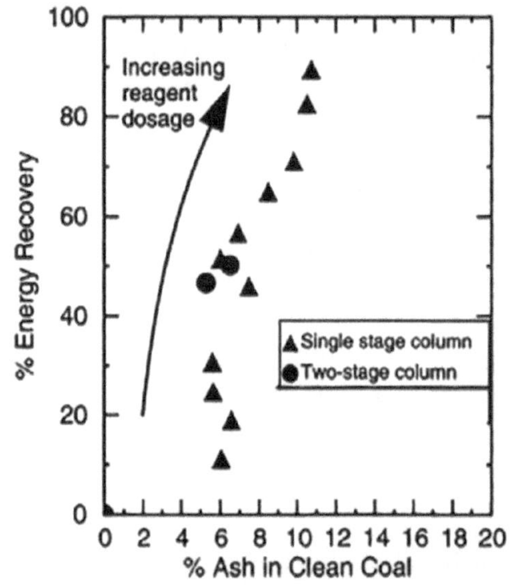

FIGURE 8.35 Grade-recovery performance of the pilot-scale column at the Empire coal processing plant, processing the fine tailings stream with all baffles installed and using only the upper bubble generator.

8.4.4.4 Baffled Column Flotation of Dewatering Screen Underflow

Continuous on-line experiments were conducted using the clean-coal dewatering screen underflow (feed stream B in Figure 8.33). These tests were mainly intended to evaluate the performance of the horizontally baffled column in steady-state operation. It should be noted that it is necessary to run a column for two to three times its residence time

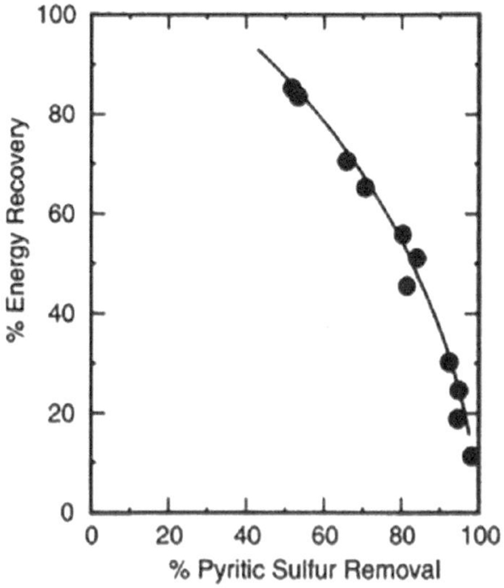

FIGURE 8.36 Comparison of energy recovery and pyritic sulfur removal in the pilot-scale column tests of the plant fine tailings stream, with all baffles installed and using only the upper bubble generator.

in order to reach steady state. This criterion was used to set a long enough time for the column to reach steady state before samples were collected for analysis.

This stream contained an average of 34.8% ash, 4.4% total sulfur, 3.1% pyritic sulfur, and 21.56 MJ/kg (9,275 BTU/lb). Its particle size distribution was 80% passing 306.7 μm and 20% passing 42 μm. This feed was coarser than the filter-press product described in the last section and contained less ash and more pyritic sulfur.

Slurry was pumped directly from the dewatering screen underflow at 10% solids and 0.5–0.8 kg of solids/min. Samples of the froth, tails, and feed were collected after allowing a full hour for the column to stabilize. The samples were taken by collecting the entire froth, tailings, and feed streams for a minimum of 45 s. A summary of the results is given in Table 8.8.

Figure 8.37 shows the grade/recovery performance of the column on-line for removing both ash and pyritic sulfur. The percent ash of the feed for each test is also plotted. From this graph, it can be seen that in spite of large variations in the feed quality, the grade of the froth product was very uniform regardless of recovery variations. The energy recovery was consistently greater than 75%, and all but two tests were above 80% energy recovery, with a high of 91.2%. The ability of the column to remove pyritic sulfur is clearly shown in Figure 8.38, which plots pyrite and ash removals against energy recovery. The removal values are the percentages of the weights of pyrite and ash originally in the feed that is rejected to the tailings. This figure shows that the column is removing between 40% and 62% of the pyritic sulfur from the coal, while simultaneously removing 80–93% of the ash. The removal of pyritic sulfur was slightly lower than that achieved with the fine tailings, both because the dewatering screen underflow was coarser and not as thoroughly liberated, and because the energy recovery was higher.

TABLE 8.8
Summary of the Froth Product Results for On-Line Tests, Using the Dewatering Screen Underflow Product

	Clean-Coal Product					
Test No.	% Pyritic S	% Ash	MJ/kg (BTU/lb)	% Energy Recovery	% Pyrite Removal	% Ash Removal
1	2.1	7.49	30.77 (13,236)	83.9	61.8	87.6
2	2.4	8.18	30.95 (13,315)	77.6	53.7	88.6
3	2.4	7.49	30.94 (13,308)	86.6	56.2	89.0
4	2.2	7.22	31.18(13,411)	90.1	48.6	80.3
5	2.6	6.78	31.29(13,460)	87.9	51.1	87.8
6	2.6	6.96	31.20 (13,420)	86.9	56.5	87.9
7	2.9	7.62	31.04(13,351)	86.7	40.3	85.2
8	2.7	6.70	31.56(13,575)	85.4	52.9	87.2
9	2.9	7.19	31.13(13,391)	79.4	54.7	88.2
10	2.8	7.84	31.08 (13,368)	91.2	49.2	84.8
11	2.4	6.71	31.42 (13,514)	88.8	56.1	87.6
12	2.7	6.90	31.40 (13,506)	91.2	49.9	87.2

Samples for each test were collected for a minimum of 45 s, sampling the entire stream. The feed quality was 31%–39% ash, 3.8%–4.8% total sulfur, and 2.2%–3.6% pyritic sulfur.

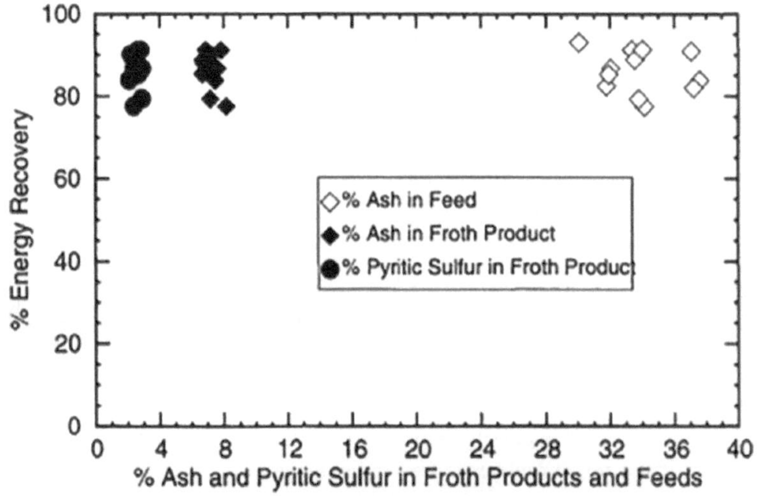

FIGURE 8.37 Grade-recovery results for the on-line continuous flotation tests, all baffles installed, with water sprays on top of the froth. The ash content of the feed for each test is also given. In spite of the wide variation in feed ash contents, the ash content of the froth product is quite constant.

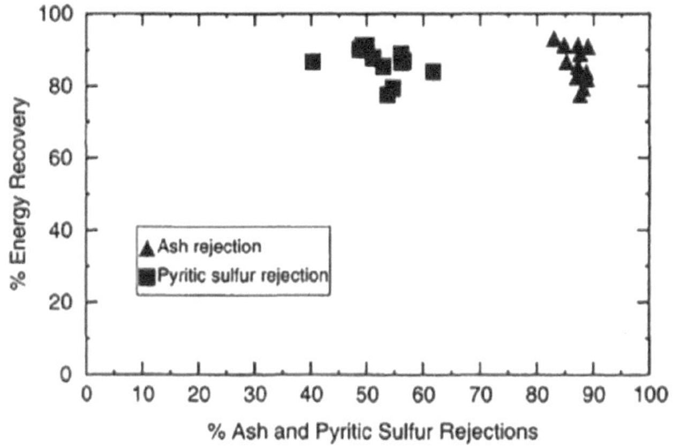

FIGURE 8.38 Pyritic sulfur and ash removals as a function of energy recovery in the online continuous tests with all baffles installed. This shows that between 40% and 65% of the pyritic sulfur is being removed by the column, even at high energy recoveries, and that over 80% of the ash is consistently being removed.

8.4.4.5 Conclusions from Baffled Column Results

From the work with the baffled column, the following conclusions were reached:

1. In conventional, open-pipe flotation columns, there is little to prevent initially small disturbances from growing into major disruptions of the column performance. As a result, even slight problems with the column operation can be rapidly amplified to the point where they degrade the column performance. Fully packed columns can prevent this from occurring, but these are prone to plugging and excessive wear. Horizontal baffling has been found to be sufficient to prevent excessive axial mixing and to prevent the formation of unusually large bubbles, and such baffles are sufficiently open and durable that plugging and rapid wear are not problems.
2. The capacity of a coal flotation column is limited by the rate at which the clean coal can be withdrawn from the froth. The horizontal baffles increase the residence time of the tailings, but they do not directly affect the froth layer, and so their effect on the capacity of coal flotation columns is minimal.
3. The column equipped with horizontal baffles could operate stably and produce a consistently high-quality product while encountering normal variations in plant feed characteristics.
4. The horizontally baffled column produced a high-quality product in a single stage from a high-clay feed coal, removing up to 87% of the ash and 50% of the pyrite while recovering 91% of the heating value.
5. The baffles were designed for use with particles as coarse as 2 mm. As a result, they never plugged in operation, and upon disassembly after several months of testwork, no evidence of plugging or rapid baffle wear was found.

Baffled columns have a promising future in coal flotation. Baffles can be installed in any column. Sometimes two or three baffles are sufficient to reduce axial mixing and improve column effectiveness.

8.5 SUMMARY

A comparison of the operational characteristics of the various types of columns is given in Table 8.9.

Column flotation is a significant improvement over conventional flotation for cleaning and desulfurization of coal. The improved selectivity of flotation columns makes it possible to efficiently produce a higher-quality product than can be produced with conventional flotation. Columns can also be used to treat coals that are so finely ground and high in impurities that they are not currently considered practical to treat by froth flotation.

There are numerous types of flotation columns available, with many of them differing only in their bubble-generating mechanism. Since columns are all very efficient, the differences in their laboratory coal desulfurization performance tend to be small. The greatest differences are seen in areas such as reagent consumption, operating stability, and coal recovery.

It has been determined that the operating stability and coal recovery of columns can be improved by the use of horizontal baffles. These baffles are used to prevent axial mixing in the column and therefore improve the separation and increase the coal recovery. This type of baffle also prevents the formation of coarse bubbles in the column, which are responsible for froth instability in conventional, unbaffled columns.

In general, open columns such as the Cominco or Microcel columns have converged over time to essentially the same design, and their performance does not vary a great deal. The main consideration in selecting an open column is the price and availability.

Packed columns produce very good metallurgical results that are generally superior to the results from open columns. However, interactions of the packing with the particles can produce operational difficulties unless the coal is ground to an extremely fine particle size.

If axial mixing and bubble coalescence are found to be problems in a particular open column installation, the problems can be corrected by retrofitting the column with horizontal baffles. Baffles can be installed above the feed inlet if the objective is to improve the product grade or below the feed inlet if the objective is to increase recovery, or in both locations to improve both the grade and the recovery.

There are many methods of classifying flotation columns. One such method is separating columns into "tall columns" and "short columns." Tall columns are typically countercurrent flotation columns which have a height generally greater than twice the diameter. Short columns refer to columns such as the Jameson cell. These are different from conventional mechanical flotation cells.

TABLE 8.9
Comparison of the Operational Characteristics of Several Types of Flotation Columns That Are Useful for Coal Flotation, Based on the Results of Testwork with Laboratory-Scale Units

	Operational Characteristics	
Column Type	**Advantages**	**Disadvantages**
Turbo-Air	Bubble generator produces fine bubbles, so air requirement is moderate. Suitable for rougher/cleaner separation.	Requires high aeration pressure (60 psi). Requires pressurized water in bubble generators. Bubble generator orifices can plug. Scale-up is limited, due to height-to-diameter ratio considerations.
Canadian	Bubble generator does not require water. Suitable for rougher/cleaner separation.	Relatively high compressed-air demand. Moderate to high frother requirements. Scale-up is limited, due to height-to-diameter ratio considerations. Bubble generators prone to plugging.
Flotaire	Bubble generator produces fine bubbles, so air requirement is moderate. Suitable for rougher/cleaner separation.	Air compressor needed. Scale-up is limited, due to height-to-diameter ratio considerations. Bubble generators prone to plugging.
Jameson	Simple to operate, with no need for compressed air. Scale-up is straightforward. Overall cell height is low. Frother demand is low. Bubble generator uses feed slurry and does not require additional water. Low washwater requirements.	Complex feed distribution system on an industrial scale. Bubble generator orifice prone to wear. High-feed solid content requires partial tailings re-circulation to maintain recovery. Feed must be injected at high pressure.
Packed	Scale-up is reportedly not subject to height-to-diameter ratio problems. Able to support very deep froths, resulting in very good refluxing action. Low aeration pressure. Bubbles generated by packing, so no bubble generator is needed.	High air requirement. High frother consumption. Tendency for wear and plugging of the packing.
Microcel	Relatively low air requirement and low frother requirement, moderate air pressure. Bubble generator uses recycled tailings stream and does not require additional water. Recycle of tailings to the bubble generator provides a scavenging action.	Static mixer is prone to wear. Scale-up is limited, due to height-to-diameter ratio considerations. Compressed air is required.
Horizontal baffles	Improves metallurgical performance. Can be retrofitted into any open column.	Additional cost of baffles.

REFERENCES

Alford, R.A. (1992), "Modeling of Single Flotation Column Stages and Column Circuits," *International Journal of Mineral Processing*, Vol. 36, pp. 155–174.

Atkinson, B.W., Conway, C.J. and Jameson, G.J. (1995), "High Efficiency Flotation of Coarse and Fine Coal," Chapter 25, *High Efficiency Coal Preparation* (Kawatra, ed.), Society for Mining, Metallurgy, and Exploration, Littleton, CO, pp. 283–294.

Boutin, P. and Wheeler, D.A. (1967), "Column Flotation Development Using an 18 Inch Pilot Unit," *Canadian Mining Journal*, Vol. 88, pp. 94–101.

Davis Jr., V.L., Stanley, F.L., Bethell, P.J., Luttrell, G.H. and Mankosa, M.J. (1994), "Column Flotation at the Middle Fork Preparation Facility," *Coal Preparation*, Vol. 14, Nos. 3–4, pp. 133–145.

Davis Jr., V.L. (1993), "Implementation of Microcel Column Flotation for Processing Fine Coal," *Proceedings of the 10th International Coal Preparation Exhibition and Conference*, Kentucky, pp. 239–250.

Davis Jr., V.L., Bethell, P.J., Stanley, F.L. and Lutrell, G.H. (1995), "Plant Practices in Fine Coal Column Flotation," Chapter 21, *High Efficiency Coal Preparation* (Kawatra, ed.), Society for Mining, Metallurgy, and Exploration, Littleton, CO, pp. 237–246.

Degner, V.R. and Person, P.L. (1991), "Leeds Column Performance Evaluation," *Minerals Engineering*, Vol. 4, Nos. 7–11, pp. 935–950.

Dell, C.C. (1976), *Froth Flotation*, British Patent No. 1,519,075.

Dell, C.C. (1985), "The Leeds Flotation Column, How It Works and What It Does," *Leeds University Mining Association Magazine*.

Dell, C.C. (1978), "Column Flotation of Coal—The Way to Easier Filtration," *Mine and Quarry*, Vol. 7, pp. 36–40.

Dell, C.C. and Jenkins, B.W. (1976), "The Leeds Flotation Column," *Seventh International Coal Preparation Congress*, Sydney, paper J3.

Dell, C.C., Bunyard, M.J., Rickleton, W.A. and Young, P.A. (1972), "Release Analysis: A Comparison of Techniques," *Transactions of the Institute of Mining and Metallurgy, Section C*, Vol. 81, pp. C89–C97.

Dobby, G.S. and Finch, J.A. (1986), "Flotation Column Scale-Up and Modeling," *CIM Bulletin*, Vol. 79, pp. 89–96.

Dobby, G.S., Amelunxen, R. and Finch, J.A. (1985), "Column Flotation: Some Plant Experience and Model Development," *International Federation of Automatic Control (IFAC)*, Vol. 18, pp. 259–263.

Eberts, D.H. (1986), "Flotation—Choose the Right Equipment for Your Needs," *Canadian Mining Journal*, Vol. 107, No. 3, pp. 25–33.

Eisele, T.C. and Kawatra, S.K. (1995), "On-Line Testing of a Horizontally-Baffled Flotation Column in an Operating Coal Cleaning Plant," Chapter 20, *High Efficiency Coal Preparation* (Kawatra, ed.), Society for Mining, Metallurgy, and Exploration, Littleton, CO, pp. 227–235.

Finch, J.A. and Dobby, G.S. (1990), *Column Flotation*, Pergamon Press, Oxford.

Groppo, J. (1986), "Column Flotation Shows Higher Recovery with Less Ash," *Coal Mining*, Vol. 23, No. 8, pp. 36–38.

Groppo, J.G., Parekh, B.K., Peters, W.J. and Kennedy, D.L. (1993), "Three years of Column Flotation Experience at Powell Mountain Coal Company," *Proceedings of the 10th International Coal Preparation Exhibition and Conference*, Kentucky, pp. 258–267.

Groppo, J.G. and Parekh, B.K. (1990), "Continuous Pilot Scale Testing of Fine Coal Flotation for Fine Coal," *Presented at the SME Annual Meeting*, Salt Lake City, UT, Preprint No. 90–153.

Harbort, G. and Clarke, D. (2017), "Fluctuations in the Popularity and Usage of Flotation Columns – An Overview," *Minerals Engineering*, Vol. 100, pp. 17–30.

Hollingsworth, C.A. (1981), "The Flotaire Flotation Cell," *SME Annual Meeting*, Chicago, February 22–26.

Honaker, R.Q. and Paul, B.C. (1994), "A Comparison Study of Column Flotation Technologies for Cleaning Illinois Coal," Final Technical Report, Illinois Clean Coal Institute, Project No. 93-1/5.1A-1P, DOE Grant No. DE-FC22-92PC92521.

Honaker, R.Q., Mohanty, M.K., and Crelling, J.C. (1996), "Coal Maceral Separation Using Column Flotation," *Minerals Engineering*, Vol. 9, No. 4, pp. 449–464.

Im, C.J. and Wolfe, R.A. (1986), "Application of Flotaire Flotation Cell in Coal Preparation Plant," *SME Annual Meeting*, New Orleans, LA, March 2–6, SME Preprint No. 86–36.

Jacobsen, P.S., Killmeyer, R.P. and Hucko, R.E. (1989), "Interlaboratory Comparison of Advanced Fine-Coal Beneficiation Processes," U.S. Department of Energy, Pittsburgh Energy Technology Center, Report No. DOE/PETC/TR–89/9.

Kawatra, S.K. and Eisele, T.C. (1987), "Column Flotation of Coal," Chapter 16, *Fine Coal Processing* (Mishra and Klimpel, eds.), Noyes Publications, Park Ridge, NJ, pp. 414–429.

Kawatra, S.K. and Eisele, T.C. (1993), "Use of Horizontal Baffles in Column Flotation," *Proceedings of the XVIII International Mineral Processing Congress,* Sydney, Australia, pp. 771–778.

Kawatra, S.K. and Eisele, T.C. (1994), "Use of Horizontal Baffles to Reduce Axial Mixing in Coal Flotation Columns," *Proceedings of the 12th International Coal Preparation Congress*, Cracow, Poland, pp. 1241–1249.

Kawatra, S.K. and Eisele, T.C. (1995a), "Laboratory Baffled-Column Flotation of Mixed Lower/Middle Kittanning Seam Bituminous Coal," *Minerals and Metallurgical Processing,* Vol. 12, No. 2, pp. 103–107.

Kawatra, S.K. and Eisele, T.C. (1995b), "Baffled-Column Flotation of a Coal Plant Fine-Waste Stream," *Minerals and Metallurgical Processing,* Vol. 12, No. 3, pp. 138–142.

Kawatra, S.K., Eisele, T.C. and Johnson, H.J. (1988), "Desulfurization of Coal by Column Flotation," Chapter 7, *Processing and Utilization of High Sulfur Coals II* (Chugh and Caudle, eds.), Elsevier, New York, pp. 61–70.

Kennedy, A. (1990), "The Jameson Flotation Cell," *Mining Magazine,* Vol. 163, No. 10, pp. 281–285.

Li, B., Tao, D., Ou, Z. and Liu, J. (2003), "Cyclo-microbubble Column Flotation of Fine Coal," *Separation Science and Technology*, Vol. 38, No. 5, pp. 1125–1140.

Luttrell, G.H. (1990), Workshop on High Efficiency Coal Preparation, Mount Vernon, 1992.

Lynch, A.J., Johnson, N.W., Manlapig, E.V. and Thome, C.G. (1981), *Mineral and Coal Flotation Circuits: Their Simulation and Control*, Elsevier, Amsterdam.

Manlapig, E.V., Jackson, B.R., Harbort, G.J. and Cheng, C.Y. (1993), "Jameson Cell Coal Flotation," *Proceedings of the 10th International Coal Preparation Exhibition and Conference,* Kentucky, pp. 203–219.

McKay, J.D., Foot, D.G. and Shirts, M.B. (1988), "Column Flotation and Bubble Generation Studies at the Bureau of Mines," *Column Flotation' 88,* SME-AIME, Littleton, CO, pp. 173–186.

Moon, K.S. and Sirois, L.L. (1983), "Column Flotation," *15th Canadian Mineral Processors Annual Meeting*, Ottawa, Ontario, Paper No. 18.

Parekh, B.K., Groppo, J.G. and Bland, A.E. (1986), "Optimization Studies of Column Flotation for Fine Coal Cleaning," *Presented at the 3rd Pittsburgh Coal Conference*, Pittsburgh, PA, September 8–12.

Peng, F.F. and Li, L. (1995), "Dispersion Characteristics in Column Flotation of Fine Coal," Chapter 18, *High Efficiency Coal Preparation* (Kawatra, ed.), Society for Mining, Metallurgy, and Exploration, Littleton, CO, pp. 207–218.

Phillips, D.I., Yoon, R., Lutrell, G.H., Fish, I. and Toney, T.A. (1997), "Installation of 4-meter diameter Microcel flotation columns at LadyDunn preparation plant," *14th International Coal Preparation Conference*, Lexington, KY, April 29–May 1, 1997

Rubinstein, J.B. (1995), *Column Flotation: Processes, Designs, and Practices*, Gordon and Breach, Basel.

Shiao, S.Y. (1994), "Removal of Pyrite and Trace Elements from Waste Coal by Dissolved CO_2 Flotation And Chelating Agents," Final Technical Report, Illinois Clean Coal Institute, pp. 1–25.

Sorokin, A.F., Filippov, A.I., Shelyakin, L.E., Medvedev, A.V. and Suslov, V.I. (1978), "Cleaning Coal Slurries in Column-Type Flotation Machines," *Koks i Khimiya*, Vol. 5, pp. 6–9.

Sutherland, K.L. and Wark, I.W. (1955), *Principles of Flotation*, Australasian Institute of Mining and Metallurgy, Melbourne.

VPI (1995), "Bench-Scale Testing of the Multi-Gravity Separator in Combination with Microcel," Final Report, prepared by Virginia Polytechnic Institute and State University, for the U.S. Department of Energy, Pittsburgh Energy Technology Center, Report No. DE-AC22–92PC92205.

Von Holt, S.T. and Franzidis, J.P. (1994), "Column Flotation of a South African Cooking Coal," *Coal Preparation*, Vol. 14, pp. 147–166.

Yang, D.C. (1984), "Static Tube Flotation for Fine Coal Cleaning," *Proceedings of the 6th International Symposium on Coal Slurry Combustion and Technology,* Orlando, FL (PETC-DOE), pp. 582–597.

Yang, D.C. (1988), "A New Packed Column Flotation System," Chapter 28, *Column Flotation* (Sastry, ed.), Society for Mining, Metallurgy, and Exploration, Littleton, CO, pp. 257–265.

Yoon, R.H. and Luttrell, G.H. (1986), "The Effect of Bubble Size on Fine Coal Flotation," *Coal Preparation,* Vol. 2, pp. 174–192.

Yoon, R.H., Luttrell, G.H. and Adel, G.T. (1990), "Advanced Systems for Producing Superclean Coal," Final Report, DOE/PC/91221-T1 (DE 91004332) August, 1990.

Zipperian, D.E. (1984), "Characteristics of Column Flotation Utilizing Aspirated Aeration," *AIME Fall Meeting*, Denver, October 24–26.

Zipperian, D.E. and Christophersen, J.A. (1985), "Plant Operation of the Deister Flotaire Column Flotation Cell," *SME Annual Meeting*, New York, NY, Preprint No. 85–128.

9 Selective Flocculation

There are several mechanisms that can be used to cause suspended particles to clump together into flocs, coagulates, or agglomerates. These mechanisms are shown in Figure 9.1. Flocculation refers to the formation of masses of particles that are held together by soluble polyelectrolyte macromolecules, such as starches, gums, and synthetic polymers, by a bridging mechanism. The flocculation is selective when the macromolecules preferentially adsorb onto a particular species (Yarar and Kitchener, 1970). Coagulation is the process of aggregation of particles by reducing the electrostatic forces between them, so that they will be held together by the London–van der Waals' interaction. In both flocculation and coagulation, the particulate masses formed are loose, fragile, and contain a considerable amount of interstitial water.

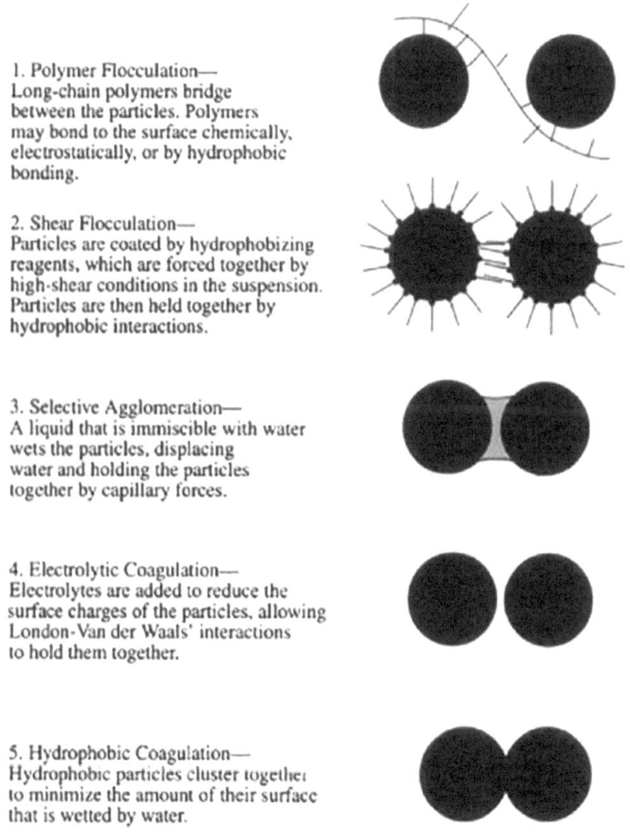

FIGURE 9.1 Schematic representations of the five basic aggregation mechanisms for flocculation of particles. Selective agglomeration is described separately in Chapter 10.

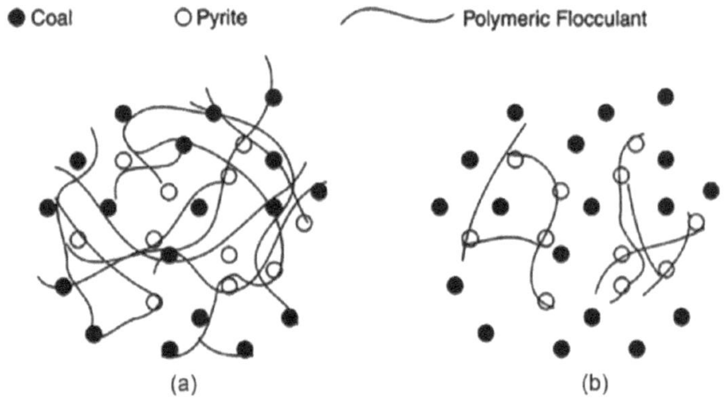

FIGURE 9.2 Flocculation of coal and pyrite based on the polymer bridge theory. (a) Bulk flocculation of coal and pyrite; (b) Selective flocculation of pyrite.

The focus of this chapter is on flocculation and coagulation, and selective agglomeration is discussed separately in Chapter 9.

The selective flocculation technique makes use of the differences in the surface chemical characteristics of coal and of ash-bearing minerals (including pyrite). The process is related to the complete flocculation that is used to remove fine solids from water in thickeners. The difference between selective flocculation and bulk flocculation is that in bulk flocculation, all of the particles present are bound together into mixed floccules, while in selective flocculation only a few selected minerals are flocculated, as illustrated in Figure 9.2.

In the fine coal-cleaning literature, considerable interest has been shown in selective flocculation because it has the potential for separating coal from ash-forming minerals at particle sizes smaller than are practical with froth flotation (Blagov, 1970; Blaschke, 1972, 1976, 1994; Hucko, 1977). These studies have concentrated on the use of polymeric flocculants to reduce the coal ash content, with only limited attention to the removal of pyrite by this means (Attia, 1985; Krishnan, 1987; Behl and Moudgil, 1992).

Selective flocculation depends broadly on two classes of factors: the surface properties of the coal and mineral particles, and the structure of the flocculant. If properly designed, the flocculants will interact chemically or physically with specific surface sites on the desired particles.

9.1 FLOCCULATING REAGENTS

Polymeric flocculants are long-chain molecules that are long enough to bridge between particles. The most common are the polyacrylamides, with molecular weights on the order of 2–10 million atomic mass units. The polyacrylamide backbone molecule is nonionic and water soluble and consists of a chain of the unit shown in Figure 9.3a. Anionic polymers can be produced by hydrolysis of the polyacrylamide, which converts some of the amide groups to carboxylic groups, which ionize in neutral or alkaline solutions to give negatively charged carboxylate

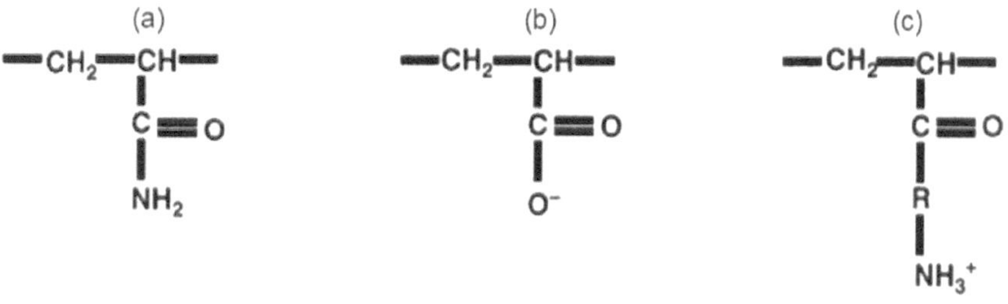

FIGURE 9.3 Basic units of polyacrylamide flocculants: (a) nonionic base unit; (b) anionic unit; (c) cationic unit. Other anionic, cationic, and nonionic groups may be substituted into the polyacrylamide chain to give specific chemical properties (Mangravite et al., 1985).

sites (Figure 9.3b). Controlled hydrolysis can therefore give the desired degree of ionic character. Cationic polymers can also be made, for example, by copolymerizing acrylamide with a suitable cationic monomer, such as the one shown in Figure 9.3c. Other polymer backbones such as polyvinyl alcohol and polyethylene oxide (nonionic), sodium polystyrene sulfonate (anionic), and polyethyleneimine or polydiallyldimethyl-ammonium chloride (cationic) are also in use (Mangravite et al., 1985).

Hydrophobic or partially hydrophobic flocculants, such as latexes and polyethylene oxide-based polymers, have been used to selectively flocculate coal particles based on their hydrophobicity. The advantages of such flocculants are that they have a high degree of selectivity toward the hydrophobic coal surfaces, and they are fully compatible with subsequent froth flotation or oil agglomeration of the coal particles.

In addition to the synthetic flocculants, there are also naturally occurring polymers such as starch, isinglass, gelatine, alginates, and other compounds that can be used for flocculation (Gregory, 1985). These are often less expensive than synthetic polymers but are also less versatile because their structures cannot easily be customized to particular applications.

Inorganic coagulants, which are mainly high-valence metal ions like Fe^{+3} and Al^{+3}, are also useful for altering the Zeta potentials of particle surfaces, promoting coagulation. These ions can either be used as simple chemicals or in a polymeric form (O'Melia, 1985).

9.2 PRINCIPLES OF SELECTIVE FLOCCULATION

The selective flocculation process can be described by the following steps, which are illustrated in Figure 9.4:

- Liberation of mineral matter from coal by comminution.
- Dispersion of coal and mineral particles in a slurry.
- Flocculation of the particles of interest.
- Floc conditioning.
- Floc/suspension separation.

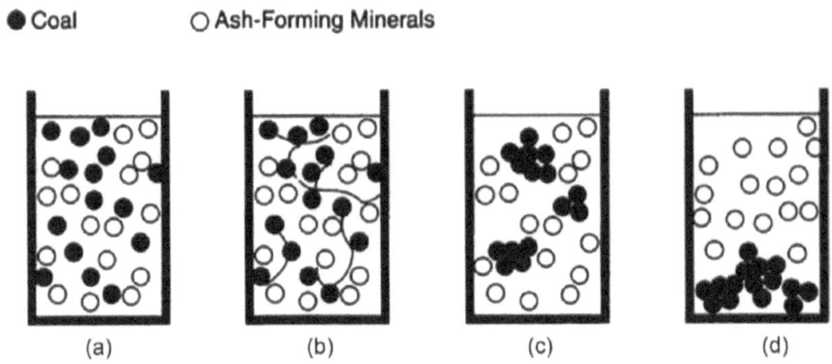

FIGURE 9.4 Four basic stages in the selective flocculation process. In this example, the coal is being flocculated while the mineral particles remain in suspension. (a) dispersion; (b) flocculation; (c) conditioning; (d) separation.

9.2.1 Liberation of Mineral Matter

As with all other physical desulfurization techniques, the success of the selective flocculation process depends on the degree of liberation of the components that are being separated. Since selective flocculation is most useful for extremely finely ground coal, a higher degree of comminution can be used than is possible for other separation techniques. The comminution has generally been carried out using conventional tumbling-media mills such as ball mills. Recently, work has been done showing that grinding to such fine sizes is more efficient when more advanced techniques such as High Pressure Roll mills (Fuerstenau et al., 1995) and Szego mills (Bajor and Trass, 1988) are used.

9.2.2 Dispersion of the Slurry

The purpose of this step is to break up any loose aggregates present in the slurry and ensure that the coal and mineral particles are suspended as individual particles (Krishnan, 1987). Dispersion is often done in the course of grinding the coal, with agitation, ultrasonics, and chemical dispersants used to keep the particles in the dispersed state. With chemical dispersants, the degree of dispersion initially increases as the dispersant dosage is increased but then gradually tapers off at higher dosages.

The surface charge of the particles in solution is very critical, because particles with a low surface charge will easily coagulate and so will be difficult to disperse. It is therefore necessary to control the surface charge at the proper level. This control is complicated by the fact that the isoelectric point (IEP) of coal depends on many factors, such as its maceral composition and degree of oxidation. In general, the IEP of coal macerals moves toward lower pH as the coal becomes more oxidized. The IEP for quartz and clay minerals is below pH 5, while for pyrite surfaces the IEP is reported to be at a pH of 5.5. In general, it is easier to maintain a well-dispersed slurry at high pH values because alkaline solutions are well away from the IEPs of the coal and mineral matter. All of the particles tend to have strong negative charges at high pH, and so they are kept from coagulating by the electrostatic repulsion between them. Anionic dispersants, such as sodium hexametaphosphate, are often used as dispersants because they adsorb onto the coal surface and increase the negative charge.

Other factors that are important for maintaining dispersion are the water quality, particle size distribution, and aging of the minerals. Factors such as water hardness, pH, ionic species in solution, dissolved organic compounds, and water temperature are all important variables that affect the surface chemistry. It has also been established that because of oxidation and hydration effects, old mineral surfaces behave differently from freshly made surfaces.

9.2.3 Flocculation

Flocculation is the most critical stage, as this is the point where gangue minerals are enclosed in the floc and where desirable particles escape flocculation. The basic principles of floc formation are shown in Figure 9.5 (Krishnan and Attia, 1985).

In the flocculation process, the flocculant first adsorbs onto the coal surface. Then, when coal particles collide with each other, the flocculant on the surface of one particle will form a bond with the second particle, and so keep them from moving apart again. This results in the formation of a floc nucleus. The nuclei continue to collide with particles and with other floc nuclei and grow into larger flocs. If flocs become large enough, shear forces that arise from fluid turbulence can cause them to break up into smaller flocs again. Depending on the type of flocculant used, the breakage of the flocs can be either reversible (allowing the flocs to re-form) or irreversible (the flocs remain small after breakage, and do not re-attach to other flocs).

For selective flocculation, it is necessary that the flocculant adsorb strongly onto one type of particle, while adsorbing very weakly onto other particle types. Depending on the type of flocculant used, it is possible to either flocculate the ash-forming minerals while leaving the coal dispersed or to flocculate the coal while leaving the minerals dispersed. To flocculate the coal, the flocculating reagent takes advantage of the natural hydrophobicity of medium- to high-rank coals. A polymeric flocculant with hydrophobic regions will selectively adsorb to the coal particles by hydrophobic bonding, while not attaching to the hydrophilic mineral particles.

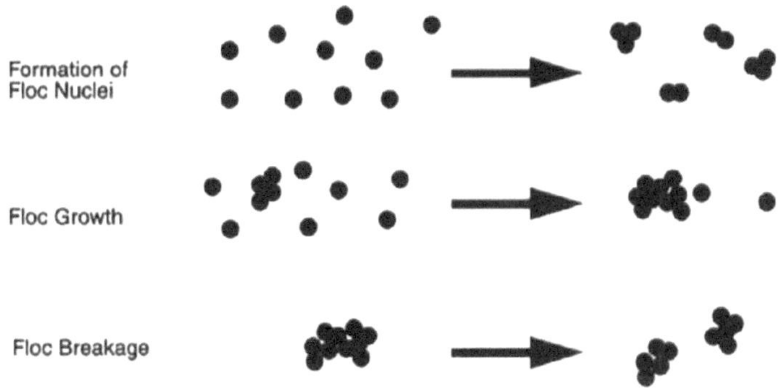

FIGURE 9.5 Formation, growth, and breakage of flocs (after Krishnan, 1987).

Mirville and Hogg (1979) determined that a nonionic polyacrylamide flocculant adsorbs onto coal at approximately ten times the level that it adsorbs onto clays and silicate minerals, and so has potential for selectively flocculating coal away from these minerals.

Polymeric flocculants can be made selective toward particular minerals by incorporating chemically active groups into the polymer chain that have a strong affinity for the mineral to be flocculated (Kitchener, 1972). For the best results, only a small proportion of the polymer chain should consist of these groups. Since pyrite has a strong affinity for xanthates while coal does not, a polymer with xanthate groups would be expected to selectively flocculate pyrite in preference to coal.

The following factors need to be considered in selective flocculation of coal:

- *Particle surface chemistry.* The effects of surface chemistry in fine coal processing are extensively discussed by Laskowski (1995). It is important to remember that the properties of coal are highly variable; therefore, the surface chemical behavior of coal particles will not be easy to control.
- *Flocculant addition.* The amount of polymeric flocculant added is critical in selective flocculation. The flocculant will always have some affinity for all of the particles in the suspension and will not restrict its action to only the particle type that is to be flocculated. If the flocculant dosage is too high, then the flocculation will become unselective and all of the particles will tend to flocculate. If the flocculant dosage is too low, then flocculation of the particles of interest will be incomplete, and many of them will not be removed from the system.
- *Percent solids.* The solids content of the suspension must be high enough that particles will be reasonably likely to collide and flocculate, but not so high that undesirable particles will be trapped into the flocs in large quantities. Increasing percent solids will also affect the reagent consumption and change the kinetics of the floc formation.

9.2.4 Floc Conditioning

There are two objectives in the floc conditioning stage. First, the flocs are being compressed and packed so that they will have sufficient strength to survive being removed from the suspension. Second, in the process of compressing the flocs, the entrapped particles that are not wanted in the flocculated product have to be freed from the flocs and returned to the suspension.

There are several sources of unwanted particles in the flocculated product:

- *Sedimentation.* Particles that are not flocculated will settle along with the flocs and be removed along with them. Although most of the particles will settle more slowly than the flocs, the coarsest particles and the particles that were near the bottom of the settling chamber will reach the bottom at the same time as the flocs and be removed. This effect can be reduced if the particles are ground to such a fine size that they settle at a negligible rate compared to the flocs.

- *Entrainment.* As the flocs form, particles are entrained in the fluid within the floc, even when the flocculant has no affinity for the trapped particles. These particles can be minimized by conditioning the flocs to minimize the amount of water that is trapped inside them.
- *Entrapment.* Unwanted particles are either included in the floc because they have some degree of affinity for the flocculant or because they have become wedged between the flocculated particles.
- *Locking.* Incompletely liberated particles are included in the floc because a portion of the particle is the desired material, and carries the undesirable material into the floc with it.

If the flocs are formed, separated from the suspension, and then re-dispersed and re-flocculated, the quality of the flocs will improve because of reductions in the amount of sedimented, entrained, and entrapped particles.

9.2.5 Separation

The separation of the flocs from the suspension needs to be done using simple, gentle techniques so that the flocs are not broken up and lost. The methods that have been reported for floc recovery include flotation, gravity decantation, and elutriation. This step is one of the primary limits on the separation efficiency because gentle separations generally do not produce a very sharp separation when compared to more aggressive separation processes.

9.3 SULFUR REMOVAL

The application of selective flocculation that is of most interest to coal desulfurization is the removal of very fine pyrite particles from finely ground coal. It is not practical to use this technique on particles that are much more than about 10 μm in diameter because they will have sedimentation speeds that are fast enough for many of the unflocculated particles to become mixed with the flocculated particles. It is generally considered to be either an alternative to flotation at these fine particle sizes or a pretreatment of material that will be re-cleaned by flotation.

Only a few investigators have reported any success in separating pyrite and coal by selective flocculation (Moudgil, 1992; Attia, 1985). These studies have been carried out only on a laboratory scale and have mainly worked with either suspensions of individual minerals or synthetic mixtures of coal and liberated pyrite. The results reported are likely to be much better than what would be seen in an industrial process, which would have to contend with locked particles.

9.3.1 Selective Flocculation

Selective flocculation of coal, leaving the pyrite dispersed in suspension, has been attempted (Attia, 1982, 1985; Attia and Fuerstenau, 1982). The method attempted was to use a xanthated polyacrylic acid (PAAX) as a dispersant for pyrite particles, while using polystyrene sulfonate (Purifloc A22) as the coal flocculant. The PAAX was produced by reacting low-molecular-weight polyacrylic acid solutions with sodium hydroxide and carbon disulfide. Tests were carried out using two different types of suspensions (pure coal and pure pyrite), both of which were 2% solids at a pH of 10. The coal and pyrite particles were reported to be finer than 37 μm. The total quantity of material used in each test was 50 mL of water and 1 g of either coal or pyrite. After conditioning and agitation, the solids were allowed to settle for 5 min before recovering the settled solids by decantation. With this procedure, the coal was found to settle completely and was not affected by the addition of PAAX. In the absence of PAAX, the pyrite also settled completely, but when the PAAX was added at dosages ranging from 50 to 400 mg/L, only 50%–55% of the total pyrite settled. Although experiments were carried out using mixtures of coal and pyrite, no quantitative results were reported (Attia and Fuerstenau, 1982; Attia, 1985).

Attia (1985) reported two experiments with a Kentucky No. 9 coal, ground to pass 37 μm and treated with 300 and 500 mg/L of PAAX before flocculating with F1029-D flocculant. The coal contained 2.5%–2.7% total sulfur, but no values were given for pyritic, sulfate, and organic sulfur. These tests were claimed to remove 9.7% and 20.3% of the total sulfur, while recovering 96.3% and 97.3% of the total feed weight. Because of the minimal characterization of the feed coal, it is difficult to determine whether this was a good separation of pyrite from the coal, or a poor one. Given the low levels of sulfur removal and the lack of follow-up in the literature, it is likely that the separation was not very effective. This is to be expected, as the high density of pyrite particles will tend to cause the coarser particles to rapidly settle along with the coal even when they are not flocculated, while the submicron pyrite will tend to remain locked to the coal and flocculate along with it, even if it is ground to very fine particle sizes.

Behl and Moudgil (1992) carried out similar studies, except that they attempted to flocculate the pyrite while leaving the coal in suspension. They used Superfloc 16 (a high-molecular-weight nonionic polyacrylamide) and Superfloc 362 (a cationic polyacrylamide) flocculants, both of which showed much greater ability to flocculate pyrite than coal in single-mineral tests. At a dosage of approximately 0.006 kg/ton and a settling time of 2 min, the Superfloc 16 could settle 100% of the pyrite and approximately 30% of the coal, while Superfloc 362 could settle 95% of the pyrite and 30% of the coal. When mixtures of 90% coal and 10% pyrite were tested, the addition of Superfloc 362 had little effect on the sulfur content of settled solids, as shown in Table 9.1 (Moudgil, 1992). The poor separation performance was credited to heteroflocculation of the pyrite to the coal particles.

Rubio and Kitchener (1977) showed that partially hydrophobic flocculants such as polyethylene oxide can flocculate inherently hydrophobic solids, such as coal. This was confirmed with unoxidized anthracite by Gochin et al. (1985), who found that upon oxidation the anthracite could no longer be flocculated by polyethylene oxide. Totally hydrophobic latexes have also been found to be suitable for selectively flocculating coal fines (Littlefair and Lowe, 1986; Attia et al., 1987; Sergeev et al., 1995).

Palmes and Laskowski (1993) have examined the behavior of three flocculants as a function of pH, using three coals that varied in their degree of hydrophobicity. The flocculants used were a fully hydrophobic latex (FR-7A, manufactured by Calgon Co.), a semihydrophilic polymer (F1029-D, manufactured by Daiichi-Kogyo

TABLE 9.1
Results Reported by Moudgil (1992) for Flocculation of Pyrite from Coal

Dosage of Superfloc 362, kg/ton	% Coal Recovery in Supernatant	% Pyritic Sulfur in Settled Solids	% Pyritic Sulfur in Supernatant
None	96.4	6.6	1.8
0.001	96.3	4.4	1.5
0.006	96.8	4.9	1.9

Settling time was 2 min. Addition of the flocculant does not appear to have improved the pyrite removal performance compared to the zero-flocculant test.

Seyaku Co.), and a hydrophilic polymer (polyacrylamide). They found that the hydrophobic latex was the most selective flocculant toward hydrophobic coal and that the more oxidized coals could not be flocculated by this reagent. The latex also required intense conditioning because it was not water soluble. The semihydrophilic polymer was also selective toward hydrophobic coal, although it did show some ability to flocculate oxidized coals. The hydrophilic polyacrylamide was essentially nonselective, flocculating all of the coal and mineral matter particles. It was also noted that there was an optimum dosage for the polyacrylamide where flocculation was most effective, while both the fully hydrophobic and semihydrophilic flocculants showed steadily increasing flocculation efficiency with increasing dosage, and did not lose effectiveness when the dosage was excessive.

Zhu et al. (2017) have used selective flocculation to separate coal from ash by enhanced gravity separation. However, this technology has a limited ability to separate sulfur from coal, with issues arising from limited floc stability during the gravity separation step and a low ability to remove organic sulfur (Zhu et al., 2017).

9.3.2 Selective Coagulation

In electrolytic coagulation, electrolytes are added to the suspension to minimize the electrostatic repulsion between particles. According to the Derjaguin, Landau, Verwey, and Overbeek (DLVO) theory, particles tend to develop an electrical charge on their surfaces from interactions with the water and the ions in solution (Verwey and Overbeek, 1947; Derjaguin and Landau, 1941; Derjaguin et al., 1954; Leja, 1982). If the proper electrolytes are added to the suspensions, they can reduce the magnitude of this charge and allow the particles to approach closely enough for the London–van der Waals' interaction to bind them together (Leja, 1982). Thus, according to the DLVO theory the electrolytic coagulation of particles is controlled by two competing effects: (1) a repulsive electrostatic interaction and (2) an attractive London–van der Waals' interaction.

A similar process, known as "shear flocculation," was developed by Warren (1975, 1977) and is more directly applicable to coal suspensions. This process was originally carried out with scheelite particles, where a collector was added, to render the scheelite particles hydrophobic. They were then agitated at high shear rates

(800–1,800 rpm) to force the particles together, with the coagulates being held together by the hydrophobic interaction between them. Since coal is naturally hydrophobic, Yoon and Luttrell (1992) coagulated coal by this means without adding any collector, using only high shear agitation. Once formed, the coal coagulates are separated from the ash minerals by sedimentation.

While it is possible to coagulate coal using only its native hydrophobicity, the resulting coagulates are not strong. Since the addition of collecting reagents such as oils will greatly increase the hydrophobicity of coal surfaces, it is obvious that the coagulation will be much more effective if such reagents are used. Yoon and Luttrell (1992) state: "There is a drawback to the hydrophobic coagulation process. The coagula formed by hydrophobic coagulation are much weaker than the flocs formed by traditional oil agglomeration processes." It may also be noted that coagulation, with its high shear rate requirement, is much slower than flocculation processes.

Yoon and Luttrell (1992) carried out continuous laboratory-scale tests of the hydrophobic coagulation process and compared the use of a sedimentation tank for coagula removal with the use of a rotating drum screen. The sedimentation tank was found to be the more efficient unit for this purpose, based on the product ash content, combustibles recovery, and separation efficiency. They achieved an ash reduction of 60%, while recovering 90% of the combustible material; however, no results for sulfur reduction were given. Parameters such as particle size were determined to be of particular importance to this process.

9.4 SUMMARY

Selective flocculation has some potential for removing fine pyrite from finely ground coal; however, it suffers from selectivity problems that appear to be inherent in the process. It is unlikely that it will become a significant means of coal desulfurization unless the process design or the reagents are greatly improved. Only a very limited amount of information is available concerning the removal of sulfur by this method. It is most likely to find use as a pretreatment to produce a rough separation before using a more selective process, such as froth flotation (Chapters 4 and 5) or selective agglomeration (Chapter 9). Because of this, it is important to be certain that the reagents used in selective flocculation will be compatible with the subsequent processing.

A large number of new reagents have been developed recently. These new flocculants have not been rigorously tested with coal specifically, but have given very promising results in other applications. It is possible that one of these new flocculants will give very good results for the selective flocculation of coal. Meanwhile, some older staples of coal flocculation, such as Superfloc 16, are not even manufactured anymore.

REFERENCES

Attia, Y.A. (1982), "Fine Particle Separation by Selective Flocculation," *Separation Science and Technology*, Vol. 17, No. 3, pp. 485–493.

Attia, Y.A. (1985), "Cleaning and Desulfurization of Coal Suspensions by Selective Flocculation," *Processing and Utilization of High Sulfur Coals* (Attia, ed.), Elsevier, New York, pp. 267–285.

Attia, Y.A. and Fuerstenau, D.W. (1982), "Feasibility of Cleaning High Sulfur Coal Fines by Selective Flocculation," *Preprints—XIV International Mineral Processing Congress*, October 17–23, Toronto, Canada, pp. 4–15.

Attia, Y.A., Yu, S. and Vecci, S. (1987), "Selective Flocculation Cleaning of Upper Freeport Coal with a Totally Hydrophobic Polymeric Flocculant," *Flocculation in Biotechnology and Separation Systems* (Attia, ed.), Elsevier, Amsterdam, p. 547.

Bajor, O. and Trass, O. (1988), "Modified Oil Agglomeration Process for Coal Beneficiation, I. Mineral Matter Liberation by Fine Grinding with the Szego Mill," *Canadian Journal of Chemical Engineering*, Vol. 66, pp. 282–285.

Behl, S. and Moudgil, B.M. (1992), "Finc Coal Cleaning by Selective Flocculation," *Coal Preparation*, Vol. 11, pp. 35–49.

Blagov, I.S. (1970), "Flocculation of Minerals and Carbon Suspensions by Polymers," *Proceedings of the 9th International Mineral Processing Congress*, Prague, Czechoslovakia, p. 166.

Blaschke, Z. (1972), "Oil Agglomeration and Selective Flocculation of Coal Slurries," *Proceedings of the 8th Mineral Processing Congress*, Cracow, Poland, pp. 763–776.

Blaschke, Z. (1976), "Beneficiation of Coal Fines by Selective Flocculation," *Proceedings of the 7th International Coal Preparation Congress*, Sydney, Australia, pp. F1–F13.

Blaschke, Z. (1994), "Oil Agglomeration and Selective Flocculation of Coal Slurries," *Proceedings of the 12th International Coal Preparation Congress*, Cracow, Poland, pp. 763–776.

Derjaguin, B.V. and Landau, L.D. (1941), "Theory of the Stability of Strongly Charged Lyophobic Sols and of the Adhesion of Strongly Charged Particles in Solutions of Electrolytes," *Acta Physiocochim, URSS*, Vol. 14, p. 633.

Derjaguin, B.V., Titievskaya, A.S., Abricossova, I.I. and Malkina, A.D. (1954), "Investigation of the Forces of Interaction of Surfaces in Different Media and Their Application to the Problem of Colloid Stability," *Discussions of the Faraday Society*, Vol. 18, pp. 18–21.

Fuerstenau, D.W., De, A., Diao, J. and Kapur, P.C. (1995), "Fine Grinding of Coal in a Two-Stage High-Pressure Roll Mill/Ball Mill Hybrid Mode," *High Efficiency Coal Preparation* (Kawatra, ed.), Society for Mining, Metallurgy, and Exploration, Littleton, CO, pp. 295–306.

Gochin, R.J., Lekili, M. and Shergold, H.L. (1985), "The Mechanism of Flocculation of Coal Particles by Polyethylene Oxide," *Coal Preparation*, Vol. 2, p. 19.

Gregory, J. (1985), "The Action of Polymeric Flocculants," *Flocculation, Sedimentation, and Consolidation* (Moudgil and Somasundaran, eds.), Engineering Foundation, New York, pp. 125–137.

Hucko, R. (1977), "Beneficiation of Coal by Selective Flocculation: A Laboratory Study," U.S. Bureau of Mines, Report of Investigations R. I. 8234.

Kitchener, J.A. (1972), "Principles of Action of Polymeric Flocculants," *British Polymers Journal*, Vol. 4, No. 3, pp. 217–229.

Krishnan, S.V. (1987), "Selective Flocculation of Fine Coal," Chapter 7, *Fine Coal Processing* (Mishra and Klimpel, eds.), Noyes Publications, Park Ridge, NJ.

Krishnan, S.V. and Attia, Y.A. (1985), "Floc Characteristics in Selective Flocculation of Fine Particles," *Flocculation, Sedimentation, and Consolidation* (Moudgil and Somasundaran, ed.), Engineering Foundation, New York, pp. 229–248.

Laskowski, J.S. (1995), "Coal Surface Chemistry and its Effects on Fine Coal Processing," *High Efficiency Coal Preparation* (Kawatra, ed.), Society for Mining, Metallurgy, and Exploration, Littleton, CO, pp. 163–176.

Leja, J. (1982), *Surface Chemistry of Froth Flotation*, Plenum Press, New York.

Littlefair, M.J. and Lowe, N.R.S. (1986), "On the Selective Flocculation Using Polystyrene Latex," *International Journal of Mineral Processing*, Vol. 17, pp. 187.

Mangravite, F.J, Leitz, C.R. and Galick, P.E. (1985), "Organic Polymeric Flocculants: Effect of Charge Density, Molecular Weight and Particle Concentration," *Flocculation, Sedimentation, and Consolidation* (Moudgil and Somasundaran, eds.), Engineering Foundation, New York, pp. 139–158.

Mirville, R.J. and Hogg, R. (1979), "Polymer Adsorption and Flocculation in the Treatment of Coal Preparation Waste Water," *Presented at the SME Annual Meeting*, February 1979, New Orleans, LA, Preprint No. 70–61.

Moudgil, B.M. (1992), "A New Approach in Ultrapurification of Coal by Selective Flocculation," Final Report, U.S. DOE, Pittsburgh Energy Technology Center, DOE/PC/88917-T8 (DE92015624).

O'Melia, C.R. (1985), "Polymeric Inorganic Flocculants," *Flocculation, Sedimentation, and Consolidation* (Moudgil and Somasundaran, eds.), Engineering Foundation, New York, pp. 159–169.

Palmes, J.R. and Laskowski, J.S. (1993), "Effect of the Properties of Coal Surface and Flocculant Type in the Flocculation of Fine Coal," *Minerals and Metallurgical Processing*, Vol. 10, No. 4, pp. 218–222.

Rubio, J. and Kitchener, J.A. (1977), "New Basis for Selective Flocculation of Mineral Slimes," *Transactions of the IMM, Section C*, Vol. 86, pp. 97–100.

Sergeev, P.V., Zalevsky, V.T., Elishevich, A.T. and Galushko, L.Ya. (1995), "Laws of the Selective Flocculation of Coals," *Solid Fuel Chemistry*, Vol. 29, No. 6, pp. 1–5.

Verwey, E.J.W. and Overbeek, J.Th.G. (1947), "Theory of the Stability of Lyophobic Colloids," *Journal of Physical and Colloid Chemistry*, Vol. 51, pp. 631–636.

Warren, L.J. (1975), "Shear-Flocculation of Ultrafine Scheelite in Sodium Oleate Solutions," *Journal of Colloidal and Interface Science*, Vol. 50, pp. 307–318.

Warren, L.J. (1977), "The Stability of Suspensions: A Guide for Beginners," *Australian Journal of Chemistry*, Vol. 44, No. 12, pp. 315–318.

Yarar, B. and Kitchener, J.A. (1970), "Selective Flocculation of Minerals," *Transactions of the IMM, Section C*, Vol. 79, p. 23.

Yoon, R.-H. and Luttrell, G.H. (1992), "Development of the Selective Coagulation Process," U.S. Department of Energy, Pittsburgh Energy Technology Center. DOE/PC/90174-T5 (DE93000484).

Zhu, X., Tao, Y. and Sun, Q. (2017), "Separation of flocculated ultrafine coal by enhanced gravity separator," *Particulate Science and Technology*, Vol. 35, No. 4, pp. 393–399.

10 Selective Agglomeration

"Agglomeration" describes processes that form balls, briquettes, nodules, flakes, or other types of particulate masses (agglomerates) from loose, fine particles (DeVaney, 1985). Selective flocculation, described in Chapter 9, can be considered a type of agglomeration process. The focus of this chapter, however, is on "selective agglomeration," which is a similar but distinct process.

Whereas selective flocculation uses small quantities of flocculants to form bridges between similar particles, selective agglomeration is performed by adding an immiscible, hydrophobic fluid. The idea is that this hydrophobic fluid will attach to the coal and displace water, as coal is more hydrophobic (water repelling) than hydrophilic (water attracting). Once this fluid has attached to the coal surface, it has the potential to form capillary bridges to other coated coal particles (Steedman and Krishnan, 1987). The formation of these capillary bridges causes the coal particles to adhere to each other. These collections of coal particles can eventually grow large enough to be separated from the bulk slurry. Since the hydrophobic fluid will not attach to hydrophilic minerals such as pyrite, this allows such minerals to be separated from the coal.

Oil is a common choice of immiscible fluid, or "agglomérant," for this purpose. As such the process is often referred to as "oil agglomeration" (Steedman and Krishnan, 1987). However, other liquids can work so long as they preferentially adhere to the coal surface, are capable of displacing water from the coal surface, and are themselves immiscible in water. The first two points are very similar conceptually, though ultimately distinct. Most nonpolar fluids have some potential for both adhering to and displacing water from a coal surface. The fluid needs to be immiscible in water so that the water phase may be completely displaced from the coal surface, rather than just diluted, and so that capillary bridges can form between coal particles in the agglomerant phase. Some liquids that have been used for selective agglomeration include hydrocarbons such as *n*-pentane, *n*-hexane; liquid carbon dioxide; or chlorofluorocarbons such as Freon-113 (Hucko and Gala, 1990). When oil is used, the quantity of oil added varies from less than 2% of the original coal weight to as much as 15% of the coal weight (Capes, 1979).

The basic principle of oil or spherical agglomeration has been known for many years, with early works dating as far back as 1921 (Perrot and Kinney, 1921) and a patent on the process as early as 1922 (Trent, 1922). Oil agglomeration was practiced commercially before froth flotation became popular, and in the 1920s four plants were in operation, with the largest producing 36,000 tons of clean coal before it was closed down for economic reasons. Conoco-Consol also commercialized an oil agglomeration process for fine coal, with a plant that processes 35–50 ton/h of feed reclaimed from a blackwater pond in southern West Virginia (Theodore, 1985).

A number of excellent articles that discuss the basic principles of selective agglomeration are available, including Steedman and Krishnan (1987), Bechtel

National, Inc. (1990), Mehrotra et al. (1983), and Keller (1995). The early works of Capes et al. (1976), Capes (1979, 1980), Armstrong et al. (1978), Labuschagne and Prinsloo (1985, 1988), and Labuschagne (1986) dealt exclusively with the recovery of normal fine coal by agglomeration. Production of ultrafine, ultraclean product coal from slurries has been dealt with by the work of Keller and Burry (1990), Keller (1995), Jha and Smit (1995), Southern Company Services (1993), and Bechtel National, Inc. (1990).

10.1 PRINCIPLES OF SELECTIVE AGGLOMERATION

Coal is primarily a hydrophobic compound, which means that the energy cost of maintaining a coal–water surface is larger than maintaining a coal-hydrophobic surface elsewhere. When a hydrophobic agglomerant such as oil is added, the coal in a coal–water slurry will be preferentially wetted by the agglomerant. This process selectively displaces the coal from the water phase, while hydrophilic minerals such as pyrite are largely unaffected. Since the agglomerant is also hydrophobic, the coal particles which are wetted by it will adhere to each other if they come into contact. This is because the formation of adhering capillaries between the coal particles minimizes the agglomerant–water surface area and the corresponding surface energy cost. In the presence of an adequate amount of oil and sufficient mechanical agitation, the coal can be formed into agglomerates that are large enough to permit separation and recovery.

The kinetics of the process depends largely on the agitation rate and the quantity of oil added. An increase in the shear rate of the suspension will increase the rate of interparticle collisions, while an increase in oil addition increases the probability that there will be oil on the colliding surfaces and that there will be sufficient oil to form a capillary bridge. At high shear rates, agglomerates form quickly, and are frequently broken, worked, and reformed, which gives rise to opportunities for entrapped mineral particles to be released. High shear rates also limit the maximum size of the agglomerates, as large agglomerates are subjected to increasingly severe forces and are more likely to be broken.

The greatest difficulty in understanding the agglomeration kinetics of coal lies in the fact that there is no simple way to quantify the ability of a specific agglomerant to displace water from the surface of a specific coal. Once the agglomerant has wetted the coal surface, the kinetics of agglomerate growth and disintegration can be quantified (Dunstan et al., 1986; Kawashima and Capes, 1974, 1975; Kawashima et al., 1989). Most of the existing kinetic models for selective agglomeration deal with coal coarser than 100 μm. However, selective agglomeration has been carried out successfully with coals with top sizes less than 1 μm (Keller and Burry, 1990). It is clear that the current kinetic models will not suffice to explain the kinetics at such small particle sizes.

In order to collect better kinetic data, Wheelock et al. (1994a,b) have developed a model mixing system that uses a photometric dispersion analyzer to measure the turbidity of the suspension. The turbidity is taken to be a measure of the degree of agglomeration, thus providing a means for continuously monitoring the progress of agglomeration as a function of time.

Thermodynamically, the wetting of coal particles by the agglomerant and their subsequent agglomeration helps to minimize the total energy of the system by reducing the contact area of dissimilar surfaces. The best performance will be achieved when the agglomerant coal interfacial energy tends towards zero, which suggests that performance can be modified by the use of surfactants or by changing the nature of the agglomerant.

10.1.1 Stages of Selective Agglomeration

The precise flowsheet used for selective agglomeration will of course depend on the feed and agglomerant characteristics and on the desired end product. Generally speaking, an agglomeration flowsheet will be similar to that shown in Figure 10.1 and will include the following steps:

- Slurry conditioning.
- Oil emulsification.
- Agglomerate formation.
- Agglomerate recovery.
- Dewatering.

10.1.1.1 Slurry Conditioning

As in the flotation process, it is important to thoroughly condition the coal slurry before agglomeration. All particles need to be thoroughly wetted by water, so that the oil does not pick up mineral particles that do not yet have water covering their surfaces. Also, any wetting or conditioning agents that are added to the process (for example, to improve the wetting behavior of oxidized coals) need to be thoroughly dispersed throughout the slurry before adding the oil. Several authors have advocated the addition of chemical reagents during this stage that make the pyrite particles more hydrophilic, which is claimed to improve the rejection of pyrite in the agglomeration step (Steedman and Krishnan, 1987; Mehrotra et al., 1983), although it is questionable whether this is really effective or necessary for most coals.

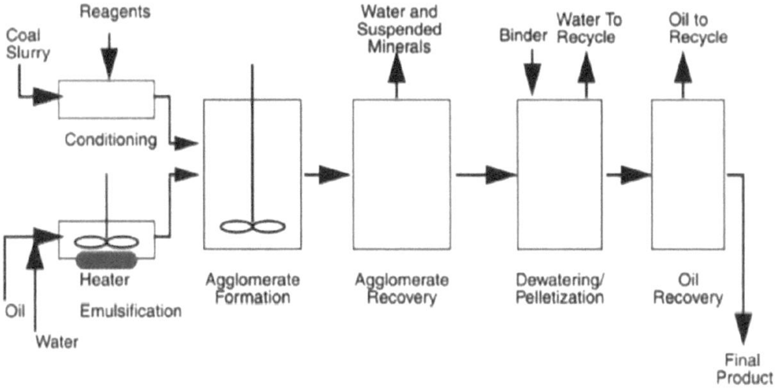

FIGURE 10.1 Generalized flowsheet for oil agglomeration (after Mehrotra et al., 1983).

10.1.1.2 Oil Emulsification

By emulsifying the oil in water before adding it to the coal slurry, the oil is dispersed so that it can more rapidly collide with and coat the coal particles. Bensley et al. (1977) studied the effects of oil emulsification in detail. Heavy oils generally produce agglomerates that are larger and stronger than agglomerates produced by light oils and are lower in cost, but these oils are difficult to disperse. Heating reduces the viscosity of heavy oils so that they are easier to disperse, and emulsification makes it possible to efficiently mix oils with the coal slurry, as shown in Figure 10.2.

10.1.1.3 Contacting and Agglomeration

In the contacting step, oil preferentially adsorbs on the coal particles. This step is carried out at high shear rates to increase the probability of collision between oil droplets and coal particles. Very often, two levels of shear rate are used, with a high shear rate during contacting and a lower level of shear during agglomerate formation. This allows the agglomerates to grow to a greater size, without sacrificing the benefits of high shear during the contacting stage.

Wheelock et al. (1994a,b) studied the effect of dispersing air in the slurry during oil agglomeration. Agglomeration was carried out with a highly hydrophobic coal and with heptane and hexadecane as agglomerants. Laboratory mixing was carried out using a flat-blade turbine agitator at a moderate shear rate. This apparatus produced nearly spherical agglomerates both with and without dispersed air. However, the addition of air increased the rate of agglomeration and produced larger agglomerates at any given oil dosage. The presence of air also improved the product quality and recovery. The effect of air dispersion was greatest at lower oil dosages.

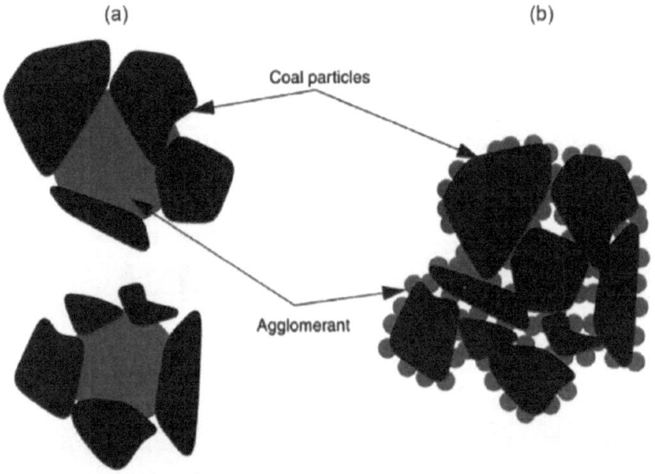

FIGURE 10.2 Effect of emulsification on the agglomeration performance of high-viscosity oils. (a) Unemulsified oils produce large droplets that have limited surface area, and do not completely cover the coal surfaces, resulting in poor agglomeration and wasted oil. (b) Emulsified oil droplets have high surface area, and completely cover the coal surfaces, resulting in good oil usage and rapid agglomeration.

10.1.1.4 Agglomerate Recovery

Once the coal/oil agglomerates are formed, they must be removed from the slurry to separate them from the unagglomerated mineral matter and water. The method used depends on the characteristics of the agglomerates. If the agglomerates are loose and fluffy, then it is most effective to either allow them to float to the surface and skim them off if they are less dense than water, or to recover them by froth flotation if they are more dense than water. If the agglomerates are compressed and durable, then they are discharged onto a screen, where the agglomerates are retained while the water and mineral matter pass through.

10.1.1.5 Dewatering

The agglomerates will contain some water that has been entrapped between particles. Since the oil displaces water from the coal surfaces, the agglomerates are much easier to dewater than the clean coal from other processes. The water will simply drain away if the agglomerates are stockpiled for a sufficient time. Alternatively, they can be dewatered using sieve bends, vibrating screens, or centrifuges.

10.1.2 Important Process Variables

The design of the selective agglomeration process depends upon the ultimate objective. The goal may range from processing refuse from coal tailings ponds to production of ultraclean fuels. The operating variables that need to be considered in planning an agglomeration process are

- Selection of agglomerant.
- Percent solids in the slurry.
- Mixing intensity.
- pH and temperature of slurry.
- Conditioning agents.
- Characteristics of coal and associated matter.

10.1.2.1 Selection of Agglomerant

A wide variety of agglomerants have been used for the selective agglomeration of coal. These agglomerants include

- Oils, such as fuel oils.
- Hydrocarbons, such as *n*-pentane and *n*-heptane.
- Liquid carbon dioxide.
- Chlorofluorocarbons, such as Freon-113.

Labuschagne (1986) has studied 41 different agglomerants, including paraffinic straight-chain hydrocarbons, olefinic straight-chain hydrocarbons, monocyclic hydrocarbons, monocyclic aromatic hydrocarbons, and dicyclic hydrocarbons. The purpose of the study was to investigate the contribution of chemical properties and structural features of agglomerants on various aspects of the selective agglomeration process. In general, it was concluded that the overall agglomeration behavior of agglomerants depends on the physical, chemical, and structural properties of these liquids. No definite conclusions could be reached, however, because of the complex relations between the various factors.

The most commonly used agglomerants are oils, which are generally of lower cost than other possibilities. Aside from the cost, it is important to select oils that will give good coal recovery while rejecting the ash and pyrite. These characteristics depend on the chemistry of oil and on its physical properties, particularly the viscosity. The best characteristics have not been fully identified because the published data is somewhat contradictory.

Sun and McMorris (1959) conducted the first study to establish the relationship between oil density/viscosity and agglomeration effectiveness. They concluded that the optimum specific gravity range was between 0.7 and 0.85, which corresponds to oils such as No. 2 fuel oil and kerosene. Oils less dense than 0.64 or denser than 0.97 were not useful agglomerants. It was claimed that the very heavy (and viscous) oils were too viscous to disperse readily, while the very light oils did not make the coal surfaces hydrophobic enough. It was reported by Sarkar et al. (1976) that the best agglomerant oils were those with surface tensions near 30 dynes/cm, which is a medium surface tension level for oils. The results of Sun and McMorris's (1959) study were disputed by Capes et al. (1976), who argued that the poor performance of high-viscosity oils was an artifact of the short mixing time used, and reported that when the mixing time was extended, even highly viscous oils could produce high coal recoveries.

A number of studies have considered the ash and sulfur rejection capabilities of various oils. It has been reported that heavy oils are less selective than lighter oils and that heavy oils produce a higher ash level in the agglomerates (Ralston, 1922; Capes et al., 1976). This effect can be seen in Table 10.1 (Tsai, 1982). Similar results were found by Elliot (1978) using emulsified oils. It has been claimed by Bensley et al. (1977), who also used emulsified oils, that increasing oil density alone did not cause a loss of ash rejection. They reported that small quantities of surface-active impurities in an oil cause the ash rejection to deteriorate and suggested that the heavy oils were most likely to contain this type of impurity. On the other hand, vegetable oils may also be easier to emulsify in water than their weight would initially suggest, as they possess functionalization that can interact with hydroxyl or fatty ester groups in water, which may be presented by the pH of the pulp or the surface of the coal (Özer et al., 2017). Vegetable oil has been found to be effective at agglomerating fine coals at very low oil concentrations with high recovery rates (Özer et al., 2017).

In general, there are apparently a number of unknown and uncontrolled variables that affect the agglomeration performance of oils, and so reported data that compares various oils for this purpose are somewhat suspect. Also, the great variability of coal properties is likely to be partly responsible for the numerous contradictory results. It has been reported, for example, that low-rank coals (sub-bituminous and lignite) only agglomerate satisfactorily with heavy oils such as coal tars, pitches, and petroleum crudes (Capes et al., 1976), which is opposite to the results reported by most investigators for higher-rank coals.

However, the nature of oil clearly has a significant impact on the agglomeration process, as it determines the character of the capillary structure between the coal

TABLE 10.1
Effect of Type of Oil Used for Agglomeration on the Beneficiation Performance

		Product			% Reduction		% Removal	
Collecting Oil	Temp., °C	% Ash	% Sulfur	% Combust. Recovery	Ash	Sulfur	Ash	Sulfur
Fuel oil	21	7.2	3.6	97	66	41	66.9	42.8
Kerosene	21	5.2	2.8	83	75	54	79.5	61.9
Kerosene	90	5.5	2.9	—	74	52	—	—
Varsol	21	5.0	2.7	91	76	56	78.4	59.7
50% Kerosene/	21	10.6	—	—	50	—	—	—
50% Bunker C	—	—	—	—	—	—	—	—
50% Kerosene/	90	14.0	4.6	92	34	25	38.9	30.6
50% Bunker C	—	—	—	—	—	—	—	—
Light coal tar	21	19.0	—	90	10	—	18.9	—
Light coal tar	90	15.7	—	90	26	—	33.0	—

Feed coal analysis (dry basis): 21.1% ash, 6.1% total sulfur, 2.3% pyritic sulfur, 2.4% sulfate sulfur, and 1.4% organic sulfur (Tsai, 1982).

particles, the degree to which mineral matter is rejected during agglomeration, and the shape, size, and strength of the final agglomerate (Özer et al., 2017).

10.1.2.2 Percent Solids in the Slurry

The pulp percent solids also affects the agglomeration performance, but this effect is not as large as many other factors. The percent solids must be high enough for the coal to contact and combine with the oil in a reasonable time, but must not be so high that the impurities become mechanically entrapped in the agglomerates. In practice, the pulp density is often in excess of 50% solids by weight (Steedman and Krishnan, 1987).

10.1.2.3 Size Distribution

The most obvious effect of particle size distribution is that the liberation improves at finer particle sizes, allowing a cleaner product to be made. Finer particle size also results in increased surface area, which in turn requires more agglomerating liquid and therefore increases the reagent costs. Finer size distributions also result in stronger agglomerates which contain moisture overall, which requires balancing the required handling properties of the agglomerates against the maximum acceptable moisture content (Özer et al., 2017).

10.1.2.4 Mixing Intensity

In general, increasing the agitator speed and mixing time tend to be beneficial in oil agglomeration. For example, Capes et al. (1971) reported that increasing the agitator speed from 3,000 to 6,000 rpm decreased the amount of time needed to complete the agglomeration from 18 to only 8 min. An increase in mixing intensity also tends to reduce the amount of agglomerant that must be added and therefore improves the overall process efficiency. The agitation intensity and mixing time needed tend to increase as the oil density and viscosity increases, as the viscous, heavy oils need more energy to thoroughly spread them over the coal surfaces (Steedman and Krishnan, 1987). Estimates of the amount of energy needed during mixing range from 10 to 40 kW/m^3 of mixer volume (Capes and Germain, 1980).

Many of the processes developed for selective agglomeration differ primarily in the intensity of mixing. The Trent process, for example, used low-intensity agitation and required agglomeration times of over 15 min (Perrot and Kinny, 1921; Batley, 1923), while more recent processes such as the Convertol and Olifloc processes use high-intensity mixers (peripheral speeds over 20 m/s) and require agglomeration times of only 15–30 s, as well as requiring much less oil (Lemke, 1954; Blankmeister et al., 1976; Steedman and Krishnan, 1987). For extremely high mixing intensity, several investigators have combined coal grinding with mixing. This has been reported to be successful using the Szego mill, with an energy consumption of approximately 30 kWh/ton (Bajor and Trass, 1988; Trass and Bajor, 1988). Emulsification rates can also be improved via ultrasonic agitation (Özer et al., 2017).

10.1.2.5 pH

The choice of pH and any chemical additives depends on the type of coal. The hydrophobicity of coal surfaces will vary with pH, so it is important that the slurry pH be in the region where the hydrophobicity of coal is high while that of other minerals is low. A particular concern is that pyrite is strongly hydrophobic at acidic pH but tends to become hydrophilic at neutral or alkaline pH (Kawatra and Eisele, 1992). It is therefore important that the agglomeration be carried out at high pH to prevent agglomeration of the pyrite along with coal.

However, excessively high pH values may reduce the agglomeration rate, as eventually, even the surfaces of the organic materials in coal will become uniformly negatively charged. This will result in them repelling each other during agglomeration, which may decrease overall agglomerate yields and a decreased recovery of organic matter. The optimal pH to perform agglomeration at is dependent on coal, and for some coals, even acidic pH values can be optimal (Özer et al., 2017).

The wetting characteristics of strongly hydrophobic coals can also be improved by increasing the ionic strength of the pulp. This destabilizes the aqueous wetting phase on the surface of the hydrophobic coal, which makes it easier for the agglomerant to form contacts with the surface (Özer et al., 2017). For less hydrophobic coals, however, the opposite effect can be observed. Perhaps this is because the presence of ions can amplify the existing hydrophilic tendencies of both the aqueous phase and the coal surface – if they are similar to begin with, then they tend to become more similar in character, but if a moderate difference in hydrophobic nature already exists, then they will tend to become more distinct overall.

10.1.2.6 Coal Characteristics and Conditioning Agents

The natural hydrophobicity of coal increases as its rank increases from lignite to anthracite. However, within anthracites the hydrophobicity may drop slightly as its specific surface area increases (Özer et al., 2017). The best agglomeration recoveries and yields are achieved when there is a drastic difference between the hydrophobicity of coal and the materials to be separated (Özer et al., 2017).

Coals lose much of their native hydrophobicity when they become oxidized, and as a result oxidized coals tend to agglomerate poorly because they are not as easily wetted by oil. The surface of the coal may undergo dry or wet oxidation over time, which generally leads to a decrease in hydrophobicity as this leads to the formation of polar ether, carboxylic, or phenolic groups. The hydrophobicity of an oxidized coal can be restored by converting these groups back into less polar groups, such as via methylation (Özer et al., 2017). Alternatively, to counteract the oxidation effect, reagents can be added to oil which improve the wetting of oxidized surfaces by oil. Dow M210 froth conditioner, for example, is a mixture of diesel oil, acetic acid, and a proprietary alkanolamide that is mixed with oils to improve their affinity for oxidized coal surfaces. Heavy oils often naturally contain partially hydrophilic compounds that can also improve wetting of oxidized coal by oil.

Coal is a complex material that is composed of several distinct, highly intermingled organic macerals, primarily including vitrinite, exinite, and inertinite. These macerals have distinct structural and surface properties that can be measured independently, but the combined wetting properties of the resulting coal have remained difficult to quantify (Özer et al., 2017).

10.1.3 Processes Based on Selective Agglomeration

Since its initial use in the Trent process in the 1920s, selective agglomeration has been used in several processes that have been developed to at least a pilot scale; however, the operating costs have been high enough that they have not been economically practical in the long run. As a result, many variants have been designed in attempts to reduce the operating costs. The following processes are examples of what has been developed, but a more complete list is available in Table 10.2:

- Trent process.
- Convertol process.
- Broken Hill Proprietary (BHP)-selective agglomeration process.
- Aglofloat process.
- Bechtel spherical agglomeration process.
- The Energy and Environmental Research Center (EERC) oil agglomeration of low-rank coals.
- Alberta Research Council (ARC) coal agglomeration process.
- Eniricerche selective agglomeration process.
- Otisca process.
- Liquid Carbon Dioxide (LICADO) process.

10.1.3.1 Trent Process

The Trent process is mainly of historical interest as the first selective agglomeration process to be used. The coal was prepared by grinding a 40% solids slurry to pass 100 mesh and was then mixed with fuel oil at a rate of 30% of the dry coal weight. The slurry was mixed with agitators rotating at 150 rpm for an extended

TABLE 10.2
Brief Overview of Industrial Coal-Agglomeration Processes Developed over the Past 100 Years (Özer et al., 2017)

Process	Year	Description
Trent Process	1920	*See Section 10.1.3.1.*
Convertol Process	1952	*See Section 10.1.3.2.*
National Research Center of Canada (NRCC) process	1982	This process involved using a Szego mill to wet-grind the coal with a light oil agglomerant. The oil was added at a high shear rate, which was then lowered to promote the formation of microagglomerates, which were transferred to a settling tank and allowed to grow. The agglomerates were then skimmed, screened, dewatered, and then grown in a pelletizing disc or drum.
OLIFLOC process	1973	This process was utilized at commercial scale to recover ultrafine coal from high-clay slimes in two plants in Germany. The process first separated the coal slurry into +0.1 and –0.1 mm sections via hydrocylcone. The fine fraction underwent agglomeration in high-energy mixers, and the coarse fraction underwent flotation. The fine fraction resulted in agglomerates with 6%–8% ash. The coarse fraction after vacuum filtration had a moisture content of less than 2%.
Broken Hill Proprietary (BHP) process	1978	*See Section 10.1.3.3.*
Shell process	1980	A two-step agglomeration process starting from a fine coal slurry. In the first step, a hydrocarbon binder and turbulence were used to initiate agglomeration. In the second step, hot water was used to soften the initial binder and turbulence, and an additional binder was used to extend agglomeration.
Central Fuel Research Institute (CFRI) process in Dhanbad, India	1979	Aimed at cleaning fines and high-ash coals having good coking properties, the feed was mixed with diesel oil and ground in a colloid mill at controlled pH. The slurry was transferred to an agitator where oil was added. Flotation was used to collect the agglomerates.
Aglofloat process	1987	*See Section 10.1.3.4.*
Agflotherm process	1987	A co-beneficiation process for low-rank coals to solid fuel and heavy oil crude, oil residue, or bitumen to synthetic crude. Run-of-mine coal was pulverized to −100 mesh and slurried with water, while the heavy oil was mixed with lighter distillates to form the agglomerant. These were mixed at high shear and separated when the agglomerates were between 0.2 and 0.5 mm. The solid fuel heating value was >22.5 GJ/t, while the synthetic crude gravity was >21° American Petroleum Institute gravity. Coal in the refuse from the flotation underflow was utilized for co-generation.

(Continued)

TABLE 10.2 (*Continued*)
Brief Overview of Industrial Coal-Agglomeration Processes Developed over the Past 100 Years (Özer et al., 2017)

Process	Year	Description
Bechtel spherical agglomeration process	1991	*See Section 10.1.3.5.*
EERC-North Dakota process	1991	*See Section 10.1.3.6.*
EPRI-ARC process	1993	*See Section 10.1.3.7.*
Eniricerche-developed selective agglomeration process	1988	*See Section 10.1.3.8.*
Otisca-T process	1981	*See Section 10.1.3.9.*
LICADO process	1988	*See Section 10.1.3.10.*

period. The agglomerates were up to 25 mm in diameter and contained only 4% ash (Mehrotra et al., 1983); however, the tremendous rate of fuel oil use made the process very expensive, and it was ultimately abandoned in favor of flotation processes.

10.1.3.2 Convertol Process

The Convertol process showed a significant drop in the oil consumption compared to the Trent process, as the amount of oil that it used was only 3%–10% of the coal weight (compared to 30% for the Trent process). This reduction in oil was achieved by much improved mixing and the use of less viscous oils. Originally, the process used "phase inversion mills" to mix the oil and coal, but in an attempt at commercial implementation, these were found to have an excessive wear rate, and so were replaced with two high-intensity mixers in series. The process had a yield of 80% and reduced the coal ash content by 50% (Lemke, 1954; Brisse and McMorris, 1958).

10.1.3.3 BHP-Selective Agglomeration Process

BHP of Australia constructed a small (5.5 ton/h) selective agglomeration plant at their John Darling Colliery in 1978, after determining on a pilot scale that it was technically feasible. The process, which is shown schematically in Figure 10.3, was successfully operated with a 40% ash feed until 1981. Ultrasonic emulsification of oil was used. The improved mixing compared to the Trent process allowed the oil addition rate to be reduced to 7%–8% of the coal weight, and the agglomerated product was between 7% and 9% ash. The weight yield was 60% of the feed coal, and the tailings ash content was approximately 85% (Nicol and Swanson, 1979).

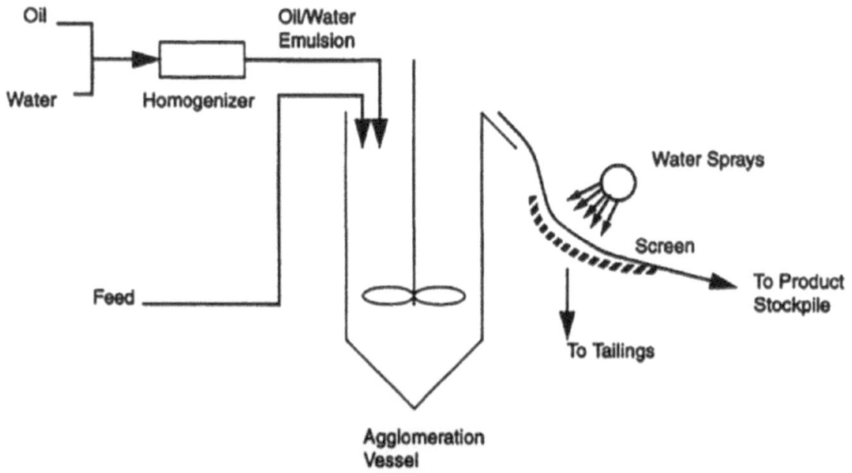

FIGURE 10.3 Schematic of the BHP oil agglomeration process.

10.1.3.4 Aglofloat Process

In the Aglofloat process, the coal was first crushed and then slurried with recycled water. The agglomerant was a mixture of crude oil with diesel fuel. The oil was mixed with the coal slurry in a high-shear mixture to form microagglomerates. Since the agglomerates were too fine to be easily removed from the slurry by screening, they were recovered by froth flotation, and further cleaned of mineral matter using a hydroseparator. The agglomerates were then transferred to a low-shear mixer, where additional oil was added and the agglomerates were enlarged to between 850 and 335 μm. This process was used to treat an Illinois No. 6 coal, containing 93.5% ash and 4.19% total sulfur. The agglomerated product recovered 89% of the combustibles, while removing 77% of the pyritic sulfur, reducing the ash content to 6.8%, and the sulfur content to 3.7% (Ignasiak et al., 1990).

10.1.3.5 Bechtel Spherical Agglomeration Process

In this process, the coal was initially agglomerated from a suspension of coal ground to less than 20 μm. The agglomerant in this first stage was heptane, which has a low boiling point that simplifies its recycle. Once the agglomerates were separated from the ash minerals, the heptane was replaced by an asphalt binder, which was used to bind together product agglomerates that were between 6.4 and 9.5 mm. The heptane was stripped from the final product by steam and was recycled to keep processing cost down. A schematic of the process is shown in Figure 10.4.

It was reported that this process could recover 98%–99% of the heating value from the coal, while rejecting over 80% of the pyrite and producing a refuse containing 85%–88% ash. The heptane recycle was quite efficient, with less than 0.2% heptane in the final agglomerates (Bechtel National, Inc. 1990; Getsoian, 1991).

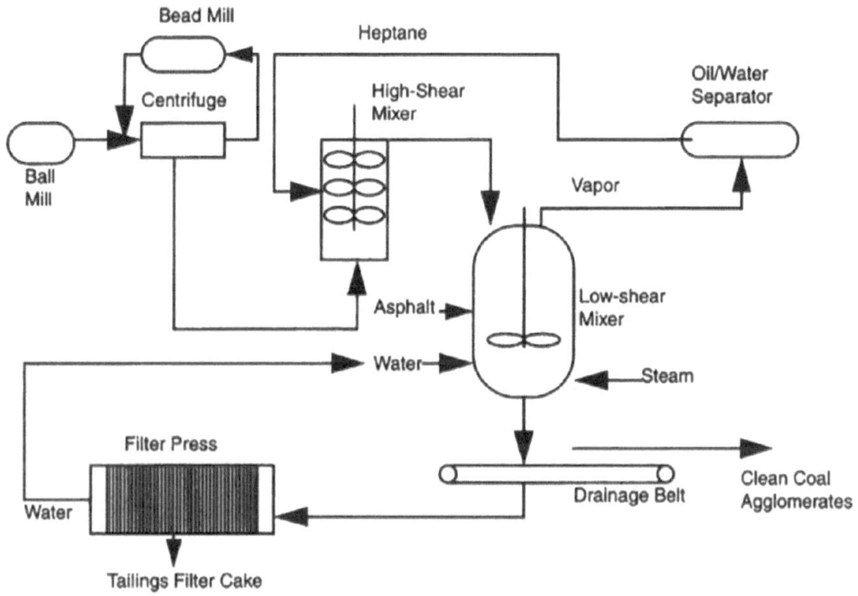

FIGURE 10.4 Schematic of Bechtel spherical agglomeration process.

10.1.3.6 EERC Oil Agglomeration of Low-Rank Coals

The EERC at the University of North Dakota has developed this process for agglomeration of low-rank coals, which are normally resistant to selective agglomeration processes. An acid-leaching step preceded the agglomerate formation, with coal stirred with a dilute acid solution in a high-shear mixer (5,500–6,000 rpm) for 1–30 min. This acid treatment made the coal more amenable to agglomeration, apparently by removing the hydrophilic compounds from the surface of the coal.

After leaching, oil was added to the coal at a rate of 70%–100% of the coal weight and agitated to form agglomerates. The agglomerates were then recovered using a 600 μm screen, and the tailings were vacuum filtered (Musich et al., 1992).

10.1.3.7 EPRI-ARC Coal Agglomeration Process

A Proof-of-Concept selective agglomeration unit was constructed by the ARC and the Electric Power Research Institute (EPRI) at the Coal Development Test Facility in Wilsonville, Alabama. A block diagram of the process used is shown in Figure 10.5. Based on laboratory-scale results, the unit used diesel oil and heptane as the agglomerants, and it was decided that diesel oil agglomeration and pelletization was the most promising route. Oil use was between 0.25% and 1.0% of the coal weight, and the small quantity of oil used made an oil recovery step unnecessary. The feed was ground to pass 150–75 μm, and in demonstration runs of several days duration the BTU recoveries were 89%–90%, while the pyritic sulfur rejections were between 60% and 80% (Southern Company Services, 1993).

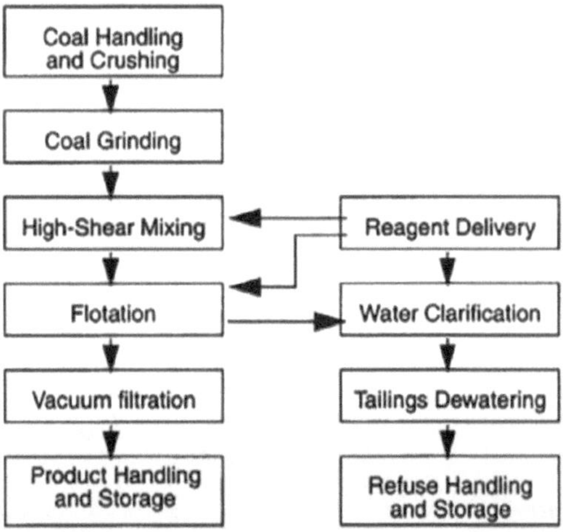

FIGURE 10.5 Block diagram of the ARC proof-of-concept selective agglomeration process.

10.1.3.8 Eniricherche Selective Agglomeration Process

This process was developed by Eniricherche (Milan, Italy), which was the corporate research arm of the Eni group. The coal was first finely ground and dispersed in water and mixed with chemicals to agglomerate the carbonaceous particles. The agglomerated coal was then recovered by filtration. The chemicals used were reported to be capable of removing both ash and pyritic sulfur. The process was claimed to produce ash reductions of 50%–75%, pyrite reductions of 50%–80%, and energy recoveries of 93%–99% (Marcotullio et al., 1989; Milana et al., 1992; Passarmi and Vettor, 1988; Staff, 1992).

10.1.3.9 Otisca Process

The Otisca process (Otisca Industries, Syracuse, NY) was originally developed to use Freon-113 as the agglomerant. Freon-113 is no longer permissible for use in most applications. This was replaced by *n*-pentane, which was lower in cost. The process proceeded in three steps:

- Raw coal was reduced to less than 500 μm in size and sometimes to as little as 5 μm.
- Coal was agglomerated from suspension.
- Agglomerant was recovered from the coal and recycled.

The energy recovery was typically near 98%, and it was reported that many coals could be reduced to 2% ash or less. Some typical results for three different coals are shown in Table 10.3.

A 15 ton/h plant, using *n*-pentane as the agglomerant, was built at Jamesville, NY. It produced over 125 tons of cleaned coal, at an estimated cost of $0.57/GJ (Blaschke, 1994; Hucko et al., 1988; Keller and Burry, 1990).

TABLE 10.3
Typical Coal Cleaning Results Using the Otisca Process

	Feed Analysis		Product Analysis	
Coal Processed	**% Ash**	**% Sulfur**	**% Ash**	**% Sulfur**
Illinois No. 6 seam	7.0–24.1	2.41–4.36	0.87	1.47
Pittsburgh No. 8 seam	11.1	3.44	1.88	2.23
Upper Freeport seam	12.3–20.0	1.4–2.2	2.06	0.62

TABLE 10.4
Reduction in Total Sulfur Content of Coal Using the LICADO Process

		% Total Sulfur	
Coal Type	**Particle Size**	**Feed**	**Clean Product**
Upper Freeport single-stage	−75 μm	1.30	0.79
Upper Freeport four-stage	−75 μm	1.30	0.52
Middle Kittanning	−250 μm	4.89	2.28
	−150 μm	4.89	1.87
	300×150 μm	6.10	1.23
	150×75 μm	4.76	1.34

10.1.3.10 LICADO Process

In the LICADO process, the agglomerating liquid was carbon dioxide. In order to liquefy the carbon dioxide at room temperature, pressures of about 6,000 kPa (870 psi) were needed. This process was not a typical agglomeration process, because it used a large excess of the water-immiscible liquid. The coal particles became preferentially attached to the CO_2/water interface and were removed as a slurry of coal in liquid CO_2. The coal and CO_2 were then rapidly separated by releasing the pressure, which allowed the CO_2 to vaporize and leave a dry coal product. The cleaned coal yield was 75%–83% with this process, and the ability to remove sulfur was good, as shown in Table 10.4.

While this process was effective for sulfur removal, it unfortunately was also of high cost, due to the expense of working at such high pressures. It was estimated that the processing cost would be at least $17.60/metric ton ($0.54/GJ), which is similar to the cost of the Otisca process (Hucko et al., 1988).

10.2 SULFUR REMOVAL

The effect of a number of processing variables on the product recovery, ash content, and sulfur content are shown in Table 10.5 for a selective agglomeration process using kerosene as agglomerant. It can be seen that increasing the mixing intensity, mixing time, pH, and kerosene all tend to increase the combustibles recovery while

TABLE 10.5
Laboratory Oil Agglomeration Test Results for a Lower Freeport Seam Coal

	Process Variables			Agglomerated Product			
				% Recovery		% Ash	% S
Agitator Speed, rpm	Slurry pH	Mixing Time, min	Kerosene Weight %	Weight	Combust.		
6,000	6.0	1	2	58.5	63.1	13.0	1.88
			4	70.5	77.6	11.2	2.19
			8	81.8	89.8	11.4	2.08
		2	2	60.4	65.3	12.8	1.96
			4	75.0	82.5	11.2	1.94
			8	82.5	91.4	10.6	2.18
	10.0	1	2	68.5	74.0	12.8	1.88
			4	78.5	85.6	12.0	1.94
			8	82.1	91.3	10.2	2.14
		2	2	66.8	73.4	11.3	1.91
			4	80.8	89.8	10.3	2.03
			8	82.7	93.0	9.2	2.22
12,000	6.0	1	2	62.0	67.6	11.6	1.76
			4	70.1	83.9	12.0	1.94
			8	82.6	91.4	10.7	2.08
		2	2	69.1	74.9	12.5	1.82
			4	77.1	83.9	12.0	1.91
			8	83.7	94.0	10.4	2.04
	10.0	1	2	68.6	74.9	10.7	1.73
			4	76.3	84.4	11.9	1.97
			8	85.0	95.5	9.4	2.15
		2	2	74.5	82.9	10.2	1.81
			4	78.5	88.7	8.8	1.93
			8	83.6	94.7	8.6	2.07

Analysis of raw coal: 19.3% ash, 1.77% pyritic sulfur, 2.50% total sulfur (Capes et al., 1976).

reducing the ash content of the product. The sulfur content of the product tends to increase with increasing oil dosage, however, and is largely insensitive to variations in the other parameters. Also, complete removal of pyrite from this coal would reduce the sulfur level to only 0.73%, but the best agglomeration product still contained 1.73% sulfur. This has been taken by many investigators to indicate that the pyrite is hydrophobic enough to have an affinity for agglomerating oils. There is little data to show that this is actually a result of the pyrite being hydrophobic. In fact, it is likely to be due simply to pyrite being more intimately locked with coal than are the other ash-forming minerals. In several of the agglomeration processes that have been discussed earlier in this chapter, the pyritic sulfur rejection was well over 50%, showing that when properly used on well-liberated coals, selective agglomeration can be an effective method for reducing the sulfur content of coal.

10.2.1 Pyrite Surface Modifiers

It has been claimed that the addition of certain reagents to the coal slurry will make the pyrite particles more hydrophilic, and therefore will improve the rejection of pyrite by oil agglomeration (Steedman and Krishnan, 1987; Mehrotra et al., 1983). This presupposes that the pyrite is hydrophobic at the beginning, and in most fresh coals, this does not appear to be the case. On the contrary, the work of Kawatra and Eisele (1992) has indicated that at alkaline pH values, the pyrite in unoxidized coal will already tend to be hydrophilic. The need for pyrite depressants is therefore questionable, unless the coal is heavily oxidized, in which case selective agglomeration will work poorly in any event.

Two approaches have been attempted to prevent agglomeration of the pyrite:

- Oxidation, using either chemical oxidants or pyrite-oxidizing bacteria (Capes et al., 1971).
- Addition of pyrite depressants such as sodium carbonate (Capes et al., 1971; Mezey et al., 1985).

While some small improvements in the pyrite rejection have been reported, these changes are generally not large enough to be significant. While there may be some tendency for the pyrite surfaces in coal to be agglomerated by oils, the success of techniques such as the Otisca process in rejecting pyrite when the coal is finely ground indicates that the major problem is locking of the pyrite to coal, and not hydrophobicity of the pyrite surface.

10.3 SUMMARY

The primary limitation on using selective agglomeration is the high usage of agglomerating liquid. This results in either high processing costs if the liquid is recovered and recycled or in high reagent costs if it remains in the final product. Selective agglomeration is also no more effective for rejecting pyrite than is froth flotation, and so the higher cost of selective agglomeration is generally not justified by any improvement in performance.

A better justification for selective agglomeration is that the agglomerating liquid displaces water from the coal surfaces, producing a lower-moisture product than is practical using froth flotation. The main industrial objection to using finely ground deep-cleaned coal is that the high surface area causes the coal to retain too much water, and drying the coal to an acceptable moisture level is expensive. Since agglomerated coals naturally contain little water, they are less expensive to dewater for marketing. In cases where the agglomerant stays in the final coal product, agglomeration also reduces or eliminates dust and converts the coal into free-flowing granules or pellets that are more convenient to handle.

The best method for gaining the maximum benefit from selective agglomeration while minimizing costs is to use an agglomerant that has a significant heating value and has a cost per gigajoule that is comparable to that of the coal being agglomerated. Such an agglomerant could be added to the coal at arbitrarily large dosages without increasing the cost per gigajoule of the coal.

REFERENCES

Armstrong, L.W., Swanson, A.R. and Nicol, S.K. (1978), "Selective Agglomeration of Fine Coal Refuse," *BHP Technical Bulletin*, Vol. 22, No. 1, pp. 37–40.

Bajor, O. and Trass, O. (1988), "Modified Oil Agglomeration Process for Coal Beneficiation 1. Mineral Matter Liberation by Fine Grinding with the Szego Mill," *Canadian Journal of Chemical Engineering*, Vol. 66, No. E, pp. 282–285.

Batley, W.W. (1923), "The Trent Process for Cleaning Coal," *Fuel*, Vol. 2, pp. 236–241.

Bechtel National, Inc. (1990), "Advanced Physical Fine Coal Cleaning: Spherical Agglomeration," Final Report to the U.S. Department of Energy, Contract No. DE-AC22-87PC79867.

Bensley, C.N., Swanson, A.R. and Nicol, S.K. (1977), "The Effect of Emulsification on the Selective Agglomeration of Fine Coal," *International Journal of Mineral Processing*, Vol. 4, pp. 173–184.

Blankmeister, W., Bogenschneider, B., Kubitza, K.H., Leininger, D., Angerstein, L. and Koehling, R. (1976), "Optimization of Dewatering of Smalls and Fines of Less than 10 mm Grain Size," *Glueckauf*, Vol. 112, No. 13, pp. 758–762.

Blaschke, Z. (1994), "Oil Agglomeration and Selective Flocculation of Coal Slurries," *Proceedings of the 12th International Coal Preparation Congress,* Cracow, Poland, pp. 763–776.

Brisse, A.M. and McMorris Jr., W.L. (1958), "Convertol Process," *Mining Engineering*, Vol. 10, pp. 258–261.

Capes, C.E., Mcllhinney, A.E., Sirianni, A.F. and Puddington, I.E. (1971), "Agglomeration in Coal Preparation," *Proceedings of the 12th Biennial Conference, Institute of Briquetting and Agglomeration,* Vancouver, BC, pp. 53–65.

Capes, C.E. (1979), "Agglomeration," *Coal Preparation*, 4th Edition (Leonard, ed.), American Institute of Mining, Metallurgical, and Petroleum Engineers, New York, pp. 10-105–10-116.

Capes, C.E. (1980), "Principles and Applications of Size Enlargement in Liquid Systems," *Fine Particle Processing* (Somasundaran, ed.), Society for Mining, Metallurgy, and Exploration, Littleton, CO, pp. 1442–1462.

Capes, C.E. and Germain, R.J. (1980), "Rejection of Suspending Liquid from Fine Particles by Agglomeration Methods," *Proceedings of the International Symposium on Drying*, 2nd July 6–9, 1980, Montreal, Quebec, Canada, Hemisphere Publishing Corp., Washington, DC, pp. 460–466.

Capes, C.E., Mcllhinney, A.E., McKeever, R.E. and Messer, L. (1976), "Application of Spherical Agglomeration to Coal Preparation," *7th International Coal Preparation Congress*, Sydney, Australia, pp. H1–H22.

DeVaney, F.D. (1985), "Agglomeration," Section 11, *SME Mineral Processing Handbook* (Weiss, ed.), American Institute of Mining, Metallurgical, and Petroleum Engineers, New York.

Dunstan, D., White, L.R. and Healy, T.W. (1986), "Kinetic Model of the Oil Agglomeration Process," *Transactions of the Institute of Mining and Metallurgy, Section C,* Vol. 95, pp. 127–132.

Elliot, R.C. (1978), *Coal Desulfurization Prior to Combustion*, Noyes Data Corp., Park Ridge NJ.

Getsoian, J.A. (1991), "Coal Treatment Process and Apparatus Therefor," U.S. Patent 5,076,812.

Hucko, R.E. and Gala, H.B. (1990), "Surface Science of Coal Preparation," U.S. DOE, Pittsburgh Energy Technology Center, DOE/PETC/TR-90/2 (DE90007161).

Hucko, R.E., Gala, H.B. and Jacobsen, P.S. (1988), "Status of DOE-sponsored Advanced Coal Cleaning Processes," *Industrial Practice of Fine Coal Processing* (Klimpel and Luckie, eds.), Society of Mining Engineers, Littleton, CO, pp. 159–210.

Ignasiak, B., Pawlak, W., Szymocha, K. and Marr, J. (1990), "Development of Clean Coal and Clean Soil Technologies Using Advanced Agglomeration Technologies, Vol. 2—Upgrading of Bituminous Coals: The Aglofloat Process," Alberta Research Council, U.S. DOE Contract DE-FG22–87-PC79865.

Jha, M.C. and Smit, F.J. (1995), "Selection of Feed Coals for Production of Premium Fuel Using Column Flotation and Selective Agglomeration Processes," *High Efficiency Coal Preparation* (Kawatra, ed.), Society for Mining, Metallurgy, and Exploration, Littleton, CO, pp. 391–400.

Kawashima, Y., Handa, T., Takeuchi, H., Niwa, T., Niwa, K., Sunada, H. and Otsuka, A. (1989), "Computer Simulation of Agglomeration by a Two-Dimensional Random Addition Model—Agglomeration Kinetics and Micromeritic Properties of Agglomerate Accompanied by Compaction Process," *Powder Technology*, Vol. 57, No. 3, pp. 157–163.

Kawashima, Y. and Capes, C.E. (1974), "Experimental Study of the Kinetics of Spherical Agglomeration in a Stirred Vessel," *Powder Technology*, Vol. 10, Nos. 1–2, pp. 85–92.

Kawashima, Y. and Capes, C.E. (1975), "Further Studies of the Kinetics of Spherical Agglomeration in a Stirred Vessel," *Powder Technology*, Vol. 13, No. 2, pp. 279–288.

Kawatra, S.K. and Eisele, T.C. (1992), "Recovery of Pyrite in Coal Rotation: Entrainment or Hydrophobicity?" *Minerals and Metallurgical Processing*, Vol. 9, No. 2, pp. 57–61.

Keller, Jr. D.V. and Burry, W.M. (1990), "The Demineralization of Coal Using Selective Agglomeration by the T-Process," *Coal Preparation*, Vol. 8, pp. 1–17.

Keller, Jr. D.V. (1995), "Variables in the Production of Premium Fuels and Ultra Clean Coal by Selective Agglomeration," *High Efficiency Coal Preparation* (Kawatra, ed.), Society for Mining, Metallurgy, and Exploration, Littleton, CO, pp. 441–52.

Labuschagne, B.S.J. and Prinsloo, J.J. (1985), "Kinetics of Coal-Oil Agglomeration Process: The Effect of Surface Variables," *Proceedings of the International Conference on Coal Science*, Pergamon Press, NY, pp. 529–533.

Labuschagne, B.S.J. (1986), "The Oil Phase in Selective Agglomeration," *Proceedings of the Pittsburgh Coal Conference*, University of Pittsburgh, Pittsburgh, PA, pp. 47–60.

Labuschagne, B.S.J. and Prinsloo, J.J. (1988), "Agglomeration Response," *Proceedings of the 13th International Coal and Slurry Tech.*, Denver, CO, pp. 471–482.

Lemke, K. (1954), "The Cleaning and Dewatering of Slurries by the Convertol Process," *Proceedings of the Second International Coal Preparation Congress*, Essen, Paper No. AIV 2.

Marcotullio, A., Passarmi, N. and Vettor, A. (1989), "Process for the Beneficiation of Coal by Selective Caking; Using a Solvent Mixture Comprising a Light Hydrocarbon, and Oil-Soluble Propoxylated Coal-Tar-Derived Phenol and a Heavy Oil," U.S. Patent No. 4,881,946.

Mehrotra, V.P., Sastry, K.V.S. and Morey, B.W. (1983), "Review of Oil Agglomeration Techniques for Processing of Fine Coals," *International Journal of Mineral Processing*, Vol. 11, pp. 175–201.

Mezey, E.J., Hayes, T.D., Mayer, R. and Dunn, D. (1985), "Application of Oil Agglomeration for Effluent Control from Coal Cleaning Plants," U.S. EPA Report No. EPA/600/7–85/042.

Milana, G., Vettor, A. and Wheelock, T.D. (1992), "Oil Agglomeration at a Moderate Shear Rate," *Proceedings of the Ninth Annual International Pittsburgh Coal Conference*, Pittsburgh, PA.

Musich, M.A., DeWall, R.A. and Timpe, R.C. (1992), "Oil Agglomeration of Low-Rank Coals," *Proceedings of the Eighth Annual Coal Preparation, Utilization and Environmental Control Contractors Conference*, Pittsburgh, PA, DOE/PETC.

Nicol, S.K. and Swanson, A.R. (1979), "Ultrafine Coal Recovery from Preparation Plant Tailings," *VIII International Coal Preparation Congress,* Donieck, Ukraine, Paper D4.

Özer, M., Basha, O.M. and Morsi, B. (2017), "Coal-Agglomeration Processes: A Review," *International Journal of Coal Preparation and Utilization*, Vol. 37, No. 3, pp. 131–167.

Passarmi, N. and Vettor, A. (1988), "Process for Beneficiating Coal by Means of Selective Agglomeration, with Blend of Light Hydrocarbons, Nonionic Surfactant, and Oils Derived from Coal Tar," U.S. Patent No. 4,776,859.

Perrot, G.S.T.J. and Kinny, S.P. (1921), "Trent Process of Cleaning Coal," U.S. Bureau of Mines Report, No. 2263, p. 18.

Ralston, O.C. (1922), "Comparison of Froth with the Trent Process," *Coal Age,* Vol. 22, No. 23, pp. 911–914.

Sarkar, G.G., Moza, A.K. and Kini, K.A. (1976), "Demineralization of Coals by Oil Agglomeration, Part 1: Studies on the Applicability of the Oil Agglomeration Technique to Various Coal Beneficiation Problems; Part 2: Basic Studies in the Mechanism of Oil Agglomeration," *Proceedings of the Seventh International Coal Preparation Congress,* Sydney, Australia.

Southern Company Services (1993), "Engineering Development of Selective Agglomeration," Final Report to U.S. Department of Energy, Contract No. DE-AC22–89PC88879.

Staff (1992), "Coal Beneficiation," Eniricherche SpA, Company Brochure.

Steedman, W.G., and Krishnan, S.V. (1987), "Oil Agglomeration Process for the Treatment of Fine Coal," *Fine Coal Processing* (Mishra and Klimpel, eds.), Noyes Publications, Park Ridge, NJ, pp. 179–204.

Sun, S.C. and McMorris III, W.L. (1959), "Factors Affecting the Cleaning of Fine Coals by the Convertol Process," *Mining Engineering*, pp. 1151–1156.

Theodore, F.W. (1985), "Oil Agglomeration for Fine Coal Recovery as Commercialized at Conoco/Consol," *4th International Symposium on Agglomeration* (Capes, ed.), Iron and Steel Society, Ontario, Canada, pp. 883–889.

Trass, O. and Bajor, O. (1988), "Modified Oil Agglomeration Process for Coal Beneficiation 2. Simultaneous Grinding and Oil Agglomeration," *Canadian Journal of Chemical Engineering*, Vol. 66, No. 2, pp. 286–290.

Trent, W.E. (1922), "Process for purifying Materials," U.S. Patent 1,420,164, June 20.

Tsai, S.C. (1982), *Fundamentals of Coal Beneficiaron and Utilization*, Elsevier, Amsterdam.

Wheelock, T.D., Drzymala, J. and Zhang, F. (1994a), "Development of a Gas-Promoted Oil Agglomeration Process; Technical Progress Report, December 1, 1993–February 28, 1994," DOE Contract FG2293PC93209, Report No. DOE/PC/93209–T2.

Wheelock, T.D., Milana, G. and Vettor, A. (1994b), "The Role of Air on Oil Agglomeration of Coal at a Moderate Shear Rate," *Fuel*, Vol. 7, No. 3, pp. 1103–1107.

11 Magnetic Separation

It is possible to separate coal from pyrite using magnetic methods. The differences in the magnetic properties of coal and pyrite are small enough, however, that either very high-intensity magnetic separators must be used or the magnetic properties of the pyrite must be enhanced. Magnetic separation can be carried out either wet or dry. Advantages of magnetic separation are its insensitivity to the coal chemistry, making it useful for oxidized coals, and its ability to remove locked coal/pyrite particles.

11.1 PRINCIPLES OF MAGNETIC SEPARATION

Magnetic fields are produced by the motion of electrical charges. On the atomic scale, electrons have magnetic moments due to their motion in the atom. These magnetic moments tend to cancel each other out in most atoms, resulting in nonmagnetic materials. For some elements, the magnetic moments of electrons are not canceled completely, and as a result, these materials have magnetic properties.

A particle placed in a magnetic field is subjected to three types of forces: magnetic, gravitational, and frictional or inertial forces. The force exerted by a magnetic field on a particle is given by the expression:

$$F = \left(K_p - K_m\right)VH\left(\frac{dH}{dx}\right) \tag{11.1}$$

where

F = force acting on the particle in the direction x;
K_p = magnetic susceptibility of the particle (volume basis);
K_m = magnetic susceptibility of the medium in which the particle is placed (volume basis);
V = volume of the particle;
H = intensity of the magnetic field acting on the particle;
dH/dx = magnetic field gradient in the direction x.

It should be noted that the force acting on a particle is a function both of the intensity of the field and of the field gradient. As a result, large magnetic forces can be produced both by using very powerful magnets and by manipulating the magnetic field so that the intensity changes very rapidly over short distances.

The important parameter in a magnetic separation is the magnetic susceptibility of the particles. The magnetic susceptibility is a measure of the degree of magnetization that is produced in a material by an applied magnetic field, and it is expressed either per unit volume (X, emu/cm^3) or per unit mass (K, emu/g) (Carmichael, 1989). In general usage, a particle is considered to be "magnetic" when it has a large, positive value for the magnetic susceptibility.

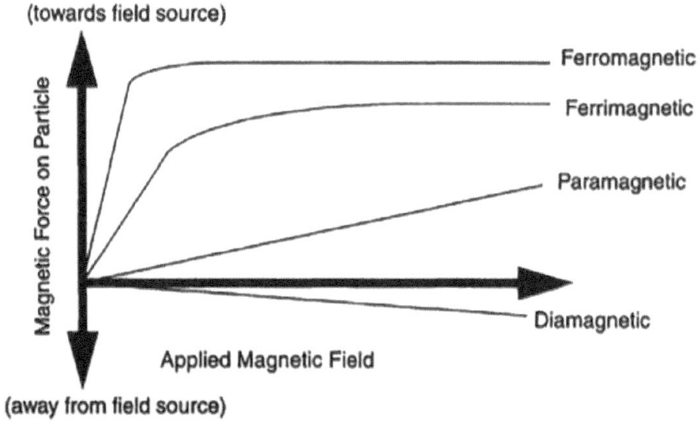

FIGURE 11.1 Responses of paramagnetic, diamagnetic, ferrimagnetic, and ferromagnetic particles to an applied magnetic field (Kolm et al., 1975). The magnetic susceptibility is the slope of the linear portion of the curve.

Magnetic materials can be classified depending on how they respond to an applied magnetic field, as shown in Figure 11.1. Ferromagnetic minerals, such as magnetite, are strongly attracted by even weak magnetic fields. This initial strong response is due to the individual magnetic "domains" that comprise these particles. The domains readily align with the applied field and so react strongly. Once the applied field is strong enough to align all of the domains, further increases in field strength do not result in an increase in the attractive force, and the material is said to be "saturated." Ferrimagnetic materials are structured differently than ferromagnetic materials, but they react in a similar way to applied fields, although with a lower saturation limit than for ferromagnetic particles.

Paramagnetic materials have a comparatively weak response to magnetic fields, but they do not reach a saturation limit even in extremely strong fields. As a result, they maintain a linear response over a very wide range of field strengths. If the applied field is strong enough, it is even possible for paramagnetic minerals to be attracted more strongly than ferromagnetic minerals. Diamagnetic minerals, such as coal, respond similarly to magnetic fields, except that they are repelled by the field instead of attracted by it. A negative value of magnetic susceptibility indicates a diamagnetic material, while a positive susceptibility indicates a paramagnetic, ferrimagnetic, or ferromagnetic material. Values of diamagnetic susceptibility are very small, and so magnetic forces exerted on diamagnetic particles are generally negligible at the field strengths and gradients present in existing magnetic separators.

It is apparent from Equation 11.1 that the force acting on a particle is proportional to the volume of the particle. The volume of a particle of constant density varies as the cube of the particle radius, while the surface area of the particle only varies as the square of the radius; therefore, as particles become smaller, the magnetic separation forces decrease more rapidly than the effects of surface area. Because of this, dry magnetic separation of particles smaller than 125 μm can be difficult because of agglomeration effects (Liu and Lin, 1976; Liu, 1982). Since fine grinding is necessary to liberate the pyrite, this limits the ability of dry magnetic separation to desulfurize coal (Burnley and Fells, 1985). Wet magnetic separation avoids the agglomeration problem, but the high surface area of finely ground coal makes it difficult to dewater. Further details of magnetic separation are discussed extensively in the literature (Gaudin, 1974; Rowson and Rice, 1989; Fletcher, 1986; Lua and Boucher, 1990; Hall and Finch, 1984; Male, 1984; Smirnova et al., 1994; Lockhart, 1984; Maxwell and Kelland, 1978; Maxwell et al., 1976; Hise, 1982; Kelland, 1982; Oberteuffer, 1973).

11.2 MAGNETIC PROPERTIES OF PYRITE AND COAL

In order to make a magnetic separation, the magnetic susceptibilities must be different for the particle types being separated. The typical ranges of magnetic susceptibilities for coal and its associated minerals are given in Table 11.1. The susceptibilities of magnetite, hematite, and pyrrhotite are included, showing that the magnetic susceptibility of pyrite and the other minerals in coal are not very high in comparison. Because of the very high magnetic susceptibility of certain iron minerals, the value of magnetic susceptibility for many rocks and minerals depends on their iron contents.

It should be noted that it is possible to convert pyrite to pyrrhotite with chemical and/or thermal treatments. This provides a method for enhancing the magnetic susceptibility of the pyrite particles so that they can be removed more easily. Pyrite is also converted during combustion to magnetite, which is then easily removed from the ash products. Converting pyrite to magnetite results in the release of the sulfur from the pyrite, and this is not relevant to coal desulfurization.

From the values given in Table 11.1, it can be seen that mineral-free coal is normally diamagnetic, while pyrite is paramagnetic. The magnetic properties

TABLE 11.1
Magnetic Susceptibilities of Coal, Pyrite, and Associated Minerals (Tsai, 1982)

Mineral	Chemical Formula	Magnetic Susceptibility, 10^{-6} emu/g
Coal		−0.1 to −0.8
Pyrite or marcasite	FeS_2	0.3–4.5
Siderite	Fe_2CO_3	331.45
Limonite	$Fe_2O_3{\cdot}3.5H_2O$	57
Calcite	$CaCO_3$	0.75
Limestone		3.8
Clay		20
Sandstone		15–20
Anhydrite	$CaSO_4$	−0.36
Gypsum	$CaSO_4{\cdot}2H_2O$	−0.38
Shale		39–45
Magnetite	Fe_3O_4	15,600
Hematite	Fe_2O_3	20.6
Monoclinic pyrrhotite	Fe_7S_8	2,800

of coal can, however, vary considerably because of the presence of paramagnetic mineral inclusions. If coal is completely demineralized by leaching, its magnetic susceptibility is reduced to less than -0.1×10^{-6} emu/g.

The removal of sulfur by magnetic separation is largely limited by the low magnetic susceptibility of pyrite. If the pyrite is treated to convert a portion of it to a more magnetic material (such as pyrrhotite or metallic iron), then it can be more easily separated. A number of studies have been carried out which attempted this.

11.2.1 Selective Microwave Heating of Pyrite

When pyrite is sufficiently heated, it is partially converted to more strong magnetic compounds, such as pyrrhotite. While early studies have investigated thermal treatments to improve the magnetic properties of coal pyrite, it was generally found to be impractical to heat the entire coal to a high enough temperature for this to occur (Liu, 1982).

When microwaves are used to heat coal, it is possible to selectively heat the pyrite to much higher temperatures than the coal matrix. This is possible because pyrite has a much greater ability to absorb microwaves than the hydrocarbon constituents of coal (Liu, 1982). It has been demonstrated that it is indeed possible to selectively heat the pyrite in coal with microwaves at a frequency of 8.3 GHz and that this heating does improve the magnetic properties (Ergun and Bean, 1968; Zavitsanos et al., 1978; Weng, 1994; Rowson and Rice, 1990a,b; Kelland et al., 1988).

11.3 MAGNETIC SEPARATORS USEFUL FOR DESULFURIZATION

The low magnetic susceptibility of pyrite makes it necessary to use powerful magnetic fields and steep field gradients to make the separation. Improvements in magnet design, such as the use of superconducting magnets, rare-earth permanent magnets, and high-gradient magnetic matrices, have made it practical to produce larger magnetic separating forces than were previously possible (Brewis, 1996; Arvidson and Henderson, 1997; Watson and Younas, 1997; Pitel and Jones, 1998; Bolt and Watson, 1998; Daugherty et al., 1996).

11.3.1 Open-Gradient Separators

In open-gradient separators, the separation is mainly dependent on the strength of the magnetic field. The magnetic field region is open and unobstructed, with no inserts to produce high magnetic field gradients. These units are easily designed for continuous operation, with magnetic particles deflected from the particle stream as the particles travel through the magnetic field. There are several different types of open-gradient separators, which are discussed below.

11.3.1.1 Falling-Stream Separator

In "falling-stream" devices, the particles fall through the field region, and the particles with the highest magnetic susceptibility are deflected, as shown in Figure 11.2. These devices are simple and can be used either for slurries or for dry powders.

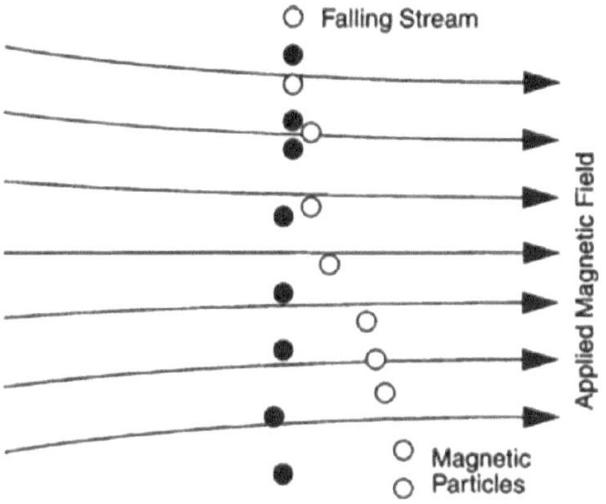

FIGURE 11.2 Principle of a falling-stream magnetic separator.

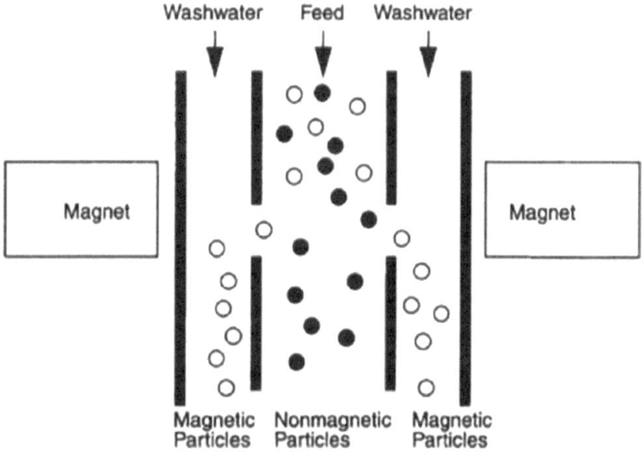

FIGURE 11.3 Side view of a falling-stream quadrupole separator. (From Aubrey et al., 1968.)

A quadrupole falling-stream separator was designed by Bethlehem Steel Corp. and is shown schematically in Figure 11.3 (Clark, 1979). The separator consists of concentric tubes, with the feed falling through the central tube. The magnets are arranged to attract the magnetic particles outward, so that they will pass through the slots cut in the inner tube. The nonmagnetic particles are unaffected by the magnetic field and continue to fall down through the central tube.

11.3.1.2 Curved-Channel Separator

The curved-channel separator is superficially similar to a spiral concentrator, with a magnetic field supplying the separating force instead of the gravitational field. A cylindrical magnet is used to produce a magnetic field on the inside edge of a curved, sloping channel, as shown in the cross-section in Figure 11.4. The feed is

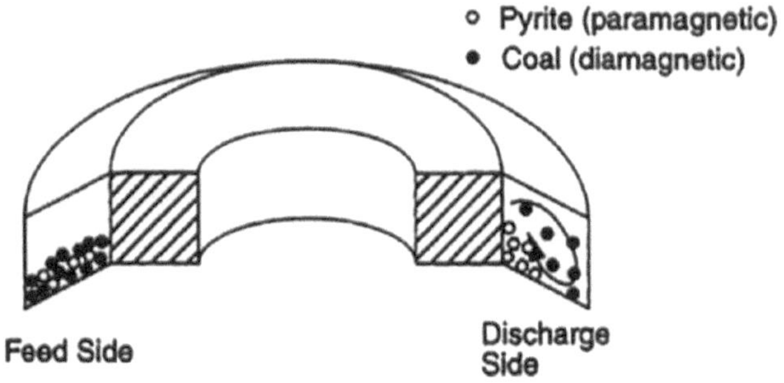

FIGURE 11.4 Cutaway view of the curved channel quadrupole magnetic separator (From Liu, 1982.)

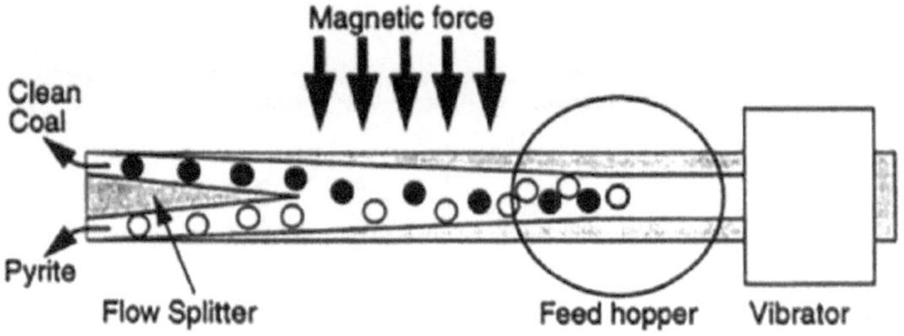

FIGURE 11.5 Frantz isodynamic laboratory separator overhead illustration. Feed coal enters through a hopper onto an inclined vibrating chute. A magnetic field applied to the chute separates coal and pyrite. (Adapted from Kester, 1965.)

conveyed into the separator as a suspension, and the nonmagnetic particles migrate to the outside of the chute, while the magnetic particles are attracted toward the center (Cohen and Good, 1976).

11.3.1.3 Frantz Isodynamic Laboratory Separator

The Frantz Isodynamic Separator is shown schematically in Figure 11.5. This is strictly a laboratory device, as its construction makes it difficult to scale up, and the vibrating-chute feeder gives it a low throughput rate. The angle of the chute and magnet to the vertical can be easily adjusted, so that the intensity of the gravitational force countering the magnetic field can be varied freely. This device is mainly useful for determining whether a magnetic separation is practical.

11.3.1.4 Sulfur Removal by Open-Gradient Separators

Kester (1965) used a Frantz separator to remove sulfur from an Upper Freeport coal, sized to 48 × 65 mesh (300 × 212 μm). The magnetic field intensity was 10,700 Gauss, and the coal feed rate was 96 g/h. The chute and magnets were inclined 5° to the side, and the chute had a downward slope of 10°. The separation performance was as shown in Table 11.2, illustrating that this type of separator could remove a significant fraction of sulfur.

TABLE 11.2
Upper Freeport Coal Cleaned by Frantz Isodynamic Laboratory Separator

Test No.	Feed % Sulfur	Clean Coal % Sulfur	Tailings % Sulfur	% Yield	% Sulfur Removal
20	3.20	1.81	10.17	83.38	52.8
21	3.09	1.79	9.29	82.67	52.1
22	3.24	1.93	12.11	87.12	48.1
23	3.10	1.67	12.67	86.98	53.2
24	3.32	1.68	14.53	87.20	56.0
Average	3.19	1.78	11.75	85.47	53.5

11.3.2 High-Gradient Separators

Since there is a limit to how strong magnets can be made, the open-gradient approach is not suitable for weakly magnetic particles because the magnetic field cannot be made sufficiently intense for open-gradient machines to work well. A different approach, which does not require continuous escalation of the magnet strength, is the introduction of extremely sharp magnetic field gradients, which strongly attract and pin magnetic particles. Such gradients can be produced by inserting ferromagnetic objects in the magnetic field. These objects act to concentrate the field, producing small volumes with a very sharp gradient, as shown in Figure 11.6.

High-gradient separators make use of this phenomenon by introducing a "matrix" of ferromagnetic material into the magnetic field. The matrix may consist of stainless steel strands, a knitted wire mesh, or an expanded metal. When a suspension carrying magnetic particles passes through the matrix, the high local-field gradients trap the paramagnetic particles, as shown in Figure 11.7. Nonmagnetic particles are unaffected by the field and pass through. The matrix therefore acts largely as a "filter" for magnetic particles.

A drawback of this type of separator is that the particles are not simply deflected into a separate stream. The paramagnetic particles are instead trapped and held in

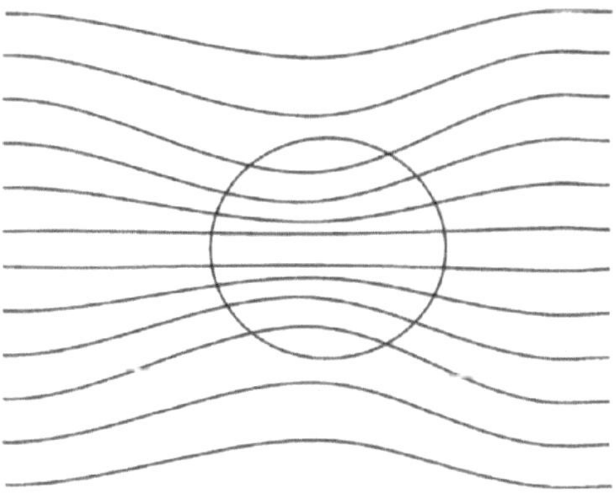

FIGURE 11.6 A ferromagnetic particle concentrates the lines of force near the particle.

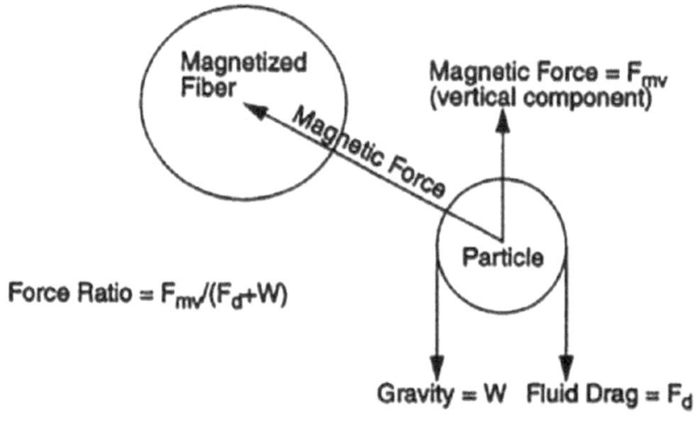

FIGURE 11.7 Forces acting on a paramagnetic particle flowing past a magnetized matrix strand (Luborsky, 1975).

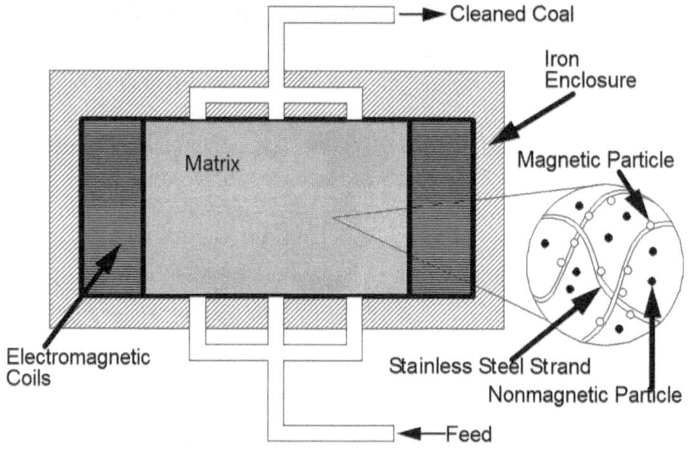

FIGURE 11.8 A batch high-gradient magnetic separator (Oder, 1976).

the matrix, while the nonmagnetic particles and water flow through. Eventually, the matrix will become saturated with paramagnetic particles. It must then be removed from the magnetic field, so that the trapped particles will be released. The various types of high-gradient magnetic separator are mainly distinguished based on how they accomplish the matrix cleaning step, as shown in the following sections.

11.3.2.1 Batch High-Intensity Separators

In the simplest type of high-intensity separator, the system is designed as shown in Figure 11.8. The unit is operated until the matrix is fully loaded with magnetic particles. The feed is then stopped, the magnet is turned off, and the magnetic particles are washed out of the matrix. This type of machine is useful for research purposes and for processes where the feed contains such a small amount of magnetic material that it takes many hours for the matrix to become loaded. Since superconducting

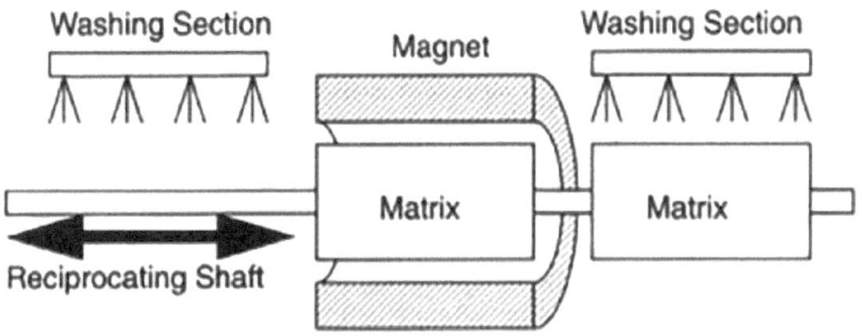

FIGURE 11.9 Reciprocating a high-gradient magnetic separator schematic diagram.

magnets should not be cycled on and off regularly, they are not suitable for this type of separator.

11.3.2.2 Reciprocating Matrix System

In this technique, shown in Figure 11.9, there are two matrix assemblies. While one matrix assembly is in the magnetic field collecting particles, the other is removed from the field and being washed. The two assemblies are swapped periodically, which makes the operation nearly continuous and makes it unnecessary to cycle the magnet on and off (Liu, 1982; Windle, 1977).

The reciprocating shaft moves identical matrices out of the magnet for cleaning and simultaneously moves a new matrix into place for separation.

11.3.2.3 Carousel Devices

This is the type of high-intensity separator that is most commonly used industrially, because it has a high capacity and operates continuously. Here, the matrix cells are arranged in a ring that rotates through the magnetic field and then rotates back out carrying the magnetic particles. The matrix is then continuously washed, as shown in Figure 11.10.

Of the various types of magnetic separators, the carousel high-gradient separator is the most nearly practical choice for coal desulfurization. These units are already in use on an industrial scale, and so machines with enough capacity for coal processing are already available.

11.3.2.4 Sulfur Removal by High-Gradient Separators

11.3.2.4.1 Pyrite Removal from Slurry Streams

Standard high-gradient magnetic separators can be effective for sulfur removal from coal slurry streams, as can be seen in Table 11.3.

In high-gradient separators that use conventional magnets, the sulfur removal is strongly dependent on the slurry flowrate. Increasing flowrates tend to increase the British Thermal Unit (BTU) recovery, while decreasing the sulfur removal. The use of superconducting magnets reduces the dependence of sulfur removal on flowrates, by allowing higher magnetic field intensities that more strongly pin the pyrite in the magnetic matrix (Liu, 1982). This allows higher BTU recoveries, although the

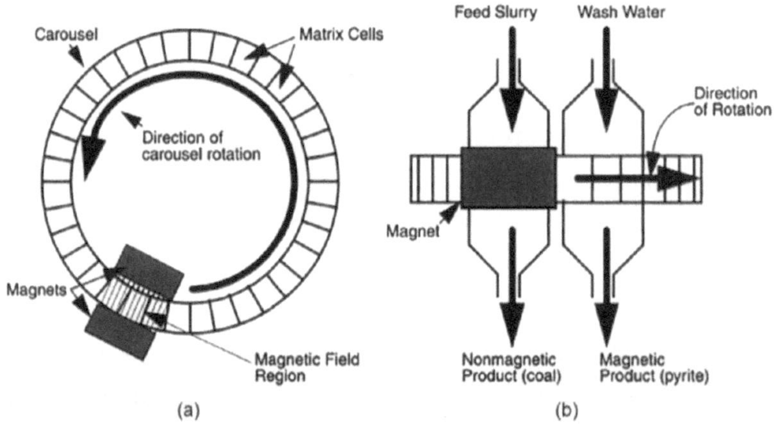

FIGURE 11.10 Illustration of a carousel high-gradient magnetic separator. (a) Top view, showing carousel motion through the magnetic field region; (b) Side view, showing feed and wash cycles. Once the matrix cells move out of the field region, the magnetic particles held on the matrix are released and flushed out by the washwater.

TABLE 11.3
Coal/Water Slurry Separation in a High-Gradient Magnetic Separator, Using Different Matrix Materials and Slurry Flow Velocities (Trindade and Kolm, 1973)

Matrix Material	% Void Volume	Slurry Velocity, cm/s	% Yield	% Sulfur Removal
Screens	87	3.9	72.6	44.5
Screens	87	0.5	49.4	63.2
Steel wool	97	1.8	75.9	58.8
Steel wool	97	0.4	48.2	79.5

cost of purchasing and cooling the superconducting magnets makes superconducting separators much more expensive than conventional separators.

11.3.2.4.1 Pyrite Removal from Dry Coal Streams

High-gradient magnetic separation can also be used for desulfurizing dry coal, which would reduce the drying costs. In the work by Oak Ridge National Laboratory and Auburn University, a dry high-gradient separator was developed that could remove a major proportion of the pyrite from coal in the 100–200 mesh size range, as shown in Table 11.4.

Zhou et al. (1996) more recently conducted extensive studies using superconducting magnets to desulfurize dry coal. Their separator was designed as shown in Figure 11.11, with the coal to be cleaned being suspended in air, and then drawn upward through the separator by a centrifugal fan. The cleaned coal was collected on a filter for analysis. The separator matrix was in the bore of a superconducting

TABLE 11.4
Magnetic Desulphurization of Pennsylvania Upper Freeport Coal by the Auburn/Oak Ridge National Laboratory Fluidized-Bed Separator, Processing a Dry Coal Stream (Liu, 1982)

	Wt %	Sulfur %	Pyrite %	% Pyrite Removed
Feed	100.00	2.123	1.519	93.17
Product	79.84	0.680	0.130	
Tailings	20.16	7.710	7.020	

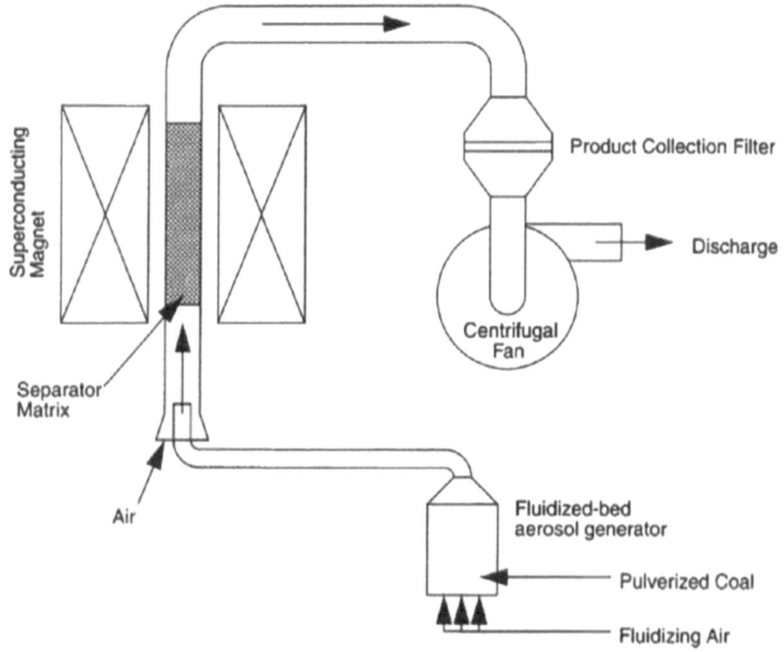

FIGURE 11.11 Dry magnetic desulfurization test rig for coal cleaning, using a superconducting magnet (Zhou et al., 1996).

magnet with a maximum rated magnetic field of 4 T, and the matrix was 250 mm in length and was made up of 0.710 mm circular stainless steel wires.

An important feature of this unit is the upward flow of the ground coal. The investigators calculated that for 100 μm diameter particles, the terminal velocity for pyrite is 1 m/s, while that of coal is only 0.35 m/s. If the vertical airflow velocity is 1.5 m/s, then the pyrite will travel through the separator matrix at only half the speed of the coal, improving the chances that the pyrite will be captured by the matrix.

The feed coal composition for the desulfurization experiments was 9.16% ash, 2.89% total sulfur, 1.79% pyritic sulfur, and 0.32% sulfate sulfur. The coal was ground and air classified to a size range of 20–105 μm, with a mean particle size of

TABLE 11.5
Effect of Applied Magnetic Field on Coal Desulfurization, Using a Superconducting Magnet to Process a Dry Coal Stream

Magnetic field, T	1.0	2.0	3.0	4.0
Magnetic material, % wt	5.22	5.74	5.78	6.80
Coal recovery, % wt	94.78	94.26	94.22	93.20
% Pyritic sulfur reduction	31.84	45.25	60.34	69.27
% Sulfate sulfur reduction	15.63	21.88	21.88	21.88
% Total sulfur reduction	42.56	44.29	49.83	56.71
% Ash reduction	13.97	28.49	20.85	33.19
% Heating value recovery	96.90	98.63	97.76	96.96
Mean particle size of magnetic fraction, pm	46.48	41.84	53.71	49.97
Pressure drop, mbar	12.00	12.30	12.30	14.00

Total sample size was 50 g, and the coal flow flux was 8.24 g/m^{2}s (Zhou et al., 1996).

TABLE 11.6
Effect of Coal Flowrate on Coal Desulfurization, Using a Superconducting Magnet to Process a Dry Coal Stream

Coal flowrate, g/min	0.906	1.10	4.00	6.43	23.66	73.16
Coal flow flux, kg/m^{2}h	6.73	8.17	29.66	47.66	175.50	542.50
Magnetic material, % wt	4.28	3.78	4.37	5.28	9.13	16.55
Coal recovery, % wt	95.72	96.22	95.63	94.72	90.87	83.45
% Pyritic sulfur reduction	56.42	55.87	57.54	56.98	59.22	53.63
% Sulfate sulfur reduction	21.88	18.75	18.75	21.88	18.75	21.88
% Total sulfur reduction	41.87	34.60	37.02	48.79	45.67	44.29
% Ash reduction	32.31	24.67	23.58	24.89	16.92	13.97
% Heating value recovery	98.83	96.25	95.66	95.17	91.37	83.99
Mean particle size of magnetic fraction, μm	53.48	49.86	42.89	46.25	37.61	40.06
Pressure drop, mbar	11.80	13.40	11.60	12.60	11.40	12.70

Total sample size was 50 g (Zhou et al., 1996).

49.73 pm. Results from the separation experiments are given in Tables 11.5–11.7. In these experiments, increasing the field strength led to improved sulfur removal without a corresponding loss in heating value recovery. As would be expected, removal of sulfur was reduced as the matrix became loaded with magnetic material.

11.3.3 Magnetic Fluids

Magnetic fluids consist of a colloidal suspension of submicron ferromagnetic particles in a carrier liquid, with a surfactant to keep them dispersed. If a proper surfactant is used (such as oleic acid), the magnetic fluid can be used to selectively

TABLE 11.7
Effect of Matrix Loading on Coal Desulfurization, Using a Superconducting Magnet to Process a Dry Coal Stream

Coal Feed, g	Pressure Drop, mbar	% Pyritic Sulfur Reduction	% Sulfate Sulfur Reduction	% Total Sulfur Reduction	% Ash Reduction
69.01	11.2	57.54	18.75	43.94	30.57
100.56	11.5	50.28	15.63	42.21	26.20
135.16	11.9	46.93	21.88	35.64	23.14
168.35	12.2	43.02	21.88	34.26	24.02
208.25	12.7	42.46	18.75	34.26	23.47
245.50	12.9	37.99	25.00	33.91	26.53
278.78	13.2	36.87	21.88	32.53	19.43
310.40	13.5	35.75	21.88	32.18	14.96

The coal flow flux was 8.24 g/m^2s, and the magnetic field strength was 3.0 T. The final coal recovery was 98.25% wt, the heating value recovery was 98.56%, and the mean particle size of the magnetic fraction was 42.89 μm (Zhou et al., 1996).

coat the pyrite and ash-forming minerals in coal. The strong magnetic properties of the magnetic fluid then make it possible to use magnetic separation to remove the particles that are coated (Sladek and Cox, 1979, 1980).

In many respects, magnetic fluid treatment is similar to froth flotation, as they both depend on the selective attachment of surface-active molecules to particular mineral species. The main difference is that the magnetic fluid method separates by magnetic forces, rather than by attachment to air bubbles. It would therefore be expected that the magnetic fluid technique will have capabilities and limitations similar to those of froth flotation.

Using magnetic fluid treatment and a Carpco high-intensity magnetic separator, an Upper Freeport Seam coal containing 17.3% ash and 2.68% sulfur was reduced to 13.3% ash and 1.17% sulfur, with a yield of 77% wt. The beneficiation performance was reported to be best at high magnetic fluid dosages, with long conditioning times (Sladek and Cox, 1980).

11.3.4 Magnex Process

A number of chemical techniques have been attempted to convert a portion of the pyrite to a more magnetic form, but only the Magnex process has been particularly successful (Liu, 1982). In this process, the coal is treated as shown in the flowsheet in Figure 11.12. The process is based on the chemical treatment of the coal with iron carbonyl vapor, which selectively deposits iron crystals on the ash-forming minerals, and reacts with the pyrite to form a material similar to pyrrhotite. Since these deposits are strongly magnetic, they make it possible to remove the ash-forming minerals and pyrite with a medium-intensity magnetic separator.

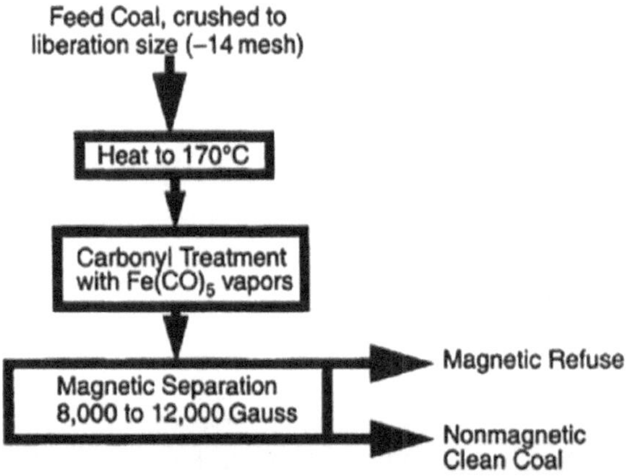

FIGURE 11.12 Flowsheet for the Magnex process, using carbonyl treatment to increase the magnetic susceptibility of pyrite before magnetic separation. The treatment enhances the magnetic susceptibility of pyrite and ash. This is a dry separation process.

TABLE 11.8
Results from Using the Magnex Process to Desulfurize Dry Pulverized Coals from Six Sources (Kindig and Geons, 1979)

				Grams of SO_2/MJ	
Coal Seam	% BTU Recovery	% Sulfur Removal	% Pyrite Removal	Feed	Clean Coal
Pittsburgh No. 8 (Green County, PA)	98.8	38.7	68.3	1.935	1.075
Allegheny Group (Mercer County, PA)	93.0	57.9	91.6	0.989	0.516
Lower Freeport (Jefferson County, OH)	87.1	43.3	62.1	2.812	1.088
Kentucky No. 11 (Muhlenberg County, KY)	85.5	22.4	56.9	2.223	1.690
Lower Freeport (Cambria County, PA)	92.5	66.5	84.4	2.060	0.542
Rosebud McKay (Big Horn County, MT)	99.0	59.6	73.5	0.679	0.297

The chemistry of the process can be summarized with the following three equations (Kindig, 1979):

$$Fe(CO)_5 \rightarrow Fe + 5CO \tag{11.2}$$

$$Fe(CO)_5 + Ash \rightarrow Fe \cdot Ash + 5CO \tag{11.3}$$

$$Fe(CO)_5 + FeS_2 \rightarrow Fe_{(1-x)}S + 5CO \quad (11.4)$$

In an industrial-scale process, the carbon monoxide that is released by each of these reactions would be captured and recycled, reacting it with a low-cost iron source (such as iron ore) to regenerate the iron carbonyl vapor.

The Magnex process has been tested in a 200 lb/h pilot plant. Typical results that have been achieved are given in Table 11.8. While the Magnex process is effective, the use of iron carbonyl presents both an explosion hazard and a toxicity problem, which has apparently prevented the process from being adopted industrially.

11.4 SUMMARY

Pyrite is sufficiently magnetic to be removed from coal by high-intensity magnetic separation, particularly when a high-gradient magnetic separator is used to provide a high value of the field gradient dH/dx in conjunction with a high value of the magnetic field intensity H. The quality of the separation can be improved, however, if the intensity of the magnetic field is made stronger than is practical using conventional electromagnets. This requires the use of advanced magnet designs, such as superconducting magnets.

High-intensity magnetic separators, particularly those using powerful superconducting magnets, have tremendous potential for use directly following coal pulverizers at electrical utility companies. These separators can remove pyrite from the coal while it is entrained in the air stream, allowing the coal sulfur content to be reduced without introducing additional complexities such as dewatering.

It is also possible to enhance the magnetic properties of pyrite, so that it can be removed without using extremely high-intensity magnetic separators. The most effective methods for doing this are by selective microwave heating and by reaction with iron carbonyl vapors.

REFERENCES

Arvidson, B.R. and Henderson, D. (1997), "Rare-Earth Magnetic Separation Equipment and Applications Developments," *Minerals Engineering*, Vol. 10, No. 2, pp. 127–137.

Aubrey Jr., W., Carpinski, J., Cahn, D. and Rauch, C. (1968), "Magnetic Separator Method and Apparatus," U.S. Patent 3,608,718, filed December 20, 1968.

Bolt, L. and Watson, J.H.P. (1998), "Magnetic Separation Using High-Tc Superconductors," *Superconductor Science and Technology*, Vol. 11, No. 1, pp. 154–161.

Brewis, T. (1996), "Magnetic Separation," *Mining Magazine*, Vol. 175, No. 4, pp. 9.

Burnley, S.J. and Fells, I. (1985), "High Gradient Magnetic Desulphurization of Coal," *Journal of the Institute of Energy*, Vol. 58, pp. 20–23.

Carmichael, R.S. (1989), *Practical Handbook of Physical Properties of Rocks and Minerals*, CRC Press, Boca Raton, FL, pp. 339–341.

Clark, N.O. (1979), "Applications of Superconducting Magnetic Separation to Clay and Mineral Processing," *Industrial Applications of Magnetic Separation* (Liu, ed.), IEEE Publication No. 78CH1447-2MAG, Institute of Electrical and Electronic Engineers, New York, pp. 62–63.

Cohen, E. and Good, J.A. (1976), "The Application of a Superconducting Magnet System to the Cleaning and Desulfurization of Coal," *IEEE Transactions on Magnetics*, Vol. MAG-12, p. 503.

Daugherty, M.A., Coulter, J.Y., Hults, W.L., Daney, D.E., Hill, D.D., McMurry, D.E., Martinez, M.C., Phillips, L.G., Willis, J.G., Boenig, H.J., Prenger, F.C., Rodenbush, A. J. and Young, S. (1996), "HTS High Gradient Magnetic Separation System," *IEEE Transactions on Applied Superconductivity, Large Scale, Proceedings of the 1996 Applied Superconductivity Conference*, Part 1 (of 3), Pittsburgh, PA, August 25–30, 1996, Vol. 7, No. 2, Part 1, pp. 650–653.

Ergun, S. and Bean, E.H. (1968), "Magnetic Separation of Pyrite from Coals," U.S. Bureau of Mines, Report of Investigations RI 7181.

Fletcher, D. (1986), "High Gradient Magnetic Filtration and Separation, Part 1," *Mine and Quarry*, Vol. 15, No. 12, pp. 35, 37–38.

Gaudin, A.M. (1974), "Progress in Magnetic Separation Using High-Intensity, High-Gradient Separators," *Mining Congress Journal*, Vol. 60, No. 1, pp. 18–21.

Hall, S.T. and Finch, J.A. (1984), "Enhanced Magnetic Desulfurization of Coal," *Minerals and Metallurgical Processing Journal*, Vol. 1, No. 3, pp. 179–184.

Hise, E.C. (1982), "Development of High Gradient and Open Gradient Magnetic Separation," *IEEE Transactions on Magnetics*, Vol. MAG-18, No. 3, pp. 847–857.

Kelland, D.R. (1982), "A Review of HGMS Methods of Coal Cleaning," *IEEE Transactions on Magnetics*, Vol. MAG-18, No. 3, pp. 841–846.

Kelland, D.R., Lai-Fook, M., Maxwell, E., Takayasu, M., Jacobs, I.S. and McConnell, M.D. (1988), "HGMS Coal Desulfurization with Microwave Magnetization Enhancement," *IEEE Transactions on Magnetics*, International Magnetics Conference—INTERMAG '88, Vol. 24, No. 6, pp. 2434–2436.

Kester, W.M. (1965), "The Demineralization of Pulverized Upper Freeport Coal in a High-Intensity Magnetic Separator," *Annual Meeting of the American Institute of Mining, Metallurgical, and Petroleum Engineers*, February 14–18, Chicago, IL.

Kindig, J.K. (1979), "The Magnex Process: Review and Current Status," *Industrial Applications of Magnetic Separation* (Liu, ed.), IEEE Publication No. 78CH1447-2 MAG, Institute of Electrical and Electronic Engineers, New York, pp. 99–104.

Kindig, J.K. and Geons, D.N. (1979), "The Dry Removal of Pyrite and Ash from Coal by the Magnex Process: Coal Properties and Process Variables," *Proceedings of the Symposium on Coal Cleaning to Achieve Energy and Environmental Goals*, Volume II (Rogers and Lemmon, eds.), EPA-600/7/79-098b, U.S. Environmental Protection Agency, Hollywood, FL, pp. 1165–1195.

Kolm, H., Oberteuffer, J. and Kelland, D. (1975), "High-Gradient Magnetic Separation," *Scientific American*, Vol. 233, No. 5, pp. 46–55.

Liu, Y.A. (1982), "High-Gradient Magnetic Separation for Coal Desulfurization," *Physical Cleaning of Coal* (Liu, ed.), Marcel Dekker, Inc., New York, Basel.

Liu, Y.A. and Lin, C.J. (1976), "Assessment of Sulfur and Ash Removal from Coals by Magnetic Separation," *IEEE Transactions on Magnetics*, Vol. MAG-12, p. 538.

Lockhart, N.C. (1984), "Dry Beneficiation of Coal," *Powder Technology*, Vol. 40, pp. 17–42.

Lua, A.C. and Boucher, R.F. (1990), "Sulphur and Ash Reduction in Coal by High Gradient Magnetic Separation," *Coal Preparation*, Vol. 8, pp. 61–71.

Luborsky, F.E. (1975), "High Gradient Magnetic Separation: A Review," General Electric Research and Development Report No. 75CRD261.

Male, S.E. (1984), "Magnetic Susceptibility and Separation of Inorganic Materials from UK Coals," *Journal of Applied Physics*, Vol. 17, pp. 155–161.

Maxwell, E. and Kelland, D.R. (1978), "High Gradient Magnetic Separation in Coal Desulfurization," *IEEE Transactions on Magnetics*, Vol. MAG-14, No. 5, pp. 482–487.

Maxwell, E., Kelland, D.R. and Akoto, I.Y. (1976), "High Gradient Magnetic Separation of Mineral Particles from Solvent Refined Coal," *IEEE Transactions on Magnetics*, Vol. MAG-12, No. 5, pp. 507–510.

Oberteuffer, J.A. (1973), "High Gradient Magnetic Separation," *IEEE Transactions on Magnetics*, Vol. MAG-9, No. 3, pp. 303–306.

Oder, R.R. (1976), "High Gradient Magnetic Separation: Theory and Applications," *IEEE Transactions on Magnetics*, Vol. MAG-12, p. 428.

Pitel, J. and Jones, H. (1998), "Computer Analysis of the 10 tesla Superconducting Magnet System of the Clarendon Laboratory for the Purposes of and Open-Gradient Magnetic Separation," *Journal of Applied Physics*, Vol. 83, No. 8, p. 4522.

Rowson, N.A. and Rice, N.M. (1989), "Desulphurisation of Coal by Magnetic Separation," *Mine & Quarry*, Vol. 18, No. 5, pp. 31–34.

Rowson, N.A. and Rice, N.M. (1990a), "Desulphurisation of Coal Using Low Power Microwave Energy," *Minerals Engineering*, Vol. 3, No. 3–4, pp. 363–368.

Rowson, N.A. and Rice, N.M. (1990b), "Magnetic Enhancement of Pyrite by Caustic Microwave Treatment," *Minerals Engineering*, Vol. 3, No. 3–4, pp. 355–361.

Sladek, T.A. and Cox, C.H. (1979), "Coal Beneficiation with Magnetic Fluids," *Industrial Applications of Magnetic Separation* (Y.A. Liu, ed.), IEEE Publication No. 78CH1447-2 MAG, Institute of Electrical and Electronic Engineers, New York, pp. 105–113.

Sladek, T.A. and Cox, C.H. (1980), "Coal Preparation Using Magnetic Separation: Vol. 4, Evaluation of Magnetic Fluids for Coal Beneficiation," EPRI Report No. CS-1517, Vol. 4, Electric Power Research Institute, Palo Alto, CA.

Smirnova, T.I., Smirnov, A.I., Clarkson, R.B. and Beiford, R.L. (1994), "Magnetic Susceptibility and Spin Exchange in Fusinite and Carbohydrate Chars," *Journal of Physical Chemistry*, Vol. 98, pp. 2464–2468.

Trindade, C.S. and Kolm, H.H. (1973), "Magnetic Desulfurization of Coal," *IEEE Transactions on Magnetics*, Vol. MAG-9, No. 3, pp. 310–313.

Tsai, S.C. (1982), *Fundamentals of Coal Beneficiation and Utilization*, Elsevier, Amsterdam, pp. 353–371.

Watson, J.H. and Younas, I. (1997), "Superconducting Discs as Permanent Magnets for Magnetic Separation," *Materials Science and Engineering B: Solid-State Materials for Advanced Technology*, Vol. B53, No. 1–2, pp. 220–224.

Weng, S. (1994), "Study of Influence of Irradiation Time on the Microwave-Magnetic Desulfurization of Raw Coal by Mossbauer Method," *He Jishu/Nuclear Techniques*, Vol. 17, No. 7, pp. 437–142.

Windle, W. (1977), "Magnetic Separation, Method and Apparatus," U.S. Patent 4,054,513, filed October 18, 1977.

Zavitsanos, P.D., Golden, J.A., Bleiler, K.W. and Kinkard, W.K. (1978), "Coal Desulfurization by Microwave Energy," Report No. EPA-600/7-78-089, U.S. Environmental Protection Agency.

Zhou, S., Garbett, E.S. and Boucher, R.F. (1996), "Dry Superconducting Magnetic Cleaning of Pulverized Coal," *AIChE Journal*, Vol. 42, No. 1, pp. 277–284.

12 Electrical Separation

Electrical separations make use of the different electrical properties of coal and mineral matter to perform a separation. The separation can be made based either on differences in particle conductivity or on differences in the electrical charge that the particles develop under certain conditions. Many electrical separation processes are carried out dry, which is advantageous for fine-coal processing because it eliminates the need for filtering and drying the clean coal product.

12.1 ELECTRICAL PROPERTIES OF PYRITE AND COAL

Pyrite and coal have significant differences in their electrical behavior. In particular, they have differences in their conductivities and electrical charging characteristics, which can be exploited by electrostatic separation techniques.

12.1.1 Conductivities of Pyrite and Coal

In general, coal particles are less conductive than mineral particles, and pyrite is the most conductive mineral that is commonly found in coal. Separation of coal and pyrite can therefore usually be carried out by methods that are based on conductivity differences. However, low-rank coals have a high content of moisture and dissolved ions, which makes these coals much more conductive (Lockhart, 1984). Because of this, conductivity separations of coal and pyrite are likely to be limited to high-rank coals.

12.2 ELECTROSTATIC FUNDAMENTALS

The concept of electric charge is considered to be undefinable in a basic sense, and instead it must be defined based on what it does. A body that exhibits a net charge will experience a force when placed in the neighborhood of other charged bodies. The magnitude of the resulting force between two charged bodies is given by the relationship:

$$F = \frac{K q_1 q_2}{d^2} \tag{12.1}$$

where q_1 and q_2 are the charges on the two bodies and d is the distance between them. The particles are attracted to each other when F is negative, and they repel each other when F is positive. The constant K is equal to $8.98755 \times 10^9 \text{N} \cdot \text{m}^2/\text{C}^2$ and is often written as

$$K = \frac{1}{4\pi\varepsilon_0} \tag{12.2}$$

where ε_0 is the permittivity of free space. Extensive discussions of electrostatic forces and their application to particle separations are given by Ralston (1961), Fraas (1962), and Lawver et al. (1986).

12.2.1 Basic Components of Electrostatic Separators

All electrostatic separation systems contain the following components (Lawver et al., 1986):

- *Charging and Discharging Mechanism*: A mechanism that will result in different particle types developing charges that (a) are of different signs, (b) are of the same sign but different magnitude, or (c) produce significantly different dipole moments.
- *External Electrical Field*: A field that can provide the necessary energy to produce separation. The field is provided by a high-voltage power source, producing an electrical potential of 10,000–100,000 V, and the electrical fields used range from 400,000 V/m up to the breakdown voltage of air, 3×10^6 V/m.
- *Nonelectrical Particle Trajectory Regulator*. A device that will send the particles on an appropriate trajectory for the separation to take place.
- *Feeding and Product Collection System*: A system to prepare the feed for separation, convey it to the separator, and to collect the separated products. Feed preparation is important, because moisture can cause serious difficulties.

12.2.2 Charging Mechanisms

There are three basic methods for producing a charge on particles, which are illustrated in Figure 12.1. These are

- *Charging by contact*: When dissimilar surfaces come in contact and are then separated, they frequently develop charges on their surfaces due to exchange of electrons and ions when they are in contact. This phenomenon is commonly referred to as triboelectrification. If one surface becomes positively charged, the other must become negatively charged, and so this is an ideal method for ensuring that two types of particles develop charges of opposite signs.
- *Charging by ion bombardment*: If a particle is bombarded with ions, such as those produced by a corona discharge, the ions will transfer charge to the particle. This is the method that produces the strongest charge on the particle surfaces, but its selectivity is practically nil, with all particles becoming strongly charged regardless of type. The ions are produced by a high-voltage electrode that ionizes the air and is designed to focus the ions onto the particles to be charged. If both conductive and nonconductive particles are charged in this manner and are on a conductive, grounded surface, then the conductive particles will redistribute their charge to the

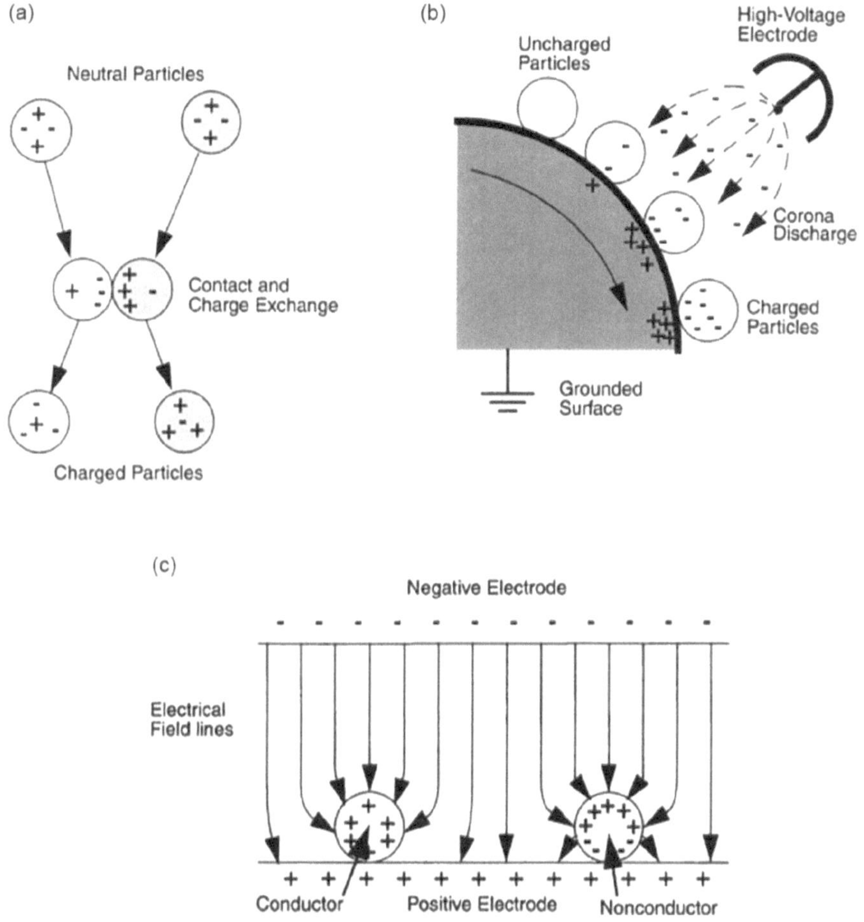

FIGURE 12.1 The three basic methods for producing electrical charge on particles: (a) contact charging, (b) ion bombardment, and (c) conductive induction.

surface almost immediately and will be free to leave the surface. In contrast, the nonconducting particles are not capable of quickly redistributing their charge, so they are held to the surface by their electrostatic attraction to the image charge that they induce on the surface.

- *Charging by conductive induction*: If a solid particle is placed on a conductive surface in the presence of an electric field, the particle will rapidly develop a surface charge by induction. Initially, the particle will polarize, and a conductive particle will then quickly become an equipotential surface and become charged to the same potential as the conductive surface it is in contact with. If the particle is then removed from the conductive surface, it will retain its charge. A particle that is a perfect insulator will not become an equipotential surface, and so it will not develop a charge. Since no real particles are either perfect conductors or perfect insulators, all particles will charge inductively to some extent; however, the speed with which they polarize and develop a charge will differ depending on a particle's electrical properties.

The charging method used depends on the nature of the material and on the nature of the separation. To separate particles based on differences in conductivity, ion bombardment or conductive induction can be used, but to separate particles with similar conductivities, contact electrification is a more promising technique.

12.2.3 Contact Charging – Triboelectrification

Triboelectrification has been found to be an effective method for charging coal and mineral matter so that the sign of the charge on the coal will be the opposite of the charge on the mineral matter (ash-forming minerals and pyrite). This charging mechanism is a result of charge transfer between two solids that have different electron free energies on their surface layers. Whenever two dissimilar materials come in contact with each other, they exchange a small amount of electrical charge. When the particles are then separated, one will have developed a positive charge while the other will have a negative charge (Mottelay, 1922; Vieweg, 1928; Shaw and Jex, 1926; Richards, 1920; Pauthenier, 1961; Crowley, 1986; Frese, 1979; Yamamoto, 1974; Cottrell et al., 1984; Tennal et al., 1997; Lindquist et al., 1995). This phenomenon, illustrated in Figure 12.2, is referred to as "triboelectrification." While the reasons for the development of charge can be given by theory (Davies, 1967), the complexity of the charge transfer in mixtures of real materials at ambient conditions makes any prediction of the polarity and amount of charge developed little more than guesswork (Inculet, 1970, 1984; Inculet and Greason, 1971; Greason, 1972; Davies, 1970; Shaw, 1917; Lowell and Rose-Innes, 1980; Lowell, 1981).

The polarity of the charge that a particle will develop depends on what material it is coming in contact with. A rule of thumb often used for predicting the sign of the surface charge is Cohen's Rule, which states that "When two dielectric materials are contacted and separated, the material with the higher dielectric constant becomes positively charged" (Knoll and Taylor, 1985; Schlichting, 1968).

The contact needed to produce triboelectrification can be obtained by two methods:

- Sliding-contact tribochargers, where the particles are forced to slide along or rub against a surface to develop a charge.
- Particle–particle contact tribochargers, where the various particle types are forced to rub against each other to develop a charge.

Sliding-contact tribochargers are very common. Charging has been accomplished by sliding feed particles down a vibrating chute (Lewowski, 1993; Masuda, 1981; Withington, 1914), through an air cyclone (Carta et al., 1976; Advanced Energy Dynamics, 1987; Ciccu et al., 1991; Masuda et al., 1981, 1983), or through a tube designed to maximize contact of particles with the tube surface (Inculet et al.,

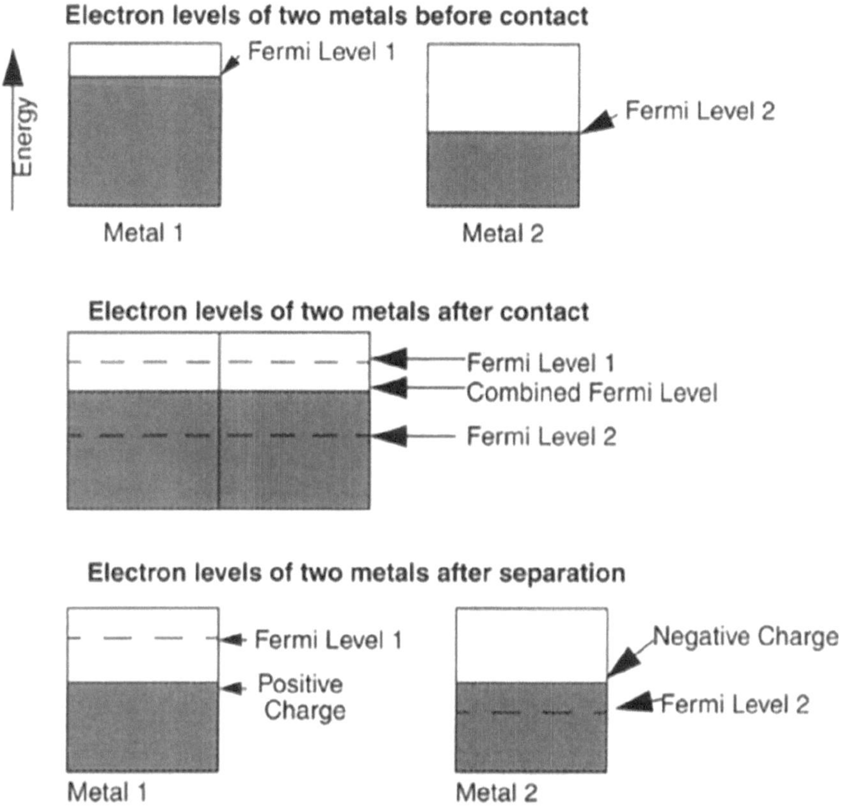

FIGURE 12.2 Illustration of the mechanism of triboelectrification (between two metals at absolute zero). When the metals come in contact, the Fermi level of one metal is lower than that of the other metal, and so electrons flow between them to produce an intermediate Fermi level midway between the two starting levels. When the two metals are separated, one will have an electron deficit (positive charge), while the other will have an electron surplus (negative charge). In real systems, with imperfect conductors and insulators at temperatures above absolute zero, the effect still occurs but there are electron backflow effects that make it difficult to predict the charges that will be developed. As a result, tribocharging effects must be determined by trial and error.

1979; Link et al., 1990; Mazumder et al., 1994; Agus et al., 1990; Finseth et al., 1993). Examples of each of these types of sliding-contact tribochargers are shown in Figure 12.3.

The charge that coal and pyrite particles will pick up in sliding-contact tribochargers depends on the type of material that the tribocharger surface is composed of. Examples of typical charges developed are shown in Table 12.1. It is desirable to have both a large difference in the charges on the coal and on the pyrite and to have one type of particle positively charged while the other is negatively charged. From the values given in Table 12.1, the tribocharger material that gives the most suitable charge to coal and pyrite is copper, and it is therefore commonly used in this type of

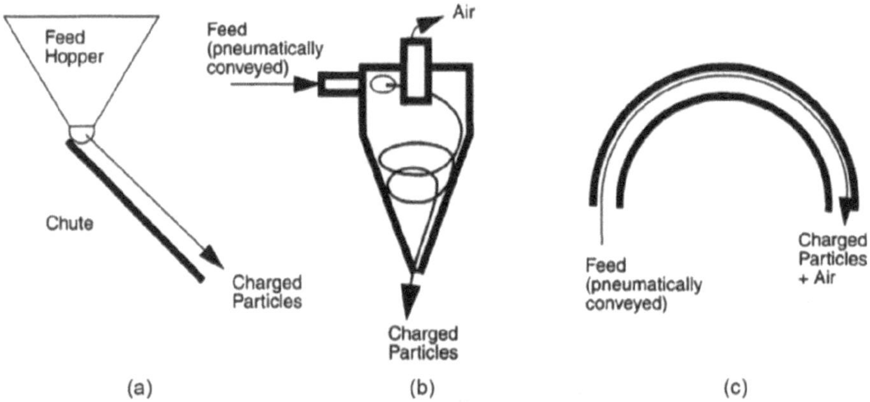

FIGURE 12.3 Examples of tribochargers where particles are charged by sliding them against a portion of the device: (a) chutes, (b) cyclones, and (c) pipe.

TABLE 12.1
Charges Developed on Coal and Pyrite Particles Sliding Inside of Tube-Type Tribochargers with Surfaces Composed of the Listed Materials (Inculet et al., 1979)

Material Used for Tribocharger Surface	Charge Developed (10^{-8}Coulombs/g/in (2.54 cm) of pipe)	
	Coal	Pyrite
Anron (PVC)	+2.00	+1.40
Galvanized steel	+0.40	−0.47
Vinyl tube	+0.22	−0.09
Copper	+2.00	−0.35
Plexiglas	−2.00	−2.30
Aluminum	−0.50	−0.31

apparatus. Steel and ceramic surfaces are also used when a higher wear resistance is needed. He et al. (2018) studied the triboelectrostatic separation of vitrinite and inertnite using aluminum, stainless steel, copper, polyethylene terephthalate (PET), polyvinyl chloride (PVC), and polyamide. Polyamide was the preferred material as it resulted in vitrinite and inertnite developing opposing charges.

Particle/particle contact tribochargers, where the coal and pyrite particles develop their charge by rubbing against each other as shown in Figure 12.4, are less commonly used. A benefit of this type of charging is that since the coal and pyrite particles are charging from contact with each other, the two types of particles will generally have opposite signs. Particle–particle tribocharging has mostly been accomplished using fluidized beds, where the particles are constantly colliding with each other (Zhou and Brown, 1988; Gidaspow et al., 1987a,b; Bergougnou et al.,

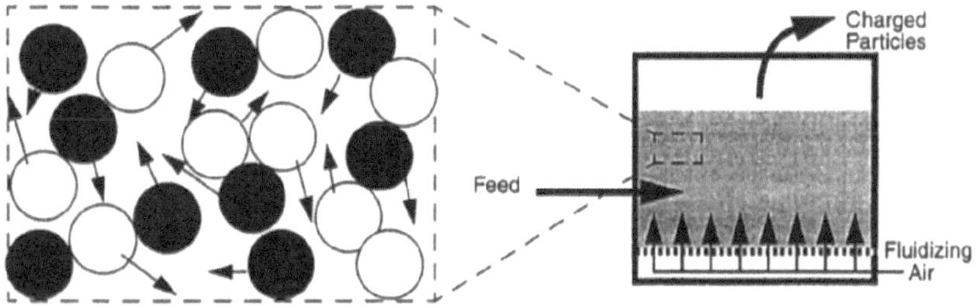

FIGURE 12.4 Charging mechanism in a fluidized-bed particle–particle contact tribocharger. As particles of one type rub against particles of another type, charge is transferred between them. This type of charging requires a high density of particles.

1977; Tauveron, 1967a,b; Morel and Tauveron, 1966; Morel, 1967; Kiewiet et al., 1978; Inculet et al., 1979, 1980; Inculet and Bergougnou, 1973; Beeckmans and Inculet, 1981; Welay and Opalinski, 1983). Advanced Energy Dynamics (1987) more recently developed a separator that uses a belt system to agitate the particles and cause them to contact each other and that can function with particles that are too fine for a fluidized bed.

12.3 ELECTROSTATIC SEPARATORS

Electrostatic separators that use triboelectrification require the following three steps, as illustrated in Figure 12.5:

- Feed preparation.
- Electrification.
- Separation.

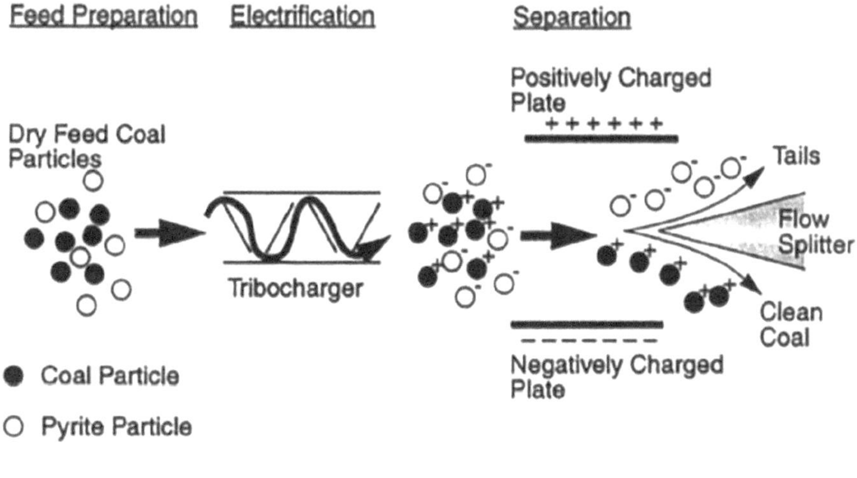

FIGURE 12.5 Basic stages of electrostatic separation, using tribocharging of the particles (Gerstler and Finseth, 1994).

12.3.1 Feed Preparation

For electrostatic separation to work effectively, the feed must first be carefully prepared so that it has the proper size distribution and is thoroughly dried. It is important to minimize the quantity of very fine particles produced during comminution (Carta et al., 1976) and to size the particles to within a specific size range (Lewowski et al., 1985; Lewowski, 1989, 1990, 1993), as the behavior of particles in the separator is strongly dependent on particle size. The moisture content has a powerful effect on the conductivities of particles, so small variations in moisture content can greatly alter their electrical properties.

Once the feed has been properly prepared, it is charged using any convenient tribocharging device and conveyed to the separation stage.

12.3.2 Plate Separators

These separators use two charged plates to separate particles based on the magnitude and polarity of the particle charges. A typical design for such a device is shown in Figure 12.6. This type of separator is best used along with a tribocharging device, so that the sign of the electrical charge on the coal can have the opposite polarity from the pyrite and ash minerals. The particles fall between the plates, with the positively

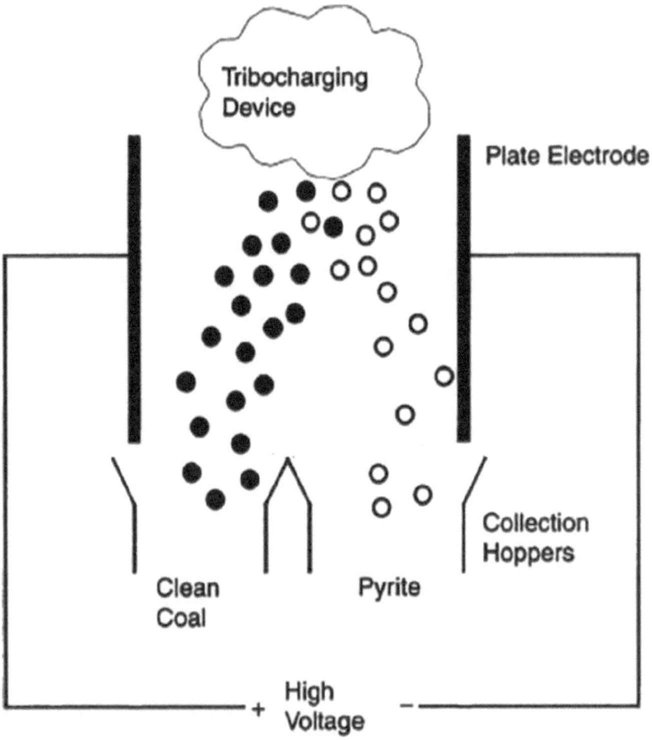

FIGURE 12.6 Basic design for an electrostatic plate separator.

charged particles diverted into one container and the negatively charged particles into the other. Since the pyrite and coal particles have opposite charge polarities, they will tend to attract one another and form agglomerates that are held together by electrostatic forces. These agglomerates have a lower net charge than the original particles, so they tend to report to the middlings. To prevent these agglomerates from forming, the separator must be operated with low particle loadings so that the particles will not interact much with each other. It has been found that the gas-to-solid mass ratio should be at least 2:1 for efficient separation, which is near the stoichiometric ratio needed for pulverized-coal burners (3:1 air/fuel ratio) (Gerstler and Finseth, 1994).

One of the difficulties with this type of separator is that very fine particles have a low falling speed in air, and so they have time to drift all the way over to one of the electrodes (Plaksin and Olofinsky, 1960, 1966). Once these particles strike an electrode, they adhere there and reduce the separator efficiency. This problem has been approached by carefully screening out the fines from the feed (Lewowski, 1993) or by using any of the several types of electrodes, designed so that the spacing becomes larger as the particles approach the bottom (Figure 12.7), preventing the fine particles from ever reaching the electrode surface. A third technique is to allow the particles to adhere to the electrodes and to rotate the electrodes past rotary scrapers to remove the particles, as shown in Figure 12.8 (Carta et al., 1976; Link et al., 1990).

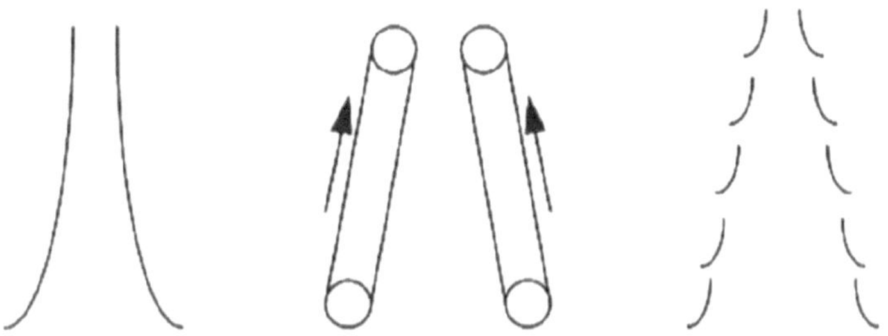

FIGURE 12.7 Plate electrode designs to prevent accumulation of fine particles on the electrode surfaces (Carta et al., 1976).

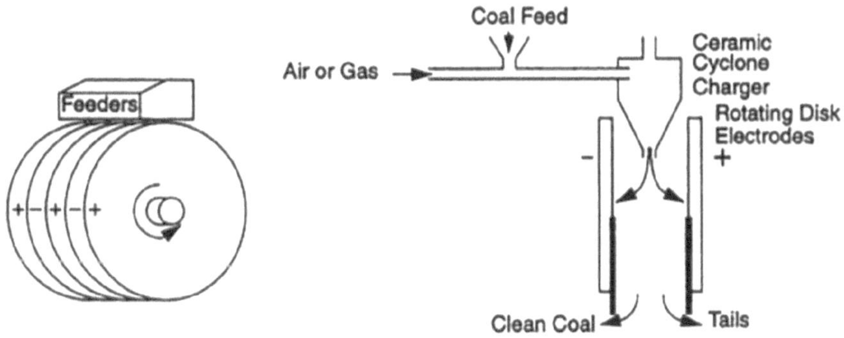

FIGURE 12.8 Disk separator. The plate electrodes are rotating disks, with scrapers to remove the fine particles that adhere to the plate.

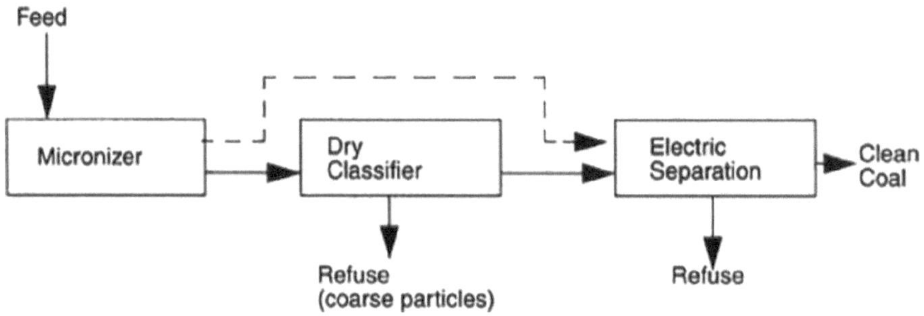

FIGURE 12.9 Selective micronization and electric separation unit used by Carta et al. (1976). The circuit could be operated to bypass the classifier if desired. Since the coal was more friable than pyrite and ash, the pyrite tended to be coarser when it left the micronizer and could be removed by the classifier based on its particle size.

Carta et al. (1976) studied the ability of a plate separator to remove pyrite from three coals, which contained pyrite in various forms. The circuit used consisted of a feed micronizer, a dry classification unit, and an electric separation as shown in Figure 12.9. It was found that when the coal was ground, the pyrite tended to be concentrated in the coarse fraction; therefore, a classifier could remove a portion of the pyrite by separating the coarse material. The circuit was designed so that the classifier could be bypassed if desired. The electric separator used a cyclone tribocharger and pneumatic transport of the solid particles. Using this unit, they obtained the results given in Table 12.2. These results show that it is possible to achieve high

TABLE 12.2
Electrostatic Separation Results for Three Coals, Using a Plate-Type Separator with a Cyclone Tribocharger (Carta et al., 1976)

Coal Type	Product	% Yield	% Pyritic Sulfur	% Pyritic Sulfur Removal
Vitrinite-rich Ruhr coal, pyrite large single grains. Combined selective comminution followed by electrostatic separation	Feed	—	0.83	73.44
	Clean coal	67.20	0.33	
	Electric separator refuse	24.57	1.91	
	Selective comminution refuse	8.23	1.75	
Ruhr coal seams, pyrite as aggregates and single grains. Electrostatic separation alone	Feed	—	0.60	79.38
	Clean coal	65.20	0.19	
	Tailings	34.80	1.37	
Pittsburgh seam coal, pyrite as large nodules and 1–15 μm grains. Electrostatic separation alone	Feed	—	2.08	65.29
	Clean coal	89.12	0.81	
	Tailings	10.88	12.48	

pyrite removals while maintaining a good coal yield. A similar system combining selective comminution and electric separation was devised by Lewowski (1993), which was also successfully used to remove pyrite from coal.

Anderson et al. (1979) used a dilute-phase loop triboelectric separator to separate coal and pyrite, as shown in Figure 12.10. The loop was constructed mainly from copper, so that the coal would develop a positive charge and the pyrite would develop a negative charge. The feed coal was ground to 100% passing 100 μm, and the separation was carried out in a dry nitrogen atmosphere to reduce the risk of explosion (Inculet et al., 1977). The products were analyzed to determine their pyrite content and petrographic composition, with the results shown in Table 12.3. These results show that while the vitrinite macerai tends to report to the clean coal product, the fusinite becomes concentrated into the reject product along with the pyrite.

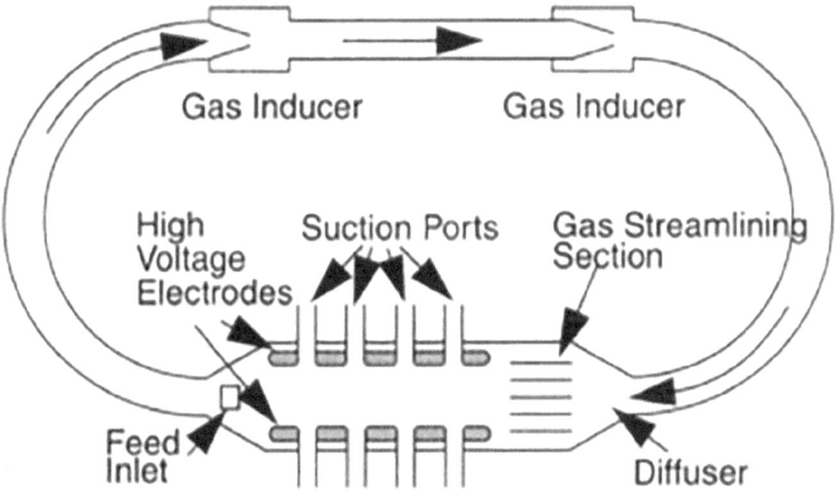

FIGURE 12.10 Electrostatic loop separator (Inculet, 1984). Gas is injected through the gas inducers and withdrawn along with separated particles through the suction ports. Particle charging occurs by contact with the curved sections.

TABLE 12.3
Electrostatic Separation of Coal Macerals Using the Electrostatic Loop Separator (Anderson et al., 1979)

Product	% Yield	% Vitrinite	% Fusinite	% Exinite	% Micrinite	% Pyrite	% Pyrite Removal
Feed	—	85.4	8.8	2.6	0.4	2.8	83.7
Product	66.5	91.1	6.3	1.6	0.5	0.5	
Tails	33.5	74.2	14.9	3.3	0.6	7.0	

12.3.3 Electrostatic Belt Separator

This separator incorporates the tribocharging mechanism with the separating mechanism. The apparatus is shown schematically in Figure 12.11 and consists of two parallel electrodes with an open-mesh belt passing between them. Feed is introduced through an opening in the center of one electrode. Once the feed particles enter the system, they are agitated by the motion of the belt, which causes the particles to rub against each other and develop electrical charges. Once the particles become charged, they are attracted to the electrode that has the opposite polarity to their own. As they approach the electrode, the belt tends to sweep them toward one end of the separator. Since the motion of the belt at each electrode is in the opposite direction from its motion at the other electrode, this results in the positively and negatively charged particles discharging from opposite ends of the separator.

The high-shear action of the belts provides good tribocharging of the particles and breaks up any particle aggregates, which reduces the problems with electrostatic agglomeration of negatively charged pyrite with positively charged coal. The countercurrent motion of the oppositely charged particles helps to prevent particles from being misplaced in the wrong product, and so the separation is more precise than that produced by other types of electrostatic separator. This unit is also able to process very fine particles, because the particles are being dragged through the units by the belt instead of falling under gravity as in most other electrostatic separators. The belt is made from high-strength woven materials such as nylon, polyester, polyolefin, coated aramids, or ultrahigh molecular weight polyethylene. The belt material must be strong and very abrasion resistant, and since the belt will develop a charge from rubbing against the particles, the belt material should be chosen so that the developed charge will not interfere with the electric fields from the electrodes.

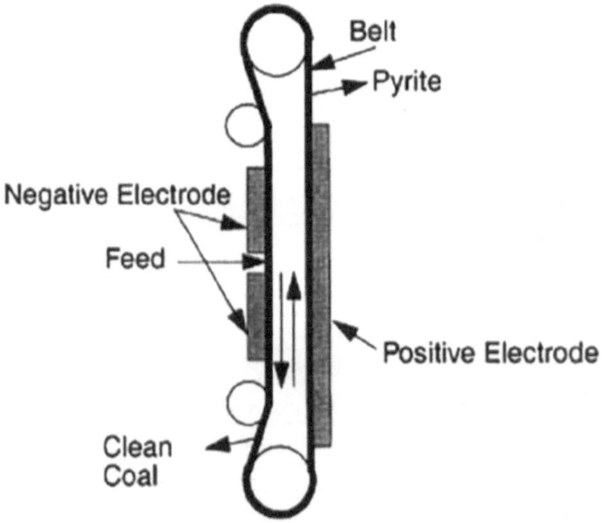

FIGURE 12.11 Vertical belt electrostatic separator (Advanced Energy Dynamics, 1987).

TABLE 12.4
Pyrite Removal Results with the Vertical Belt Electrostatic Separator (Advanced Energy Dynamics, 1987)

Feed Coal	Product	% Yield	% Ash	% Pyritic Sulfur	% Pyrite Removal
Lower Kittanning seam	Feed	—	10.07	2.88	88.9
Mean diameter 15 μm	Clean coal	62.9	3.23	0.51	
	Reject	37.1	21.21	6.90	
Illinois No. 6 seam	Feed	—	8.41	0.63	75.1
94% passing 45 μm	Clean coal	54.5	3.57	0.35	
	Reject	45.5	13.38	1.04	
Upper Freeport seam	Feed	—	11.64	1.20	71.4
90% passing 45 μm	Clean coal	55.6	8.54	0.83	
	Reject	44.4	19.75	1.93	

A 3.04 m long unit of this type was used to clean three coals that have been ground to fine particle sizes to liberate pyrite and ash. The results of this separation are shown in Table 12.4.

It was reported by the investigators that the various coal macerals behaved in different ways in this separator. In particular, it was found that in first-stage separations, where there was a large amount of vitrinite present, the fusinite would charge negatively due to contact with the vitrinite particles and report to the reject product along with the pyrite and ash. If the reject product was re-treated, however, the fusinite would mainly tribocharge by contact with the pyrite and ash particles and would therefore charge positively and report to the clean coal product.

12.3.4 Commercial Application of Electrostatic Separation

At present, there are no known commercial installations that use electrostatic separation with triboelectric charging to process coal, although as of 1996 an 18 ton/h electrostatic belt separator was being used commercially to remove unburned carbon from fly-ash (Tondu et al., 1996). There are two reasons for this lack of application:

- The capacity of a triboelectric separator is low compared to other common separation techniques, due to the difficulty of charging large quantities of particles in existing tribochargers. For a throughput of 40 kg/h, approximately 1 m^2 of floor space is needed for conventional parallel-plate units (Lewowski, 1993). It should be noted that electrostatic belt separators have considerably higher capacity per unit volume than traditional electrostatic separators, so this problem may become much less serious in the near future.
- The selectivity of the separation for removing ash and pyrite from coal is low, due to effects such as electrostatic agglomeration of particles with opposite charges and disturbances of falling particles by air currents.

An additional consideration is that in order for triboelectrification to work, the coal must be finely ground and dried, both of which would add a considerable cost in a coal preparation plant. The conditions for electrostatic separators are much better at the utility plant, where the coal must be dried and pulverized before it is conveyed by turbulent airflow to the combustor. The conditions during pulverization and conveying are exactly those that would be expected to triboelectrically charge particles; therefore, there is good potential for including electrostatic separation in the plant immediately before combustion. To be able to implement an electrostatic separator in this situation, knowledge is needed of the sign and magnitude of the charges developed on the coal, ash, and pyrite particles.

Schaefer et al. (1995, 1996) conducted studies of coal charging in pulverizers in two utility plants. In the first plant (Spurlock Station, Maysville, KY, using a pressure-fed pulverizer), the charge developed on the particles was in the range of 32×10^{-6} to 53×10^{-6} coulombs/kg, and at the second plant (Widows Creek Station, Stevenson, AL, using a suction-fed pulverizer), the charge was in the range of -59×10^{-6} to -108×10^{-6} coulombs/kg. It should be noted that while the magnitudes of the charges developed were roughly similar, the polarity of the charge was positive at Spurlock Station, but it was negative at Widows Creek. Diagrams of pressure-fed and suction-fed pulverizer circuits are shown in Figure 12.12.

The charge measured on the particles after they leave the pulverizer is the net charge. In theory, it would be possible for the total charge on the coal particles to have the same magnitude as the total charge on the ash and pyrite particles, but the opposite sign, and as a result the net charge would be zero. In practice, the quantities of the different types of particles and the charges that they develop will be different, and so they will not cancel out completely, leading to a net charge on the particles. The observed values of the net charge indicate that the charges on individual particles are at least that high, and potentially even higher, so a high degree of electrostatic separation is possible.

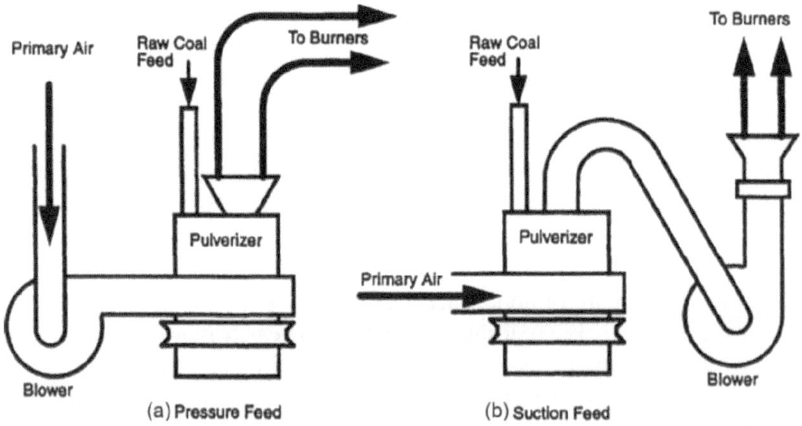

FIGURE 12.12 Schematics of pressurized and suction pulverizer systems, such as those studied by Schaefer et al. (1996). Particles are subjected to considerable friction inside pulverizers, and can therefore develop a considerable electrical charge.

Once it had been determined that particles were becoming triboelectrically charged in coal pulverizers, Schaefer et al. (1995, 1996) conducted further experiments to determine whether this charge could be used to remove mineral matter from the coal. A portion of the pulverizer product was diverted through a small parallel-plate separator with an 8 cm plate spacing, operated at 7.5–30 kV. At Spurlock Station, this unit was able to achieve 76%–80% combustibles recovery while removing 43%–51% of the ash minerals. When used at Widows Creek Station, the combustibles recovery was 81%–86% while removing 21%–44% of the ash minerals. A properly optimized separator could be expected to give even better results (Stencel et al., 1998).

12.4 ELECTRODYNAMIC (CORONA) SEPARATION

Electrodynamic separation is also commonly known as "high-tension" separation, but the latter term should be discouraged because many other electrical separation techniques also use high voltages. This technique makes use of ion bombardment to charge particles, with the ions produced by a corona discharge. The separation is based on the rate at which the charge deposited on a particle can be conducted away again. Since the charge is flowing through the particles, it is not correct to refer to this type of separation as being "electrostatic."

Most early designs for electrodynamic separators are now obsolete, and current designs are almost invariably based on a rotating, grounded drum and a small diameter or pointed electrode to provide a corona discharge, as shown in Figure 12.13. Feed particles enter at the top of the drum, and are charged by bombardment with ions from the corona electrodes. Once charged, the particles become "pinned" to the surface by electrostatic attraction. As the drum rotates, it carries the particles out of the corona region. The charge then flows from the particles into the grounded drum. When enough charge has flowed from the particle, it is no longer pinned to the drum, and it falls off. Charge flows away from the conductors faster than from the

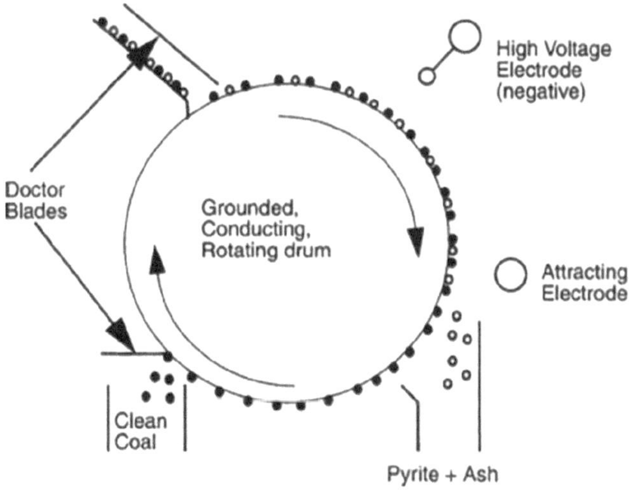

FIGURE 12.13 Schematic of an electrodynamic drum separator.

nonconductors, so the conductors fall off of the drum first and are separated. Modem machines usually include a larger diameter nonionizing electrode after the corona electrode, which lifts and deflects the conductive particles, thereby improving the selectivity of the separation. Drum separators are limited to particles coarser than 38 µm, both because of capacity limitations and because the air currents produced by the rotating drum and the corona discharge tend to interfere with the separation of particles finer than this size.

The rotating-drum electrostatic separator has been widely used to separate a wide range of particle mixtures, including metal from glass, plastics from paper, heavy minerals from beach sands, etc. There has been relatively little scientific work with this technique since the publication of Ralston's book in 1961, and there would seem to be scope for research to improve the effectiveness of the process. One suggestion for extending the range of particle conductivities that can be separated is to superimpose a radio-frequency electric field onto the direct field, so that the effective conductivity of particles will include dielectric loss, allowing some insulators to be separated based on their dielectric constants (Horowitz, 1966; Szczerbinski, 1983).

12.4.1 Electrodynamic Separation of Coal

Most of the early work in this area is reviewed by Ralston (1961) and Fraas (1962). One exception is the work done by Gray and Whelan (1956), who studied the effects of rank, particle size, and moisture content for a variety of coals, ranging in size from 2 mm to 125 µm. They used several different electrode types, various drum materials, and a range of drum speeds and operating voltages, and identified a wide range of conditions that produced good separation of coal and shale.

Experiments have been conducted for all ranks of coals ranging from German brown coals to Pennsylvania anthracite, both to produce low-ash and low-sulfur coals, and to fractionate the coal macerals (Ralston, 1961). Work has also been done to classify coal by size, with beneficiation occurring as an incidental benefit (Olofinskii, 1957). In experiments using Japanese coals that contained 6.5%–10% ash, and were sized to 1,700 × 400 µm, a product that was enriched in vitrinite and contained only 2.5%–3.0% ash was produced. The reported recoveries were 96%–98% of the recoveries that were achieved by sink–float analysis at the same ash content (Mukai et al., 1963; Mukai, 1964). The removal of pyrite by corona separation has also been reported (Lockhart, 1984; Abel et al., 1972, 1973). In the work by Abel et al., the electrodynamic separation followed a classifier, and the combination was used to process a Pittsburgh seam coal. By use of staged grinding, this circuit could remove nearly 90% of the pyrite from the coal (Abel et al., 1973).

At present, there are no known applications of electrodynamic separation of coal, although a pilot plant was extensively studied in Germany in the 1940s on coal feeds 2.0–0.1 mm in size, and a larger plant was constructed in the Ruhr Valley. This plant never operated due to bomb damage in the war (Lockhart, 1984).

Extensive tests have been carried out on pulverized coals by Advanced Energy Dynamics (1987), using a standard Carpco pilot-scale machine, which was essentially identical to that shown in Figure 12.13. Dry coal particles ranging in size from 250 to 38 µm were processed, with the mineral matter acting as a conductor while the coal acted as a semiconductor or nonconductor. A10 ton/h commercial-scale unit was tested at the Picway Station Power Plant in Columbus, OH, where it was found that it could not handle particles finer than 38 mm and had a lower-than-expected capacity. Operation of this unit was therefore suspended in 1985.

12.5 ELECTROPHORETIC SEPARATION

In addition to the dry electrical separation techniques described earlier, it is also possible to desulfurize coal by a wet electrical separation, based on thc Zeta potentials of the various particles. The Zeta potential is the charge that develops around the surface of a particle when it is immersed in a liquid such as water, and the surface selectively exchanges ions with the solution. The electrophoretic separation process was studied by the Bureau of Mines on a laboratory scale, but the complexity of the equipment and the power consumption of the process were considered to make the technique impractical on a large scale (Miller and Baker, 1974). A variation on the technique has recently been tried, where electrophoresis was used to accelerate the separation rate of an oil/water system (Mang and Oder, 1990). While the electrophoretic process is normally considered to be distinct from other types of electrical separation, it has a number of similar features that make it appropriate to discuss here. The schematic diagram of the Bureau of Mines electrophoretic separator is shown in Figure 12.14.

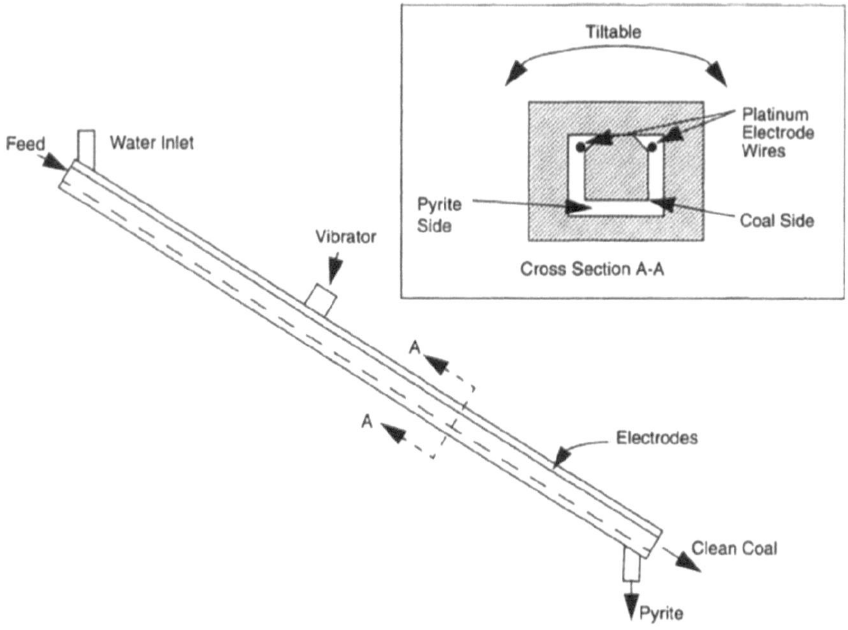

FIGURE 12.14 Schematic diagram of the Bureau of Mines electrophoretic separator (Miller and Baker, 1974). The column could be tilted around its longitudinal axis so that gravity could assist in the separation.

12.6 SUMMARY

A tremendous amount of research has gone into the study of electrical separation of coal, but there are difficulties involved in applying it at an industrial scale. The major challenges are

- Maintaining high processing rates;
- Processing very fine particles;
- Avoiding undesirable agglomeration in the feed;
- Acquiring a sharp separation so that high grades can be achieved without greatly sacrificing recovery.

However, many of these problems are at this point simply engineering problems which have been investigated thoroughly. Industrial triboelectric separators are beginning to become useful for coal separations.

Electrostatic and electrodynamic separators require particles to hold charge on their surfaces for some period of time. This requires that the particles be in an insulating medium such as dry air, so these processes must be applied to dry powders by their very nature. This requirement for dry powders can be attractive for a number of reasons:

- The product remains dry, and so filtering and drying costs are eliminated.
- There are many coal-producing areas that are very arid, and water is difficult to obtain in large enough quantities for conventional wet-coal processing. Dry processing is therefore preferred in these areas when possible.
- In regions where the temperature is below freezing for much of the year, operations are greatly simplified if no water is involved.
- A process that removes sulfur from dry coal would be suitable for use on-site with pulverized-coal burners. The coal could be pulverized, cleaned, and then burned immediately at the power plant, without the need for shipping, storage, or excessive handling of finely ground coal.

In particular, it is very expensive to dewater coal that has been ground to the very fine particle sizes needed to liberate the fine pyrite grains. There would therefore be significant benefits to coal desulfurization processes that do not need to suspend the coal in water; therefore, dry separation techniques would have certain advantages even if their separation efficiency is less than the corresponding wet separations. It should be noted, however, that coal dust suspended in air can be a serious fire and explosion hazard, and this hazard is even more severe when high voltages are present, which can cause electrical sparks. Every precaution must be taken against fire and explosion when processing dry coal electrically.

While most electrostatic separators are basically similar to the types that have been used for many years, there are a few recent developments that may make these separators more practical on an industrial scale. In particular, the Vertical Belt Separator is a significant departure from traditional designs that partially overcomes the capacity and particle size limitations, and it may be useful for treating fine coal at higher throughputs.

Triboelectrification and the resulting separation techniques are very important technologies that can be immediately applied to power plants burning coal. In a utility plant, when coal is injected into the burner after comminution, it is very dry and ground to a very fine size. An electrode there will attract pyrite. While it will require periodic cleaning, this simple application can reduce the amount of SO_2 being released into the atmosphere.

In addition to the dry separation technologies, it is also possible to carry out electrophoretic separations in aqueous suspensions. Electrophoretic separation has a very high power demand, and since it is a wet process, it does not have the advantages of the dry electrical separation techniques. As a result, it is only of academic interest and there is no industrial interest in the process.

REFERENCES

Advanced Energy Dynamics, Inc. (1987), "Advanced Physical Fine Coal Cleaning: Final Report." U.S. Department of Energy, Report No. DOE/PC/81211-T17 (DE88004959).

Abel, W.T., Zulkoski, M. and Gauntlett, G.J. (1972), "Dry Separation of Pyrite from Coal," *Industrial Engineering Chemistry Product Research and Development*, Vol. 11, No. 3, pp. 342–347.

Abel, W.T., Zulkoski, M., Brady, G.A. and Eckerd, J.W. (1973), "Removing Pyrite from Coal by Dry-Separation Methods," U.S. Bureau of Mines Report of Investigations RI 7732, 26 pages.

Agus, M., Carbini, P., Ciccu, R. and Ghiani, M. (1990), "Triboelectric Coal Cleaning and Desulphurization with the Turbocharger Separator," *Processing and Utilization of High-Sulfur Coals III* (Markuszewski and Wheelock, eds.), Elsevier Science Publishers B.V., Amsterdam, pp. 311–320.

Anderson, J.M., Perobek, L., Bergougnou, M.A. and Inculet, I.I. (1979), "Electrostatic Separation of Coal Macerals," *IEEE Transactions on Industry Applications*, Vol. IA-15, No. 3, pp. 291–293.

Beeckmans, J.M. and Inculet, I.I. (1981), "Electrostatic Method and Apparatus for Sorting Fluidized Particulate Material," U.S. Patent 4,274,947, June 23, 1981.

Bergougnou, M.A., Inculet, I.I., Anderson, J. and Parobek, L. (1977), "Electrostatic Beneficiation of Coal in a Fluidized State," *Powder and Bulk Solids Technology*, Vol. 1, No. 3, pp. 22–26.

Carta, M., Del Fa, C., Ciccu, R., Curreli, L. and Agus, M. (1976), "Technical and Economical Problems Connected with the Dry Cleaning of Raw Coal and in Particular with Pyrite Removal by Means of Electric Separation," *Proceedings of the 7th International Coal Preparation Congress,* Sydney, Australia, Paper K.2, 35 p.

Ciccu, R., Ghiani, M., Peretti, R., Serci, A. and Zucca, A. (1991), "Tribocharging Studies of Ground Coal Matter," *Processing and Utilization of High-Sulfur Coals IV* (Dugan, Quigley, and Attia, eds.), Elsevier Science Publishers B.V., Amsterdam, pp. 223–237.

Cottrell, G.A., Hatto, C.E., Reed, C. and Rose-Innes, A.C. (1984), "Contact Charging of Ideal Insulators: Experiments on Solidified Rare Gases," *Journal of Physics-D: Applied Physics,* Vol. 17, pp. 989–1005.

Crowley, J.M. (1986), *Fundamentals of Applied Electrostatics*, John Wiley and Sons, New York.

Davies, D.K. (1967), "The Generation and Dissipation of Static Charge on Dielectrics in a Vacuum," *Proceedings, 2nd Conference on Static Electrification, Institute of Physics,* London, Conference Series No. 4, pp. 29–36.

Davies, D.K. (1970), "Charge Generation on Solids," *Proceedings, 1st International Conference on Static Electricity,* European Federation of Chemical Engineering, Vienna, pp. 10–21.

Finseth, D., Newby, T. and Elstrodt, R. (1993), "Dry Electrostatic Separation of Fine Coal," *Processing and Utilization of High-Sulfur Coals V* (Parekh and Groppo, eds.), Elsevier Science Publishers B.V., Amsterdam, pp. 91–98.

Fraas, F. (1962), "Electrostatic Separation of Granular Materials," U.S. Bureau of Mines Bulletin No. 603, 165 pages.

Frese, Jr. K.W. (1979), "Simple Method for Estimating Energy Levels of Solids," *Journal of Vacuum Science Technology*, Vol. 16, No. 4, pp. 1042–1046.

Gerstler, W.D. and Finseth, D.H. (1994), "Development of the Tribo-electrostatic Fine Coal Separator," *Proceedings of the Tenth Annual Coal Preparation, Utilization, and Environmental Control Contractors Conference*, July 18–21,1994, Pittsburgh, PA, pp. 125–132.

Gidaspow, D., Shih, Y.T. and Wasan, D. (1987a), "Hydrodynamics of Electrofluidization: Separation of Pyrites from Coal," *American Institute of Chemical Engineering Journal*, Vol. 33, No. 8, pp. 1322–1333.

Gidaspow, D., Gupta, R., Mukherjee, A. and Wasan, D. (1987b), "Separation of Pyrites from Illinois Coals Using Electrofluidized Beds and Electrostatic Sieve Conveyors," *Processing and Utilization of High-Sulfur Coals II* (Chugh and Caudle, eds.), Elsevier Science Publishers B.V., Amsterdam, pp. 271–281.

Gray, V.R. and Whelan, P.F. (1956), "Electrostatic Cleaning of Low Rank Coal by the Drum Separator," *Fuel*, Vol. 35, p. 184.

Greason, W.D. (1972), "Effect of Electric Fields and Temperature on the Electrification of Metals in Contact with Insulators and Semi-Conductors," Ph.D. Thesis, The University of Western Ontario, Canada.

He, X., Sun, H., Zhao, B., Chen, X., Zhang, X. and Komarneni, S. (2018), "Tribocharging of Macerals with Various Materials: Role of Surface Oxygen-Containing Groups and Potential Difference of Macerals," Fuel, Vol. 233, No. 1, pp. 759–768.

Horowitz, A.S. (1966), "Differential Separation of Particulates by Combined Electrostatic and Radio-Frequency Means," U.S. Patent No. 3,293,789.

Inculet, I.I. (1984), *Electrostatic Mineral Separation*, Research Studies Press, Ltd, Letch-worth.

Inculet, I.I. (1970), "Influence of Electric Fields on Friction Electrification Between Metals and Borosilicate Glass," *Journal of Colloid and Interface Science*, Vol. 32, No. 3, pp. 395–400.

Inculet, I.I. and Greason, W.D. (1971), "Effect of Electric Fields and Temperature on Electrification of Metals in Contact with Glass and Quartz," *Static Electrification, Proceedings of the Institute of Physics,* London, Conference Series, No. 11, pp. 23–32.

Inculet, I.I. and Bergougnou, M.A. (1973), "Electrostatic Beneficiation of Fine Mineral Particles in a Fluidized Bed," *Proceedings of the 10th International Mineral Processing Congress*, London, England, Paper 11, pp. 1–14.

Inculet, I.I., Bergougnou, M.A. and Brown, J.D. (1977), "Electrostatic Separation of Particles Below 40 μm in a Dilute Phase Continuous Loop," *IEEE Transactions on Industry Applications*, Vol. IA-13, No. 4, pp. 370–373.

Inculet, I.I., Bergougnou, M.A., Quigley, R.M. and Brown, J.D. (1979), "Fluidized Electrostatic Removal of Mineral Matter from Coals Mined at Hat Creek, British Columbia, Canada," *Conference Record 1979, 14th Annual Meeting IEEE Industry Applications Society*, Cleveland, pp. 112–116.

Inculet, I.I., Quigley, R.M., Bergougnou, M.A., Brown, J.D. and Faurschou, D.K. (1980), "Electrostatic Beneficiation of Hat Creek Coal in the Fluidized State," *CIM Bulletin*, Vol. 73, No. 822, pp. 51–61.

Kiewiet, C.W., Bergougnou, M.A., Brown, J.D. and Inculet, I.I. (1978), "Electrostatic Separation of Fine Particles in Vibrated Fluidized Beds," *IEEE Transactions on Industry Applications*, Vol. IA-14, No. 6, pp. 526–530.

Knoll, F.S. and Taylor, J.B. (1985), "Advances in Electrostatic Separation," *Mining, Metallurgy & Exploration*, Vol. 2, No. 2, pp. 106–114.

Lawver, J.E., Taylor, J.B. and Knoll, F.S. (1986), "Laboratory Testing for Electrostatic Concentration Circuit Design," *Design and Installation of Concentration and Dewatering Circuits* (Anderson and Mular, eds.), SME-AIME, Littleton, CO, pp. 454–177.

Lewowski, T., Grodzicki, A. and Matz, T. (1985), "Electrostatic Separator," Polish Patent No. 142955.

Lewowski, T. (1989), "Szkie historii i perspektywy elektrostatycznego odsiarczania wegla (The History and prospects of hard coal electrostatic desulphurization)," *Przeglad Górgniczy*, Vol. 45, No. 1, pp. 24–29.

Lewowski, T. (1990), "O mozliwosci obnizenia zawartosci pirytu w niektórych weglach energetyczynch (Possibilities of reducing pyrite content in some steam coals)," *Przeglad Górgniczy*, Vol. 46, No. 1, pp. 16–19.

Lewowski, T. (1993), "Electrostatic Desulphurization of Polish Steam Coals," *Coal Preparation*, Vol. 13, pp. 97–105.

Lindquist, D.A., Mazumber, M.K., Tennal, K.B., McKendree, M.H., Kleve, M.G. and Scruggs, S. (1995), "Electrostatic Beneficiation of Coal," *Proceedings of the MRS Fall Meeting*, Vol. 371, pp. 459–463.

Link, T., Killmeyer, R., Elstrodt, R. and Haden, N. (1990), "Initial Study of Dry Ultrafine Coal Beneficiation Utilizing Triboelectric Charging with Subsequent Electrostatic Separation," U.S. Department of Energy, Report No. DOE/PETC/TR-90/11 (DE91000943).

Lockhart, N.C. (1984), "Dry Beneficiation of Coal," *Powder Technology*, Vol. 40, pp. 17–2.

Lowell, J. and Rose-Innes, A.C. (1980), "Contact Electrification," *Advances in Physics*, Vol. 29, pp. 947–1023.

Lowell, J. (1981), "The Effect of an Electric Field on Contact Electrification," *Journal of Physics-D: Applied Physics*, Vol. 14, pp. 1513–1522.

Mang, J.T. and Oder, R.R. (1990), "Coal Beneficiation by Electrostatic Coalescence," *Processing and Utilization of High-Sulfur Coals III* (Markuszewski and Wheelock, eds.), Elsevier Science Publishers B.V., Amsterdam, pp. 341–350.

Masuda, S. (1981), "Industrial Applications of Electrostatics," *Journal of Electrostatics*, Vol. 10, pp. 1–15.

Masuda, S., Toraguchi, T., Takahashi, T. and Haga, K. (1981), "Beneficiation of Coal, Using a Cyclone Tribocharger," *Conference Record IEEE/IAS Annual Meeting*, Philadelphia, PA, pp. 1001–1005.

Masuda, S., Toraguchi, T., Takahashi, T. and Haga, K. (1983), "Electrostatic Beneficiation of Coal Using a Cyclone-Tribocharger," *IEEE Transactions on Industrial Applications*, Vol. LA-19, No. 5, pp. 789–793.

Mazumder, M.K., Tennal, K.B. and Linquist, D. (1994), "Electrostatic Beneficiation of Coal," *Proceedings of the Tenth Annual Coal Preparation, Utilization, and Environmental Control Contractors Conference,* July 18–21, 1994, Pittsburgh, PA, pp. 111–116.

Miller, K.J. and Baker, A.F. (1974), "Evaluation of a Novel Electrophoretic Separation Method to Remove Pyritic Sulfur from Coal," U.S. Bureau of Mines Report of Investigations RI 7960.

Morel, R. (1967), "Electrostatic Sorting in a Fluidized Bed," *Proceedings of the Static Electrification Conference*, London, England, May, pp. 165–168.

Morel, R. and Tauveron, P. (1966), "Method and Apparatus for the Electrostatic Sorting of Granular Materials," Canadian Patent No. 746,343, Nov. 15, 1966.

Mottelay, P.F. (1922), *Bibliographical History of Electricity and Magnetism*, Charles Griffin & Co., London.

Mukai, S., Wakamatsu, T. and Shida, Y. (1963), "Investigation of the Electrostatic Concentration of Coal with a Low Iron Content in Corona Discharge Fields," *Suiyokaishi,* Vol. 15, pp. 51–56.

Mukai, S. (1964), "Electrostatic Enrichment of Coals with Low Ash Content in the Corona Field," *Freiberger Forschungsh.,* A326, pp. 161–165.

Olofinskii, N.F. (1957), *Electric Corona Separation of Coal Fines and Certain Minerals*, Moscow, 1957, Translated by Israel Program for Scientific Translations, 1969.

Pauthenier, M. (1961), "Lois de Charge des Particules Sphéiques Conductrices dans un Champ Electrique Bi-ionisé," *La Physique des Forces Electrostatiques et Leurs Applications—Colloques Internationaux du Centre National de la Recherche Scientifique*, pp. 279–287.

Plaksin, I.N. and Olofinsky, N.F. (1966), "Review of Electrical Separation Methods in Mineral Technology," *Institution of Mining and Metallurgy Transactions, Section C*, Vol. 75, No. 712, pp. C57–C64.

Plaksin, I.N. and Olofinsky, N.F. (1960), "Dedusting and Classifying Minerals in the Fine Size Range with Corona-Type Separators," *Bulletin of the Institution of Mining and Metallurgy*, Vol. 69, No. 645, pp. 613–626.

Ralston, O.C. (1961), *Electrostatic Separation of Mixed Granular Solids*, Elsevier, Amsterdam.

Richards, H.F. (1920), "Electrification by Impact," *Physics Review, Series 2*, Vol. 16, pp. 290–304.

Schaefer, J.L., Ban, H., Saito, K. and Stencel, J.M. (1995), "Quantifying Pulverizer Induced Charge in a Utility Boiler," *Proceedings of the 12th International Pittsburgh Coal Conference*, Pittsburgh, PA, pp. 320–325.

Schaefer, J.L., Ban, H., Saito, K. and Stencel, J.M. (1996), "Pulverizer Induced Charge: Comparison of Separate Utility Pulverizer Configurations," *Proceedings of the 13th International Pittsburgh Coal Conference*, Pittsburgh, PA, pp. 867–872.

Schlichting, H. (1968), *Boundary Layer Theory*, 6th edition, McGraw-Hill Inc., New York.

Shaw, P.E. (1917), "Experiments on Tribo-Electricity," *Proceedings of the Royal Society*, Vol. 94, p. 16.

Shaw, P.E. and Jex, C.S. (1926), "Triboelectricity and Friction," *Proceedings of the Royal Society,* Vol. 111, pp. 339–355.

Stencel, J.M., Shaefer, J.L., Ban, H., Li, T.-X. and Neathery, J.K. (1998), "Triboelectrostatic Cleaning of Coal In-Line Between Pulverizers and Burners at Utilities," *Coal Preparation,* Vol. 19, Nos. 1–2, pp. 115–129.

Szczerbinski, M. (1983), "Electrodynamic Conductance Separation with Use of Alternating Fields," *Journal of Electrostatics,* Vol. 14, No. 2, pp. 175–186.

Tauveron, P. (1967a), "Method and Apparatus for Separating Materials," Canadian Patent No. 766,113, Aug. 29, 1967.

Tauveron, P. (1967b), "Triage Electrostatique en Bain Fluidisé," *Revue de l'Industrie Minérale,* pp. 442–449.

Tennal, K.B., Mazumder, M.K., Lindquist, D., Zhang, J. and Tendeku, F. (1997), "Triboelectric Separation of Granular Materials," *Proceedings of the 1997 IEEE Industry Applications Conference,* 32nd IAS Annual Meeting, Part 3 of 3, Oct. 5–9, New Orleans, LA, pp. 1724–1729.

Tondu, E., Thompson, W.G., Whitlock, D.R., Bittner, J.D. and Vailiauskas, A. (1996), "Commercial Separation of Unburned Carbon From Fly Ash," *Mining Engineering*, Vol. 48, No. 6, pp. 47–50.

Vieweg, H.F. (1928), "Frictional Electricity," *Journal of Physical Chemistry*, Vol. 30, pp. 865–889.

Welay, A. and Opalinski, I. (1983), "Electric Charge Neutralization by Addition of Fines to a Fluidized Bed Composed of Coarse Dielectric Particles," *Journal of Electrostatics*, Vol. 14, pp. 279–289.
Withington, H.P. (1914), "Cleaning Anthracite by the Huff Electrostatic Process," *Coal Age*, Vol. 5, No. 8, pp. 452–153.
Yamamoto, S. et al. (1974), "Work Functions of Binary Compounds," *Journal of Chemical Physics*, Vol. 60, p. 10.
Zhou, G. and Brown, J.D. (1988), "Coal Surface Conditioning for Electrostatic Separation," *Canadian Journal of Chemical Engineering*, Vol. 66, pp. 858–863.

13 Combination Techniques

13.1 ADVANTAGES OF COMBINED PROCESSES FOR DESULFURIZATION

It has often been found that a combination of two different processes gives better desulfurization results than multiple stages of a single type of desulfurization process. This is because different types of separators will be most effective for removing particular kinds of sulfur-bearing particles. For example, if a pyrite particle has a thin coating of coal on its surface, it will behave as a coal particle in column flotation and will report to the clean coal product. The same particle will be removed by a density-based separation because its high pyrite content will cause it to have a high density. Conversely, a very fine particle of pure pyrite is likely to be too fine to be removed by a density-based separator but can be rejected by, for example, froth flotation. It should also be remembered that fine pyrite particles without coal inclusions can also become hydrophobic upon oxidation.

Column flotation is now in commercial use for removing ash-forming minerals from fine coal and has been studied for pyrite removal as well. According to release analysis tests that have been carried out on many high-sulfur coals, the maximum possible sulfur removal by column flotation is substantially less than the sulfur removal predicted from float/sink washability tests. It is generally accepted that release analysis represents the best possible flotation separation, while float/sink washability results are the best possible using density-based separations.

There are many possible combinations of desulfurization processes that would perform better than either process alone. The most basic approach is to divide the feed coal stream into size fractions, and then to process each stream by the most efficient process for that size. For example, Pittston Coal Management Company (USA) has constructed the Pittston Moss 3 preparation plant, which has a capacity of 800 tons/h and replaces a previous 35-year-old facility. This plant incorporates a heavy-media cyclone/spiral/froth flotation circuit, which produces a product that is 2% lower in ash than was produced by the old heavy-media bath/Deister table/froth flotation plant, without a loss in yield. Adjacent to this plant is the Middle Fork Pond Recovery operation, which replaced conventional flotation with Microcel column flotation to meet coal quality requirements. The Microcel columns are combined with improved desliming of the spiral product, which has reduced product ash and improved overall recovery (Stanley and Bethell, 1993).

While many plants utilize combinations of several different desulfurization processes, relatively few use these processes in series, with the same coal passing through two different types of separation. This is the type of combined operation that is most capable of producing a synergistic effect, where the product quality is higher than is possible with either process alone. Many of the possible process combinations have not yet been attempted. Others, such as the combination of selective comminution/classification with electrostatic separation, have been used in the laboratory but have not been used on a larger scale. There are many examples of such combinations, which are too numerous to be within the scope of this book. The combinations discussed will be those that are currently used in industrial practice or those that have a significant potential for commercialization.

13.2 FLOTATION AND GRAVITY

13.2.1 Gravity Separator Followed by Flotation

Several examples of gravity separators followed by froth flotation have been described in the literature. A selection of case studies is presented below.

13.2.1.1 Case Study No. 1: Heavy-Media/Water-Only Cyclone, Followed by Column Flotation with Re-grind

A circuit was set up at the Ohio Coal Testing and Development center (OCTAD) as shown in Figure 13.1. This circuit used heavy-media cyclones and water-only cyclones to preclean the feed to an advanced column flotation cell. The circuit was operated at a feed rate of 2–3 tons/h and was used to process coals from the Pittsburgh No. 8, Upper Freeport, and Illinois No. 6 seams (ICF Kaiser, 1993, 1994; Ferris and Bencho, 1995).

The flowsheet was configured so that the raw coal was crushed to pass 630 μm and then screened on a deslime sieve at 300 μm. The material coarser than 300 μm was fed to a 31.8 cm diameter heavy-media cyclone. The heavy-media refuse was dewatered and disposed of, while the heavy-media cleaned coal was crushed to pass 300 μm. This recrushed coal was combined with the minus 300 μm material from the deslime

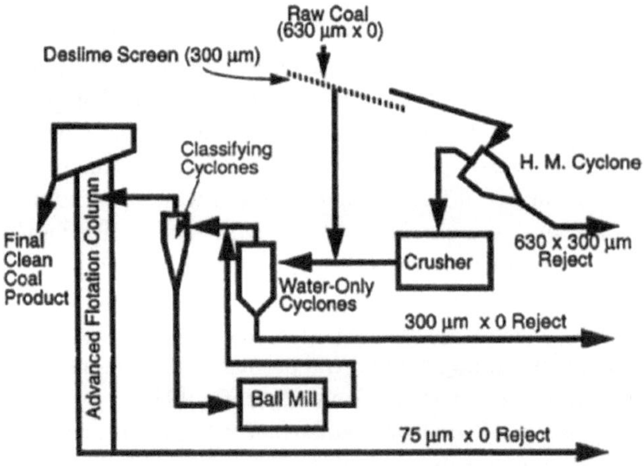

FIGURE 13.1 Flow diagram for the combined gravity/column flotation circuit tested at the OCTAD facility.

screen and pumped to two 15.24 cm diameter water-only cyclones. The water-only cyclone refuse was sent to the refuse thickener, while the clean coal overflow was directed to classifying cyclones. The classifying cyclone underflow (300 × 75 μm) was re-ground in a 1.83 m diameter ball mill, which was operated in closed circuit with the classifying cyclones. The classifying cyclone overflow, consisting of coal that had been precleaned, crushed, and ground to pass 75 μm, was used as feed to a 1.83 m diameter advanced flotation cell. The column produced a final clean coal product that was thickened and dewatered using a solid-bowl centrifuge.

The performance of this circuit in processing a Pittsburgh No. 8 seam coal is given in Table 13.1, which shows the distribution of total weight, heating value, and total pyrite among the reject products. Overall, the circuit showed a 48% yield at 89.5% heating value recovery, while removing 73.6% of the pyrite. A summary of the results obtained for all three coal seams processed is given in Table 13.2. These results indicate that for all three coals, it was possible to reject over 73% of the pyritic sulfur while recovering over 85% of the total heating value. This separation is largely due to the ability of the flotation column to reject the fine pyrite that was liberated during the crushing and grinding operations. This pyrite could not have been removed by the heavy-media or water-only cyclones because it was too fine for these units to process.

TABLE 13.1
Summary of the Results of Processing a Pittsburgh No. 8 Seam Coal, Using the OCTAD Process Flowsheet Shown in Figure 13.1

Circuit Product	% wt	% Ash	% Pyritic Sulfur	% of Feed Heating Value	% of Feed Pyrite Content
Raw feed	100.0	46.05	2.04	100.0	–
1/4 in × 48 mesh reject	31.1	—	—	3.8	44.3
48 mesh × 0 reject	4.5	—	—	4.0	16.6
200 mesh × 0 reject	16.2	—	—	2.7	12.7
Total refuse	51.8	81.77	2.90	10.5	73.6
Final clean coal	48.2	7.65	1.12	89.5	26.4

TABLE 13.2
Summary of the Results of 24-h Demonstration Runs of the OCTAD Circuit Shown in Figure 13.1, for Coal from Three Different Seams

Coal Seam Processed	% Heating Value Recovery	% Pyrite Removal
Pittsburgh No. 8 seam	89.5	73.6
Upper Freeport seam	87.5	76.2
Illinois No. 6 seam	85.8	79.4

13.2.1.2 Case Study No. 2: Column Flotation Followed by Multi-gravity Separator

The Mozley multi-gravity separator (MGS) is adversely affected by large quantities of ultrafine minerals (such as clays) in the feed, so pilot-scale studies were carried out using a Microcel flotation column to preclean coal before using the MGS (Venkatraman et al., 1995). When the MGS was used alone to treat Pittsburgh No. 8 seam coal, it could remove 70% of the pyritic sulfur but only 40% of the ash at 95% British Thermal Unit (BTU) recovery. When the flotation column was used to pretreat the MGS feed, the overall removals of both ash and pyritic sulfur were 75%–85%, at heating value recoveries of 85%–90%.

13.2.1.3 Case Study No. 3: In-Plant Studies of Floatex Hydrosizer/Falcon Concentrator/Jameson Cell

Honaker et al. (1995, 1996a,b) studied the performance of the Floatex Hydrosizer followed by a Falcon Concentrator and a Jameson Flotation Cell. This circuit was installed at Kerr-McGee's Galatia preparation plant, and the performance was compared with the existing circuit that consisted of spirals and a flotation bank. For a low-sulfur Illinois No. 5 coal, the Floatex/Falcon/Jameson circuit reduced the pyritic sulfur content from 0.67% to 0.34%, at 93.2% combustibles recovery. The ash content was simultaneously decreased from 27.6% to 5.84%. Use of this circuit to treat a high-sulfur Illinois No. 5 coal resulted in rejection of 72.7% of the pyritic sulfur and 82.3% of the ash, while recovering 81% of the combustibles. Subsequently, the cleaned product was pulverized and re-treated in the Falcon concentrator, followed by flotation in a packed column, which reduced the pyritic sulfur content of the cleaned coal from 2.43% to 0.30%. At the same time, the pyritic sulfur and ash rejections increased to 89% and 93%, respectively. The entire process reduced the sulfur dioxide emissions from 2.67 g SO_2/MJ to only 0.75 g SO_2/MJ, which is Phase 1 compliance coal. Comparison of the Floatex/Falcon/Jameson circuit with the existing circuit showed a 10%–20% improvement in the concentrate yield while improving ash and pyritic sulfur rejection.

13.2.2 Air-Sparged Hydrocyclone (ASH)

The air-sparged hydrocyclone (ASH) is an attempt to combine the hydrocyclone with froth flotation. It can therefore be considered as a combination of density separation and flotation separation being carried out simultaneously (Miller and Van Camp, 1981). A schematic of an ASH is shown in Figure 13.2.

In ASH flotation, the slurry is fed into the device through a standard hydrocyclone header. The swirling slurry enters a porous tube where air bubbles are sparged into the slurry. The slurry is then discharged through an annular opening at the base of the cyclone. The feed rate and the diameter of the annular opening are adjusted to maintain a particular thickness of slurry on the cyclone wall (called the swirl-layer thickness). The air that is sparged into the cyclone is sheared into small bubbles at the cyclone wall, and the bubbles then migrate toward the center axis. As the bubbles travel, they attach to hydrophobic particles and carry them to the froth phase. The froth

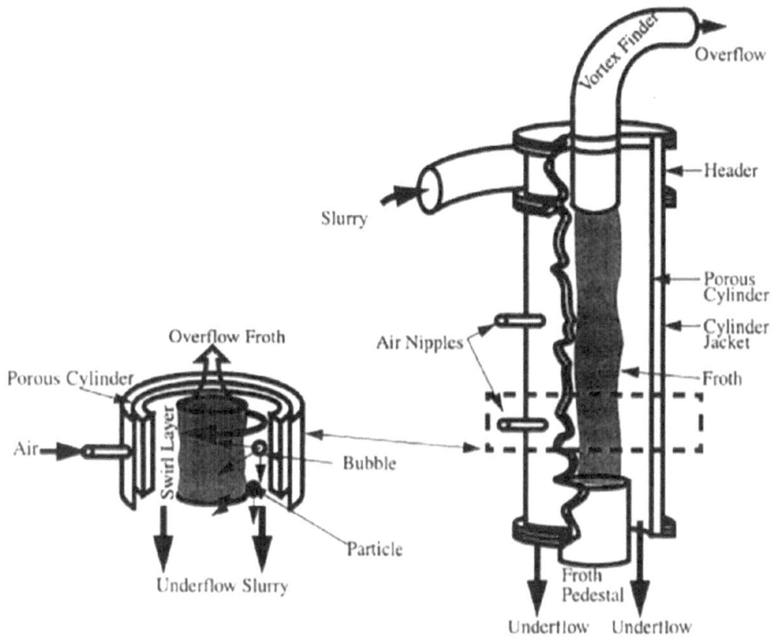

FIGURE 13.2 Cutaway diagram of an ASH.

phase forms on the axis of the cyclone and is supported by the froth pedestal, which forces the froth to discharge through the vortex finder. As in conventional flotation, the hydrophilic particles remain in suspension and are discharged with the slurry phase. The flow patterns that occur in an ASH are shown in Figure 13.3.

Conventional flotation separations become less effective as the particle size is reduced because of the lower probability of collision between particles and air bubbles at fine particle sizes. In the air-sparged cyclone, the application of a centrifugal force increases the probability that fine particles will interact with coal particles.

Miller and Kinneberg (1984) studied the effects of frother dosage on the bubble size in the air-sparged cyclone. As the level of methyl isobutyl carbinol (MIBC) frother was increased from 0 to 120 ppm, the bubble size remained essentially constant. Frother addition in the air-sparged cyclone is therefore required only to stabilize the froth phase.

The primary advantage of ASH is that it has a higher capacity per unit volume than conventional froth flotation machines, with up to 300 times the capacity of conventional flotation machines. The high capacity is a direct result of the short residence time of slurry in the cyclone, which is only a few seconds. The major drawback of air-sparged cyclones in coal flotation is that they require a considerably increased collector dosage compared to conventional cells, without a corresponding improvement in grade or recovery (Breed and Franzidis, 1995). Another considerable drawback of the design in Figure 13.2 is that the porous cylinder can become blocked quite readily. It is generally easier to produce bubbles externally to the cyclone, as is often done in column flotation.

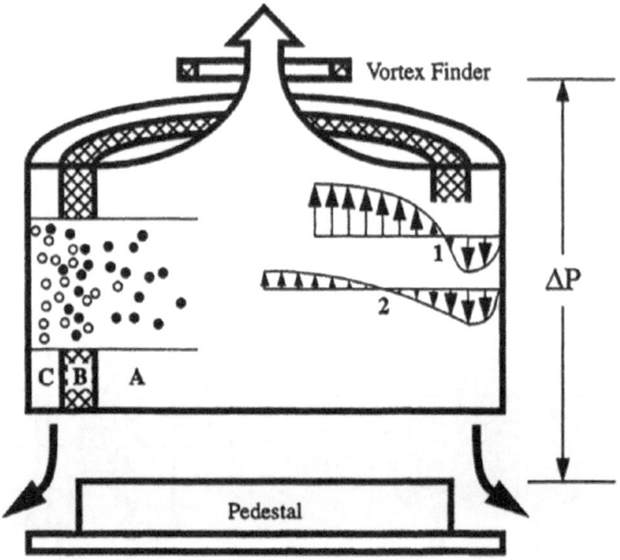

FIGURE 13.3 Flow profiles in an ASH. As the diameter of the underflow opening is decreased, the magnitude of the vertical flows in the cyclone is altered as shown.

13.2.2.1 Froth Structure

In ASH, the air bubbles and attached hydrophobic particles lose their rotational velocity and move radially to the center of the cylinder, forming the froth core. The size of the froth core and its motion are dependent on the cyclone operating conditions and feed characteristics. The froth core is constrained on the sides by the swirl layer and is prevented from dropping out of the bottom of the cyclone by the froth pedestal. The froth is therefore forced to move upwards, discharging through the vortex finder (Stoessner, 1990).

13.2.2.2 Particle Size Effects

Since it is a modified hydrocyclone, the particles move through ASH in a manner similar to their motion in a conventional hydrocyclone. The coarser and denser particles tend to report to the underflow, where they discharge through the annular opening. The fine particles and low-density particles tend to be removed through the vortex finder, along with the froth product (Stoessner, 1990).

13.2.2.3 Operating Variables

A number of variables that are important for conventional hydrocyclones also influence the performance of ASHs. These include cyclone length, vortex finder diameter, size of underflow discharge, slurry inlet pressure, and feed percent solids. The operation of the air-sparged cyclone is also affected by the variables that influence flotation, including collector and frother type and dosages, particle size, and water quality. In addition, there are variables that are specific to the air-sparged

TABLE 13.3
Pyrite Removal Results Reported for a 5.08 cm (2 in) Diameter ASH (Stoessner, 1990)

Feed Coal	Feed % Pyrite Sulfur	Feed Pressure, PSI	Air Flow Rate, slpm	Clean Coal % Pyrite Sulfur	Yield	% Pyrite Sulfur Removal	% Coal Recovery
Upper Freeport	0.31	5	150	0.21	74.6	49.2	83.2
seam	0.33	20	450	0.28	57.2	51.9	65.4
Pittsburgh No.	1.57	5	150	0.59	58.4	77.9	73.1
8 seam	1.84	20	450	0.52	35.5	90.1	46.7
Illinois No. 6	1.11	5	150	1.08	37.0	64.2	63.1
seam	1.94	20	450	1.60	34.0	71.4	56.0

cyclone, such as the pore diameter of the porous wall and the air flow rate. These numerous variables make the operation of the air-sparged cyclone complex, and its operation in a plant environment can be very difficult.

Stoessner (1990) has discussed the effects of these variables on the operation of 5.08 and 12.7 cm diameter air-sparged cyclones. Even in laboratory studies, the "fine" and "medium" grade porous tubes that sparge the air into the cyclones had to be replaced every 5–6 h of operation because of plugging. The "coarse" grade of porous tubes had a longer lifetime, in excess of 120 h.

13.2.2.4 Processing Results

A 5.08 cm diameter air-sparged cyclone was tested at the Electric Power Research Institute/Coal Quality Development Center (EPRI/CQDC) test facility, processing coals from the Upper Freeport, Pittsburgh No. 8, and Illinois No. 6 seams. Each coal was sized to pass 150 μm. Testing of the Pittsburgh No. 8 and Illinois No. 6 coals at this facility was not successful, with the failure attributed to water quality (Stoessner, 1990); however, successful flotation of these coals was reported by the University of Utah. The reported results are given in Table 13.3.

Tests have been carried out using larger-diameter ASHs (12.7 and 38.1 cm in diameter), but pyrite removal results for these units are not available.

13.3 FLOTATION AND AGGLOMERATION

Froth flotation and agglomeration are both dependent on particle hydrophobicity and can readily be used in combination. Typically, the coal is first ground to the pyrite liberation size and agglomerated with an appropriate agglomerating agent. The agglomerates are then recovered by froth flotation.

One combined agglomeration/flotation process, called aglofloat, has been developed by Ignasiak et al. (1991). In this process, run-of-mine coal was first crushed to pass 600 μm and suspended in a slurry. A mixture of diesel oil and crude oil was added to the slurry to act as a bridging liquid, and the slurry was conditioned in a high-shear mixer to produce microagglomerates about 212 μm in size. These agglomerates were then recovered in a conventional froth flotation cell. After flotation, the agglomerates were washed using a hydroseparator to remove ash and pyrite. The washed agglomerates were then transferred to a low-shear mixer where additional bridging oil was added and the agglomerates were enlarged. A schematic of the process is shown in Figure 13.4.

Some work has also been done using a two-stage process where the intermediate products are re-ground to promote liberation (Pawlak et al., 1990). Using the two-stage process, run-of-mine Illinois No. 6 coal (39.5% ash and 4.19% total sulfur) was cleaned to produce a product of 6.8% ash and 3.62% total sulfur, at 89% combustibles recovery. Unfortunately, the additional grinding stage tends to make the entire process unattractive due to the increased cost and operational complexity.

Chander et al. (1994) carried out agglomeration directly in a flotation cell, with the objective of improving the selectivity and recovery for small particles. The performance of froth flotation degrades at very fine sizes, due to overloading of the froth phase, excessive carryover of water, and poor contact efficiency between fine particles and air bubbles. Agglomeration can be used to increase the effective particle size of the coal, which improves the flotation performance. In their agglomeration/flotation experiments, Chander et al. used polyethylene oxide/polypropylene oxide (PEO/PPO) triblock copolymers in conjunction with insoluble oil collectors. When the insoluble oil was used alone, the agglomeration tended to be nonselective and ash reduction was poor. The addition of PEO/PPO block copolymers modified the coal surfaces to reduce nonselective agglomeration, and the polymers also acted as emulsifying agents for the insoluble oil. The result was the formation of smaller more hydrophobic agglomerates with a lower ash content than those produced without the polymers.

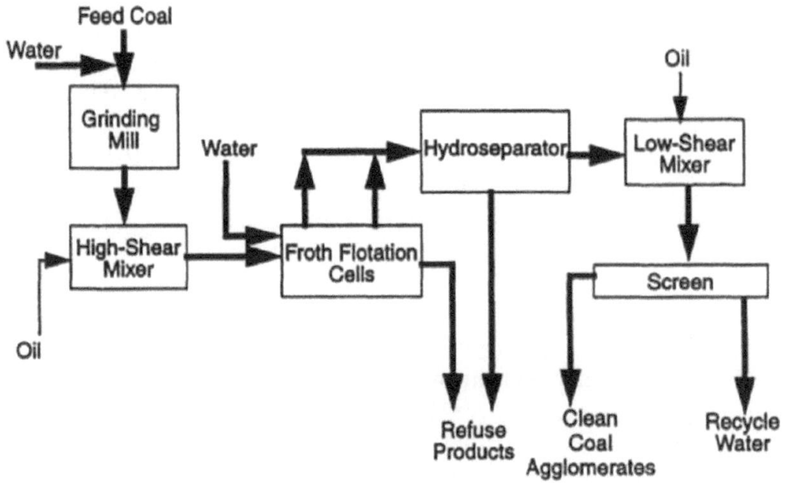

FIGURE 13.4 Schematic of the single-stage aglofloat process.

13.4 GRAVITY AND MAGNETIC SEPARATION

Gravity separation processes, such as heavy-media separation, can be used in combination with magnetic separation (Liu, 1982). The advantage of this process is that fine grinding to liberate the smallest pyrite inclusions can be delayed until after the gravity separation, which saves the cost of grinding large amounts of gangue minerals that can be rejected at a coarse size. The removal of the bulk of the mineral matter before it reaches the magnetic separator also decreases the demands on this separator, allowing it to operate more efficiently. One of the potential disadvantages of this technique is that the magnetite heavy media that is commonly used in heavy-media separators can be carried over into the magnetic separator, interfering with its performance.

High-gradient magnetic separation (HGMS) has been studied in conjunction with heavy-media separation, using a pulverized Pennsylvania coal from the Sewickley seam (Weschler et al., 1980). The coal was ground to 90% finer than 74 μm (200 mesh), and a portion of it was cleaned by heavy-media separation, using zinc chloride solution at a specific gravity of 1.6 as the dense medium. Using a magnetic field intensity of 17.1 kilo-oersteds and a slurry velocity of 18–20 mm/s, the combined process could produce a product with a total sulfur content of 1.03%, compared to 1.33% sulfur when the magnetic separator was used alone and 1.63% sulfur when the heavy-media separation was used alone.

13.5 FLOTATION AND MAGNETIC SEPARATION

Flotation can be readily used in conjunction with HGMS because both processes work well for fine coal particles (Liu, 1982). Flotation can remove the bulk of the impurities, including the clays with low magnetic susceptibilities, and the magnetic separator can then readily remove the residual pyrite.

Experiments were conducted by the US Department of Energy to compare the performance of a combination of flotation and HGMS with the performance of two-stage coal-pyrite flotation (Hucko, 1979). The magnetic field used was 2.16 T, and coals ground to pass 500 and 75 μm were studied. For the −500 μm coal, the flotation/HGMS combination could remove 76% of the pyritic sulfur at a clean product yield of 65%. When the coal was ground to pass 75 μm, 77% of the pyritic sulfur was removed at a product yield of 69%. The performance was approximately equal to that of two-stage coal-pyrite flotation at −500 μm and was superior to the coal-pyrite flotation at −75 μm.

Pilot-scale tests of combined froth flotation/HGMS have been carried out for pulverized Pittsburgh No. 8 and Illinois No. 6 coals (Freyberger et al., 1979). A rod-mill/ball-mill/cyclone circuit was used to grind the coals to 75% passing 45 jam, and they were then treated by rougher-cleaner froth flotation. The cleaner froth was then screened at 75 μm to remove coarse particles that could plug the magnetic separator. The objective was to demonstrate the performance of the process during continuous operation in 15-h operating shifts. With the Illinois No. 6 coal, pyrite removal was 73%–76% at a clean coal yield of 85%.

TABLE 13.4
Results from Processing a Micronized Lower Freeport Middling Coal Using Agglomeration Followed by Froth Flotation (Oder, 1991)

Product	% Combustibles Recovery	% Ash	% Total Sulfur	% Pyritic Sulfur
Feed coal	100.0	15.37	1.78	1.11
Agglomeration clean coal product	99.1	4.69	0.98	0.35
Agglomeration + magnetic separation clean coal product	76.5	1.65	0.76	0.18

13.6 AGGLOMERATION AND MAGNETIC SEPARATION

The combination of agglomeration and high-intensity magnetic separation has been studied on a bench scale. Agglomeration was used first to remove the bulk of the (largely nonmagnetic) mineral matter, and it was then followed by HGMS to remove the pyrite particles not fully removed by the agglomeration stage.

This approach has been studied by Oder (1991) in bench-scale experiments, using Pittsburgh and Lower Freeport seam coals wet-ground to mean particle sizes of 7–9 μm. The agglomerant used was perchloroethylene (PCE), added at dosages ranging from 10% to 150% of the coal weight. Agglomeration was carried out in a 14 speed Hamilton Beach blender, with the shear rate varied by adjusting the blender speed. Agglomerates were screened at 75 μm to separate them from the water and dispersed mineral matter. The agglomerated material was then treated in a high-intensity magnetic separator at magnetic field strengths up to 15 T (150,000 gauss).

Using the above process, a Lower Freeport middling coal was treated, producing the results given in Table 13.4. These results show that by using this combination of processes, it is possible to produce a superclean coal. The excellent ash and pyrite removal is made possible by the ability of the agglomeration stage to process very finely ground and well-liberated coal and of the magnetic stage to remove the remaining locked coal/pyrite particles.

13.7 SUMMARY

It has been shown that in many cases, treating a coal using two different processes in series provides better desulfurization results than can be achieved with either process alone. This is possible because the types of sulfur-bearing particle that are poorly removed by one type of separator can be readily removed by another. This is particularly true of combinations of processes based on surface effects (such as froth flotation or oil agglomeration) with processes based on volume effects (such as density separators or magnetic separators). The main drawback to such combinations is that it is necessary to pay for installing and operating two different separation processes instead of only a single one.

If all of the processes in a combination are used to treat slurries, such as combined gravity/flotation separations, then it is most reasonable for all of the separation to be done in the coal preparation plant. The equipment for these processes is available, and the coal in the plant is already in slurry form, making it straightforward to use them in series.

If a combination of a wet separation process with a dry separation process is used, such as froth flotation/electrostatic separation, it is more sensible to carry out the wet process at the preparation plant and the dry process at the power plant just prior to combustion. This will work well because the power plant will pulverize the dry coal before combustion, improving the liberation of the pyrite and ash, and converting the coal to a form that is ideal for electrostatic separation and other dry separation processes.

REFERENCES

Breed, A.W. and Franzidis, J.-P. (1995), "A Factorial Design Investigation of the Parameters Affecting the Performance of an Air-Sparged Hydrocyclone (ASH) Treating a South African Coal," *Coal Preparation*, Vol. 16, No. 1–2, pp. 81–102.

Chander, S., Polat, H. and Polat, M. (1994), "Mechanism of Coal Flotation by Insoluble Collectors in the Presence of PEO/PPO Triblock Co-Polymers," *Changing Scopes in Mineral Processing—6th International Mineral Processing Symposium* (Kemal, Arslan, Akar and Canbasoglu, eds.), Kusadasi, Turkey, pp. 461–467.

Ferris, D.D. and Bencho, J.R. (1995), "Engineering Development of Advanced Physical Fine Coal Cleaning Technologies," *Proceedings of the 11th Annual Coal Preparation, Utilization and Environmental Control Conference*, U.S. Department of Energy, Pittsburgh Energy Technology Center, Pittsburgh, PA, pp. 9–16.

Freyberger, W.L., Keck, J.W., Spottiswood, D.J., Solem, N.D. and Doane, V.L. (1979), "Cleaning of Eastern Bituminous Coals by Fine Grinding, Froth Flotation, and High Gradient Magnetic Separation," *Proceedings of the Symposium on Coal Cleaning to Achieve Energy and Environmental Goals* (Rodgers and Lemmon, eds.), U.S. Environmental Protection Agency, EPA-600/7-79-098a, Hollywood, FL, pp. 534–567.

Honaker, R.Q., Reed, S. and Ho, K. (1995), "A Fine Coal Circuitry Study Using Column Flotation and Gravity Separation," Final Technical Report, Illinois Clean Coal Institute, Project No. 94-1/1.1A-1P.

Honaker, R.Q., Reed, S., Mohanty, M.K. and Ho, K. (1996a), "In-Plant Testing of a Novel Coal Cleaning Circuit Using Advanced Technologies," Final Technical Report, Illinois Clean Coal Institute, Project No. 95-1/1.1 A-IP.

Honaker, R.Q., Mohanty, M.K., Wang, D. and Ho, K. (1996b), "Fine Coal Circuitry Using Advanced Physical Cleaning Processes," *Proceedings of the 13th International Pittsburgh Coal Conference* (S.H. Chiang, ed.), September 3–7, Pittsburgh, PA, pp. 1375–1380.

Hucko, R.E. (1979), "DOE Research in High Gradient Magnetic Separation Applied to Coal Beneficiation," *Industrial Applications of Magnetic Separation* (Liu, ed.), IEEE publication no. 78CH1447-2MAG, Institute of Electrical and Electronic Engineers, New York, pp. 62–63.

ICF Kaiser (1993), "Engineering Development of Advanced Physical Fine Coal Cleaning Technologies," 18th Quarterly Progress Report, U.S. Department of Energy, Pittsburgh Energy Technology Center, Project No. DE-AC22-88-PC88881.

ICF Kaiser (1994), "Engineering Development of Advanced Physical Fine Coal Cleaning Technologies," 25th Quarterly Progress Report, U.S. Department of Energy, Pittsburgh Energy Technology Center, Project No. DE-AC22-88-PC88881.

Ignasiak, B., Pawlak, W. and Szymocha, K. (1991), *Development of Clean Coal and Clean Soil Technologies Using Advanced Agglomeration Techniques*, Volume 2; Upgrading of Bituminous Coals by the Aglofloat Process, Alberta Research Council, Devon, AB.

Liu, Y.A. (1982), "High-Gradient Magnetic Separation for Coal Desulfurization," *Physical Cleaning of Coal* (Liu, ed.), Marcel Dekker, Inc., New York, Basel.

Miller, J.D. and Van Camp, M.C. (1981), "Fine Coal Flotation in a Centrifugal Force Field with an Air Sparged Hydrocyclone," *SME-AIME Fall Meeting*, Chicago, IL, Preprint No. 81-360.

Miller, J.D. and Kinneberg, D.J. (1984), "Fast Flotation in an Air-Sparged Hydro-cyclone," *Proceedings of MINTEK 50*, South Africa, Vol. 1, p. 353.

Oder, R.R. (1991), "Separation of Pyrites by Combined Agglomeration and High Field Magnetic Separation," *Processing and Utilization of High Sulfur Coals IV* (Dugan, Quigley, and Attia, eds.), Elsevier, Amsterdam, pp. 491–502.

Pawlak, W., Szymocha, K., Riker, Y. and Ignasiak, B. (1990), "High-Sulfur Coal Upgrading by Improved Oil Agglomeration," *Processing and Utilization of High-Sulfur Coals III* (Markuszewski and Wheelock, eds.), Elsevier, Amsterdam, pp. 279–288.

Stanley, F.L. and Bethell, P.J. (1993), "Process Modifications at Moss 3 Preparation Plant," *Proceedings of the 6th Australian Coal Preparation Conference* (Davis and MacKay, eds.), Mackay, Queensland, pp. 20–31, paper A2.

Stoessner, R.D. (1990), "Air-Sparged Hydrocyclone/Advanced Froth Flotation Fine Coal Cleaning," Final Report, U.S. Department of Energy, Pittsburgh Energy Technology Center, Report No. DOE/PC/88853-T2 (DE90016410).

Venkatraman, P., Luttrell, G.H. and Yoon, R.-H. (1995), "Fine Coal Cleaning Using the Multi-Gravity Separator," *High-Efficiency Coal Preparation* (Kawatra, ed.), Society for Mining, Metallurgy, and Exploration, Littleton, CO, pp. 109–117.

Weschler, I., Doulin, J. and Eddy, R. (1980), "Coal Preparation Using Magnetic Separation," EPRI Report No. CS1517, Vol. 3, Electric Power Research Institute, Palo Alto, CA.

14 Microbial Processing

There has been considerable interest in using microorganisms to remove sulfur from coal. Certain types of microorganisms are known to dissolve pyritic sulfur without the need for large quantities of expensive reagents, and there are organisms under study that can solubilize some coals, and others that may be able to remove some of the organic sulfur. The current limitations on microbial processing of coal are the time required for pyrite dissolution, the early stage of development of specialized strains for this purpose, and the engineering difficulties of scaling up microbial processes to handle the necessary tonnages at a practical rate. The solution to these problems will require a better understanding of how the microorganisms interact with coal and genetic engineering to improve their performance in industrial applications. While microbiological coal desulfurization has a great deal of promise, there is also a considerable amount of research to be done before such processes will be industrially useful.

14.1 MICROBIOLOGY BACKGROUND

Before discussing the applications of microorganisms to coal desulfurization, it is important to understand the basic terminology of microbiology.

14.1.1 Types of Microorganisms

The term "microorganism" is an extremely wide one and simply refers to the fact that these organisms are too small to be seen with the naked eye. A broad classification of microorganisms is given in Table 14.1. In coal desulfurization, the organisms that are of most interest are the bacteria and fungi, as it is these organisms that are most able to break down either mineral matter or complex organic material such as the molecules that comprise coal.

It has been known for many years that certain bacteria can metabolize inorganic matter as their sole source of energy (Colmer and Hinkle, 1947). Certain of these organisms are responsible for acid mine drainage from coal mines, by oxidizing pyrite to iron sulfate and sulfuric acid. Such organisms are referred to as *chemoautotrophs*, meaning that they get their metabolic energy from inorganic reactions (chemotrophy), and use carbon dioxide as their carbon source for growth (autotrophy). This is in contrast to *photoautotrophs* such as blue-green algae and plants (which also obtain carbon from carbon dioxide, but extract their energy from light), and *chemoheterotrophs* (which include most other organisms, and use organic compounds both as a source of energy and as a source of carbon). These fundamentals of microbiology are discussed in great detail in standard texts, such as Brock et al. (1984). The most common types of microorganism metabolisms are given in Table 14.2. The chemoautotrophs that oxidize pyrite are typically also *acidophilic*, meaning that they grow best in a highly acidic environment.

TABLE 14.1
Basic Types of Microorganisms (Brock et al., 1984)

Organism Type	Description and Significance
Bacteria	Cells small (generally on the order of 1–10 µm), relatively simple, cell shape maintained by a rigid cell wall, have no distinct nucleus or internal compartmentalization. Found in all environments that can support life. Vary greatly in their sources of energy and food, making them of great interest in coal desulfurization.
Algae	Cells complex, have distinct nucleus and internal structure, shape maintained by a rigid cell wall, use sunlight as an energy source and carbon dioxide as a carbon source for growth. Generally not of interest in coal desulfurization, except as a possible supplemental food source for other organisms that are useful for desulfurization.
Fungi	Cells complex, have distinct nucleus and internal structure, shape maintained by a rigid cell wall, decompose organic matter as food source. Mainly of interest for breaking down the coal structure into liquid or gaseous fuels.
Protozoa	Cells complex, have distinct nucleus and internal structure, no cell wall, obtain food by ingesting other organisms. Not useful for coal desulfurization, but some species have been reported to prey on desirable organisms in the culture tanks, greatly reducing their numbers and effectiveness.

TABLE 14.2
Basic Classifications of Microorganism Metabolisms

Metabolic Type	Energy Source	Carbon Source	Examples
Chemoautotrophs	Inorganic chemical reactions	Carbon dioxide	*T. ferrooxidans, Sulfolobus* species
Photoautotrophs	Sunlight	Carbon dioxide	Photosynthetic bacteria, algae
Chemoheterotrophs	Organic chemical reactions	Organic compounds	Most bacteria, all fungi, and protozoans
Photoheterotrophs	Sunlight	Organic compounds	Some species of purple and green sulfur bacteria

There are four basic ways that microorganisms can be used to desulfurize coal. These are

- Oxidize and dissolve the pyritic sulfur from the coal.
- Selectively attack the molecules that contain the organic sulfur in the coal.
- Break the coal down into liquids, gases, or water-soluble compounds to produce liquid or gaseous fuels, so that normal chemical processing methods can be used to remove sulfur from these fuels.
- Alter the properties of either the coal or the sulfur so that other sulfur-removal processes will be more effective.

These actions can either be done directly by the microorganisms (cellular material) or by chemicals that they produce, which are either naturally released into solution or are extracted from the organisms (acellular material). The acellular materials that are of most interest are *enzymes*, which is a general term for highly specific catalysts that the cells produce. If an organism secretes the appropriate enzyme for a particular desulfurization reaction, it can accelerate the rate of the reaction by several orders of magnitude and make a reaction practical that would not otherwise even be worth considering. Enzymes are a particular class of *proteins*, which are an extremely versatile class of compounds used by microorganisms (and indeed, all life) for a wide variety of chemical and structural purposes. In some cases, proteins other than enzymes can be useful in coal desulfurization.

To use microorganisms or their products effectively, it is first necessary to grow them in sufficient numbers. The most straightforward means of growing microorganisms is in a simple batch culture. This consists of a suitable vessel, filled with a solution that contains all of the necessary nutrients and trace elements that the organisms need for growth. This vessel is then *inoculated* by adding a small quantity of the organisms to be grown. A major consideration in any type of bioprocessing is that, unlike chemical reagents, the cells will reproduce rapidly when conditions are favorable but will die when they are not. It is therefore necessary to more carefully control the conditions in a bioreactor than in a reactor that uses only chemical reagents. In general, a bacterial population in a batch culture will behave as shown in Figure 14.1, assuming that the initial conditions are suitable for growth.

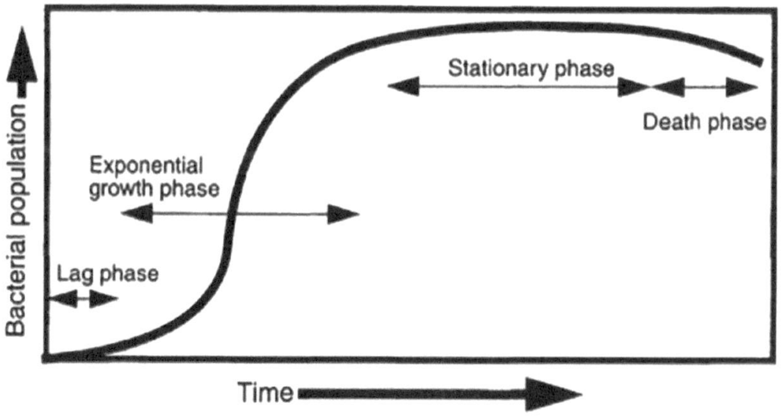

FIGURE 14.1 Growth curve for microorganisms in a batch culture. Initially growth is slow, as the organisms adapt to the prevailing conditions. Growth then becomes exponential as the bacteria divide and multiply, until nutrients become depleted. The cell numbers then go through a plateau where the reproduction rate equals the death rate. Ultimately, toxins build up or food supplies are depleted, and the cells then begin to die off. For the bacterium *T. ferrooxidans*, an organism that is commonly studied for use in coal desulfurization, this curve would typically extend over several weeks, and the peak bacterial population would be on the order of 10^9 cells/mL.

For a short period after inoculation, the organisms will not reproduce, as they must first adapt their metabolisms to the conditions in the culture. Once this "lag phase" is completed, the organisms enter the exponential growth phase, where they are actively growing and reproducing and very few are dying. The number of cells increases exponentially during this phase because, as the organisms reproduce, the number of cells able to reproduce further is steadily increasing. Ultimately, a point will be reached where some factors such as limitations of some trace element, accumulations of toxins in the growth media, or simple crowding, prevent the organisms from reproducing further. They then enter what is termed the stationary phase, where the number of living cells is no longer increasing. While some cells may still reproduce under these conditions, others are dying rapidly enough to compensate for this. Ultimately, conditions will reach the point where the death rate exceeds the reproduction rate. When this happens, the number of living cells in the culture falls off as it enters the death phase. The stationary phase can last for a long time, because many bacteria remain viable for long periods following entry into stationary phase or after starvation.

The activity of the cells is highest when they are in the exponential growth phase, and so normally it is desirable to maintain the cells in this condition as much as possible in a continuous operation. While it is not possible to maintain a batch culture in a particular phase indefinitely, this can be done with a continuous culture, where cells and used nutrient media are constantly drawn off to keep the cell numbers from becoming too high, while at the same time adding fresh nutrient media so that the remaining microorganisms can continue to grow. Since high cell numbers are also desirable, it is preferred that the continuous culture be maintained with high enough cell numbers that it is on the boundary between the exponential growth and stationary phases. Unfortunately, this is the point that is most difficult to maintain, because small changes in the absolute growth rate can easily tip the culture into the stationary phase. Particular care is therefore needed in control of such processes.

Once the organisms have grown, they can be either introduced directly to the coal to be processed or chemicals that they produce can be extracted from the cells and added to the coal. While it is possible in many cases for the culture vat to also serve as the desulfurization reactor, with coal added directly to the growing microorganisms, it is not necessary to do this. In some cases, growing the microorganisms separately from the desulfurization reactor will make the process easier to control. In these cases, the simplified control may make the use of extra equipment worthwhile.

14.1.2 Genetic Engineering

While bacteria found in nature can be quite effective for coal desulfurization, it would be helpful to be able to modify the genes of the organisms to improve their leaching performance, ability to tolerate toxins, growth rate, and other characteristics. This can be done to some extent by adaptation of a bacterial strain to grow well in particular conditions, but there is a limit to what can be done by adaptation, because it depends on natural, random mutations.

Genetic engineering provides a means for producing artificial mutations at any desired point in the microbial genome. The advantage is that the desired mutation can be made at any time, rather than waiting for them to occur naturally. This also allows the possibility of producing multiple simultaneous mutations, which has essentially zero probability of happening by chance. Genetic engineering also allows genetic material to be taken from other, unrelated species and used to modify the target organism in a way that would never occur in nature.

A typical genetic engineering strategy can be divided into the following ten basic steps (Holmes and Yates, 1990).

1. Isolate DNA from the organism of interest (in this case, *Thiobacillus ferrooxidans*).
2. Break down the DNA into gene-sized pieces, using restriction enzymes.
3. Isolate a single gene-size piece of interest. This step is quite complex, due to the large number of genes present in a given microorganism, but there are a number of molecular biology techniques that can be used, which are described in the standard texts of the field. Once the gene is isolated, it is incorporated into a small circle of DNA (a *plasmid)* that was taken from well-characterized bacteria such as *Escherichia coli.*
4. The plasmid containing the gene is then reinserted into *E. coli.* This organism is commonly used for this purpose because it is thoroughly characterized, and techniques for handling, manipulating, and cultivating it are very well-developed.
5. The modified *E. coli* is cultivated, producing billions of bacteria from the original modified cell. Each bacteria carries a copy of the plasmid with the inserted gene, so billions of copies of the gene are produced as well.
6. Plasmids bearing the gene of interest are isolated from the host *E. coli,* providing enough material for geneticists to study and manipulate.
7. The gene of interest is sequenced to determine its structure. In the past, this was a laborious process, but in recent years it has become simpler and more rapid. Further advances are constantly being made that will continue to reduce the cost and difficulty of this step.
8. The region of the gene is identified where the desired mutation can be made. This region is then cut out using restriction enzymes and replaced with a chemically synthesized strand of DNA with the desired base sequence.
9. The mutated gene is copied by repeating steps 4–6.
10. Finally, the mutated gene is returned to the organism that it was originally taken from. Four requirements must be met for this step to be successful. First, the plasmid carrying the gene must enter the bacterial cell. Second, it must stably replicate inside the cell so that the daughter cells will receive copies of the plasmid. Third, the host cell must be able to interpret the gene properly. Fourth, it must be possible to select the modified cells from the unmodified cells. This is necessary because successful transformation is a rare event, and modified cells are always far less numerous than unmodified cells. A typical method for making the selection is to include a gene for resistance to a particular antibiotic in the plasmid. Then, when the bacterial culture is treated with the antibiotic, only those that successfully took up the plasmid will survive.

So far, *T. ferrooxidans* genes have been taken through this process as far as Step 6, with successful isolation of replicated plasmids from *E. coli*. Work is ongoing to sequence and mutate these genes, and reinsert functioning copies into *T. ferrooxidans*. Engineered strains of this and other sulfur-dissolving organisms are not yet available but are likely to be produced in the near future.

14.2 BACTERIAL LEACHING OF PYRITIC SULFUR FROM COAL

14.2.1 Pyrite Dissolution Reactions

Of the two primary classes of sulfur in coal, the pyritic sulfur is the most amenable to bacterial leaching and has been quite thoroughly studied. The ability of microorganisms to actually participate in the dissolution of pyrite was first proposed by Zarubina et al. (1959) and Silverman et al. (1963). The leaching of pyrite is based on the fact that pyrite can be oxidized in air, according to the reaction:

$$2FeS_2 + 7O_2 + 2H_2O \rightarrow 2FeSO_4 + 2H_2SO_4 \tag{14.1}$$

The resulting ferrous sulfate and sulfuric acid are highly soluble in water, where the original pyrite was almost completely insoluble. Removal of the sulfur from the coal is therefore straightforward once it has been oxidized. This reaction is extremely slow, however, and is not practical without some sort of catalyst. One such catalyst is ferric iron (Fe^{+3}), which will react with pyrite according to the equations

$$FeS_2 + 7Fe_2(SO_4)_3 + 8H_2O \rightarrow 15FeSO_4 + 8H_2SO_4 \tag{14.2}$$

$$FeS_2 + Fe_2(SO_4)_3 \rightarrow 3FeSO_4 + 2S \tag{14.3}$$

Depending on the conditions, the pyrite can be either completely oxidized to ferrous sulfate and sulfuric acid or only partially oxidized to ferrous sulfate and elemental sulfur. While these reactions proceed quite rapidly, the ferric iron must then be regenerated by oxidizing the ferrous sulfate. Otherwise, the reaction will stop when the Fe^{+3}/Fe^{+2} ratio becomes less than about 2:1 (Singer and Stumm, 1970). Another consideration is that ferric sulfate is much less soluble than ferrous sulfate and will not dissolve in water unless it is more acidic than pH=4. Since the ferric iron has to be in solution before it can reach the pyrite, the reaction will have to be carried out in acidic solution.

The rate-limiting step in pyrite oxidation is the reoxidation of ferrous iron to ferric iron. Since $FeSO_4$ is relatively stable to chemical oxidation in the acid solution necessary to keep $Fe_2(SO_4)_3$ dissolved, some sort of catalyst is needed to accelerate the reoxidation step. Otherwise, the reaction will be far too slow to be practical, with the iron requiring weeks to reoxidize. The primary action of chemoautotrophic bacteria in pyrite dissolution is that they catalyze the ferric iron regeneration reaction (Hoffmann et al., 1981):

$$4FeSO_4 + O_2 + 2H_2SO_4 \rightarrow 2Fe_2(SO_4)_3 + 2H_2 \tag{14.4}$$

When the proper chemoautotrophic bacteria are present, they can accelerate this reaction by as much as a factor of 10^6 (Singer and Stumm, 1970), so instead of requiring weeks, the regeneration can be carried out in a matter of hours, depending on the solution pH and the quantity of bacteria available to carry out the reoxidation. The benefit to the bacteria of catalyzing this reaction is that they can extract energy from it. They accomplish this by attaching the dissolved iron to molecules in their cell membranes and extracting energy from the transport of electrons and hydrogen across the membrane as the iron oxidation proceeds. Ultimately, the complete pyrite oxidation reaction when catalyzed by bacteria is

$$4FeS_2 + 15O_2 + 2H_2O \rightarrow 2Fe_2(SO_4)_3 + 2H_2SO_4 \quad (14.5)$$

Another manner in which some bacteria can oxidize pyrite is to attach to the surface and, by more complex mechanisms, directly oxidize the sulfur (Singer and Stumm, 1970). This tends to be much slower than the reaction with ferric iron but is important when the leaching process is just getting started and there is not yet a significant amount of iron in solution. It is also an important mechanism for dealing with the elemental sulfur that can be formed during the pyrite oxidation. Since elemental sulfur is not soluble in water and can also introduce other processing problems, it must be oxidized to sulfate and removed.

14.2.2 Pyrite-Oxidizing Organisms

Several different organisms are able to metabolize pyrite and sulfur compounds, some of which are listed in Table 14.3. Of these, only *T. ferrooxidans*, *Thiobacillus thiooxidans*, and *Sulfolobus acidocaldarius* have been much studied for coal desulfurization. *Thiobacillus thiooxidans* has not been conclusively shown to be capable of oxidizing pyrite independently, but it is very useful in mixed cultures with *T. ferrooxidans*, where it rapidly deals with the elemental sulfur that forms during oxidation by the dissolved iron. The other organisms have the potential advantage of higher leaching rates because they grow best at higher temperatures which make dissolution of pyrite more rapid (Huber et al., 1986). High-temperature organisms of many types are regularly isolated from geothermal sites, and many species are still being characterized (Kelly and Deming, 1988).

14.2.2.1 *Thiobacillus ferrooxidans*

This microorganism was among the first of the pyrite oxidizers to be discovered and isolated, and appears to draw its energy almost entirely from the oxidation of dissolved iron. It is extremely common, found in virtually all exposed deposits of pyrite that are in contact with water, and is the primary cause of acid mine drainage. It requires a pH of less than 4, and grows best at a pH of about 2 and a temperature of approximately 30°C. When this microorganism is used to leach pyrite from coal, it will generally require several weeks for complete leaching, as shown in Figure 14.3.

The rate of leaching by these bacteria is limited by various factors, such as their metabolic rate, the availability of oxygen, the availability of trace minerals

TABLE 14.3
Microorganisms That Are Capable of Oxidizing and Dissolving Pyritic Sulfur from Coal

Organism	Notes
Thiobacillus ferrooxidans	pH 1–4, 4°C–40°C (optimum 35°C) iron oxidizer
Thiobacillus thiooxidans	Sulfur oxidizer, best mixed with *T. ferrooxidans*, grows under the same conditions
Thiobacillus acidophilus	Sulfur oxidizer, grows under same conditions as *T. ferrooxidans*
Leptospirillum ferrooxidans	Iron oxidizer, elongated shape may allow better penetration into coal pores
Sulfolobus acidocaldarius	pH 0.5–6, 60°C–90°C (optimum 75°C)
Sulfolobus brierleyi	pH 1–5, 60°C–90°C (optimum 75°C)
Sulfolobus solfataricus	pH 1–5, 70°C–90°C (optimum 87°C)
Acidianus infernus	pH 1–5, 65°C–95°C (optimum 90°C)

and carbon dioxide, and the absolute numbers of bacteria in the suspension. It is difficult to grow cultures of *T. ferrooxidans* to numbers much more than 1×10^9 cells/mL, which places a cap on the maximum leaching rate with these microbes. When growing these organisms in the laboratory in the absence of pyrite, the nutrient media compositions given in Table 14.4 are frequently used.

In these nutrient media, the ferrous sulfate is provided as the energy source for the organisms, while the other components are trace nutrients necessary for growth. The pH of the solution is critical, since the organisms will die if the pH becomes higher than about 4. These media are highly selective for *T. ferrooxidans*, since

TABLE 14.4
Compositions of Common Nutrient Media for *T. ferrooxidans* Cultures

Compound	T&K	9K	Low-P
Ammonium sulfate, $(NH_4)_2SO_4$	0.4 g	3.0 g	0.4 g
Potassium chloride, KCl	—	0.1 g	—
Potassium phosphate, K_2HPO_4	0.4 g	0.5 g	0.1 g
Magnesium sulfate, $MgSO_4{\_}7H_2O$	0.4 g	0.5 g	0.6 g
Calcium nitrate, $Ca(NO_3)_2$	—	0.01 g	—
Ferrous sulfate, $FeSO_4{\_}7H_2O$	33.3 g	44.5 g	33.4 g
Distilled, sterile water	1,000g	1,000g	1,000g
Sulfuric acid to pH = 2.0			

The T&K media was formulated by Tuovinen and Kelly (1973), while the 9K media (where the 9K signifies that the medium contains 9,000 ppm dissolved iron) was formulated by Silverman and Lundgren (1959). The Low-P (low-phosphorus) medium (McCready et al., 1986) was formulated because some strains of *T. ferrooxidans* grow more slowly when excess phosphate is present.

other organisms will not be able to metabolize the iron, and are unlikely to survive under such acid conditions. As a result, contamination by other organisms is less likely to be a problem than is normally the case in microbiological processes. There has been a report, however, that there is at least one species of protozoan that preys on *T. ferrooxidans* and can be introduced into cultures with unsterilized water (McCready, 1990). These predators are capable of reducing *Thiobacillus* populations from 10^8 cells/mL to only 10^2 cells/mL in only a few hours, and are therefore capable of bringing pyrite dissolution activity to a halt very rapidly. These organisms are most likely to infest a culture when the pH rises to near 4, and the best defense against them is careful sterilization of all feed materials, and maintaining the pH at 2.5 or below.

When no other organisms are present, *T. ferrooxidans* has difficulty in initially metabolizing pyrite because it is not well-adapted for direct oxidation of sulfur and needs the dissolved iron to grow well. There is a related organism, *T. thiooxidans*, which grows under similar conditions and is primarily adapted for sulfur oxidation and not for iron oxidation. A mixed culture of these two organisms is therefore more effective for dissolving pyrite than either organism alone, since *T thiooxidans* can attack the sulfur in the pyrite and release the iron, and *T. ferrooxidans* can then oxidize the iron and further dissolve the pyrite indirectly.

14.2.2.2 *Sulfolobus acidocaldarius*

Bacteria of the *Sulfolobus* genus are not only acidophilic like *T. ferrooxidans* but also thermophilic, growing best in the temperature range of 50°C–80°C. These organisms are unusual in this respect, as most other organisms will be rapidly killed by such high temperatures. These bacteria are native to sulfurous hot springs worldwide (Brock, 1978). This ability to live at high temperature is their chief advantage, since at higher temperatures, the relevant chemical reactions run faster and so the times needed for leaching are reduced. As a result of this, *S. acidocaldarius* dissolves pyrite at about twice the rate of *T. ferrooxidans* (Dugan, 1986). Since the use of elevated temperatures requires a heat source, the use of *S. acidocaldarius* or other thermophilic organisms to leach coal will be most practical at the power plant, where waste heat from the boilers is available to heat the process water. Otherwise, the fuel costs for heating the slurry could be prohibitive.

In addition to faster leaching, some investigators have also reported that *S. acidocaldarius* can dissolve some of the organic sulfur as well as the pyrite. This would be a significant benefit if it were true, since organic sulfur is much more troublesome to remove than pyritic sulfur. This has been claimed based on the fact that *S. acidocaldarius* can metabolize elemental sulfur and that when the ASTM International analysis method is used to determine organic sulfur, any elemental sulfur present will be recorded as organic sulfur. These organisms do seem to reduce the ASTM organic sulfur in some coals by a small amount (Kargi and Robinson, 1986). There is considerable doubt, however, whether the organisms are actually having any effect on the organic sulfur, because the magnitude of the organic sulfur reduction is comparable to the magnitude of the uncertainty in the organic sulfur determination. In any case, the effect is small enough that it is unlikely to be industrially significant.

14.2.2.3 Other Iron-Oxidizing Bacteria

Other organisms that can oxidize dissolved iron have also been discovered, such as *Leptospirillum ferrooxidans*. These organisms will therefore also be able to indirectly oxidize pyrite (Norris et al., 1986). However, they are likely to lack the capability of *T. ferrooxidans* to directly oxidize pyrite in the absence of dissolved iron, and so will need some initial level of dissolved iron to start leaching. This need for dissolved iron to already be present makes their cultures difficult to establish from scratch in many cases. This disadvantage may be partially countered by a more rapid rate of metabolizing iron for some species.

14.2.3 Factors Influencing Pyrite Removal

Once living organisms become involved in a process, the number of variables that can strongly affect that process becomes very large. This is a result of both the great sensitivity of life to small changes in conditions and to differences between various species and strains of organisms. The most important factors that affect pyrite dissolution are shown in Figure 14.2 (Olson and Brinckmann, 1986). Because there are so many possible factors that affect microbial pyrite dissolution, it is difficult to compare the results from various researchers.

A standard ASTM method (designation E1357–90) has been developed for comparing various strains of *T. ferrooxidans,* which should be used for evaluating bacterial strains whenever practical. The basic procedure is to suspend 1.0 g of pyrite in 50 mL of 1/10-strength iron-free 9K salts, in 250-mL sterilized conical flasks. Bacteria for the test are grown in 9K media, harvested by centrifugation, washed in 0.01 M sulfuric acid, and suspended at an appropriate concentration to provide an inoculum of 1.0×10^7 to 5.0×10^7 cells/mL in the test flasks. The cultures are then incubated at 28°C on a gyratory shaker at 200 rpm, with liquid samples removed

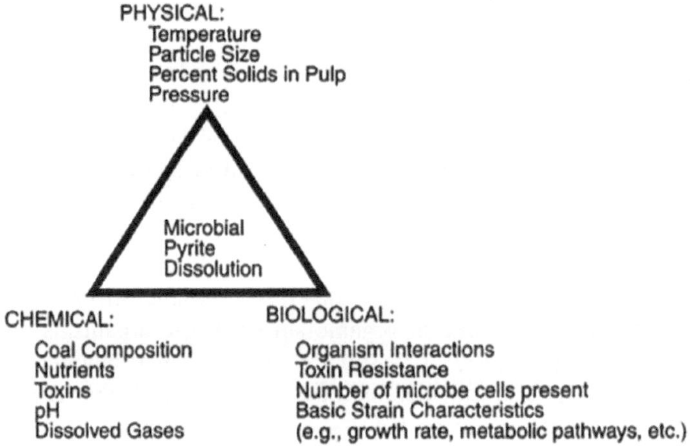

FIGURE 14.2 Factors that affect the rate and completeness of microbial dissolution of pyrite (Olson and Brinckmann, 1986).

periodically for determination of total soluble iron. The rate of leaching is determined from the slope of the soluble iron vs. time curve, with the slope of the linear part of the curve being taken as the active leaching rate for comparison with other cultures. In interlaboratory comparisons using this method, the linear portion of the curve occurred at times as short as 1–6 days and as long as 12–21 days (Olson, 1991). This illustrates that even under comparable conditions, the characteristics of different strains of the same organism can vary markedly.

14.2.3.1 Physical Factors

The physical factors mainly control the speed of the bacterial metabolism and the rate at which they can get at and react with the pyrite. Increasing the temperature increases the metabolic rate of the bacteria up to the point where the heat starts to coagulate the cell proteins, which kills the cells. This upper temperature limit will vary depending on the organism. Particle size controls the amount of pyrite surface that is exposed to bacterial action, with smaller particles increasing the maximum rate of dissolution and also allowing more complete dissolution. Increasing the percent solids increases the quantity of coal that can be treated per unit reactor volume, and increasing the pressure allows more oxygen and carbon dioxide to be dissolved in the water, both of which the bacteria need to dissolve pyrite.

14.2.3.2 Chemical Factors

The chemical factors determine whether a given bacterial strain will be able to survive at all, and what the maximum number of cells will be that can be maintained. Coal composition will determine whether all of the nutrients and trace minerals that the bacteria need are already present or must be added separately. The coal composition will also determine whether there are toxins, such as heavy metals, that will tend to impair the bacteria. The pH is also a critical factor, since if it leaves the tolerable range the bacteria will die. Since coals differ in their content of alkalies such as carbonates, the pH must be controlled carefully, and if the pH leaves the optimum range, the leaching kinetics will slow considerably (Bos, 1990; Bos et al., 1992). The composition of the gases dissolved in the solution can also have an effect. While oxygen is needed to oxidize the pyrite, carbon dioxide is also necessary so that the bacteria can use it as a carbon source while growing. Since the concentration of carbon dioxide in the atmosphere is quite low, time is required for CO_2 to diffuse to the cell to be used to produce cell mass. The use of supplemental carbon dioxide to increase the CO_2 concentration has been reported to improve the growth of *T. ferrooxidans* (Kargi, 1982a,b), and hence to increase the rate of pyrite leaching.

14.2.3.3 Biological Factors

The biological factors determine how effective a given species or strain of microorganism will be for desulfurization under a particular set of physical and chemical conditions. These factors are the most difficult to accurately determine and control, because there is no guarantee that they will remain the same indefinitely. As the organisms adapt to conditions or mutate, their tolerance for various factors or their nutritional requirements can change noticeably. Different strains of *T. ferrooxidans*, under apparently the same conditions, are known to leach pyrite at different rates as

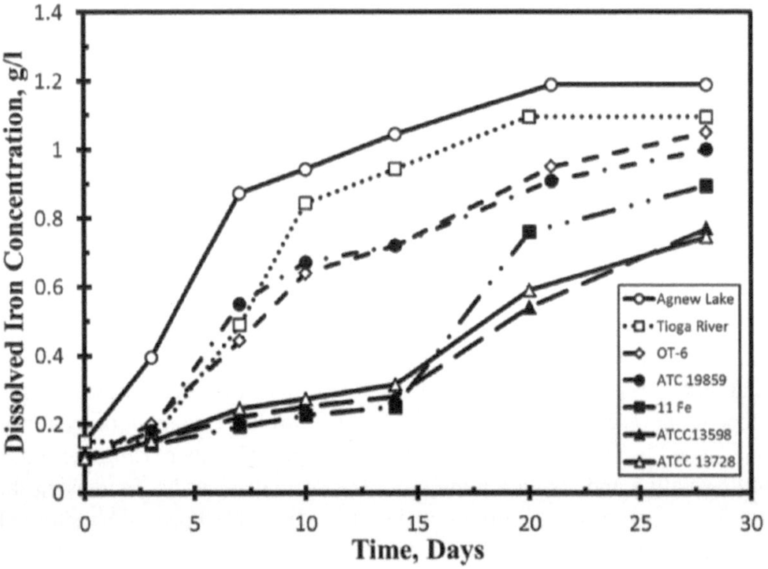

FIGURE 14.3 Variation in the pyrite-dissolution rate for several strains of *T. ferrooxidans*, all treating coal from the same source under the same conditions. The dissolved iron concentration is an indicator of pyrite dissolution, with higher concentrations indicating more dissolution.

shown in Figure 14.3 (Hoffmann et al., 1981), although the precise reasons for the differences are not well-understood (Olson and Kelly, 1990). In particular, there is considerable variation in the degree of heavy-metal tolerance, which has a profound and variable effect on bacterial growth. Since coals vary a good deal in their metal contents, bacteria that are optimum for one coal may not be suitable for another (Hoffmann et al., 1981). Table 14.5 shows the effects of some of the biological and physical parameters on the rate and extent of coal desulfurization.

14.3 ORGANIC SULFUR REMOVAL

Removal of organic sulfur by microbial leaching is in a fairly primitive state, as no organisms have yet been discovered that can dependably dissolve this form of sulfur from anything but a very narrow range of coals. Some of the microorganisms that have been investigated for this purpose are listed in Table 14.6.

Certain organisms have been found to metabolize some common, stable sulfur-containing organic compounds, such as dibenzothiophenes (DBTs). The organisms are not using these compounds as a source of energy, but rather are removing the sulfur as an essential nutrient. Compounds of this type account for a substantial fraction of the organic sulfur in coal, and so these organisms have some potential for removing much of this sulfur. Most of the work done to date has been conducted with pure model compounds such as DBT, to reduce the uncertainty involved in determining just how much sulfur the organisms were removing (Kilbane and Maka, 1989).

TABLE 14.5
Effects of Some Parameters on the Rate and Extent of Coal Desulfurization (Monticello and Finnerty, 1985)

Cultures	pH	Temp °C	Coal Source	Size, μm	Slurry % Solids	% Pyrite Removal	Time, Days
T. ferrooxidans (ATCC 19859)	2.5	25	Eastern US	43–74	≤20	90–98	8–12
T. ferrooxidans	2.0	35	Ohio	<37	25	77	N.R.
isolated from acid mine water	2.0	35	New Mexico	<37	25	83	N.R.
S. acidocaldarius (strain 98-3)	2.5	70	Pennsylvania	100–149	5	30	10
S. acidocaldarius	1.5	70	Pennsylvania	<48	20	90	N.R.
S. acidocaldarius and	1.5–2	28	Illinois No. 2	<74	20	90	14
Ferrolobus species	1.5–2	60	Illinois No. 2	<74	20	90	6
T. ferrooxidans and *T.*	2.5	25	West Virginia	<74	20	90	14
thiooxidans	2–2.5	~25	N.A.	74–300	20	97	5

It should be noted that there is no uniform method used by all investigators for handling, adapting, or monitoring these organisms, and so comparisons of results are difficult.
N.R., not recorded; ATCC, American Type Culture Collection, Rockville, MD; N.A., not available.

TABLE 14.6
Organisms Studied for Removing Organic Sulfur from Coal (Arctech, 1988; Kilbane and Maka, 1989; Ma and Li, 2010)

Organism Notes	Notes
Acinetobacter species	CB-2, mutant from soil bacteria. Apparently genetically unstable.
Pseudomonas species	CB-1, mutant from soil bacteria. Genetically unstable.
Sulfolobus acidocaldarius	Questionable effectiveness, difficult to distinguish organic S
Sulfolobus brierleyi	removal from pyrite removal or elemental S removal.
Methylomonas albus	Model compounds only, not coal.
IGTS4–7	Mixed cultures, soil organisms from various sources, used by the Institute of Gas Technology.
Rhodococcus species	Mechanism for removal of DBT and BT explained by Ma and Li (2010).

There is some evidence that the ability of *Pseudomonas* species to oxidize DBT is encoded in a plasmid, making it possible to exchange the necessary genetic information between organisms (Monticello et al., 1985).

To date, many organisms and mixed cultures have been isolated that have been claimed to remove at least some of the organic sulfur from coal, but results have been poorly reproducible (Kilbane and Maka, 1989). For example, the bacteria strain CB-1 was reported to be capable of oxidizing the sulfur from DBT without breaking the carbon ring structure, unlike other organisms, which tend to oxidize the entire molecule (Isbister and Kobylinski, 1985). In a few cases, these bacteria were reported to remove up to 47% of the organic sulfur from coal, but they do not seem to work very well on coals in general. Work with other organisms rarely showed organic sulfur removals much above the uncertainty in the pyritic sulfur assay, and even the organisms that were effective tended to be genetically unstable or require very long leaching times with very high cell concentrations (Kilbane and Maka, 1989).

In addition to oxidation of the organic sulfur by *aerobic* organisms (which require oxygen to survive), some work has been done to use *anaerobic* bacteria (which do not require oxygen) to metabolize sulfur. Some of these organisms are capable of metabolizing model sulfur compounds, but have not yet been successful in desulfurizing coal.

Organisms that remove organic sulfur are generally unable to remove pyrite from coal, because they cannot live at the low pH where pyrite leaching is carried out. Similarly, they cannot be used in mixed culture with pyrite-oxidizing bacteria. It is extremely unlikely that an organism will be discovered that can remove both organic sulfur and pyrite, and so producing such an organism will require some rather sophisticated genetic engineering to insert organic-sulfur-removal genes into pyrite-removing organisms such as *T. ferrooxidans.*

An important point in microbial processing is that a microorganism can only be depended on to do those things that it must do to survive. While a bacterium may be capable of removing organic sulfur from coal, it will not do this unless it derives some benefit from doing so, such as obtaining energy or a vital nutrient. If a mutant strain is developed that carries out an action that is industrially useful but of no benefit to the bacterium, then the mutation will probably not be stable, since the organism will be expending its energies to do something that it need not do to survive. Such organisms will be outcompeted by new mutant strains that do not carry out this action, and the original, desirable organisms will ultimately be replaced by the undesirable strain. Organisms that attack organic sulfur fall into this class, since there are both better sources of sulfur (dissolved sulfates from oxidized pyrite) and better sources of energy (the coal matrix itself) than the organic sulfur compounds. It is therefore extremely unlikely that a dependable organism will be either discovered or developed for leaching the organic sulfur from coal without some profound advances in genetic engineering.

Mineral bioprocessing research and development needs have been described by Hanumantha Rao et al. (2010). Some of the most promising and well-understood organisms remain a laboratory curiosity. Despite the desulfurization mechanism of *Rhodococcus* being thoroughly analyzed and understood, it has never been used in an industrial application (Ma and Li, 2010). However, this mechanism provides insight into the removal of organic sulfur by other means.

Rhodococcus promotes two major desulfurization mechanisms: one that removes DBT and the other that removes benzothiophene (BT). The DBT pathway, also known as the "4S" pathway, produces four distinct intermediate chemicals as it cleaves the C-S bonds and oxidizes the sulfur to a sulfate (Ma and Li, 2010). The intermediates of the DBT pathway are DBT-sulfoxide, DBT-sulfone, 2-(2-hydroxyphenyl) benzenesulfinate, and finally 2-hydroxybiphenyl (Ma and Li, 2010). The 2-hydroxybiphenyl product returns to the oil/coal phase, preserving the overall fuel value of the coal while removing sulfur (Ma and Li, 2010).

The BT pathway, on the other hand, is comparatively novel as few bacteria are known to desulfurize BT, which is common in gasoline (Ma and Li, 2010). The specific pathway *Rhodococcus* undergoes depends on species, but it is known that the first steps are the conversion of BT to BT-sulfoxide, BT-sulfone, and then 2-(2'-hydroxyphenyl)ethen 1-sulfinate (Ma and Li, 2010). The sulfinate may then be oxidized to remove the sulfate to form 2-(2'-hydroxyphenyl)ethan-1-al, or it may be desulfinated directly to form o-hydroxystyrene (Ma and Li, 2010).

It would be nice if there were a bacterium that could solubilize organic sulfur. In that case, it may be possible to extract the plasmid and add it to a bacteria that solubilizes inorganic sulfur as well. Such a bacteria could solubilize inorganic and organic sulfur, simplifying the desulfurization process immensely.

Initially, it is best to isolate a bacteria from a coal dump of the plant where the bacteria is to be used. Six or seven strains should be isolated, and the one which is most successful at surviving should form the basis of the bacterial desulfurization process. The most successful organism can be specifically isolated, multiplied, and applied at the full plant scale. Commercial strains are typically less suitable for any given plant.

Bacterial leaching has been successful for processing copper and gold, but has been mostly a laboratory curiosity for coal (Smith and Miettinen, 2006). Hopefully, new technology will be developed, which will make it more practical to remove otherwise difficult-to-remove sulfur from coal using bacteria. So far, it seems that bacterial technology will be required to remove the finest of organic sulfur compounds from coal, as other techniques have consistently been met with very little success. The removal of pyritic sulfur is comparatively quite easy using bacteria or techniques described earlier in this book (Patra and Natarajan, 2003).

14.4 COAL-SOLUBILIZING ORGANISMS

It has recently been discovered that some microorganisms, particularly fungi and bacteria that decompose wood, are capable of converting some coals into liquid- or water-soluble forms (Klein et al., 1991). Some of these are listed in Table 14.7. The greatest effect is seen for low-rank, highly oxidized coals, which are chemically most similar to wood. Research in this area is not yet advanced, and results on lignite and sub-bituminous coals are spotty and somewhat unpredictable. At present, higher-rank coals are impervious to microbial solubilization, although it may be possible to pretreat them to make them susceptible.

The products produced are generally water-soluble, highly aromatic, high-molecular-weight compounds that are as yet poorly characterized. While there is no direct use for these compounds as yet, their solubility in water raises the possibility that they can be used as chemical feedstocks. The relevance to coal desulfurization is that after solubilizing the coal, the sulfur can be removed from the other organic molecules by conventional chemical engineering techniques.

TABLE 14.7
Some Representative Coal-Solubilizing Organisms (Scott et al., 1986; Yen, 1988; Rayner and Boddy, 1988)

Organism	Notes
Pseudomonas species	Some bacterial species in this genus known to decompose hydrocarbons. Ambient temperature and near-neutral pH. Very widespread and tenacious, and resistant to a large number of chemicals
Corynebacterium species	Bacteria
Micrococcus species	Bacteria
Streptomyces species	Filamentary soil bacteria
Polyporus versicolor *Trametes versicolor* *Poria placenta* *Penicillium waksmanii* *Candida* species *Aspergillus* species *Sporothrix* species *Paecilomyces* species	Fungi that are responsible for white rot (selective delignification) in wood. The same enzymes that degrade lignin apparently also degrade low-rank coals. These fungi need highly oxidizing conditions and a supplemental food source to degrade lignin, which is not sufficient by itself. They grow at approximately neutral pH and ambient temperature

The microorganisms apparently need coals that contain a large number of oxygenated functional groups, which can be considered to give the organisms a "handle" to break up the molecular structure of the coal. Low-rank coals are higher in these groups than higher-rank coals, and so are easier for the organisms to attack. The coal can also sometimes be made more susceptible by treating it with concentrated alkali, which changes the coal structure enough for the microorganisms to be more active.

The organisms that were originally seen to decompose coal were fungi, which are quite complex organisms. While they can use complex reaction schemes to break down compounds that other organisms have difficulty with, they unfortunately also grow much slower than the much simpler bacteria. So, if possible, bacteria will be preferred for this type of processing simply to increase the rate of the process, although they will probably not decompose the coal as thoroughly.

14.5 MICROBIAL LEACHING REACTORS

Using bacterial leaching requires the following things:

- The leaching reactor must have enough residence time (typically several days) for the dissolution to be carried out.
- The suspension pH and temperature must be carefully maintained at the optimum level for microorganism growth.
- Supplemental nutrients, oxygen, and a carbon source must be provided for the organisms.
- Care must be taken to prevent contamination with other organisms, as these foreign organisms can outcompete the desired culture for nutrients to the

point where they can no longer survive. In some cases, the foreign organism may even actively prey on them. Such predation on *T. ferrooxidans* has been reported in the literature (McCready, 1990).

- The coal must be reduced to a fine particle size to maximize both the leaching rate and the ultimate removal of sulfur.
- The microorganisms must have sufficient time to grow to a useful population, and there should be sufficient backmixing in the reactor to prevent the organisms from being "washed out" by fresh feed. Alternatively, fresh microorganism can be added with the feed, but in this case, a sufficient number must be grown in an external fermentor to keep the process well-supplied.
- Toxins must be prevented from building up in the reactor sufficiently to harm the microorganisms.

For the organisms to be most effective in continuous leaching, they should be maintained in conditions of exponential growth, as they are then multiplying and metabolizing sulfur at their maximum rate (Murphy et al., 1985). The difficulty is that during exponential growth, the number of organisms in the reactor changes rapidly, and so a small change in an operating parameter can cause large changes not only in the activity of the bacteria but also of their numbers. This causes disproportionately large changes in the leaching rate from small changes in the reactor conditions. If the bacteria grow too rapidly, their numbers will outstrip the rate that a new feed is being introduced, and they will overpopulate the reactor, toxins will accumulate, and they will begin to die back. If they do not grow rapidly enough, then they will be flushed out of the reactor with the product faster than they can replace themselves. While preventing either of these results would be reasonably straightforward if both growth and activity were linear, the fact that they are not linear makes control extremely difficult.

Another difficulty is determining how many active cells are actually present in a reactor, which is a parameter that should be monitored for control purposes. The obvious method is simply to examine a sample of slurry under a microscope and count the number of bacteria in a known volume. This technique has two problems: first, it is not practical to distinguish living cells from dead cells by this method; and second, since a slurry is being treated, the suspended solid particles make it difficult to see all of the organisms, and may be mistaken for organisms themselves. A related problem is that bacteria tend to attach themselves to solid surfaces, such as the suspended particles, and are not counted. Similarly, many of the other common cell enumeration techniques (such as optical density measurement) are also unusable. Methods that provide a value for the number of live cells in the suspension are available, but these take several days (or in the case of *Thiobacillus*, weeks) to be completed, and so are not useful for control purposes. The reactors must therefore be controlled based entirely on the progress of leaching.

14.5.1 Types of Microbial Leaching Reactors

There are numerous reactor designs for coal leaching with bacteria, several of which are described below.

- *Stirred Tanks*: As shown in Figure 14.4, a stirred-tank reactor is simply a corrosion-resistant vessel with agitators to keep the solids in suspension. The agitation of the slurry will keep it aerated to some extent, but air spargers will improve the aeration a good deal. The advantage of this type of reactor is that it resists plugging and can be easily covered and heated for leaching with *Sulfolobus* or other thermophiles. In this case, the reactor should be located at the power plant where the coal will be burned, so that the waste heat from the boilers can be used to heat the reactor. The fact that these tanks do not require air to operate also makes them suitable for use with anaerobic bacteria, assuming that any such strains are ever developed for use in coal desulfurization. This type of reactor, however, is more expensive to install and operate than pachuca tanks, due to having moving parts in contact with the abrasive, usually corrosive slurry. These are most suitable for thermophilic organisms such as *Sulfolobus* species and for leaching with organic-sulfur-removing organisms that grow at near-neutral pH and are therefore subject to contamination.
- *Pachuca Tanks*: These consist of a tank with a riser tube, as shown in Figure 14.5. Air is bubbled into the bottom of the tank. As the bubbles rise, they carry slurry up the riser tube with them, producing a vigorous vertical mixing action. Since the slurry is in intimate contact with air bubbles in the riser, the aeration of the slurry is excellent. Since the tank has no moving

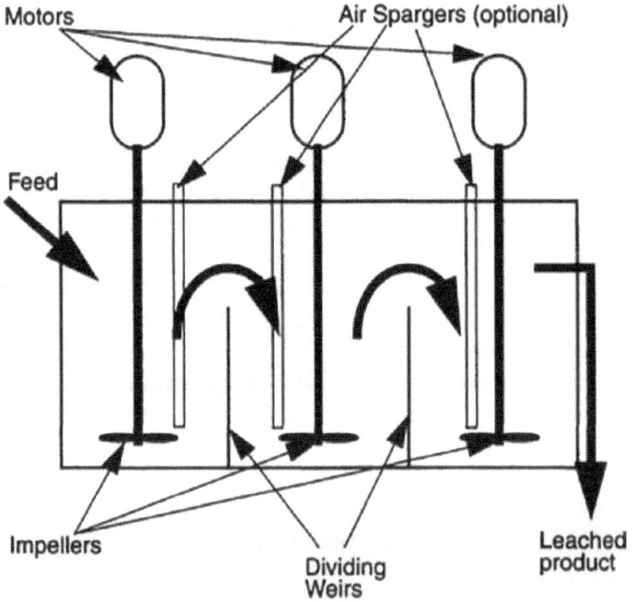

FIGURE 14.4 Stirred-tank reactor. The mechanical agitation makes it practical to seal and heat this type of reactor, making it most useful for thermophilic microorganisms. The weirs help prevent short-circuiting and backmixing, so that all of the solids remain in the reactor for approximately the same length of time.

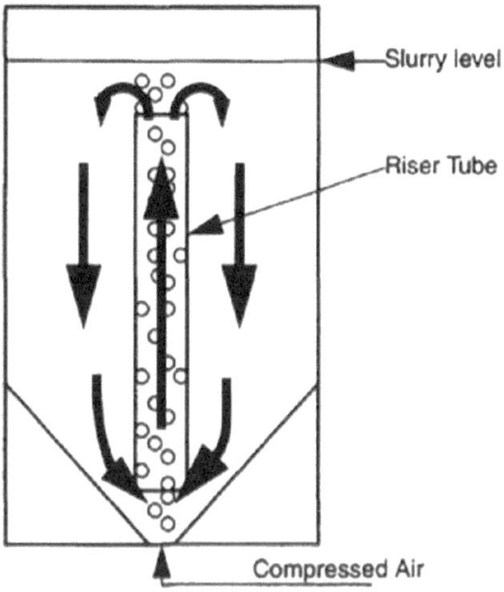

FIGURE 14.5 Schematic representation of a pachuca tank. The rising air bubbles carry slurry upward in the riser tube, which produces vigorous vertical mixing while thoroughly aerating the slurry. Since there are no moving parts in contact with the slurry, these tanks have less tendency to abrade particles to fines than many other reactor types do. Air can be injected through the base, as shown, or through a bubbler inserted from the top of the tank. The conical base prevents particles from settling out and accumulating in the corners.

parts, it can be made wear- and corrosion-resistant and has a lower tendency to abrade particles and make slimes. If the coal contains very much clay, these tanks are prone to gradual plugging, as the clay accumulates around the base of the riser. The need for high volumes of air flow makes them difficult to seal from the atmosphere, and so contamination by wild strains of microorganisms is difficult to prevent. Pachuca tanks can be used either for batch processing or in series for continuous processing. These are most suitable for *Thiobacillus* leaching, which grow at near-ambient temperatures. The excellent aeration in pachuca tanks can markedly improve the leaching rate of pyrite, compared to the rate in shake-flasks, as is shown by the results in Figure 14.6. In these experiments, coal from the Meigs Creek seam, OH, was used, with parallel experiments using both laboratory-scale pachuca tanks and mechanically shaken Erlenmeyer flasks as reactor vessels. These experiments were batch tests rather than continuous tests, to avoid the problems with controlling continuous cultures. The shake-flasks are comparable to a mechanically stirred reactor. In these experiments, it was found that the bacteria completed their lag phase and began rapidly leaching the pyrite approximately 3 days sooner in the pachuca tanks than in the shake-flasks and also show some increase in the initial leaching rate (Kawatra et al., 1987, 1989).

- *Rotating Drums*: These reactors consist of a very long, slowly rotating drum partly filled with slurry, as shown in Figure 14.7. As the drum rotates, it mixes and aerates the slurry. The major advantage is that such a reactor can

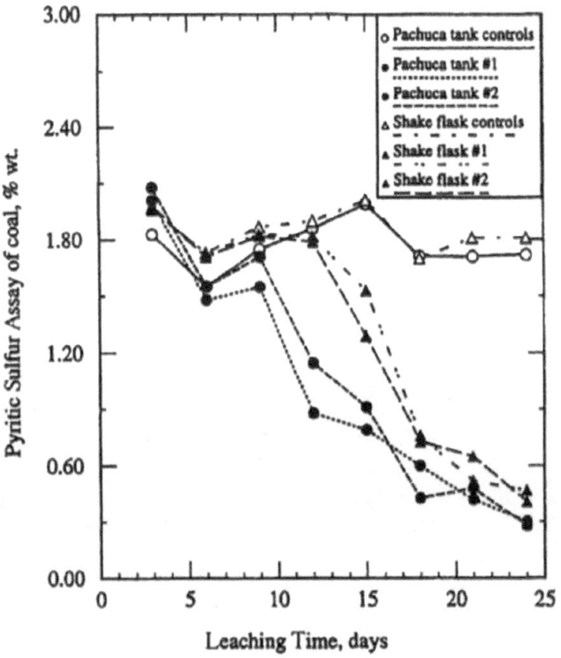

FIGURE 14.6 Leaching of coal with *T. ferrooxidans*, using both air-agitated reactors and continuously shaken flasks. The reaction initially starts slowly due to low numbers of bacteria and a lack of dissolved iron, but then accelerates as the bacteria numbers increase. The greater aeration of the air-agitated reactor also acts to increase the leaching rate over the shake-flasks. The coal was from the Meigs Creek seam; contained 2% pyritic sulfur, 4.8% total sulfur, and 18% ash; and had a gross heating value of 11,406 British Thermal Units per pound (BTU/lb). The coal was crushed and ground to 98% passing 150 μm. The pachuca tanks had a working volume of 1.25 L, and the shake-flasks had a working volume of 125 mL. A total of 24 shake-flasks and 24 pachuca tanks were used, 16 of each with growing bacteria and 8 of each poisoned with mercuric chloride to prevent bacterial growth and act as controls. (Kawatra et al., 1987, 1989).

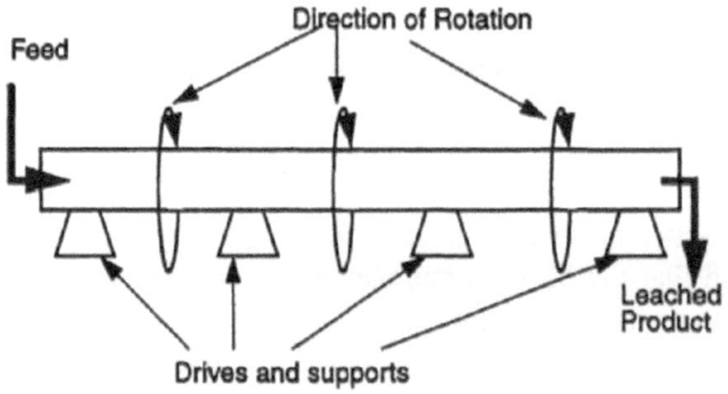

FIGURE 14.7 Rotating drum reactor. As slurry flows down the length of the drum, the drum slowly rotates to keep the slurry agitated, aerated, and suspended. The length:diameter ratio for such a reactor can be as great as 50:1.

operate with a higher percentage of solids in the slurry than either stirred tanks or pachuca tanks. Again, the capital costs are high due to the need for moving parts, and sparging of air into the drum is somewhat difficult. A drum reactor is well-suited for continuous processing, with all of the solids spending approximately the same amount of time in the reactor; however, the lack of backmixing makes it more prone to washout of bacteria than the other types. The best solution is to grow the bacteria in a separate fermentor and add them along with the feed. This type of reactor is moderately easy to seal against contamination, but it is not as easy to heat as a stirred-tank reactor. It is therefore most suitable for room-temperature processing.

- *Pipeline Leaching*: Since there is some interest in transporting coal from the mine to the consumer in slurry pipelines, it has been suggested that the coal be bacterially leached at the same time. The slurry would be inoculated with the appropriate bacteria at the start of the pipeline and then leached during the several days that it spent traveling down it. The coal would then be dewatered and burned at the far end. This would obviously require a very long pipeline to have enough retention time, so applications would be somewhat limited. Another difficulty is that the corrosiveness of a leaching slurry would cause pipeline maintenance problems.
- *Heap-Leaching*: If the coal is piled onto a watertight surface and the acid solution is trickled over it, as shown in Figure 14.8, then the pyrite is dissolved by the bacteria and carried away by the solution. This is a batch process, with a heap being made, leached as far as possible, and then removed and replaced. Very large quantities can be treated in a single heap, requiring only a large, flat area for the watertight impermeable pad. This has the advantage of being extremely cheap, since no energy needs to be expended to keep solids in suspension, there are no moving parts, and water pumps are much easier to operate than air compressors. Since the solution is typically sprayed over the surface, it is well-aerated. Unfortunately, there are a number of severe problems that make this method unsuitable as a sole method for

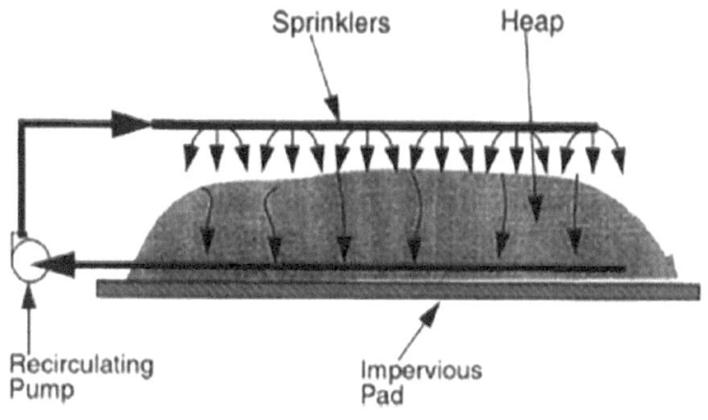

FIGURE 14.8 Heap-leach. The coarse solids rest on an impermeable pad, which collects the runoff for recycle. The particles must be coarse enough for the solution to percolate through easily and to keep from being washed away, which limits the proportion of the pyrite that can be removed. The sprinkler system provides excellent aeration for the solution.

desulfurizing coal. Most important is that the coal in the heap must be coarse enough to keep from being washed away, which means that the coal is too coarse to be leached effectively. The leaching solution also has a tendency to "channel" through the heap, so that some areas are not in contact with it and are poorly leached. In general, a heap-leach of coal would only remove about 60% of the pyrite, which is not sufficient. In coal desulfurization, a heap-leach would probably be most useful as a source of bacteria for other types of reactors or as a pretreatment stage before grinding the coal finely enough for a second stage of leaching. If microbial coal liquefaction becomes practical, heaps could be the best choice, because particle size limitations will be far less important if the entire coal matrix is being dissolved rather than just the sulfur-bearing parts. Since coal is prone to spontaneous combustion, and microbial action generates a great deal of heat, the heating of the center of a coal heap-leach will be a hazard that will have to be watched closely.

14.6 BACTERIALLY ASSISTED COAL CLEANING

The main reason bacteria are not yet used in industrial coal processing is that they are too slow, and the cost of the necessary reactor volume is therefore prohibitive. Other techniques for cleaning coal are much faster but have difficulty removing very fine pyrite; however, it is sometimes possible to use the bacteria for a relatively rapid pretreatment, which will improve the efficiency of subsequent processing (Capes et al., 1973; Kempton et al., 1980). The bacteria in this case are acting as surfactants (surface-active agents) and can be treated as if they were chemical reagents. This will of course require that the bacteria be cultured separately, so that they can be metered into the process as desired. Otherwise, the bacteria would not have time to multiply to sufficient numbers during the pretreatment stage. This technique is also limited to removing pyritic sulfur, as this is the only sulfur form that is amenable to physical separation.

While bacteria take a considerable amount of time to dissolve pyrite, they can attach to and alter the surface of pyrite very rapidly, in a matter of minutes (Ohmura et al., 1993). By changing the surface chemistry, the ability of some processes to reject pyrite can be improved. As discussed in Chapter 5, pyrite can, under certain circumstances, become hydrophobic due to formation of reduced sulfur species on its surface, and it is then recovered along with coal in surface-chemistry-based separations such as froth flotation, selective flocculation, and oil agglomeration.

14.6.1 Mechanisms of Pyrite Surface Alteration by Microorganisms

There are three different ways that microorganisms have been hypothesized to alter the pyrite surface. One is by oxidizing the pyrite surface, the second is by physically attaching to the pyrite, and the third is by coating it with released proteins or other organic compounds. The oxidation of the surface can only be accomplished by the chemoautotrophic bacteria like *T. ferrooxidans*, but physical coating can be done by any number of bacteria, yeasts, and other organisms.

14.6.1.1 Surface Oxidation

When chemoautotrophic bacteria oxidize pyrite, the oxidation obviously has to start at the mineral surface. In the course of oxidation, the pyrite does not go directly to iron sulfate, and the final iron sulfate will take some time to dissolve. There will therefore be some partially oxidized compounds on the pyrite surface, some of which are strongly hydrophilic. If the bacteria can be made to preferentially produce these hydrophilic compounds, then the surface chemistry of the pyrite will become less similar to coal; however, the oxidation can also produce compounds similar to elemental sulfur. These compounds are hydrophobic, and will cause pyrite to be recovered by froth flotation and oil agglomeration. So, forming these compounds is counterproductive and must be avoided.

In order for the oxidation to be helpful, the following conditions are needed:

- The suspension must be suitable for the organisms to be alive and active. For *T. ferrooxidans*, this means pH less than 4, about 30°C, and a considerable amount of Fe^{+3} in solution.
- The hydrophilic species should not dissolve immediately from the pyrite surface. Since this will tend to happen in acid solution, the pH should not be allowed to become too low. Otherwise, the pyrite will end up coated with elemental sulfur and will become strongly hydrophobic, and the entire purpose of pyrite surface oxidation will be lost.

These two requirements are obviously contradictory, so the conclusion is that under normal circumstances, bacterial oxidation of the pyrite surface will not make it more hydrophilic, and so will not be beneficial for surface-chemistry-based separations. It would be possible to simply use bacteria to produce dissolved iron in a separate reactor, then add the dissolved iron to the coal to be processed, and raise the pH enough for the iron to precipitate on the pyrite surface. This is, however, indistinguishable from simple chemical pyrite depression (Baker and Miller, 1971).

14.6.1.2 Bacterial Attachment

Many bacteria tend to attach themselves to solid surfaces whenever possible. This may be to obtain nutrients from the solid surface or simply to keep from being swept by flowing currents into an environment that would kill them. Since bacterial cell walls tend to be hydrophilic, the pyrite can be made hydrophilic as well if the bacteria selectively attach to it and cover it completely. *T. ferrooxidans*, *S. acidocaldarius*, and related organisms have a strong tendency to selectively attach to pyrite because it is their food source. Since they grow so slowly, it would be preferable to use other organisms, some of which can grow much faster. The appropriate pH conditions will have to be used since the bacteria have to be alive to attach themselves to surfaces. For *T. ferrooxidans*, this again means using an acidic solution, but most other organisms will need near-neutral pH. This has not yet been thoroughly studied for nonchemoautotrophic organisms, so the degree of selectivity between coal and pyrite is not well-known.

14.6.1.3 Coating by Proteins and Other Compounds

Microorganisms release a variety of organic compounds, enzymes, waste products, and other substances into solution. Some of these compounds are hydrophilic and selectively coat pyrite rather than coal. They therefore can give an effect very similar to direct attachment by the organisms. The main difference is that these compounds can attach to the pyrite regardless of whether the organisms that made them are alive or not. In fact, when microorganisms die, the cells *lyse*, or break open, and release their contents into the solution (Brock et al., 1984). This releases far more compounds than would be provided by living cells, and so coating is probably most effective when the cells are dead. This mechanism is the one that is probably most common and effective. It is similar to adding organic colloids as pyrite depressants.

14.6.2 Applications of Bacterial Surface Modification

There are no industrial operations that intentionally use bacteria to improve the removal of pyrite from coal, although it is possible that some may do it by accident if the coal is stockpiled for long periods. It is more likely in these cases that the bacteria will be generating elemental sulfur (Chou, 1990), and the surface modification will be a problem rather than a benefit. Work has been done to use bacteria as reagents in oil agglomeration (Capes et al., 1973; Kempton et al., 1980) and in froth flotation (Attia and Elzeky, 1985; Dogan et al., 1985). In both processes, the bacteria have been claimed to improve pyrite rejection. Work has also been done to use the bacteria to selectively coat and flocculate the coal, and so replace synthetic flocculants for selective flocculation (Misra et al., 1992).

14.6.2.1 Oil Agglomeration

Oil agglomeration is particularly prone to pick up pyrite along with the coal because the large quantities of oil used will tend to agglomerate all particles that are even slightly hydrophobic. *Thiobacillus ferrooxidans* cultures have been claimed to improve the pyrite rejection during oil agglomeration for some coals (Capes et al., 1973).

14.6.2.2 Froth Flotation

Bacterial modification is not as beneficial for froth flotation as for oil agglomeration, because the flotation process is more selective against pyrite.

Originally, work was done using *T. ferrooxidans* (Elzeky and Attia, 1987; Atkins et al., 1987; Townsley et al., 1987), but it was later found that other organisms were effective as well (Stainthorpe, 1989; Townsley and Atkins, 1986; Atkins, 1990; Eisele and Kawatra, 1994). In fact, under the conditions where pyrite is most likely to be recovered by froth flotation, the nonchemoautotrophic bacteria are more effective than *Thiobacillus*, although *T. ferrooxidans* was still superior to flotation with no organisms at all, as shown in Table 14.8.

Because organisms other than *T. ferrooxidans* are killed in acid solution, the depression by these organisms appears to be due to the release of the contents of dead cells and not to any activity of the living organisms. There are two reasons for the lower effectiveness of *T. ferrooxidans* strains. First, the fact that they were still

TABLE 14.8
Results of Flotation of the Pure Rico Pyrite with Various Microorganisms at a pH of 2

Organisms	Percentage of Pyrite Reporting to the Froth
Pseudomonas fluorescens	2.0 ± 0.25
Lactobacillus acidophilus	2.4 ± 1.40
Staphylococcus epidermis	2.3 ± 0.60
Klebsiella terrigena	4.4 ± 0.40
Saccharomyces cerevisiae	1.8 ± 0.02
Thiobacillus ferrooxidans	9.4 ± 1.80
T. ferrooxidans strain PA-1	39.9 ± 2.70
Distilled water controls	76.5 ± 2.50

The two *T. ferrooxidans* strains were obtained from EG&G, ID, and the other organisms were taken from stock cultures maintained by the Biological Sciences department at Michigan Technological University. Tests were run in triplicate and the results were averaged. The range of uncertainty is equal to the difference between the highest and lowest values for each set of three tests. The microorganism inoculum for each test was held constant at 3.2×10^9 cells/g of pyrite, at 5.88% solids. The microorganisms for each test were centrifuged, washed, and resuspended to prevent side effects from the growth medium.

alive kept their cell contents from being dispersed, and so each cell covered less area. Second, they were oxidizing the pyrite in an acid solution. This generated elemental sulfur on the pyrite surface, tending to counter the effect of the attached cells. Judging from these results, the effect of released cell products is a far more important effect than either live-cell attachment or surface oxidation.

The ability of microorganisms to depress the flotation of pure mineral pyrite does not necessarily mean that they will be able to actually improve pyrite removal from coal. For example, Table 14.9 shows flotation results for a coal that was treated with bacteria for extended periods. For a 2-h contact time, there was no significant improvement in the pyrite removal by adding the bacteria. While the pyrite removal did improve for the 23-h contact time, this was accompanied by a corresponding loss in the weight recovery, and so the improvement in the quality of the froth product was quite small.

The differences in response of mineral pyrite and coal pyrite to bacterial treatment are shown in Figures 14.9 and 14.10, and show that the response to both pH changes and microorganism additions is radically different for the two types of pyrite (Kawatra and Eisele, 1993; Eisele and Kawatra, 1994). The mineral pyrite is strongly floatable at acid pH, but its floatability drops to nearly zero at neutral or alkaline pH. When the pyrite is floatable, the microorganisms can depress it, but obviously they have little effect when the pyrite is not floatable. In contrast, the coal pyrite is most floatable at neutral pH, and the microorganisms have a much smaller effect on its floatability. Figure 14.10 also shows that while the yeast reduce the recovery of the pyrite, they also reduce the total weight recovery to the same extent, which indicates

TABLE 14.9

Flotation Results for Coal from the Meigs Creek Seam after Conditioning with *T. ferrooxidans* for Various Times in Pachuca Tanks at a pH of 2

Contact Time, h	Collector, kg/metric ton	% Weight Recovery	Froth % Ash	% Ash Removal	Froth % Pyritic S	% Pyritic S Removal
2 (control)	0.22	72.8	9.8	56.5	1.5	45.4
	0.43	78.8	10.0	52.1	1.4	44.8
	0.86	86.9	10.9	43.4	1.5	34.8
2	0.22	75.8	9.8	54.0	1.5	43.2
	0.43	83.6	10.2	45.6	1.5	37.3
	0.86	87.8	11.0	40.0	1.6	29.8
23	0.22	56.8	9.2	66.9	1.6	54.6
	0.43	76.4	9.4	54.5	1.4	46.5
	0.86	85.5	9.7	46.1	1.4	40.1

Flotation pH was between 5.5 and 6, and the coal contained 18% ash, 4.8% total sulfur, 2.0% pyritic sulfur, and 2.8% organic sulfur. No detectable amount of sulfur was dissolved by the bacteria in the course of the test. The "control" tests used no bacteria, but were otherwise run identical to the 2-h contact time tests. While the increase in pyritic sulfur removal after 23 h is large at the lowest collector dosage, the loss of coal recovery due to reduced floatability is equally large. Increasing collector dosage improves the recovery but reduces the pyritic sulfur removal. The froth % pyritic sulfur is only reduced by 0.1% sulfur by the bacteria after 23 h, which is a barely detectable improvement.

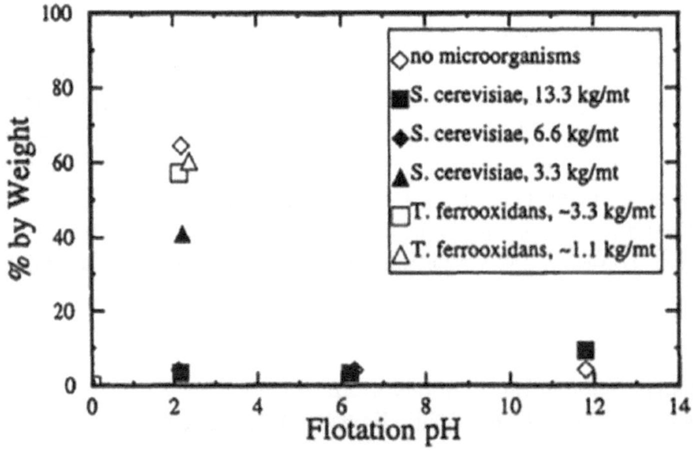

FIGURE 14.9 Flotation of 150 × 200 mesh mineral pyrite over a wide pH range, with varying inoculum sizes of both *T. ferrooxidans* and *Saccharomyces cerevisiae*. The *T. ferrooxidans* inoculum size was estimated from cell counts and is only approximate. A high concentration of microorganisms is required to depress flotation at a pH of 2. At higher pH, the floatability of this pyrite is lost and little effect is seen from microorganisms.

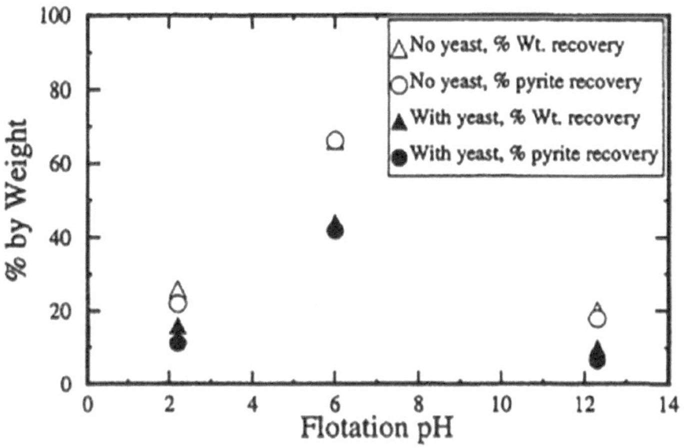

FIGURE 14.10 Flotation of 150 × 200 mesh Empire coal pyrite, with and without added yeast (*S. cerevisiae*). The yeast produces a general depression of all components and is not selective toward pyrite, as is shown by the close similarity of the values for total weight recovery and pyrite recovery.

that the associated coal is depressed as much as the coal pyrite. Based on this, the yeast appear to be a general flotation depressant and not a truly selective depressant for coal pyrite. While microbiological depression of pyrite flotation may be of some utility for some mineral applications, it does not appear to be a highly effective technique for desulfurizing coal.

It should be noted that the changes in flotation behavior of the coal pyrite with changing pH was very similar to the pH response that has been reported for coal (Zimmerman, 1979). The behavior of the coal pyrite is consistent with pyrite floating because the particles have surfaces that are at least partially made up of hydrophobic coal, and not because the pyrite itself is hydrophobic. If the actual pyrite surfaces were contributing to the hydrophobicity, then the flotation of pyrite should have been considerably enhanced at a pH of 2, as was seen for the mineral pyrite. The concept of using bacteria as a pyrite depressant for coal flotation is therefore flawed, as the coal pyrite is floating due to locking with coal particles, not because the pyrite itself is particularly hydrophobic. Bacterial pyrite depressants thus have the same inherent limitation as all of the other pyrite depressants that were described in Chapter 6 and are equally unlikely to find industrial application in the coal industry (Kawatra and Eisele, 1997).

14.6.2.3 Selective Flocculation

Selective flocculation using bacteria is a different proposition than either froth flotation or oil agglomeration. In this case, the bacteria are used to selectively attach to the coal, rather than to the pyrite (Misra et al., 1992; Raichur et al., 1995). The bacteria are an integral part of the process, as they are responsible for the flocculation of the coal particles. The organism used is *Mycobacterium phlei*, which is one of the few bacteria that is strongly hydrophobic. The hydrophobic nature of the cell causes it to selectively attach to multiple coal particles, binding them together into flocs that can be separated from pyrite and ash minerals by their higher settling rate.

Work in this area is in extremely early stages, and so it is difficult to say whether such processing will ever be practical on a large scale.

14.7 SUMMARY

Several organisms exist that are very effective for removing pyritic sulfur from coal. These organisms use the oxidation of pyrite to iron sulfates as their source of energy, in a solution that is acidic enough to dissolve the sulfates. Since the bacteria need to oxidize pyrite to live, they are very dependable. The main drawback of these organisms is that they are very slow-growing, and so removal of 80%–90% of the pyrite from coal requires about 1–3 weeks.

The various forms of organic sulfur in coal present a much greater problem, since there are no organisms that break down sulfur–carbon compounds as a vital part of their metabolism. Organisms have been isolated that can remove organic sulfur from coal, but only when they cannot obtain sulfur elsewhere. These organisms are typically not genetically stable. They tend to revert back to forms that do not attack sulfur compounds whenever sulfur is available from other sources.

Some microorganisms are capable of breaking down coal into water-soluble organic liquids. These organisms are typically those that decompose wood, and they are only effective on low-rank coals. Chemical pretreatment improves the effectiveness of this process. While there is no current use for the organic liquids produced, they may be useful at some point as chemical feedstocks.

Numerous reactor types are available that are suitable for bacterial leaching. The type of reactor selected depends on the oxygen and temperature needs of the organism and on the particle size of the coal. Pachuca tank and rotating drum reactors are well-suited for pyrite-oxidizing bacteria, with the rotating drum being superior for coarser particles and higher percent solids slurries. Stirred-tank reactors are superior when the reactor must be heated, or oxygen is not critical for the organism in use. Heap-leaching has a considerable advantage over the other types of reactors in its low capital and operating costs, but because the heap must use relatively coarse coal, the leaching is not complete. A heap-leach would therefore be most suitable as an adjunct to other leaching processes if it were used in coal desulfurization at all.

The use of various microorganisms to assist certain physical processes for removing pyrite has also been investigated. The goal of this is to modify the surface of pyrite so that it can be more easily removed by froth flotation, oil agglomeration, or selective flocculation. In general, the greatest effect is seen when the organisms have died and lysed, with the released proteins and other organic compounds acting as surfactants in the process.

While there has been a great deal of research done to use microorganisms in coal desulfurization, actual application of such methods are conspicuously absent in the coal industry. This lack of industrial application is due mainly to the long times required to produce results, as well as to the basic unpredictability of processes that use living organisms of any sort. Unless there are major breakthroughs in this area, such as could result from successful genetic engineering to greatly increase the growth rate and effectiveness of microorganisms, then this entire area is likely to remain a laboratory curiosity as far as the coal industry is concerned. To date, everything in coal processing that could be done by microorganisms can still be done more cheaply and reliably by other means.

The ability to remove organic sulfur from coal would be very beneficial to the coal processing industry. Sulfur removal from copper and gold with bioprocessing methods has already been successful, and we hope that there is similar success in the near future for coal.

REFERENCES

Arctech, Inc. (1988), "Microbially Mediated Removal of Organic Sulfur from Coal," DOE/PC/81207-T1 (DE89001572).

Atkins, A.S. (1990), "Developments in the Biological Suppression of Pyritie Sulfur in Coal Flotation," *Bioprocessing and Biotreatment of Coal* (D.L. Wise, ed.), Marcel Dekker, New York, pp. 507–548.

Atkins, A.S., Bridgewood, E.W., Davis, A.J. and Pooley, F.D. (1987), "A Study of the Suppression of Pyritic Sulfur in Coal Froth Flotation by *Thiobacillus ferrooxidans*," *Coal Preparation*, Vol. 5, pp. 1–13.

Attia, Y.A. and Elzeky, M.A. (1985), "Biosurface Modification in the Separation of Pyrite from Coal by Froth Flotation," *Coal Science and Technology*, Vol. 9, pp. 673–682.

Baker, A.F. and Miller, K.J. (1971), "Hydrolyzed Metal Ions as Pyrite Depressants in Coal Flotation: A Laboratory Study," Bureau of Mines Report of Investigations RI7518.

Bos, P. (1990), "Advantages and Disadvantages of Microbial Coal Desulfurization," *Proceedings of the 7th Annual International Pittsburgh Coal Conference*, Pittsburgh, PA, pp. 605–611.

Bos, P., Boogerd, F.C. and Kuenen, G.J. (1992), "Microbial Desulfurization of Coal," *Environmental Microbiology* (Mitchell, ed.), Wiley, New York, pp. 375–403.

Brock, T.D., Smith, D.W. and Maidgan, M.T. (1984), *Biology of Microorganisms*, 4th edition, Prentice-Hall, Englewood Cliffs, NJ.

Brock, T.D. (1978), *Thermophilic Organisms and Life at High Temperatures*, Springer-Verlag, New York.

Capes, C.E., Mcllhinney, A.E., Sirianni, A.F. and Puddington, L.E. (1973), "Bacterial Oxidation in Upgrading Pyritic Coals," *CIM Bulletin*, Vol. 66, No. 739, pp. 88–91.

Chou, C.L. (1990), "Geochemistry of Sulfur in Coal," *Geochemistry of Sulfur in Fossil Fuels*, ACS Symposium Series No. 429, American Chemical Society, Washington, DC.

Colmer, A.R. and Hinkle, M.E. (1947), "The Role of Microorganisms in Acid Mine Drainage: A Preliminary Report," *Science*, Vol. 106, pp. 253–256.

Dogan, M.Z., Ozbayglu, G., Hicyilmaz, C., Sarikaya, M. and Ozcengiz, G. (1985), "Bacterial Leaching versus Bacterial Conditioning and Flotation in Desulfurization of Coal," *15th Congrès International de Minéralurgie, Cannes, France*, Vol. 2, pp. 304–313.

Dugan, P.R. (1986), "Microbial Desulfurization of Coal and its Increased Monetary Value," *Biotechnology and Bioengineering Symposium No. 16*, John Wiley & Sons, Troy, NY, pp. 227–237.

Eisele, T.C. and Kawatra, S.K. (1994), "Use of Fungi and Bacteria for the Depression of Mineral and Coal Pyrite," *Reagents for Better Metallurgy* (Mulukutla, ed.), Chapter 11, Society for Mining, Metallurgy, and Exploration, Littleton, CO, pp. 91–100.

Elzeky, M. and Attia, Y.A. (1987), "Coal Slurry Desulfurization by Flotation Using Thiophilic Bacteria for Pyrite Depression," *Coal Preparation*, Vol. 5, pp. 15–37.

Hanumantha Rao, K., Vilinska, A. and Chernyshova, I.V. (2010), "Minerals bioprocessing: R & D needs in mineral biobeneficiation," *Hydrometallurgy*, Vol. 104, pp. 465–470.

Hoffmann, M.R., Faust, B.C., Panda, F.A., Koo, H.H. and Tsuchiya, H.M. (1981), "Kinetics of the Removal of Iron Pyrite from Coal by Microbial Catalysis," *Applied and Environmental Microbiology*, Vol. 42, pp. 259–271.

Holmes, D.S. and Yates, J.R. (1990), "Basic Principles of Genetic Manipulation of *Thiobacillus ferrooxidans* for Biohydrometallurgical Applications," *Microbial Mineral Recovery* (Ehrlich and Brierley, eds.), McGraw-Hill, New York, pp. 29–54.

Huber, G., Huber, H. and Stetter, K.O. (1986), "Isolation and Characterization of New Metal-Mobilizing Bacteria," *Biotechnology and Bioengineering Symposium No. 16*, Troy, NY, John Wiley & Sons, pp. 239–251.

Isbister, J.D. and Kobylinski, E.A. (1985), "Microbial Desulfurization of Coal," *Coal Science and Technology*, Vol. 9 (Processing and Utilization of High-Sulfur Coals), pp. 627–641.

Kargi, F. (1982a), "Enhancement of Microbial Removal of Pyritic Sulfur from Coal using Concentrated Cell Suspensions of *T. Ferrooxidans* and an External Carbon Dioxide Supply," *Biotechnology and Bioengineering*, Vol. 24, No. 3, pp. 749–52.

Kargi, F. (1982b), "Microbiological Coal Desulfurization," *Enzymatic and Microbial Technology*, Vol. 4, No. 1, pp. 13–19.

Kargi, F. and Robinson, J.M. (1986), "Removal of Organic Sulphur from Bituminous Coal: Use of the Thermophilic Organism Sulfolobus acidocaldarius," *Fuel*, Vol. 65, No. 3, pp. 397–399.

Kawatra, S.K. and Eisele, T.C. (1993), "Depression of Pyrite Flotation by Microorganisms as a Function of pH," *Processing and Utilization of High-Sulfur Coals V* (Parekh and Groppo, eds.), Elsevier, Amsterdam, pp. 139–148.

Kawatra, S.K. and Eisele, T.C. (1997), "Pyrite Recovery Mechanisms in Coal Flotation," *International Journal of Mineral Processing*, Vol. 50, No. 3, pp. 187–201.

Kawatra, S.K., Eisele, T.C. and Bagley, S. (1987), "Coal Desulfurization by Bacteria," *Mineral and Metallurgical Processing*, Vol. 4, No. 4, pp. 189–192.

Kawatra, S.K., Eisele, T.C. and Bagley, S.T. (1989), "Studies of Pyrite Dissolution in Pachuca Tanks and Depression of Pyrite Flotation by Bacteria," *Biotechnology in Minerals and Metal Processing* (Scheiner, Doyle and Kawatra, eds.), Society of Mining Engineers, Littleton, CO, pp. 55–62.

Kelly, R.M. and Deming, J.W. (1988), Extremely Thermophilic Archaebacteria: Biological and Engineering Considerations," *Biotechnology Progress*, Vol. 4, No. 2, pp. 47–62.

Kempton, A.G., Moneib, N., McCready, R.G.L. and Capes, C.E. (1980), "Removal of Pyrite from Coal by Conditioning with *Thiobacillus Ferrooxidans* followed by Oil Agglomeration," *Hydrometallurgy*, Vol. 5, pp. 117–125.

Kilbane, J. and Maka, A. (1989), "Microbial Removal of Organic Sulfur from Coal," Department of Energy Report No. DOE/PC/81210-T10 (DE89009049).

Klein, J., Van Afferden, M., Beyer, M., Pfeifer, F. and Schact, S. (1991), "Coal in Biotechnology," *Genetic Engineering and Biotechnology*, Vol. 11, No. 3, pp. 11–17.

Ma, T. and Li, S. (2010), "The Desulfurization Pathways in Pathways in *Rhodococcus*" *Biology of Rhodococcus*. Microbiology Monographs, (H. Alverz, eds.), Vol. 16, Springer, Berlin.

McCready, R.G.L. (1990), "Microbiological Studies on High-Sulfur Eastern Canadian Coals," *Bioprocessing and Biotreatment of Coal* (D.L. Wise, ed.), Marcel Dekker, New York, pp. 685–732.

McCready, R.G.L., Wadden, D. and Marchbank, A. (1986), "Nutrient Requirements for the in-place leaching of uranium by *Thiobacillus ferrooxidans*," *Hydrometallurgy*, Vol. 17, pp. 61–71.

Misra, M., Smith, R.W., Dubel, J. and Chen, S. (1992), "Selective Flocculation of Fine Coal with Hydrophobic *Mycobacterium Phlei*," *SME Annual Meeting*, Feb. 24–27, 1992, Phoenix, AZ, Preprint No. 92–87.

Monticello, D.J. and Finnerty, W.R. (1985), "Microbial Desulfurization of Fossil Fuels," *Annual Review of Microbiology*, Vol. 39, pp. 371–89.

Monticello, D.J., Bakker, D. and Finnerty, W.R. (1985), "Plasmid-Mediated Degradation of Dibenzothiophene by *Pseudomonas* Species," *Applied and Environmental Microbiology*, Vol. 49, No. 4, pp. 756–760.

Murphy, J., Riestenberg, E., Möhler, R., Marek, D., Beck, B. and Skidmore, D. (1985), "Coal Desulfurization by Microbial Processing," *Coal Science and Technology*, Vol. 9 (Processing and Utilization of High-sulfur Coals), pp. 643–652.

Norris, P.R., Parrott, L. and Marsh, R.M. (1986), "Moderately Thermophilic Mineral-Oxidizing Bacteria," *Biotechnology and Bioengineering Symposium, No. 16* (Biotechnology in Minerals and Metals-refining, and Fossil Fuel Process Industries), Troy, NY, pp. 253–262.

Ohmura, N., Kitamura, K. and Saiki, H. (1993), "Mechanism of Microbial Flotation Using *Thiobacillus ferrooxidans* for pyrite suppression," *Biotechnology and Bioengineering*, Vol. 14, pp. 671–676.

Olson, G.J. and Brinckmann, F.E. (1986), "Bioprocessing of Coal," *Fuel*, Vol. 65, No. 12, pp. 1638–1646.

Olson, G.J. and Kelly, R.M. (1990), "Microbial Processing of Coal," *Bioprocessing and Biotreatment of Coal* (D.L. Wise, ed.), Dekker, New York, pp. 465–86.

Olson, G.J. (1991), "Rate of Pyrite Bioleaching by *Thiobacillus Ferrooxidans*: Results of an Interlaboratory Comparison," *Applied and Environmental Microbiology*, Vol. 57, No. 3, pp. 642–644.

Patra, P. and Natarajan, K.A. (2003) "Microbially-Induced Flocculation and Flotation for Pyrite Separation from Iron Oxide Gangue Minerals," *Minerals Engineering*, Vol. 16, pp. 965–973.

Raichur, A.M., Misra, M. and Smith, R.W. (1995), "Differential Adhesion of Hydrophobic Bacteria onto Coal and Associated Minerals," *Coal Preparation*, Vol. 16, pp. 51–63.

Rayner, A.D.M. and Boddy, L. (1988), *Fungal Decomposition of Wood: Its Biology and Ecology*, John Wiley & Sons, New York.

Scott, C.D., Strandberg, G.W. and Lewis, S.N. (1986), "Microbial Solubilization of Coal," *Biotechnology Progress*, Vol. 2, No. 3, pp. 131–139.

Silverman, M.P., Rogoff, M.H. and Wender, I. (1963), "Removal of Pyritic Sulfur from Coal by Bacterial Action," *Fuel*, Vol. 42, pp. 113–124.

Silverman, M.P. and Lundgren, D.G. (1959), "Studies on the Chemoautotrophic Iron Bacterium *F ferrooxidans* I. An Improved Medium and Harvesting Procedure for Securing High Cell Yields," *Journal of Bacteriology*, Vol. 77, pp. 642–647.

Smith, R.W. and Miettinen, M. (2006), "Microorganisms in Flotation and Flocculation: Future Technology or Laboratory Curiosity?" *Minerals Engineering*, Vol. 19, pp. 548–553.

Stainthorpe, A.C. (1989), "An Investigation of the Efficacy of Biological Additives for the Suppression of Pyritic Sulfur During Simulated Froth Flotation of Coal," *Biotechnology and Bioengineering*, Vol. 33, pp. 694–698.

Singer, P.L. and Stumm, W. (1970), "Acidic Mine Drainage: The Rate Determining Step," *Science*, Vol. 167, p. 1121.

Townsley, C.C. and Atkins, A.S. (1986), "Comparative Coal Fines Desulphurization Using the Iron Oxidising Bacterium *Thiobacillus ferrooxidans* and the Yeast *Saccharomyces cerevisiae* During Simulated Froth Flotation," *Process Biochemistry*, Vol. 21, No. 6, pp. 188–191.

Townsley, C.C., Atkins, A.S. and Davis, A.J. (1987), "Suppression of Pyritic Sulfur During Flotation Tests Using the Bacterium Thiobacillus ferrooxidans," *Biotechnology and Bioengineering*, Vol. 30, p. 108.

Tuovinen, O.H. and Kelly, D.P. (1973), "Studies of the Growth of *Thiobacillus ferrooxidans,*" *Archiv für Mikrobiologie*, Vol. 88, pp. 285–298.

Yen, T.F. (1988), "Microbial Screening Test for Lignite Degradation," DOE/PC/70809T15 (DE89008184).

Zarubina, Z.M., Lyalikova, N.N. and Shmunk, E.I. (1959), "Investigation of Microbiological Oxidation of Coal Pyrite," *Izvest. Akad. Nauk S.S.S.R.*, Otdel. Tekh. Nauk, Met. i Toplivo, No. l, pp. 117–119.

Zimmerman, R.E. (1979), "Froth Flotation," *Coal Preparation*, 4th Edition (J.W. Leonard, ed.), AIME, New York, Chapter 10, Part 3.

15 Chemical Processing

Many coals contain as much as 50% organic sulfur, which cannot be removed by physical coal-cleaning methods (Morrison, 1981; Meyers, 1977; Abdul et al., 1992; Yaman et al., 1995). Removal of the organic sulfur requires chemical desulfurization techniques, many of which can also remove inorganic sulfur and ash-forming minerals. Recent research in this area has laid the foundations for chemical desulfurization processes capable of reducing the sulfur content of high-sulfur coal by as much as 90%, thus meeting New Source Performance Standards (NSPS) for control of sulfur oxides (Nowak and Meyers, 1993). Although chemical desulfurization of coal is not profitable under current economic conditions, it may become economical in the future as regulations of SO_2 emissions are tightened and low-sulfur coal reserves are depleted. Furthermore, future development of coal-oil, coal-water, and coal-methanol fuel applications may require precombustion desulfurization to levels only achievable by chemical processing.

An important point to keep in mind when reviewing the chemical desulfurization literature is that simply reporting the organic sulfur as a percent weight of the treated product can be very misleading. The reason for this is that chemical reactions can break down and remove a portion of the coal along with the associated organic sulfur, while leaving inert ash-forming minerals behind. When this happens, the sulfur content expressed as weight percent of the product will decrease because of increased dilution of the carbonaceous material with ash-forming materials, but the actual quantity of sulfur present per unit energy content of the product might be the same as or even higher than in the original feed. In order to be sure that a process is truly selectively removing organic sulfur, it is necessary to perform a complete materials balance (including gaseous and dissolved products) to account for the dilution effect. At a minimum, the sulfur content of the desulfurized product must be reported as sulfur per unit heat content and not simply as sulfur per unit weight. If this is not done, sulfur removals reported on a mass basis are almost meaningless (Meyers, 1996, Personal Communication). This effect is much more of a problem for chemical desulfurization than for physical desulfurization, for two reasons:

1. Physical separations leave all components of the coal in solid forms that are easily captured and measured, while chemical desulfurization reactions can convert coal to gaseous or water-soluble components that may be difficult to capture and analyze properly.
2. The clean-coal product from physical coal cleaning is lower in ash than the feed coal, and so ash dilution is not a problem.

15.1 FUNDAMENTALS OF CHEMICAL DESULFURIZATION

Based on physical and organic chemistry theory and on work done by oil refineries for desulfurization of petroleum, Meyers (1977) proposed the following six types of reaction by which organic sulfur may be chemically removed from coal prior to combustion:

1. *Solvent Partitioning*: Treatment with a solvent that dissolves all or part of the organic portion of the coal. In order for desulfurization to take place, the dissolved sulfur-bearing species must then be separated from the low-sulfur species.
2. *Thermal Decomposition*: Heating the coal in the absence of oxygen, to decompose the sulfur-containing compounds so that they can escape as gases (H_2S and/or SO_2). Since other compounds in the coal are also prone to decomposition at relatively moderate temperatures, a catalytic approach is needed to encourage the sulfur-containing compounds to decompose first.
3. *Acid–Base Neutralization*: Mercaptans and thiols are weakly acidic, and so they can be dissolved in aqueous solutions of weak bases; however, sulfides, disulfides, and aromatic sulfur compounds cannot be removed in this way.
4. *Sulfur Reduction*: Reaction of the sulfur-containing compounds with a reducing agent (such as hydrogen) to chemically reduce the sulfur to H_2S gas. For example:

$$R-S-R'+2H_2 \rightarrow R-H-H_2S+R'-H$$

If the temperature is above 400°C, this reaction also results in the production of liquid and gaseous products, and the process becomes "coal liquefaction."

5. *Sulfur Oxidation*: Reaction of sulfur-bearing compounds with oxidizing agents to convert them to water-soluble sulfates. For example:

$$R-SH+2O_2+H_2O \rightarrow R-OH+H_2SO_4$$

6. *Nucleophilic Displacement*: Replacement of the sulfur atom in a molecule with another species that bonds into the molecule more strongly than sulfur. The displacement takes place without any oxidation or reduction occurring. For example:

$$R-S-R'+2NaOH \rightarrow R-O-R'+Na_2S+H_2O$$

In practice, the most successful processes developed to date for removal of organic sulfur from coal utilize oxidation or displacement reactions (Morrison, 1981).

Chemical processes can also be used to convert pyritic sulfur into soluble forms. This is of particular value for pyrite which is too finely distributed to be removed by physical separation. The possible chemical reactions of pyrite are numerous, and include (Meyers, 1977)

1. *Displacement Reactions*: Sulfur may in theory be displaced from pyrite by a nucleophilic species (OH^-), without any reduction or oxidation reaction occurring, although in practice the sulfur and iron in pyrite generally change their oxidation states during chemical reactions. For example:

$$2OH^- + FeS_2 \rightarrow Fe(OH)_2 + S_2^{-2}$$

2. *Acid–Base Neutralization*: Iron disulfide (pyrite) is a weak alkali and can react with strong acids to produce hydrogen sulfide, as follows:

$$4H^+ + FeS_2 \rightarrow Fe^{+2} + 2H_2S$$

3. *Oxidation Reactions*: The disulfide in pyrite can be oxidized to elemental sulfur or to more oxidized sulfur forms such as sulfate. Many oxidizing agents are suitable for pyritic sulfur oxidation, such as $KMnO_4$, NaOCl, H_2O_2, Cl_2, NO_2, HNO_3, SO_2, O_2, and $HClO_4$. Similarly, oxidation can be performed by salts of metals in a high-valence state, such as Fe^{+3}. For example:

$$[O] + FeS_2 \rightarrow Fe^{+2} + 2S + \left[O^{-2}\right]$$

4. *Reduction Reactions*: Pyrite can react with reducing agents (such as hydrogen) to produce lower-order sulfides or metallic iron and hydrogen sulfide, as shown below:

$$H_2 + FeS_2 \rightarrow FeS + H_2S$$

$$H_2 + FeS \rightarrow Fe + H_2S$$

Oxidation reactions are primarily effective for pyritic sulfur removal. Oxidation-based processes need to be selective for sulfur to be practical, as otherwise oxidation of the coal results in losses of heating value.

Both pyritic and organic sulfur can be removed by treatment with a strong base, either aqueous or molten, such as NaOH, KOH, or $Ca(OH)_2$. While reactions with a strong base are known to work well, the mechanisms and products of the reaction are complex and not well-characterized (Morrison, 1981).

15.2 CAUSTIC TREATMENTS

Removal of sulfur with strong alkali is essentially an example of a displacement reaction. Treatment of pyrite with a base such as sodium hydroxide, whether molten or aqueous, converts pyritic sulfur (FeS_2) to iron oxide and water-soluble sulfides and sulfates through any of several chemical reactions, including

$$8FeS_2 + 30NaOH \rightarrow 4Fe_2O_3 + 14Na_2S + Na_2S_2O_3 + 15H_2O$$

$$35FeS_2 + 140NaOH \rightarrow 16Fe_2O_3 + 64Na_2S + 64H_2O + Fe_3O_4 + 6Na_2SO_4 + 6H_2$$

Organic sulfur can undergo a displacement reaction with alkali, in which the sulfur atom is replaced by an oxygen atom (Tsai, 1982). For example:

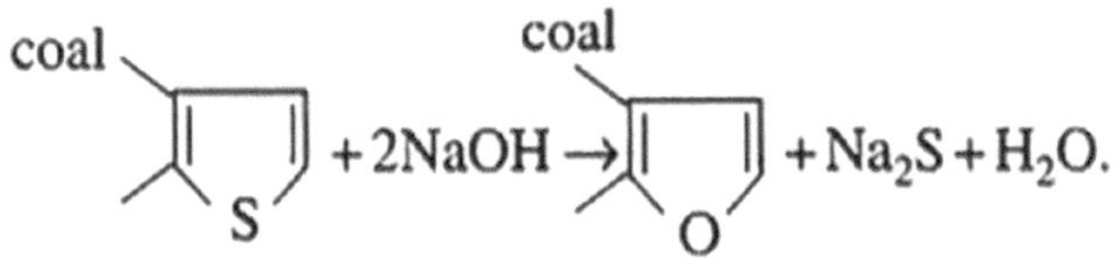

After these reactions are completed, alkali can be regenerated from the sodium sulfide by reaction with carbon dioxide and lime, as shown below:

$$Na_2S + CO_2 + H_2O \rightarrow H_2S + Na_2CO_3$$

$$Na_2CO_3 + CaO + H_2O \rightarrow CaCO_3 + 2NaOH$$

Since strong alkalis can remove both inorganic and organic sulfur, they are good candidates for chemical desulfurization processes. Several processes are based on these types of reactions and are described in the following sections.

15.2.1 Molten Caustic Leaching

One of the most effective methods for chemical desulfurization of coal is treatment by molten caustic leaching (MCL), which uses a strong base in the absence of water at high temperature to remove ash as well as inorganic and organic sulfur. To date, this is one of only two chemical desulfurization processes that have been successful enough to warrant pilot-scale studies, the other being the Meyers ferric salt leaching process (Meyers, 1996, Personal Communication).

The caustic used most often for MCL is sodium hydroxide, although the more reactive and more expensive potassium hydroxide is sometimes used in place of up to half of the NaOH. One MCL process, known as the Gravimeli process, shown in Figure 15.1, has been tested on a pilot plant scale (10–20 lb/h) by Thompson Ramo Woolridge (TRW) in California (Nowak and Meyers, 1993). In the Gravimelt process, anhydrous NaOH is mixed with pulverized dry coal in a ratio of 2.5:1 and heated to about 400°C for 1–2 h. This mixture is then rinsed with 2–10 parts water by countercurrent flow. Finally the coal is washed with H_2SO_4, rinsed, and dewatered. The Gravimelt process is capable of removing more than 90% of the ash and sulfur, including organic sulfur, from both bituminous coals and brown coals. MCL removes all of the pyritic and sulfidic sulfur and about 90% of the thiophenic organic sulfur, while producing small amounts of elemental and sulfatic sulfur (Huffman et al., 1995). When used to treat brown coals, MCL reduces the inherent moisture content as well as the ash content, and so it increases the heating value of brown coal, sometimes by as much as a factor of 2 (Valasek et al., 1992; Rondio et al., 1993). It is estimated that with an optimized design and economies of scale, the Gravimelt process could produce desulfurized coal at a cost of approximately $51/ton (Nowak and Meyers, 1993).

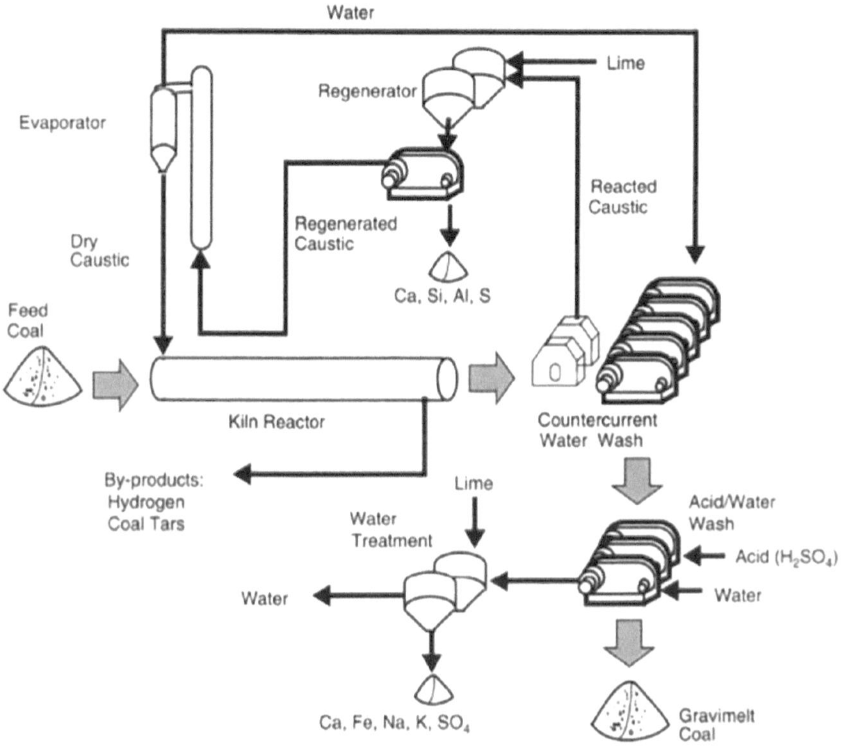

FIGURE 15.1 Schematic of Gravimelt molten caustic leaching process, with the approximate quantities of material that report to various products, expressed as percentages of the coal feed weight (Nowak and Meyers, 1993; Anastasi et al., 1990).

The MCL treatment is a harsh process, and it results in a partial conversion of the coal to volatiles and produces changes in the coal structure. The volatiles are rich in combustible materials, and they can be used as fuels themselves. The reduced volatiles content of the solid product makes the MCL product more difficult to ignite than the original coal, but once ignited, the burning characteristics are similar to those of the parent coal. The ash fusion temperature of the ash from MCL products is very similar to that of the ash from the parent coal, indicating that the slagging tendencies will be similar. There is some possibility, however, that MCL products will show more severe fouling tendencies, as a result of their higher Na_2O contents. In any case, the very low ash contents of the MCL products are expected to outweigh any increased tendency for slagging and fouling (Mcllvried et al., 1990).

The effect of using KOH as a fraction of caustic in MCL was investigated by Chriswell et al. (1991) at the Ames Laboratory in Iowa, USA. They found that for medium-rank coals such as Illinois No. 6, NaOH–KOH mixtures were generally less effective than NaOH alone at high temperatures (i.e., above 350°C). Below 350°C, KOH–NaOH mixtures were more effective. Since high-sulfur reductions are only achieved at high temperatures, the use of NaOH alone is recommended for medium-rank coals. For high-rank coal from the Upper Freeport seam, the Ames Laboratory found that sulfur reductions were greater at all temperatures when KOH was added, compared to use of NaOH alone.

There have been reports that some low-rank coals, such as Hanna 80, are easily dissolved by MCL, even at low temperatures, resulting in very poor coal yields (Chriswell et al., 1991). In tests with some European coals, it was also demonstrated that MCL was effective for removing sulfur and ash, but that the coal rank strongly influenced the overall yield. Heating value losses were greatest for the low-rank Sulcis coal at 38%, and least for the higher-rank Ruhr coal at only 2.2% (Carbini et al., 1990).

A modified MCL process has been proposed where coal was washed in boiling water for 1 h, followed by a float/sink separation in 50% NaOH. The low-density product of this separation was then brought to 390°C for 15 min. This procedure was reported to allow the use of a 1:1 ratio of NaOH to coal, while still achieving results comparable to those obtained at higher NaOH:coal ratios (Akhtar and Chriswell, 1993).

The extent of organic sulfur removal by MCL as a function of time is dependent on temperature. Pyritic and aliphatic sulfur are removed at low temperatures, while the thiophenic compounds are removed at higher temperatures. It has been determined that to minimize loss of heating value while maximizing sulfur removal, it is preferable to process for a longer time rather than to process at a higher temperature. Molten caustic treatment causes particles to fragment at low temperatures, but at high temperature, the cross-linking of carbon into aromatic structures helps to fuse the particles back together (Majchrowicz et al., 1991).

15.2.2 Aqueous Base Leaching

An early alkali-leaching method was the Battelle Hydrothermal Coal Process. This consisted of first grinding the coal to 100% passing 28 mesh (600 μm) and 70% passing 74 μm (200 mesh). The ground coal was then mixed with an aqueous solution containing 10% NaOH and 2%–3% $Ca(OH)_2$, and heated in an autoclave at 250°C–350°C and 4,140–17,200 kPa (600–2,500 psi) for 10–30 min to extract sulfur and leachable mineral matter. After leaching, the slurry was cooled and decanted to separate the leachant from the coal. The coal was dried without attempting to wash off the residual alkali, as it was expected that any alkali remaining on the coal surface would absorb a portion of the remaining sulfur during combustion (Stambaugh, 1977; Stambaugh et al., 1975). The leachant was then regenerated by first treating the solution with carbon dioxide to drive off the sulfur as H_2S, followed by addition of lime to convert sodium carbonate to sodium hydroxide. A schematic diagram of the process is shown in Figure 15.2.

It was found that for a wide variety of US coals, 90%–99% of the pyrite could be removed and 20%–70% of the organic sulfur could be removed through the Battelle process. Heating value losses were between 5% and 10%. Unfortunately, when using strongly alkaline aqueous solutions, severe corrosion problems are almost unavoidable at temperatures above 250°C. It should be noted that while the hydrothermal process requires temperatures somewhat lower than for the MCL process, it also

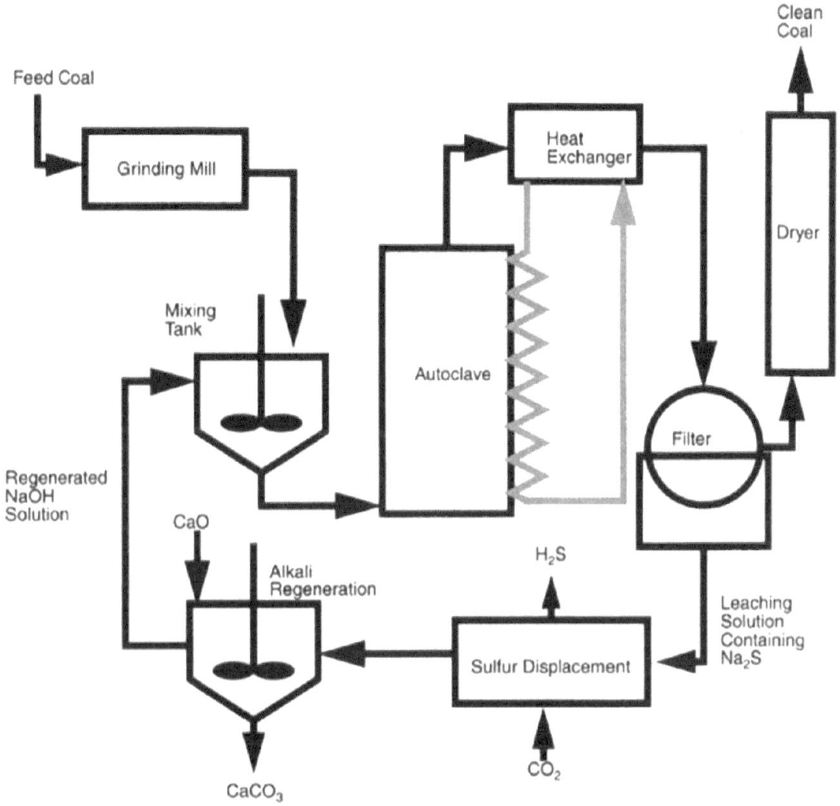

FIGURE 15.2 Schematic diagram of the Battelle Hydrothermal Coal Process.

requires an autoclave and a finely ground feed. The equipment and feed preparation are therefore more expensive for the hydrothermal process than for MCL. As a result, work on the hydrothermal process has been suspended. It should also be noted that the organic sulfur content per unit heat content was never determined for the Battelle process, and so it is possible that much of the reduction in sulfur was actually due to ash dilution (Meyers, 1977).

15.2.3 Microwave Desulfurization

The objective of microwave desulfurization of coal is to selectively heat the pyrite particles and aqueous solution at a high rate, which is possible due to their greater ability to absorb microwave energy than the rest of the coal matrix (Yang and Ren, 1988; Tsai, 1982; Yang and Wi, 1987; Bluhm et al., 1986; Kusakabe et al., 1988). This supplies the energy needed for the chemical desulfurization reactions directly at the locations in the coal where it is needed. The intent is that the microwaves will penetrate the coal quickly and bring the high-sulfur regions to reaction temperature. This can ultimately accomplish the reaction more rapidly than would be possible with conventional heating. It should be noted that microwave heating is more expensive per unit of energy delivered than conventional heating. It is only beneficial if very rapid or selective heating is necessary (Meyers, 1998, Personal Communication).

An early attempt at microwave desulfurization was devised by General Electric (GE). The basic process was first to grind the coal to between 600 μm (28 mesh) and 74 μm (200 mesh) in size and produce a slurry of the coal with sodium hydroxide and water. The slurry was then dewatered, producing a moist coal cake with a layer of sodium hydroxide solution coating the coal particles. This cake was irradiated with microwaves for 1 min at a power level of 1 kW, and then washed first with water, followed by a wash with dilute hydrochloric acid, and dried (Zavitsanos et al., 1982). Unfortunately, the GE reports did not specify how long the coal and alkali were in contact before microwave treatment, or the HCl concentration in the final wash stage (Norton et al., 1991). GE researchers reported that the best results were obtained with 0.6–1.0 g of NaOH/g of coal. When the treatment was repeated twice on a single coal sample, up to 89% of the sulfur could be removed, along with 40% of the ash-forming minerals.

In the time since GE discontinued work on the project, investigators have been unable to duplicate their results consistently. For example, Norton et al. (1991) found that longer irradiation times (at least 3 min) or extra repetitions of the microwave-cleaning process were required to reach the desulfurization levels claimed by GE. The experimentally determined best conditions for microwave desulfurization in caustic were a forward (incident) power level of 1.5 kW for 2.0 min irradiating 50 g of coal, at a NaOH:coal ratio of 1:1 using −600 μm (28 mesh) coal. Using these conditions, an Illinois No. 6 coal precleaned by heavy-media sink/float was treated three times for an overall sulfur reduction of 83% and an ash reduction of 87%. The Space and Technology group of TRW conducted 95 bench-scale tests using Kentucky No. 11, Pittsburg No. 8, and Middle Kittanning coals. They achieved sulfur reductions of only 32%–40% and ash reductions of 60%–86% using a double treatment method similar to the GE procedure (Rhee and Wan, 1987); however, Norton et al. (1991) listed several important factors that may have hindered the TRW desulfurization experiments, including use of a nonresonant applicator cavity and a metal conveyor belt.

Microwaves have also been considered in conjunction with MCL. Simple pretreatment before MCL did not improve the desulfurization process any more than pretreatment by conventional heating (Norton et al., 1991). It has been reported, however, that the use of microwaves to heat the alkali/coal mixture could significantly increase the rate of coal desulfurization. This improvement in rate is believed to be due to the localized microwave heating improving the contact between the coal and the molten caustic, which increases the mass-transfer rate (Hayashi et al., 1990).

In addition to use with alkali-leaching processes, microwaves have also been examined as a means for accelerating other reactions. For example, microwaves can selectively heat pyrite to speed its reaction with hydrogen (Tsai, 1982):

$$H_2 + FeS_2 + (\text{microwaves}) \rightarrow FeS_{1-x} + H_2S$$

This releases a portion of the sulfur from pyrite particles and also converts the pyrite to pyrrhotite, which is magnetic. The magnetic pyrrhotite can then be removed physically using magnetic separation. Microwaves can also be used to promote liquefaction of coal with hydriodic acid and tetrahydrofuran (Andres et al., 1998) and to thermally convert pyrite to pyrrhotite ($Fe_{1-x}S$) or troilite (FeS), both of which are soluble in HCl solutions (Weng and Wang, 1992; Wang et al., 1990).

While microwave desulfurization has shown some good results, the performance is variable and difficult to reproduce. The main difficulty seems to be that microwave processing requires a good understanding not only of the chemical reactions involved but also of the physics of microwave irradiation.

15.3 OXIDATIVE TREATMENTS

Many different reagents have been used to attempt to selectively oxidize the sulfur in coal without excessively oxidizing the coal. Oxygen can be used but requires high temperature and pressure to obtain significant conversions. Other oxidizing agents can be used in milder conditions, such as metallic salts, chlorine gas, nitrogen or sulfur dioxides, peroxides, organic peroxyacids, sodium hypochlorite, ozone, or potassium permanganate. These other oxidizing agents are more expensive than oxygen, and their consumption by the process is a major operating expense. Most of these processes do not have supporting data to show that they actually reduce the sulfur content per unit of heating value.

15.3.1 Metallic Salts

Many metallic salts in aqueous solution are good oxidizing agents for pyritic and/or organic sulfur in coal. Typically these salts are able to reduce inorganic sulfur by 70%–90%. These reactions generally proceed faster for finer coal particles and higher temperatures (Prasassarakich and Pecharanond, 1992).

15.3.1.1 Meyers Process

The most successful example of a metallic salt sulfur oxidation method is the Meyers process, which used ferric iron to selectively oxidize pyrite (Meyers et al., 1972). This is one of only two chemical desulfurization processes to be used to date on a pilot scale (Meyers, 1996, Personal Communication). In the Meyers process, coal is leached in an $Fe_2(SO_4)_3$ solution at atmospheric pressure and 90°C–130°C for 4–6 h (Meyers et al., 1979; Morrison, 1981). The reaction equation is shown below:

$$5FeS_2 + 23Fe_2(SO_4)_3 + 24H_2O \rightarrow 51FeSO_4 + 24H_2SO_4 + 4S$$

The ferrous sulfate is then reoxidized by reaction with oxygen, as follows:

$$4FeSO_4 + 2H_2SO_4 + O_2 \rightarrow 2Fe_2(SO_4)_3 + 2H_2O$$

The coal is then filtered and washed. Since the sulfates are water soluble, they are carried away in the leach solution. The elemental sulfur is then removed by acetone extraction.

Before recycling the leach solution, it is neutralized and the sulfate concentration brought down to its original value by the addition of lime, as shown below:

$$Fe(SO_4)_3 + CaO \rightarrow Ca(SO_4) + Fe_2O_3$$

$$FeSO_4 + CaO \rightarrow Ca(SO_4) + FeO$$

The Meyers process can remove 83%–98% of the pyritic sulfur with little coal loss, and it is an excellent choice for removing pyritic sulfur that is too finely divided for removal by physical separation. Unfortunately, it is not effective for organic sulfur removal. Because of the inability of the Meyers process to remove organic sulfur, development has not continued.

15.3.1.2 Ferric Chloride

Some recent work has centered on the use of iron (III) or copper (II) chloride solutions, which was claimed to be capable of removing both organic and inorganic sulfur. Ferric chloride ($FeCl_3$) has been shown to be more effective than copper chloride for removal of organic sulfur (Fan et al., 1987). Fan et al. (1987) treated six high organic sulfur coals sized to 74×38 μm for 1 h in a 10% (by weight) ferric chloride solution at 300°C, followed by washing in boiling 2M hydrochloric acid for 30 min. This treatment was reported to remove essentially all of the pyritic sulfur and 35%–50% of the organic sulfur in 1 h, with British Thermal Unit (BTU) recoveries of 74%–82%. Testing on model organic sulfur compounds showed this procedure to be ineffective for oxidation of the benzothiophene and dibenzothiophene forms of organic sulfur, although it was effective for breaking down thiophenol, phenyl disulfide, benzyl methyl sulfide, benzyl phenyl sulfide, benzyl disulfide, phenyl sulfide, and 2,5-dimethylthiophene. It should be noted that $FeCl_3$ becomes unstable at approximately 300°C and forms an insoluble basic iron oxide that will precipitate on the coal. This insoluble material would dilute the sulfur, making it appear that more was removed than was actually the case. It would also increase the ash content of the coal.

15.3.2 Oxydesulfurization

Oxydesulfurization is the use of oxygen (or air) to desulfurize coal, usually at very high temperatures and pressures. The process may be carried out in either wet or dry states, depending on the pressures and temperatures used (Mishra et al., 1995). Pulverized coal is first slurried with water, sometimes in the presence of acid or base, pressurized with oxygen, and then heated to the desired reaction temperature. The oxydesulfurization can be carried out in a fluidized bed, which is suitable for high-sulfur caking coals that contain a significant amount of finely dispersed pyritic sulfur. Losses of heating value during this process can be as high as 40% for some coals (Anonymous, 1989), and it has been reported that the porosity of the coal is increased by this treatment, making it more reactive (Akhtar and Cliffe, 1990). Oxydesulfurization is most effective for removal of pyritic sulfur, by conversion to soluble sulfates with dissolved oxygen as shown in the equations below (Tsai, 1982):

$$2FeS_2 + 7O_2 + 2H_2O \rightarrow 2FeSO_4 + 2H_2SO_4$$

$$4FeSO_4 + O_2 + 2H_2SO_4 \rightarrow 2Fe_2(SO_4) + 2H_2O$$

$$Fe_2(SO_4)_3 + 3H_2O \rightarrow Fe_2O_3 + 3H_2SO_4$$

These reactions are similar to those used by the Meyers process, with ferric iron from the pyrite catalyzing the oxidation of additional pyrite. The main difference from the Meyers process is that the ferric iron must be generated from the pyrite, and so the reaction can take some time to get started. It has been reported that maintaining a leaching solution pH greater (more alkaline) than 0.5 prevents the formation of elemental sulfur, which is not soluble in water and therefore would be difficult to remove.

The use of alkaline solutions during oxydesulfurization, while enabling removal of some of the organic sulfur, generally results in higher carbon losses than desulfurization in acidic solution (Tsai, 1982). Temperatures and pressures vary from process to process. Some examples are shown in Table 15.1 (Morrison, 1981).

The reagents used are among the most inexpensive available. Drawbacks to oxydesulfurization methods include increased coal oxidation, loss of heating value, and high expense due to the use of high temperature and pressure. It is worth noting that the use of oxalic acid as a promoter allows the process to operate at milder temperatures and pressures with better oxygen selectivity and higher BTU recovery, as was done in the Atlantic Richfield Company (ARCO) promoted oxydesulfurization process. Development of the ARCO oxydesulfurization process has been discontinued, because organic sulfur removals were found to be insignificant in further Department of Energy (DOE)-sponsored testing (Meyers, 1998, Personal Communication). It has been reported that the use of ammonia solution for extraction in oxydesulfurization increases the efficiency of the process (Yaman and Kucukbayrak, 1996).

TABLE 15.1
Comparison of Several Oxydesulfurization Processes (after Morrison, 1981)

Process	Temperature, °C	Pressure, atm	Time, h	Chemicals Used	% Sulfur Removal	
					Pyritic	Organic
Pittsburgh Energy Technology Center (PETC)	180–200	34–68	<1	Air, lime	95	13
Ames	150 240	14	1	Oxygen, sodium carbonate	80–90	0 <20
Ledgemont oxygen leaching (LOL)	130	10–20	1–2	Oxygen, lime, ammonia	80–90	<20
ARCO promoted	120	2	1	Oxygen, lime,	88–98	
Oxydesulfurization	350		1	Oxalic acid	65–89	23–30

15.3.3 Chlorinolysis

A process for desulfurization of coal by chlorinolysis was developed by the Jet Propulsion Laboratory at the California Institute of Technology (Kalvinskas and Hsu, 1979; Morrison, 1981). This process uses coal pulverized to less than 94 μm slurried in 1,1,1-trichloroethane (TCE) or water in a coal to solvent ratio of 2:1 by weight. This slurry is then exposed to chlorine gas for 45 min at temperatures of 60°C–130°C and pressures of 1–5 atm. The reaction of chlorine with pyrite produces iron chloride and S_2Cl_2 which, in the presence of water at temperatures above 50°C, is converted to sulfuric and hydrochloric acid:

$$FeS_2 + 2Cl_2 \rightarrow FeCl_2 + S_2Cl_2$$

$$S_2Cl_2 + 8H_2O + 5Cl_2 \rightarrow 2H_2SO_4 + 12HCl$$

Reaction of chlorine with organic sulfur yields sulfenyl chlorides (RSCl), which react with water to form sulfonates and sulfuric acid:

$$R - S - R' + Cl_2 \rightarrow R - SCl + R'Cl$$

$$R - S - S - R + Cl_2 \rightarrow R - SCl + R'Cl$$

$$R - SCl + 2Cl_2 + 3H_2O \rightarrow R - SO_3H + 5HCl$$

$$R - SCl + 3Cl_2 + 4H_2O \rightarrow R - Cl + H_2SO_4 + 6HCl$$

After reaction, the TCE solvent can be recovered by steam distillation. The wet coal is then filtered, washed, and brought to 400°C in a stream of inert gas. This drives off much of the chlorine as HCl gas, which may possibly be reconverted into chlorine gas using a modified Deacon process. If water is used in place of TCE, the process is simplified and total sulfur reductions are comparable (45%–66% for water versus 52%–63% for TCE). However, organic sulfur reductions drop from 46% to 89% for TCE to 0%–24% for water, although water is more effective for removing pyritic sulfur. One major problem with the chlorinolysis process is the corrosive environment created by the use of Cl_2 and HCl. Other drawbacks of the chlorinolysis process include poor coal recovery from the dechlorination step (88%) and high chlorine contents (0.23%–5.1%) of the product coal.

15.3.4 KVB Method

The KVB process utilizes NO_2 to oxidize pyrite and a hot caustic solution to displace organic sulfur from coal (Guth, 1979; Morrison, 1981). The chemical reaction for pyrite oxidation is shown below:

$$FeS_2 + 6NO_2 \rightarrow FeSO_4 + SO_2 + 6NO$$

In the KVB process, dry coal, pulverized to 14–28 mesh is treated with a gas stream of 5%–10% NO_2 (in a mixture of NO, NO_2, O_2, and N_2) at 93°C for 30–60 min.

The nitrogen dioxide oxidizes the pyritic sulfur to iron sulfate, and gaseous sulfur dioxide and nitric oxide are given off. The nitric oxide is then reoxidized to nitrogen dioxide to continue the process. The treated coal is washed with hot water to remove the iron sulfate for a pyrite reduction of 60%–100%. The coal is then washed in a caustic solution at 25°C–100°C for 30–60 min, giving a 30%–50% reduction in organic sulfur. The entire process is carried out at atmospheric pressure (Tsai, 1982).

15.3.5 Autoxidation of Sulfur Dioxide

This process is based on the production of peroxide species as transients during the autoxidation of sulfur dioxide. The basic reactions are as follows, where M represents a transition metal such as iron, and the transient peroxide species are underlined:

$$SO_2 + H_2O \rightarrow HSO_3^- + H^+ \tag{15.1}$$

$$M^{(n+1)+} + HSO_3^- \rightarrow M^{n+} + HSO_3^{\bullet} \tag{15.2}$$

$$HSO_3^{\bullet} + O_2 \rightarrow \underline{HSO_5^{\bullet}} \tag{15.3}$$

$$HSO_5^{\bullet} + HSO_3^- \rightarrow \underline{HSO_5^-} + HSO_3^{\bullet} \tag{15.4}$$

$$HSO_5^- + HSO_3^- \rightarrow 2HSO_4^- \tag{15.5}$$

$$2M^{n+} + O_2 + H^+ \rightarrow 2M^{(n+1)+} + \underline{H_2O_2/HO_2^-} \tag{15.6}$$

The transient species $HSO_5^{\bullet}$, HSO_5^-, H_2O_2, and HO_2^- are all able to oxidize additional sulfur from coal. Iron is a suitable metal ion to catalyze the reactions and is particularly well-suited because it is naturally present when pyrite is being dissolved. It is reported that this method could achieve 73% conversion of pyrite in 22 h at 67°C, using air with 0.07 vol% SO_2 as the source of both sulfur dioxide and oxygen. In comparison, only 20% of the pyrite was converted when SO_2 was absent from the air (Bronikowski et al., 1990).

15.4 OTHER METHODS

In addition to processes that use alkali-leaching and oxidative processes, a number of other techniques have been attempted for coal desulfurization, which are described in the following sections.

15.4.1 Pyrolysis

When coals are heated in a vessel where oxygen is excluded, a portion of the sulfur is released by thermal decomposition; however, this requires temperatures high enough that much of the coal is broken down as well, converting it to coal gases and a relatively unreactive "char."

In experiments with Polish coals of various ranks, coals were pyrolyzed at 1,000°C under an atmosphere of the pyrolysis gases. It was found that the proportion of sulfur removed by this method decreased as the coal rank increased. This change appeared to be due to the increasing proportion of thiophenic sulfur in high-rank coals, which is more stable and resistant to pyrolysis (Gryglewicz, 1996). It has also been reported that at temperatures above 500°C, the H_2S released by decomposing pyrite can be taken up by the coal char as organic sulfur, rather than being removed. Because of this effect, it is important to remove the pyrite as thoroughly as possible before attempting desulfurization by pyrolysis (Ibarra et al., 1994). Mild pyrolysis at 475°C has been reported to remove up to 33.2% of the total sulfur in coal in 6 min, in a reactor designed to capture the evolved H_2S with a CaO absorber (Lin et al., 1997).

15.4.2 IGT Hydrodesulfurization

The Institute of Gas Technology (IGT) in Chicago developed a two-step process for coal desulfurization that was a combination of oxydesulfurization and hydrogenation treatments (Fleming et al., 1977; Morrison, 1981). Specifically, pulverized coal was first contacted with air in a fluidized-bed reactor at 400°C and atmospheric pressure, removing 25%–30% of the sulfur, destroying 8%–12% of the coal, and generating steam and a low BTU gas. Although this step was necessary for later removal of organic sulfur, it destroyed the tendency of coal to cake upon heating.

The remaining coal was then reacted with hydrogen in a second fluidized-bed reactor at 800°C and atmospheric pressure, removing the sulfur as H_2S. Sulfur acceptors, such as calcium or iron oxide, could be used to reduce hydrogen consumption. The useful products were light oil, fuel gas, and char that cannot be directly substituted for coal.

A preliminary process design estimated that for bituminous coal on a commercial scale, 83% BTU recovery could be achieved if all products were recovered and steam was generated from energy lost due to coal consumption.

15.4.3 Magnex Process

The Magnex process is unique in that it converts weakly magnetic pyrite and nonmagnetic mineral matter into paramagnetic material, which may be easily removed by low-intensity magnetic separation (Kindig and Turner, 1976; Kindig and Goens, 1979; Porter and Goens, 1979; Morrison, 1981). The process involves heating coal ground to less than 1.41 mm to 170°C, using 100 kg steam/ton of coal. The steam heating removes volatile compounds and elemental sulfur, which impede the paramagnetic conversion. Next the coal is treated with iron pentacarbonyl ($Fe(CO)_5$) vapor, which decomposes and reacts with the pyrite and mineral matter surfaces.

$$FeS_2 + Fe(CO)_5 \rightarrow 2Fe_{(1-x)}S + 5CO$$

$$\text{Minerals} + Fe(CO)_5 \rightarrow Fe \cdot \text{Minerals} + 5CO$$

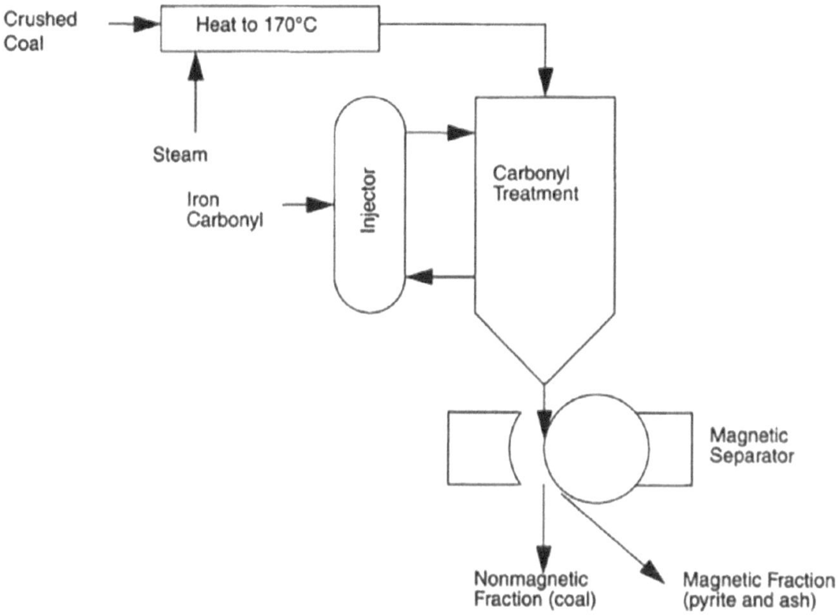

FIGURE 15.3 Schematic diagram of the Magnex process.

This forms metallic iron on the surface of the mineral matter and converts the surface of the pyrite particles into pyrrhotite. Since both metallic iron and pyrrhotite are strongly magnetic, these particles are then removable using low- to medium-intensity magnetic separators. A basic flowsheet of the Magnex process is shown in Figure 15.3.

Laboratory-scale testing of seven coals showed pyritic sulfur removal to vary between 57% and 92% when treated with 16 kg of $Fe(CO)_5$/ton coal followed by magnetic separation. Mineral matter removal was 7%–71%. BTU recoveries ranged from 86% to 96%. On a pilot plant scale a bituminous coal was treated with 10 kg $Fe(CO)_5$/ton coal followed by magnetic separation, achieving 85% pyritic sulfur removal and 86% BTU recovery; however, only pyritic sulfur was removed in this process.

15.4.4 Chemical Comminution

Chemically initiated fracture of coal was used in place of crushing and grinding in a process developed by Syracuse Research Corporation (Howard and Datta, 1977; Datta, 1977; Morrison, 1981). The feed coal was first conventionally sized to between 380 and 0.149 mm, and the undersize material was sent directly to a physical cleaning step. The remaining +0.149 mm coal was then reacted with ammonia vapor at 0.91 MPa gauge pressure until it had disintegrated to a top size of less than 10 mm, which took about 2 h, with the time depending on the type of coal. This coal was then physically cleaned, removing the liberated pyrite.

The advantage of chemical comminution is that breakage tends to occur along internal phase boundaries, with the result that mineral particles are more thoroughly liberated from the coal than they would be by conventional crushing and grinding. Lower-rank coals tend to respond more readily to this treatment than do high-rank coals, and it has been reported that chemical comminution is more effective for pyrite liberation than for ash mineral liberation (Tsai, 1982). The primary drawbacks are the long times required for fracture and the cost of the chemical comminution agent.

It is estimated that 50%–70% of the pyritic sulfur can be liberated by this method with 60%–90% product recovery (Contos et al., 1978; Morrison, 1981). A combination of chemical swelling agents with conventional comminution has recently been shown to reduce the grinding energy requirements for coal and to improve liberation and grinding efficiency (Wang et al., 1996).

15.5 LABORATORY SULFUR EXTRACTION METHODS

In addition to processes which are intended for industrial use, there are a number of desulfurization methods that are not expected to be practical on a large scale, due to high reagent costs. These processes are mainly useful for laboratory studies of the forms of sulfur in coal (Calkins, 1994; Bottrell et al., 1994).

15.5.1 Peroxyacetic Acid

Peroxyacetic acid was originally tested as a single-stage sulfur extraction technique, with disappointing results. It was found, however, to convert sulfur compounds into forms that were more amenable to other desulfurization processes, and it was therefore studied as a pretreatment stage for alkali extraction. The two-stage process consisted of coal dispersal in glacial acetic acid (CH_3COOH), warming to the desired temperature (21°C–104°C), and then mixing with 30% hydrogen peroxide (H_2O_2) in a 1:3 (vol.) ratio. The coal was then filtered and reacted with sodium bicarbonate in methanol at 350°C–450°C. When used in combination with treatment by base in an organic solvent, peroxyacetic acid removed 85%–95% of the sulfur from Illinois No. 6 and Indiana No. 5 coals, while producing a yield of about 80% and no loss in heat content (Palmer et al., 1995).

15.5.2 Sodium Hypochlorite

Sodium hypochlorite (NaOCl) selectively oxidizes sulfur in coal, especially sulfatic and organic sulfur. In one study, 20 g each of three –150 μm (100 mesh) Illinois coals were treated in 100 mL of 5.25 wt% NaOCl for 1 h at room temperature and atmospheric pressure, followed by a 0.3 M sodium carbonate wash at 80°C for 1 h (Brubaker and Stoicos, 1985). This reduced inorganic sulfur by 15%–25% and organic sulfur by 11%–42%. Repeating these two treatment steps resulted in a 26%–49.5% overall sulfur reduction, with 81%–95% heating value recovery. However, the product coal contained between 2% and 2.7% chlorine, which would need to be removed before combustion. It was found that primarily sulfatic and organic sulfur forms were removed, with pyrite removal being very low.

15.5.3 Potassium Permanganate

Potassium permanganate ($KMnO_4$) can reduce both organic and pyritic sulfur levels in coal by the process shown in Figure 15.4. A laboratory demonstration of the process consisted of first precleaning the coal by heavy-media separation using TCE as the heavy liquid, followed by grinding to –74 μm (200 mesh), and oxidizing the coal with potassium permanganate three times, with washing between the oxidation stages (Attia and Lei, 1987; Attia and Fung, 1993). Each oxidation stage consisted of treatment with 6% $KMnO_4$ for 1.5 h at room temperature and atmospheric pressure followed by a water wash at 80°C for 1 h. Direct comparison of potassium permanganate with sodium hypochlorite on identical coal samples under similar conditions showed the permanganate treatment to be superior for organic sulfur removal (Attia and Lei, 1987). Originally, the procedure used only water washing after each oxidation stage, but Attia and Fung (1993) modified this procedure by acid washing with 16% HCl after each step, and 15 min of ultrasonic treatment during the third step to help break up the oxidation product layer. Using the modified procedure on a Canadian Nova Scotia coal gave organic sulfur reductions of 27% and 38%, and total sulfur reductions of 54% and 63%, for –74 μm (200 mesh) and –38 μm (400 mesh) coal sizes, respectively. Pyritic sulfur reductions of up to 97% were achieved. Sulfur reductions were highest for an initial $KMnO_4$ leach at pH 7.

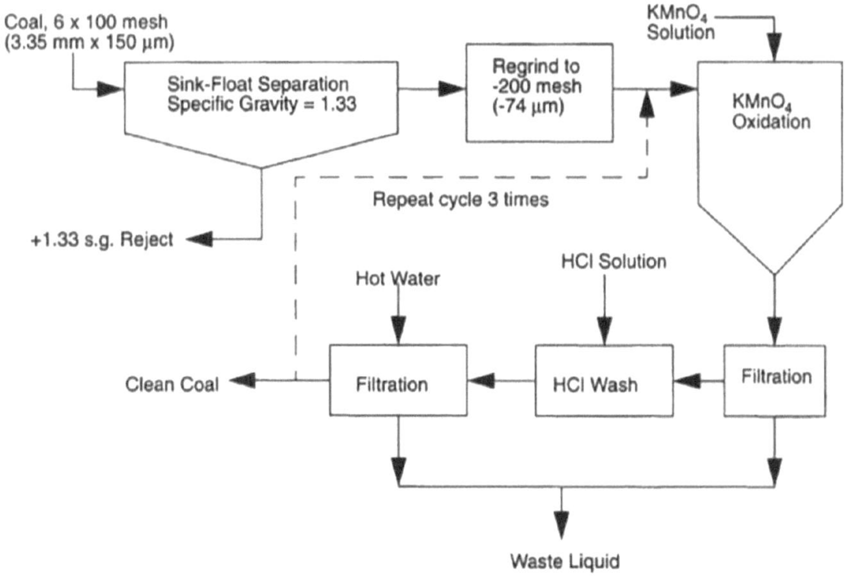

FIGURE 15.4 Basic procedure for $KMnO_4$ desulfurization of coal (Attia and Fung, 1993).

15.5.4 Supercritical Fluid Extraction

Supercritical fluid extraction (SFE) of sulfur from coal was performed on a very small scale with either CO_2–10% methanol or pure CO_2 under supercritical conditions (54.72 MPa, 482°C). When used on IBC-101 coal, the CO_2–methanol mixture primarily extracted only thiophenic sulfur, while the pure CO_2 removed 50% of the total sulfur, including pyritic, sulfatic, and thiophenic forms in only 30 min. When the coal was "spiked" with phosphoric acid before treatment, the sulfur removal increased to 82%. (Louie et al., 1994). It should be noted that supercritical extraction requires very high pressures and high temperatures, which tends to elevate the cost of the process. Similar processes use supercritical alcohol and alcohol/water solutions (Li and Guo, 1995; Duty and Penrod, 1994; Yueruem and Tugluhan, 1990).

15.5.5 Hydride Reduction, SET, and Lochmann's Base

Stock and Chatterjee (1993) have studied the use of three methods for the selective removal of various forms of sulfur in coal using highly selective reagents. Their methods are primarily intended to be used for characterizing sulfur forms in coal, and therefore make use of expensive specialized reagents.

Hydride reduction consists of treatment of −150 μm (100 mesh) coal with lithium aluminum hydride in tetrahydrofuran at 67°C for 24 h to selectively and completely reduce the pyrite. This treatment has been used on Illinois No. 6 coal to obtain a pyrite-free product as feed for organic sulfur removal experiments such as the SET and BASE treatments described below.

In single electron transfer, or SET, coal is treated with potassium naphthalene (−1) in tetrahydrofuran at 67°C for 24 h (Huffman et al., 1995; Chatterjee et al., 1990). SET gives organic sulfur conversions of 33%–50% for Illinois No. 6 coal, preferentially removing aromatic and heterocyclic sulfur.

Lochmann's base, or BASE, is a very reactive mixture of *n*-butyllithium and potassium *t*-butoxide in heptane, which preferentially removes sulfidic organic sulfur. In BASE, coal is added to the reaction mixture and allowed to react at 98°C for 6 h (Huffman et al., 1995; Chatterjee and Stock, 1991). Application of the BASE method to Illinois No. 6 coal gave 25%–40% organic sulfur reduction. The use of SET and BASE sequentially reduced the organic sulfur content of a coal from 1.77% to between 0.84% and 0.72%. The remaining sulfur was about half elemental and oxidized sulfur forms, which are more easily removed from the coal than organic sulfur.

15.5.6 Extraction with Chlorinated Solvents

It has been reported that hot perchloroethylene (PCE) is effective in a two-stage process for removing both pyritic sulfur and organic sulfur from coal, while losing less than 1% of the coal heating value (Fullerton et al., 1991; Midwest Ore Processing Co., 1992). This treatment removes 90%–99% of the pyritic sulfur and 30%–70% of the organic sulfur. The process is a combination of chemical displacement reactions and solvent partitioning, and the response of coals to the process varies depending on their content of impurities that promote extraction (Lee et al., 1993). The pyritic and organic sulfur are removed in separate steps, with the nature and extent of the reaction controlled by adjusting the reaction time and temperature. Desulfurization is most efficient when the organic sulfur is extracted before the pyritic sulfur, because the presence of mineral matter promotes the extraction of the organic sulfur (Vishnubhatt et al., 1993).

The time needed for the extraction varies depending on the type of coal. If excessive time is allowed, the extracted sulfur can re-react with the coal, which impairs the process efficiency (Vishnubhatt and Lee, 1993). The most recent work with this technique has shown that TCE as well as PCE can be used in this process (Lee et al., 1994). Due to solubility considerations, the extraction should be carried out at or near the boiling point of the solvent. Since TCE boils at only 74°C, compared to 121°C for PCE, the TCE extraction is carried out at a lower temperature than PCE extraction, and is therefore milder and more selective than PCE extraction (Lee et al., 1994). These solvents tend to leave chlorine in the coal, which can be removed by solvents such as supercritical CO_2 (Kocher et al., 1994).

It should be noted that this process has been viewed with extreme skepticism by other workers in the area, who have apparently been unable to duplicate the reported results (Wheelock, 1998, Personal Communication; Vorres, 1990).

15.6 SUMMARY

Many of the chemical coal desulfurization processes are able to remove both pyritic and organic sulfur, unlike the physical desulfurization techniques, which can only remove pyritic sulfur. The major limitation of the chemical processes is that they are too expensive to use at current coal prices. As environmental regulations become more stringent, however, it is possible that these technologies will become economical on an industrial scale. Further work is needed to improve the performance and economics of chemical coal desulfurization.

A summary of the performances of the various chemical desulfurization technologies is given in Tables 15.2 and 15.3, which illustrate the range and capabilities of the various processes. These results should be interpreted cautiously, because many studies did not provide mass balances, or even calculations of sulfur per unit heat content in the feed and final product. They therefore may not have distinguished between actual selective sulfur removal and simple dilution effects. Out of all of these processes, only two have been sufficiently promising or sufficiently well advanced to be tested on a pilot scale: the MCL process and the Meyers process. It is most likely that any full-scale chemical desulfurization plants will use some variation of the MCL process, which is unique in that it produces a high-quality, low-ash, low-sulfur solid product efficiently with a high yield.

As of yet no coal preparation facility is known to use chemical desulfurization methods on a full plant scale. The costs remain too high compared to other coal-cleaning methods, and the price of competing fuel sources is often too low to justify the expense at all. Without major breakthroughs to decrease the processing cost or greatly increase the efficiency of these methods, it is unlikely that these methods will ever be practical at a full plant scale.

TABLE 15.2
Summary of Chemical Desulfurization Processes

	% Sulfur Reduction						
Process	**Pyritic**	**Organic**	**% BTU Recovery**	**% Ash Reduction**	**Scale**	**Problems**	**Source**
TRW molten caustic leaching (Gravimelt)	~100	>70	87	38–97	10–20 lb/h integrated process	High temperature (400°C), some loss of volatiles/caking properties, poor yield for low-rank coals	Nowak and Meyers (1993); Chriswell et al. (1991)
Battelle hydrothermal	90–95	20–70	90–95	N/A	10 kg/h continuous	High temperature (250°C–350°C) and pressure (4–17 MPa)	Stambaugh et al. (1975, 1977)
GE microwave	90	50–70[a]	~100	N/A	10–500 g batch	Results not reproducible on large scale, high energy cost	Zavitsanos et al. (1978)
ARCO promoted oxydesulfurization	88–98	0 (one stage)	95	Dec-50	1.2 kg/h continuous	Moderate temperature, moderate pressure	Beckberger et al. (1979)
		20–30 (two stage)				Slagging problems on combustion	
TRW Meyers	84–99	0	~100	04-Feb	8 tons/day pilot	Corrosion, only removes pyrite	Meyers et al. (1979)
Iron chloride	~100	30–50	74–82	N/A	Bench	Poor oxidation of benzothiophenes	Fan et al. (1987)
Jet Propulsion Laboratory (JPL) chlorinolysis	71–95[b]	46–98[b]	Lost up to 3 MJ/kg	04-Jan	2 kg/h continuous integrated	Up to 5% chlorine retention, very corrosive reagents	Kalvinskas et al. (1980)

(Continued)

TABLE 15.2 (*Continued*)
Summary of Chemical Desulfurization Processes

Process	% Sulfur Reduction		% BTU	% Ash	Scale	Problems	Source
	Pyritic	Organic	Recovery	Reduction			
KVB	60–100	30–50	74–82	N/A	50 g batch	Waste disposal, high NO_x product	Guth (1979)
IGT hydrodesulfurization	90% total Sulfur		55–84	N/A	45 kg/h Continuous	High temperature (800°C), product is not coal	Fleming et al. (1977)
Magnex	57–92	0	86	Jul-71	91 kg/h continuous integrated	Only removes pyrite and some ash minerals, needs magnetic separator	Porter and Goens (1979)
Chemical comminution	50–70	0	60–90	50–60	23 kg batch	Only liberates pyrite, requires subsequent physical cleaning step	Contos et al. (1978); Howard and Datta (1977)

Source: Adapted and extended from Morrison (1981).

[a] Organic sulfur removal for two reaction steps.

[b] JPL process using TCE.

TABLE 15.3
Summary of Recent Bench-Scale Chemical Desulfurization Studies

Process	% Sulfur Reduction		% BTU	% Ash	Problems	Source
	Pyritic	Organic	Recovery	Reduction		
Peroxyacetic acid + sodium carbonate	85%–95% total sulphur		<80%	N/A	High temperature, long residence time	Palmer et al. (1995)
Sodium hypochlorite	15–25	Nov-42	81–95	N/A	Long residence time, retains 2% Cl	Brubaker and Stoicos (1985)
Potassium permanganate	Up to 97	20–48	92	Poor	Moderately expensive reagent	Attia and Fung (1993); Attia et al. (1990)
Supercritical fluid extraction	80% total sulfur		N/A	N/A	Designed for small scale, relatively undeveloped	Louie et al. (1994)
Hydride reduction + SET + BASE	100	63	N/A	N/A	Very expensive reagents, reaction times of 24 h for each step	Stock and Chatterjee (1993)

REFERENCES

Abdul, A., Srivastava, S.K. and Haque, R. (1992), "Chemical desulfurization of high sulfur coals," *Fuel*, Vol. 71, pp. 24–29.

Akhtar, S.S. and Chriswell, C.D. (1993), "Hydrothermal Pretreatment of Coal Before Molten Caustic Leaching," *Processing and Utilization of High-Sulfur Coals V* (Parekh and Groppo, eds.), Elsevier Science Publishers, Amsterdam, pp. 288–290.

Akhtar, S.S. and Cliffe, K.R. (1990), "Effect of Oxydesulfurization on the Properties of the Cleaned Coal," *Fuel Processing Technology*, Vol. 24, No. 1–3, pp. 437–443.

Anastasi, J.L., Barrish, E.M., Coleman, W.B., Hart, W.D., Jones, J.F., Ledgerwood, L., McClanathan, L.C., Meyers, R.A., Shih, C.C. and Turner, W.B. (1990), "Molten Caustic Leaching (Gravimelt Process) Integrated Test Circuit Operation Results," *Processing and Utilization of High-Sulfur Coals III* (Markuszewski and Wheelock, eds.), Elsevier, Amsterdam, pp. 371–378.

Andres, J.M., Ferrando, A.C. and Ferrer, P. (1998), "Liquefaction of Low-Rank Coals with Hydriodic Acid and Microwaves," *Energy and Fuels*, Vol. 12, No. 3, pp. 563–569.

Anonymous (1989), "Selective Oxidation Desulfurization Process: Bench-Scale Studies," Electric Power Research Institute Report No. EPRI ER #6366, 64 pp.

Attia, Y.A. and Fung, A. (1993), "Chemical Desulfurization of Canadian High Sulfur Coal with Potassium Permanganate," *Processing and Utilization of High-Sulfur Coals V* (Parekh and Groppo, eds.), Elsevier Science Publishers, Amsterdam, pp. 263–282.

Attia, Y.A. and Lei, W. (1987), "Removal of organic sulfur from high sulfur coals by mild chemical oxidation using potassium permanganate," *Processing and Utilization of High-Sulfur Coals II* (Chugh and Caudle, eds.), Elsevier, Amsterdam, pp. 202–212.

Attia, Y.A., Lei, W. and Carlton, R.W. (1990), "Removal of reaction products from chemical oxidation of coal using potassium permanganate for desulfurization," *Processing and Utilization of High-Sulfur Coals III* (Markuszewski and Wheelock, eds.), Elsevier, Amsterdam, pp. 391–404.

Beckberger, L.H., Burk, E.H., Grossboll, M.P. and Yoo, J.S. (1979), "Preliminary Evaluation of Chemical Coal Cleaning by Promoted Oxydesulfurization," EPRI-EM-1044, April 1979, Atlantic Richfield Company, Harvey, IL, 128 pp.

Bluhm, D., Richardson, C., Markuszewski, R., Fanslow, G., Durham, K., and Green, T. (1986), "Effect of Microwave Irradiation on the Desulfurization and Deashing of Caustic-Treated Coal," *Journal of Microwave Power and Electromagnetic Energy*—International Microwave Power Symposium, Vol. 21, No. 2, pp. 75–76.

Bottrell, S.H., Louie, P.K.K., Timpe, R.C. and Hawthorne, S.B. (1994), "The use of stable sulfur isotope ratio analysis to assess selectivity of chemical analyses and extractions of forms of sulfur in coal," *Fuel*, Vol. 73, No. 10, pp. 1578–1582.

Bronikowski, T., Pasiuk-Bronikowska, W. and Ulejczyk, M. (1990), "Coal Desulphurization Facilitated by Coupled Autoxidation of Sulphur Dioxide," *Processing and Utilization of High-Sulfur Coals III* (Markuszewski and Wheelock, eds.), Elsevier, Amsterdam, pp. 425–435.

Brubaker, I.M. and Stoicos, T. (1985), "Precombustion Coal Desulfurization with Sodium Hypochlorite," *Processing and Utilization of High-Sulfur Coals* (Attia, ed.), Elsevier, Amsterdam, pp. 311–326.

Calkins, W.H. (1994), "The Chemical Forms of Sulfur in Coal: A Review," *Fuel*, Vol. 73, No. 4, pp. 475–484.

Carbini, P., Curreli, L., Ghiani, M. and Satta, F. (1990), "Desulphurization of European Coals Using Molten Caustic Mixtures," *Processing and Utilization of High-Sulfur Coals III* (Markuszewski and Wheelock, eds.), Elsevier, Amsterdam, pp. 361–369.

Chatterjee, K. and Stock, L.M. (1991), "Toward Organic Desulfurization: The Treatment of Bituminous Coals with Strong Bases," *Energy and Fuels*, Vol. 5, No. 5, pp. 704–707.

Chatterjee, K., Wolny, R. and Stock, L.M. (1990), "Coal Desulfurization by Single Electron Transfer Reactions," *Energy and Fuels*, Vol. 4, No. 4, pp. 402–406.

Chriswell, C.D., Markuszewski, R. and Norton, G.A. (1991), "Use of NaOH Alone vs. NaOH–KOH Mixtures for the Removal of Sulfur and Ash from Coal by the Molten Caustic Leaching Process," *Processing and Utilization of High-Sulfur Coals IV* (Dugan, Quigley, and Attia, eds.), Elsevier, Amsterdam, pp. 385–397.

Contos, G.Y., Frankel, I.F. and McCandless, L.C. (1978), "Assessment of Coal Cleaning Technology: An Evaluation of Chemical Coal Cleaning Processes," PB 289 493, August 1978, U.S. Environmental Protection Agency, Washington, DC, 299 pp.

Datta, R.S. (1977), "Chemical Fragmentation of Coal," *Proceedings of the Third Symposium on Coal Preparation*, CONF-7710113, October 18–20, 1977, Louisville, KY; U.S. National Coal Association, Washington, DC, pp. 22–31.

Duty, R.C. and Penrod, J.M. (1994), "Desulfurization of Illinois No. 6 Bituminous Coal via Reductive Carboxylation in Absolute Ethanol," *Energy and Fuels*, Vol. 8, No. 1, pp. 234–238.

Fan, C.W., Dong, G.W., Markuszewski, R. and Wheelock, T.D. (1987), "Coal Desulfurization by Leaching with Ferric or Cupric Salt Solutions," *Processing and Utilization of High-Sulfur Coals II* (Chugh and Caudle, eds.), Elsevier, Amsterdam, pp. 172–182.

Fleming, D.K., Smith, R.D. and Aquino, M.R.Y. (1977), "Hydrodesulfurization of Coals," *Coal Desulfurization, Chemical and Physical Methods*, ACS Symposium Series, Vol. 64 (Wheelock, ed.), American Chemical Society, Washington, DC, pp. 267–279.

Fullerton, K.L., Lee, S. and Kulik, C.J. (1991), "Process Engineering Studies of the Perchloroethylene Coal Cleaning Process," *Fuel Science and Technology International*, Vol. 9, No. 7, pp. 873–888.

Gryglewicz, G. (1996), "Effectiveness of High Temperature Pyrolysis in Sulfur Removal from Coal," *Fuel Processing Technology*, Vol. 46, No. 3, pp. 217–226.

Guth, E.D. (1979), "Oxidative Coal Desulfurization Using Nitrogen Oxides: The KVB Process," *Proceedings of the Symposium on Coal Cleaning to Achieve Energy and Environmental Goals*, PB-299 384, September 1978, Hollywood, FL; April 1979, U.S. Environmental Protection Agency, Washington, DC, pp. 1141–1164.

Hayashi, J.-I., Oku, K., Kusakabe, K. and Morooka, S. (1990), "Role of Microwave Irradiation in Coal Desulphurization with Molten Caustics," *Fuel*, Vol. 69, No. 6, pp. 739–742.

Howard, P.H. and Datta, R.S. (1977), "Chemical comminution: A Process for Liberating the Mineral Matter from Coal," *Coal Desulfurization, Chemical and Physical Methods*, ACS Symposium Series, Vol. 64 (Wheelock, ed.), American Chemical Society, Washington DC, pp. 58–69.

Huffman, G.P., Shah, N., Huggins, F.E., Stock, L.M., Chatterjee, K., Kilbane II, J.J., Chou, M.M. and Buchanan, D.H. (1995), "Sulfur Speciation of Desulfurized Coals by XANES Spectroscopy," *Fuel*, Vol. 74, No. 4, pp. 549–555.

Ibarra, J.V., Palacios, J.M., Moliner, R. and Bonet, A.J. (1994), "Evidence of Reciprocal Organic Matter-Pyrite Interactions Affecting Sulfur Removal During Coal Pyrolysis," *Fuel*, Vol. 73, No. 7, pp. 1046–1050.

Kalvinskas, J., Grohmann, K., Rohatgi, N., Ernest, J. and Feller, D. (1980), "Coal Desulfurization by Low Temperature Chlorinolysis, Phase II," Final Report, January 15, 1980, Jet Propulsion Laboratory, California Institute of Technology, Pasadena, CA, 150 pp.

Kalvinskas, J. and Hsu, G.C. (1979), "JPL Coal Desulfurization Process by Low Temperature Chlorinolysis," *Proceedings of the Symposium on Coal Cleaning to Achieve Energy and Environmental Goals*, PB-299 384, September 11–15, 1978, Hollywood, FL; U.S. Environmental Protection Agency, Washington, DC, pp. 1096–1140.

Kindig, J.K. and Goens, D.N. (1979), "The Dry Removal of Pyrite and Ash from Coal by the Magnex Process, Coal Properties and Process Variables," *Proceedings of the Symposium on Coal Cleaning to Achieve Energy and Environmental Goals*, PB-299 384, September, 1978, Hollywood, FL; U.S. Environmental Protection Agency, Washington DC, pp. 1165–1196.

Kindig, J.K. and Turner, R.L. (1976), "Dry Chemical Process to Magnetize Pyrite and Ash for Removal from Coal," *Preprint of paper presented at 1976 SME-AIME Fall Meeting and Exhibit*, September 1–3, 1976, Denver, CO, Preprint No. 76-F-306; Society of Mining Engineers of AIME, Salt Lake City, UT, 18 pp.

Kocher, B.S., Azzam, F.O. and Lee, S. (1994), "Removal of Residual Chlorine from Indiana 5 Coal by Supercritical Carbon Dioxide Extraction," *Fuel Science and Technology International*, Vol. 12, No. 2, pp. 303–313.

Kusakabe, K., Morooka, S. and Aso, S. (1988), "Chemical Coal Cleaning with Molten Alkali Hydroxides in the Presence of Microwave Radiation," *Fuel Processing Technology*, Vol. 19, No. 3, pp. 235–242.

Lee, S., Vishnubhatt, P., Berthinee, E.A. and Kulik, C.J. (1993), "Cobeneficiation of Coals in Perchloroethylene Extraction Desulfurization," *Fuel Science and Technology International*, Vol. 11, No. 5–6, pp. 831–844.

Lee, S., Vishnubhatt, P. and Kulik, C.J. (1994), "Selective Removal of Organic Sulfur from Coal by Trichloroethane Extraction," *Fuel Science and Technology International*, Vol. 12, No. 2, pp. 211–228.

Li, W. and Guo, S. (1995), "Removal of Organosulfur by Supercritical Extraction of Coal with Alcohol," *Dalian Ligong Daxue Xuebao/Joumal of Dalian University of Technology*, Vol. 35, No. 3, pp. 325–329.

Lin, L., Khang, S.J. and Keener, T.C. (1997), "Coal Desulfurization by Mild Pyrolysis in a Dual-Auger Coal Feeder," *Fuel Processing Technology*, Vol. 53, No. 1–2, pp. 15–29.

Louie, P.K.K., Timpe, R.C., Hawthorne, S.B. and Miller, D.J. (1994), "Sulfur Removal from Coal by Analytical-scale Supercritical Fluid Extraction (SFE) Under Pyrolysis Conditions," *Fuel*, Vol. 73, No. 7, pp. 1173–1178.

Majchrowicz, B.B., Franco, D.V., Yperman, J., Reggers, G., Gelan, J., Martens, H., Mullens, J. and Van Poucke, L.C. (1991), "Investigation into the Changes of Structure and Reactivity during Desulphurization of Bituminous Coals," *Fuel*, Vol. 70, No. 3, pp. 434–441.

McIlvried, T.S., Smouse, S.M., Ekmann, J.M. (1990), "Evaluation of combustion performance of molten-caustic-leached coals," *Processing and Utilization of High-Sulfur Coals III* (Markuszewski and Wheelock, eds.), Elsevier, Amsterdam, pp. 379–390.

Meyers, R.A. (1977), *Coal Desulfurization*, Marcel Dekker, New York, 254 pp.

Meyers, R.A., Hamersma, J.W., Land, J.S. and Kraft, M.L. (1972), "Desulfurization of Coal," Science, Vol. 177, pp. 1187–1188.

Meyers, R.A., Koutsoukos, E.P., Santy, M.J. and Orsini, R. (1979), "Bench-Scale Development of Meyers Process for Coal Desulfurization," Final Report, PB-290 515, January, 1979, TRW Systems Group, Redondo Beach, CA, 104 pp.

Midwest Ore Processing Co. (1992), "Coal Desulfurization by Perchloroethylene Processing," Final Report, Vol. 1–3, EPRITR-100551, Project 3027-1.

Mishra, V.S., Mahajani, V.V. and Joshi, J.B. (1995), "Wet Air Oxidation," *Industrial and Engineering Chemistry Research*, Vol. 34, No. 1, pp. 2–48.

Morrison, G.F. (1981), *Chemical Desulfurization of Coal*, IEA Coal Research, London.

Norton, G.A., Bluhm, D.D., Markuszewski, R. and Chriswell, C.D. (1991), "Application of Microwave Energy to Caustic Cleaning of Coal," *Processing and Utilization of High-Sulfur Coals IV* (Dugan, Quigley and Attia, eds.), Elsevier, Amsterdam, pp. 425–138.

Nowak, M.A. and Meyers, R.A. (1993), "Molten-Caustic Leaching (MCL) process Integration," *Processing and Utilization of High-Sulfur Coals V* (Parekh and Groppo, eds.), Elsevier Science Publishers, Amsterdam, pp. 305–315.

Palmer, S.R., Hippo, E.J. and Dorai, X.A. (1995), "Selective Oxidation Pretreatments for the Enhanced Desulfurization of Coal," *Fuel*, Vol. 74, No. 2, pp. 193–200.

Porter, C.R. and Goens, D.N. (1979), "Magnex Pilot Plant Evaluation—A Dry Chemical Process for the Removal of Pyrite and Ash from Coal," *Mining Engineering*, Vol. 31, No. 2, pp. 175–180.

Prasassarakich, P. and Pecharanond, P. (1992), "Kinetics of Coal Desulfurization in Aqueous Copper (II) Sulfate," *Fuel*, Vol. 71, pp. 929–933.

Rhee, K.H. and Wan, E.I. (1987), "Overview of Chemical and Microbial Coal-Cleaning Activities," *Processing and Utilization of High-Sulfur Coals II* (Chugh and Caudle, eds.), Elsevier, Amsterdam, pp. 224–231.

Rondio, K., Breiter, B., Meyers, R.A., McClanathan, L., Hardgrove, J., Storsul, S. and Meland, P. (1993), "Polish Gravimelt Coal Refinery Project," *Presented at the Polish-American Conference on Reconstruction and Modernization of Heat and Power Generating Plants*, June 15–17, Warszawa, Poland.

Stambaugh, E.P. (1977), "Hydrothermal Coal Process," *Coal Desulfurization—Chemical and Physical Methods*, ACS Symposium Series No. 64 (T.D. Wheelock, ed.), American Chemical Society, Washington, DC, pp. 198–205.

Stambaugh, E.P., Miller, J.F., Tam, S.S., Chauhan, S.P., Feldmann, H.F., Carlton, H.E., Foster, J.F., Nack, H. and Oxley, J.H. (1975), "Hydrothermal Process Produces Clean Fuel," *Hydrocarbon Processing*, Vol. 54, No. 7, pp. 115–116.

Stock, L.M. and Chatterjee, K. (1993), "Organic Desulfurization of Illinois No. 6 Coal," *Proceedings of the 5th International Conference on Processing and Utilization of High Sulfur Coals*, Elsevier Science Publishers, Amsterdam, p. 291.

Tsai, S.C. (1982), *Fundamentals of Coal Beneficiation and Utilization*, Elsevier, Amsterdam, pp. 353–371.

Valasek, V., Roucek, V., Meyers, R.A., Hardgrove, J.M., McClanathan, L., Storsul, S. and Meland, P. (1992), "Gravimelt Coal Refinery Project in Czechoslovakia," *Presented at the International Conference on the Clean and Efficient Use of Coal*, February 24–27, Budapest, Czechoslovakia.

Vishnubhatt, P. and Lee, S. (1993), "Effect of Filtration Temperature on Organic Sulfur Removal from Coal by Perchloroethylene Coal Cleaning Process," *Fuel Science and Technology International*, Vol. 11, No. 8, pp. 1081–1093.

Vishnubhatt, P., Thome, T. and Lee, S. (1993), "Effect of Pyritic Sulfur and Mineral Matter on Organic Sulfur Removal from Coal," *Fuel Science and Technology International*, Vol. 11, No. 7, pp. 923–936.

Vorres, K.S. (1990), "Sulfur Species in Perchloroethylene and Other Coal Extracts," *American Chemical Society Division of Fuel Chemistry Preprints*, Vol. 35, No. 2, pp. 523–529.

Wang, X.-H., Guo, Q., Yingling, J.C. and Parekh, B.K. (1996), "Improving Pyrite Liberation and Grinding Efficiency in Fine Coal Comminution by Swelling Pretreatment," *Coal Preparation*, Vol. 17, No. 3–4, pp. 185–198.

Wang, J., Yan, J. and Wong, S. (1990), "Changes of Iron-Sulfur Compounds in the Process of Coal Microwave Desulfurization," *Huadong Huagong Xueyuan Xuebao/Joumal of East China Institute of Chemical Technology*, Vol. 16, No. 1, pp. 45–49.

Weng, S. and Wang, J. (1992), "Exploration on the Mechanism of Coal Desulfurization Using Microwave Irradiation/Acid Washing Method," *Fuel Processing Technology*, Vol. 31, No. 3, pp. 233–240.

Yaman, S., Ersoy-Mericboyu, A. and Kucukbayrak, S. (1995), "Chemical Coal Desulfurization Research in Turkey," *Fuel Science and Technology International*, Vol. 13, No. 1, pp. 49–58.

Yaman, S. and Kucukbayrak, S. (1996), "Oxydesulfurization of a Turkish Hard Lignite with Ammonia Solutions," *Energy Sources*, Vol. 18, No. 6, pp. 677–683.

Yang, J. and Ren, J. (1988), "Coal Desulfurization by Microwave and Its Relation with Sample Dielectric Property," *Journal of East China Institute of Chemical Technology*, Vol. 14, No. 6, pp. 713–718.

Yang, J.-K. and Wi, Y.-M. (1987), "Relation between Dielectric Property and Desulphurization of Coal by Microwaves," *Fuel*, Vol. 66, No. 12, pp. 1745–1747.

Yueruem, Y. and Tugluhan, A. (1990), "Supercritical Extraction and Desulphurization of Beypazari Lignite by Ethyl Alcohol/NaOH Treatment 2. Kinetics," *Fuel Science and Technology International*, Vol. 8, No. 3, pp. 221–240.

Zavitsanos, P.D., Golden, J.A., Bleier, K.V. and Jacobs, I.S. (1982), "Coal Desulfurization by a Microwave Process," DOE Contract No. DE-AC22-80PC30142, Progress Reports through January 1982.

Zavitsanos, P.D., Golden, J.A., Bleiler, K.W. and Kinkead, W.K. (1978), "Coal Desulfurization Using Microwave Energy," PB-285 880, U.S. Environmental Protection Agency, Washington, DC, 79 pp.

16 CO_2 Sequestration

Burning coal releases carbon dioxide (CO_2). As of 2016, approximately 36 billion tons of CO_2 are emitted around the world by human sources. Of these emissions, about 15.5 billion tons of CO_2 are attributable to the combustion of coal. Around the world, coal is responsible for 43%–44% of all human CO_2 emissions (Janssens-Maenhout et al., 2017; EIA, 2018).

CO_2 is believed to be a greenhouse gas and contribute to global warming, and as such many countries are seeking to regulate CO_2 emissions. No single fuel will be impacted by these actions more than coal. The potential for CO_2 to be emitted by coal is an inherent result of the chemistry of combustion and the composition of coal, but it is possible to mitigate or eliminate CO_2 emissions by preventing them from reaching the atmosphere. If legislation restricting CO_2 emissions come to pass, coal users will need to adapt.

The capture and sequestration technologies presented in this section have been developed for the future of coal, the future of fossil fuels, and the future of the world as a whole. Initial investments into CO_2 capture and sequestration were stymied in many locations due to the cost at the time and the lack of a marketable use for the captured CO_2. However, CO_2 capture can be made highly efficient, and there are sequestration opportunities that can result in a net profit. This section aims to provide a guideline of the technologies available and the development of the capture and sequestration of CO_2 from fossil fuels. While there is a focus on coal, these techniques are typically applicable to any fuel in any large-scale process.

It will be best if we can sequester CO_2 while simultaneously removing SO_2 and NO_x compounds (Kawatra and Eisele, 2001). It is necessary to develop a scrubber that will remove SO_2, NO_x, and CO_2 simultaneously. Methane is another important pollutant to consider during the mining of coal, but at a power facility it simply serves as another fuel and can be removed by combusting it. However, managing methane emissions from a mine is outside the scope of this chapter.

16.1 CO_2 SEQUESTRATION OVERVIEW

There are three primary methods that have been considered in much detail for sequestering, or permanently removing, CO_2 from the atmosphere or atmospheric emissions. These methods are

- Geological sequestration: Storing the CO_2 in underground reservoirs that are believed to be capable of permanently containing it, even under great pressures and in large quantities (Bachu, 2015)
- Oceanic sequestration: Reacting the CO_2 with mildly alkaline oceanic waters to form carbonates or bicarbonates, which are nonvolatile and largely unreactive

- Mineral carbonation: Reacting CO_2 with alkaline minerals or mineral-bearing wastes to form carbonates, which again are nonvolatile and largely unreactive

These three methodologies are in development because the amount of CO_2 produced is enormous. While it would absolutely be preferable to reuse the CO_2 in a more practical, economically beneficial process than pumping it underground or dissolving it in the ocean, there do not yet exist sufficient uses for CO_2 to use even a meaningful fraction of it. In comparison, these sequestration techniques are expected to have the theoretical capacity to capture the majority or even entirety of human CO_2 emissions.

The infrastructure to actually capture 36 billion tons of CO_2, of course, does not yet exist. Developing this infrastructure is the largest hurdle to mitigating these CO_2 emissions, and is at least as important as sequestering or utilizing the CO_2. However, as of 2017, there were approximately 31 million tons of active, large-scale CO_2 capture and sequestration projects in the world (Global CCS Institute, 2017). Many of these projects are enhanced oil recovery (EOR) projects, wherein CO_2 is used to displace oil from partially depleted oil reservoirs. These reservoirs are expected to meet the requirements for geological sequestration and contain the CO_2 indefinitely into the future. The oil recovered in this process is then a profitable incentive to continue the operation. However, not all oil reservoirs are suitable for EOR, the specific details of which will be discussed in Chapter 17.

Mineral carbonation occurs naturally as well, but is typically quite slow. The process can be made more practical by increasing the surface area of the minerals and adding a catalyzing or pseudo-catalyzing step (Wang and Maroto-Valer, 2011; Yuen et al., 2016).

Each of these methods will be discussed in detail in Sections 16.2.4 and 16.3 later.

For each of these methods, it is ideal to have a CO_2 stream that is as concentrated in CO_2 as is practical. This typically involves separating it from other gases, especially nitrogen in air or flue gases. Several separation methodologies for this problem are described in Section 16.2.

An example of the distribution of CO_2 emissions by sector in the United States is provided in Figure 16.1.

16.2 CAPTURING CO_2

Carbon capture is the process of removing CO_2 from a gas stream that would ultimately be released to the atmosphere. The separated CO_2 is typically purified and then sequestered or utilized elsewhere. There are four major strategies as to where to start with carbon capture.

- *Post-combustion capture*: remove CO_2 from an existing flue gas stream at a facility by adding a scrubber or adsorber loop. Useful for retrofitting onto existing plants, but with the lowest overall efficiency.
- *Pre-combustion capture*: remove CO_2 from the fuel directly by gasifying the fuel into hydrogen and carbon dioxide. The CO_2 can then be removed

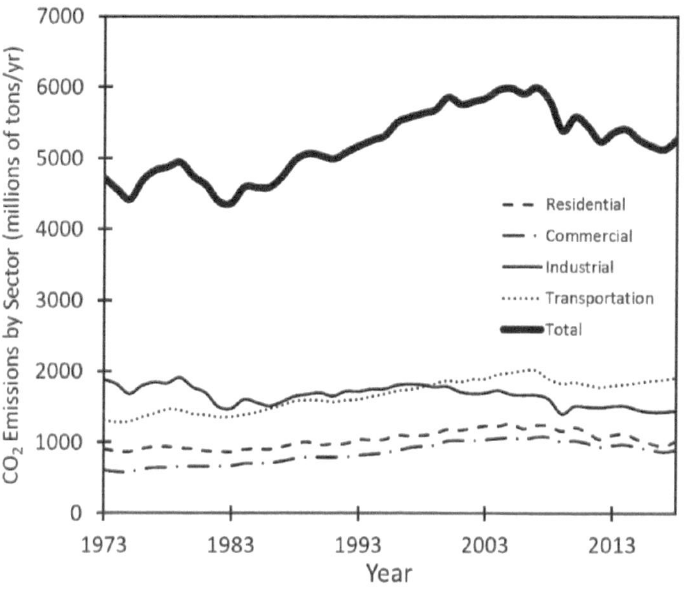

FIGURE 16.1 A breakdown of CO_2 emissions by each labeled sector in the US. In recent years transportation has overtaken industrial CO_2 emissions. Aside from transportation, the majority of CO_2 emissions can also be traced back to fossil-fuel based electric power generation. (Data from EIA, 2019.)

directly via scrubbing, ultrafiltration, or other methods. Typically requires a complete process redesign, as the fuel of combustion is now hydrogen.

- *Oxy-combustion capture*: perform the fuel combustion in an oxygen-rich atmosphere, so that the resulting flue gas is a nearly pure CO_2 stream to begin with. Does not require changing the theoretical unit operations, but the limits of the equipment must be taken into consideration as combustion temperatures in pure oxygen can be dramatically higher than in air.
- *Chemical looping capture*: use various reagents to selectively transfer oxygen from an air reactor to a combustion reactor. This avoids needing to perform air separation elsewhere.

The simplest method is *post-combustion capture*. Once combustion proceeds, simply separate the CO_2 from the flue gases before they are released from the atmosphere. The largest advantage of post-combustion capture is that it does not require a major redesign of existing facilities, and can be installed as a simple tail unit. The disadvantage of post-combustion capture is that it must take place somewhat close to atmospheric pressure, working with relatively dilute CO_2 streams. The stripper column and subsequent compression step typically require a significant portion of a power plant's output. In the case of a coal-fired facility the worst-case scenario may be as much as one-third of the total power output, resulting in a 70% increase in electricity costs.

A typical post-combustion CO_2 scrubber process is setup along the lines of Figure 16.2. Amine solutions are predominantly used as the capture solution in industry today (Rochelle, 2009; Spigarelli and Kawatra, 2013), but aside from bicarbonates any alkaline compound has the potential to capture CO_2. In particular, sodium hydroxide (NaOH) and sodium carbonate (Na_2CO_3) have been found to be effective for capturing CO_2 (Kawatra et al., 2011). Bicarbonates (HCO_3^-) are excepted because they are already saturated with CO_2: there is no intermediate compound that can be formed by an interaction between a bicarbonate and carbonic acid.

A typical CO_2 inlet stream for such a scrubber is around 16% CO_2, with the remainder as mostly nitrogen. In coal-fired facilities, this typically requires that SO_x and NO_x compounds be removed first, as they will compete with, inhibit, or bypass the CO_2 capture process, depending on the capture solution and regeneration step. The capture solution is alkaline in nature, capturing CO_2 as described in Section 16.2.1 (Yeh et al., 2005).

However, the fundamental chemistry of post-combustion CO_2 capture is very similar to the post-combustion capture of SO_2 and NO_x compounds. Perhaps it would be possible to design a post-combustion capture system that overcomes these problems.

CO_2 capture can proceed at moderately high temperatures, between 40°C and 70°C, but lower temperatures near ambient are generally preferable. This is partially because the solubility of CO_2 in water decreases as temperature increases, but also because above 70°C most practical capture solutions begin to release their captured CO_2 and regenerate. Thus, hot flue gases should be cooled before the scrubber column.

In the stripper column, the most typical stripping gas is steam. Steam is compatible with aqueous solutions, but it will slightly dilute the overall solution. Between this and the potential loss of alkaline solute during regeneration, this process will also require a bleed stream to extract the excess liquid and a fresh feed stream to replace any missing alkaline solute. Besides steam, hot CO_2 gas will be compatible

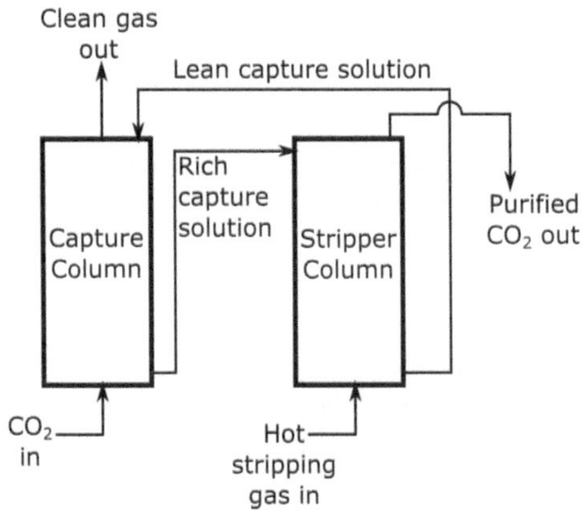

FIGURE 16.2 Generic post-combustion CO_2 scrubber. The CO_2 stream input is cleaned in the capture column, while the capture solution is regenerated in the stripper column. The most common stripping gas is water.

with all thermally stable capture solutions as a stripping gas. However, if the alkaline solute is, for example, a volatile amine or the regeneration needs to take place near or above water's boiling point, it is still possible for various contaminants to get into the purified CO_2 stream. These potential impurities can be removed with a condenser or water wash.

The most common capture solutions used in industry today are amine chemistries, but there has also been significant investigation into hydroxide chemistries, carbonates (Spigarelli et al., 2013), and various mineral slurries (Rochelle, 2009; Boot-Handford et al., 2014).

In all cases, most of the cost of post-combustion capture is in energy input into the stripping step. Careful management of the heat exchange between hot and cold streams in these processes is vital to maintaining low operating costs (Boot-Handford et al., 2014).

The direct alternative is *pre-combustion capture*. Instead of combusting a coal or hydrocarbon fuel, the fuel is gasified first to form hydrogen gas and CO_2 (Spigarelli and Kawatra, 2013). The CO_2 is then separated before the hydrogen is sent off for combustion. As hydrogen gas contains no carbon, its combustion does not result in the formation of CO_2, so the carbon has been captured before combustion takes place. The advantage of this strategy is that the CO_2 content and partial pressures are quite a bit higher, allowing for better reaction kinetics during the capture step.

However, it is not usually practical to retrofit pre-combustion capture onto an existing facility – pre-combustion typically takes place at pressures of 15–40 bars, which typically requires a complete redesign of all of the involved equipment. Furthermore hydrogen's high-energy density presents rather different difficulties as a fuel than any hydrocarbon. Pre-combustion capture is far more enticing in the design of new facilities, where the total cost of electricity production is expected to increase by less than 25% compared to a baseline of no-capture. This setup is presented in Figure 16.3.

A middle ground of these two is *oxy-combustion* (Spigarelli and Kawatra, 2013). In most processes, most of the gas flowing through any combustion system is actually nitrogen, due to its ubiquitous presence in atmospheric air. By separating that out and using pure oxygen for combustion instead, the starting CO_2 product can be

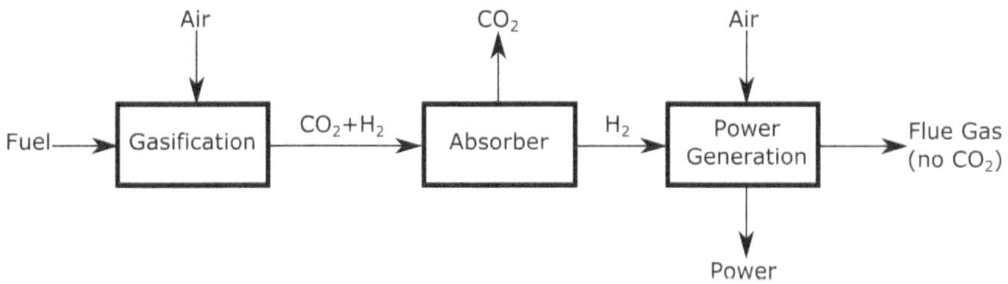

FIGURE 16.3 Generic pre-combustion capture CO_2 scrubber. Pre-combustion capture is specifically intended as an alternative to the combustion of other fuels – the final power generation step may actually be performed in fuel cells. The common absorber technology is membrane separation.

made significantly purer – in a stoichiometric and complete combustion followed by a simple water wash, it is possible that the flue gas may in fact be pure CO_2. If high purities must be guaranteed, oxy-combustion processes may not completely suffice, since stoichiometric complete combustion is difficult to guarantee. The CO_2 from the flue gas may be looped back through the combustion to achieve higher CO_2 purities. This strategy is shown in Figure 16.4.

Chemical looping combustion is one particular implementation of the oxy-combustion strategy worth to note. By using metals that can be oxidized alongside suitably reducing fuels, it is possible to cycle oxygen, and only oxygen, from an air vessel to a fuel reactor (Spigarelli and Kawatra, 2013). This avoids many of the typical difficulties and hazards of working with pure oxygen. This strategy is outlined in Figure 16.5.

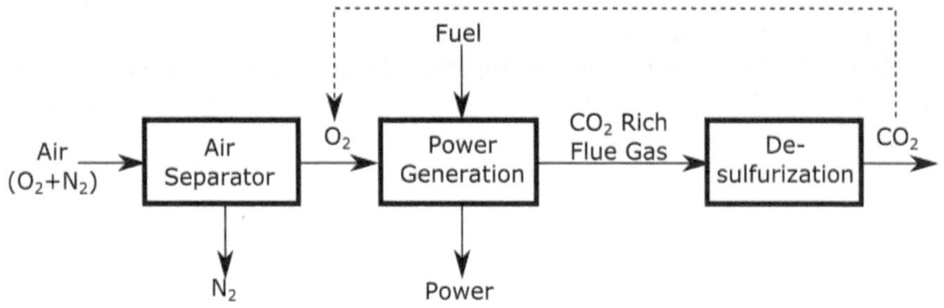

FIGURE 16.4 Generic oxy-combustion process. By performing combustion in pure oxygen, energy is not spent heating or removing nitrogen, and higher CO_2 purities can be achieved. Air separation can be achieved with membrane separation technologies, while desulfurization processes have been discussed earlier in this book. CO_2 may be recycled back to the power generation step to increase total pressure, if needed.

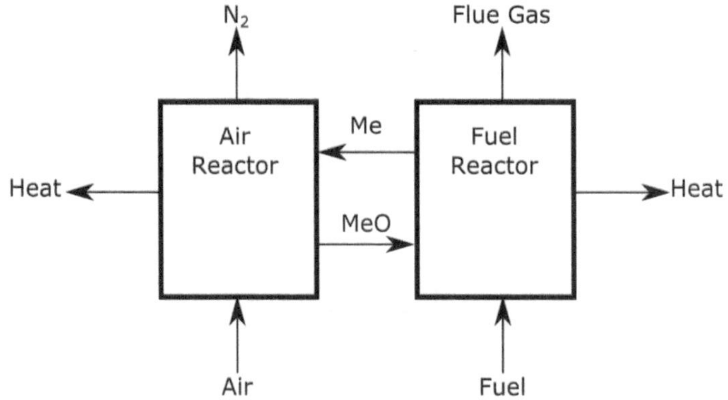

FIGURE 16.5 Generic chemical looping process using metal oxides. A metal is used to selectively transfer oxygen from the air reactor to the fuel reactor. The metal oxide may be an iron, nickel, copper, manganese, or other oxide. The flue gas will be rich in CO_2, as no nitrogen is present during combustion. Heat can typically be recovered from both reactors.

TABLE 16.1
Comparison of CO_2 Capture Chemistries

Chemistry	Typical Use Case	Advantages	Disadvantages
Hydroxides (e.g. NaOH)	Solid sorbents Direct air capture	Very good capture kinetics Very good capture capacity Thermally stable	Difficult to regenerate Very corrosive
Simple amines (e.g. MEA)	Post-combustion capture	Easily regenerated Good capture capacity	Thermally unstable Toxic Somewhat corrosive
Hindered amines (e.g. KS-1, PZ)	Post-combustion capture	Easily regenerated Good capture kinetics More stable during regeneration	Toxic
Carbonates (e.g. Na_2CO_3, K_2CO_3)	Post-combustion capture	Easily regenerated Thermally stable during regeneration	Slow capture kinetics
Alkaline minerals (e.g. serpentinite, olivine)	Sequestration as mineral carbonates	Simple sequestration case Depending on location, reagents may be abundant	Very slow capture kinetics Difficult regeneration
Activated carbon	Air filtration	Does not require energy Consistent capture capacity	Limited capture capacity
MOFs	High-efficiency capture and storage	Very good capture capacity Low energy cost of regeneration	Temperature sensitive Expensive Not in industrial use as of 2018

In all of these capture strategies, it is typically necessary to separate the CO_2 from a stream containing potentially many compounds. There are several chemistries available for this, which will be discussed in detail starting at Section 16.2.1, but are outlined in Table 16.1.

Once CO_2 has been effectively captured, the question becomes what should be done with it. In this chapter, we will discuss various options of sequestration – fixing the CO_2 in a stable, environmentally-friendly location or form. It is possible to store CO_2 underground or in the ocean – geological and oceanic sequestration – but there are other possibilities as well. These include forestation, ocean fertilization, photosynthesis, and mineral carbonation. These options will be discussed in Section 16.3.

16.2.1 Chemistry of Capturing CO_2

CO_2 is soluble in and is hydrolyzed by water. These equilibrium reactions are shown in Equations 16.1 and 16.2 below, forming carbonic acid (H_2CO_3) at the end (Lower, 1999).

$$CO_2(g) \rightleftharpoons CO_2(aq) \tag{16.1}$$

$$CO_2(aq) + H_2O(l) \rightleftharpoons H_2CO_3(aq) \tag{16.2}$$

The equilibrium constant of Equation 16.2 is relatively unfavorable towards the formation of carbonic acid. In particular, the equilibrium is given by Equation 16.3 below, where square brackets indicate the concentration of the given component: (Lower, 1999)

$$K_{H_2CO_3} = \frac{[H_2CO_3]}{[CO_2(aq)]} \approx \frac{1}{650} \tag{16.3}$$

Equation 16.3 assumes that the water is free of other impurities, which may shift the equilibrium in either direction. Note however that it is common for claimed concentrations of carbonic acid to be the combined concentrations of the true carbonic acid and the dissolved by still hydrated CO_2 (Lower, 1999).

The equilibrium between the dissolved gas and the proper gas phase is given by Equation 16.4, based on Henry's law:

$$P_{CO_2} = K_H[CO_2(aq)] \tag{16.4}$$

K_H is typically around 30 atm/(mol/L) and depends strongly on temperature. As a gas, the solubility of CO_2 in water decreases as temperature increases, meaning that K_H increases as temperature increases (Lower, 1999).

All of these things together mean that CO_2 over water will diffuse into the water, where it will begin to behave as an acid. This is the mechanism that every post-combustion CO_2 capture process designed to date relies on. The unstable carbonic acid can be neutralized using alkaline compounds to stabilize the carbon compounds and shift the equilibrium so that more CO_2 may dissolve.

In the presence of an alkaline compound, the acid–base reactions that capture the CO_2 as carbonates or bicarbonates tend to be extremely fast. The kinetics of this interaction is thus primarily limited by the dissolution and hydrolysis of CO_2. The dissolution kinetics can be improved by increasing the interfacial area between the gas and liquid phases. In a column, this can be accomplished with packing or by decreasing the bubble size of the injected CO_2 (Aaron and Tsouris, 2005).

Typical scrubbing solutions prefer solutions with high heats of absorption for CO_2. This is because these compounds exhibit a stronger shift in relative pressures of CO_2 vs. H_2O at the solution surface at elevated temperatures and pressures. Thus, at high temperatures and pressures, these solutions end up requiring less energy overall to regenerate. At low pressures and vacuum systems, the difference in relative pressures is not as relevant – solvents with low heats of absorption are preferable for isothermal systems. However, as the CO_2 being pulled off in these scrubbers is typically immediately heading to a compressor, working at higher pressures is more preferable than not (Oyenekan and Rochelle, 2007; Boot-Handford et al., 2014).

At higher temperatures, such as those required in pre-combustion capture processes, it is sometimes ideal to forgo the liquid phase entirely. Solid sorbent systems can achieve high CO_2 capture capacities – lithium hydroxide is likely the extreme example, with a capacity of approximately 6.8 mol CO_2/kg sorbent. In comparison, most aqueous capture solutions manage less than 1 mol CO_2/kg solute. With respect to coal processing, these sorbents are primarily useful when combined with coal gasification techniques or oxycombustion techniques. They are less commonly used for post-combustion capture, as the temperature requirements to achieve sorbent regeneration are quite a bit higher and the heat management is significantly trickier with solids than liquid streams. Due to the elimination of water and other carrier species, however, the heat required overall is not too much higher, despite the increased temperatures (Satyapal et al., 2001; Li et al., 2013; Boot-Handford et al., 2014).

CO_2 sorbents are also the most practical way of capturing CO_2 without any associated power cost. A sorbent at an exothermic heat of absorption can capture CO_2 with no outside energy source – so long as regeneration is not required. For this reason, solid sorbents are common in CO_2 scrubbers designed for submarines and spaceships (Satyapal et al., 2001).

Sorbent technologies are primarily based on compounds with high thermal stability, such as alkaline hydroxides and zeolites. There has been development of amine-based sorbents for space missions which can be regenerated at more reasonable temperatures and does not need to be thermally stable at elevated temperatures. Such sorbents may also be useful in undersea vessels, as it is less caustic than traditional hydroxide sorbents, and potentially presents a less immediate hazard in the case of a hull breach (Satyapal et al., 2001; Li et al., 2013; Boot-Handford et al., 2014).

16.2.2 Amine Scrubbing Systems

Amines have been found to be quite effective at capturing CO_2 in aqueous solutions. Measuring by weight of CO_2 captured per weight of amine, the most effective amine for CO_2 capture is ammonia, which has the added benefit of being particularly inexpensive. However, ammonia is significantly volatile, and up to 30% of it may be removed from the capture solution during stripping. This can typically be recovered by a condenser or water wash, but it adds to the overall heat load of the process. However, ammonia does not thermally degrade at temperatures sufficient to regenerate it, which may potentially provide a cost saving over heavier amines (Yeh et al., 2005).

Current industrial practice favors these heavier amines – the most commonly discussed amine in literature for CO_2 scrubbing is monoethanolamine (MEA). However, most industrial processes rely on proprietary amines such as KS-1 or PSR solvents, or recently developed solutes such as piperazines. KS-1 and PSR are both sterically hindered amines which are intended to provide better regeneration characteristics than traditional amines (Aaron and Tsouris, 2005; Rochelle, 2009; Spigarelli and Kawatra, 2013; Boot-Handford et al., 2014).

These amines are typically regenerated with steam at stripping temperatures between 100°C and 120°C, followed by a water wash to create a high-purity CO_2 product. The CO_2 is then typically compressed for shipment, storage, or sequestration, often to 100–150 bar. The stripping should ideally take place at the highest temperature allowed by the thermal degradation of the solute, so that CO_2 pressures at the compressor inlet are as high as possible, minimizing compression work (Rochelle, 2009; Boot-Handford et al., 2014).

Common concentrations of MEA in the scrubbing solution are 20%–30%. This concentration is mostly to provide CO_2 scrubbing capacity – capture rates of 90%

can be achieved with much lower concentrations of MEA if the CO_2 flow rate is low enough to avoid overloading it. Higher capture rates can be achieved with amine scrubbers, but it is expected that the maximum energy efficiencies occur around 90% (Rochelle, 2009).

MEA is subject to thermal degradation, but it can be minimized by stripping at lower temperatures (i.e. 100°C) and MEA is one of the least expensive amines. Costs from MEA replacement are expected to remain under \$5/ton CO_2 captured. (Rochelle, 2009)

The proprietary amines are harder to make general statements. They are typically sterically hindered amines, such as KS-1, designed to improve the thermodynamics of CO_2 capture. These sterically hindered amines are easier to thermally regenerate, requiring less energy overall per ton of CO_2 captured – 2.7 MJ/kg CO_2 for KS-1 vs. 3.7 MJ/kg CO_2 for MEA (MacDonald, 2008; Rochelle, 2009).

Piperazine is another amine alternative that is receiving significant investigation. It has been found to be comparatively thermally stable, allowing for stripping to occur efficiently at elevated pressures. Piperazine concentrations are typically around 40 wt%. Piperazine has also been found to have very good absorption kinetics of amines tested so far, as compared to hindered amines that typically have poor kinetics overall. High pressure stripping environments may also open the door to using organic solvents instead of water for the solvent phase – these solvents typically have lower heat capacities than water, and may represent a significant energy savings. However, at lower pressures, they are too volatile to be worthwhile (Aaron and Tsouris, 2005; Boot-Handford et al., 2014).

MEA can also be degraded by SO_x, NO_x, fly-ash, free oxygen, chloride, and gypsum. SO_2 neutralizes MEA, resulting in less capture, while NO_2 results in carcinogenic nitrosamines. SO_3 and fly-ash can aerosolize the amines and cause equipment fouling. Chloride results in corrosion, while gypsum can cause precipitation in the solvent itself (Rochelle, 2009; Boot-Handford et al., 2014; Rochelle, 2014).

Oxidative degradation of MEA is primarily catalyzed by dissolved iron and manganese, leading to the formation of various nuisance chemicals. There are concerns that these products may not be environmentally tolerable at large concentrations that may be formed with large-scale plants, but they have been worked around at small scales so far. Some amines, notably tertiary amines, piperazine, and hindered amines (presumably including KS-1), are more resistant to this oxidation. These oxidations can only occur in the presence of oxygen, however, and taking measures to remove the oxygen will correspondingly mitigate this degradation. It can be further mitigated by performing the stripping at lower temperatures, when possible (Boot-Handford et al., 2014).

SO_2 and SO_3 removal has been discussed in prior chapters, but may not be sufficient to remove all SO_x compounds of relevance to CO_2 scrubbing. It has been found that extremely fine H_2SO_4 aerosols and fly-ash may persist even beyond those pre-treatments, which may have a negative impact on amine capture efficiency (Boot-Handford et al., 2014).

NO_x removal is a different issue, but is also important as that can also cause problems in an amine scrubber. In particular, NO_2 will react with secondary amines to produce nitrosamines. These present environmental risks and are significantly toxic.

Even in systems where no secondary amines are added, nitrosamines may eventually emerge – the thermal degradation of many amines (including MEA) results in the formation of secondary amines, eventually leading to the formation of nitrosamines. It is believed that UV radiation may be an effective treatment method for decomposing the nitrosamines into less dangerous chemicals, and thus air emissions of nitrosamines should produce little long-term environmental risk. However, in the short term they are undesirable (Boot-Handford et al., 2014).

Amine aerosols can form from SO_3 or small particles in the system, and cannot easily be removed in a condenser or water wash. Instead, filtration with a fiber filter is required, with a significant associated pressure drop (150–250 mmHg). This can be completely resolved by using amines with little to no volatility, but most practical scrubbing amines are at least somewhat volatile (Boot-Handford et al., 2014).

Some amines, especially MEA, can cause problems with equipment corrosion at the concentrations used, but are still gentler than processes using hydroxides as the capturing solute. The corrosion caused by amines can be mitigated with certain additives, and are practically nonexistent with KS-1 (Aaron and Tsouris, 2005; MacDonald, 2008).

16.2.3 Alkaline Scrubbing Systems

A major, categorical alternative to amines in scrubbing columns is the use of inorganic alkaline compounds. These include but are not limited to sodium hydroxide, lithium hydroxide, potassium hydroxide, sodium carbonate (Kawatra et al., 2011), potassium carbonate, and others. These can typically be divided into strongly alkaline compounds (e.g. hydroxides), and weakly alkaline compounds (e.g. carbonates).

The only alkaline compound that could be considered categorically incapable of capturing CO_2 is bicarbonates with HCO_3^- anion. As previously stated, this is because there is no intermediate compound between carbonic acid and bicarbonate anion. Bicarbonates however can typically be easily regenerated into carbonate compounds at relatively low temperatures of around 80°C. At much higher temperatures, they may be calcined to hydroxides.

Strongly alkaline scrubbing systems have the potential to result in precipitation, as carbonate compounds are formed and have limited solubility in water. In particular, the presence of divalent metal cations practically guarantees the formation of insoluble carbonate and bicarbonate salts. In many locations this means that water softening and treatment is a requirement before an inorganic alkaline CO_2 scrubber to avoid equipment fouling. However, as hard water is almost always responsible for some level of equipment fouling, especially in power facilities, it is likely that the required soft water infrastructure is already present in any existing plant.

Weakly alkaline scrubbing systems based on carbonate chemistries are, perhaps surprisingly, not nearly as subject to this effect. This is because as capture proceeds, carbonate anions are transformed into bicarbonate anions, and the total concentration of compounds which are likely to be insoluble typically decreases. During regeneration, when the opposite happens, it is practically guaranteed that the final regenerated solution will contain no more carbonate than the initial solution, and thus is also not subject to precipitation. These systems simply do not contain anions which can capture more than the solvent's maximum loading. Water softening is still a requirement, as the initial carbonate loading will precipitate any divalent metal cations quite readily, but it is unlikely for any carbonates to precipitate in the scrubber itself during operation.

The strongly alkaline compounds typically have significant corrosion issues in long-term usage. The weakly alkaline chemistries, however, aside from any potential equipment fouling, are typically noncorrosive to most steels (Spigarelli and Kawatra, 2013). One other attraction to carbonate chemistries is that they are largely harmless to humans and typically pose little environmental risk (Boot-Handford et al., 2014).

Carbonate chemistries typically have a low heat of absorption, and hydroxide chemistries ultimately behave quite similarly to carbonate chemistries during regeneration. The complete regeneration of a hydroxide compound typically requires extremely elevated temperatures in calcination.

One other difference with inorganic alkaline scrubber solutes is that they are occasionally practical as solid-phase adsorbents for CO_2 as well.

16.2.3.1 Alkaline Hydroxides

The advantages of sodium hydroxide are that it is relatively inexpensive and very effective for capturing CO_2. The downsides of sodium hydroxide include difficulty in fully regenerating the capture solution and in the cost of the equipment required to handle the corrosion that it causes.

Aqueous sodium hydroxide scrubbers are not typically considered for high-capacity usage such as flue gas of a coal-fired power plant. Almost all literature focuses on the use of amines or carbonates, or mixtures of the two. Sodium hydroxide can certainly achieve similar capture rates from flue gases as MEA scrubbers, but since the complete regeneration takes place at temperatures around 800°C–1,000°C, its use with aqueous solutions in recycle loops is highly impractical (Oyenekan and Rochelle, 2007; Rochelle, 2009; Boot-Handford et al., 2014).

However, sodium hydroxide has been investigated for scrubbers designed to work with atmospheric air. This is likely to take advantage of its rapid capture kinetics with CO_2 – the hydroxide anion can directly bind with aqueous CO_2, bypassing the formation of carbonic acid entirely. This, combined with an exceptionally surface area provided by extremely fine spray nozzles is expected to allow for an acceptable capture rate from gas streams as dilute in CO_2 as atmospheric air. The downsides of such a system are a considerable water usage, the cost of replacing the sodium hydroxide, and the energy costs of the fans and pumps required to maintain the relatively large flow rates through the column for the small amount of CO_2 being captured (Stolaroff et al., 2008; Boot-Handford et al., 2014).

Sodium hydroxide can also be made effective as a solid sorbent for capturing CO_2 while it is still at high temperatures. This may be practical as a part of a pre-combustion capture process, such as coal gasification. In particular, a mixed sodium hydroxide-calcium oxide sorbent was developed, which could capture CO_2 at 315°C and could be regenerated at 700°C. Solid sorbents can achieve high CO_2 capture, in this case 3 mol CO_2/kg sorbent. This exceeds the capacity of all practical amine chemistries in use today (Siriwardane et al., 2007; Boot-Handford et al., 2014).

Sodium hydroxide is also reasonably common in minerals used for mineral carbonation, a specific form of CO_2 sequestration. In this case, there is no intention to regenerate the sodium hydroxide, and the final product is typically less corrosive than the starting mineral, negating the major disadvantages of sodium hydroxide for CO_2 scrubbing (Boot-Handford et al., 2014).

Chemically, lithium hydroxide behaves quite similarly to sodium hydroxide, but was used in space travel for quite some time due to its lower molar mass and relatively high capture capacity of almost 6.8 mol CO_2/kg sorbent (Satyapal et al., 2001).

16.2.3.2 Case Study: Optimizing Alkaline Scrubber Kinetics

At Michigan Technological University, a set of experiments were carried out with a pilot-scale CO_2 scrubber column to determine the optimal conditions for capturing CO_2. The scrubber column was a 4 ft packed bed column with a diameter of 4 in. Two gallons/min of scrubbing solution was input into the top, while approximately 45 L/min of a simulated flue gas was bubbled up from the bottom of the column. This column utilizes a sodium carbonate-based scrubbing solution, developed and patented at Michigan Technological University (Kawatra et al., 2011).

The capture solution was varied over numerous experiments with a baseline case of 2 wt% sodium carbonate in water. This was compared to 2 wt% sodium hydroxide, 1 wt% sodium carbonate, 3 wt% sodium carbonate, and 2 wt% sodium carbonate with a handful of additional modifiers. The water chemistry was taken into account by testing capture efficiency using distilled, softened, and city tap water.

The simulated flue gas was a CO_2 mixture with compressed air. The baseline CO_2 composition was chosen to be 16%, for the sake of simulating a typical flue gas CO_2 concentration. In the interests of evaluating the effects of a real flue gas on CO_2 capture, arrangements were also made with the Michigan Technological University steam plant, run by the university to provide steam and heating to the campus, to tap into their real flue gas to test the scrubbing column there as well. It was found that the steam plant's flue gas was closer to 8% CO_2, and so comparison studies between work in the laboratory and the steam plant used CO_2 concentrations near 8% instead.

The scrubbing column, as installed at the Michigan Technological University steam plant, is pictured in Figure 16.6.

Despite the relatively low concentrations of the capturing reagent in the scrubbing solutions, it was found that in many cases a significant portion of CO_2 could be captured in a single pass through the scrubbing column. Even pure water was modestly effective at collecting the CO_2 from the gas stream, as shown in Figure 16.7 compared to sodium carbonate and sodium hydroxide. Two weight % sodium carbonate alone captured between 20% and 60% of the incoming gas in all water chemistries tested, as shown in Figure 16.8. Capture performance was consistently higher with distilled water than city tap water, likely due to the high concentration of Ca^{2+} and Mg^{2+} ions in the tap water. These ions would result in carbonate precipitating, resulting in a lower overall concentration of capture solution.

In an attempt to improve scrubber capture efficiency, frothers were added to stabilize the CO_2 bubbles. The idea is that if CO_2 absorption is primarily limited by the mass transfer of CO_2 from the bubbles to the aqueous solution, then increasing the surface area by decreasing the bubble size should increase the adsorption rate.

FIGURE 16.6 The alkaline scrubber column tested at Michigan Technological University, as installed in the on-campus steam plant. Scrubbing solution is added at point A and removed at point D, and the flue gas is added at point C and removed at point B.

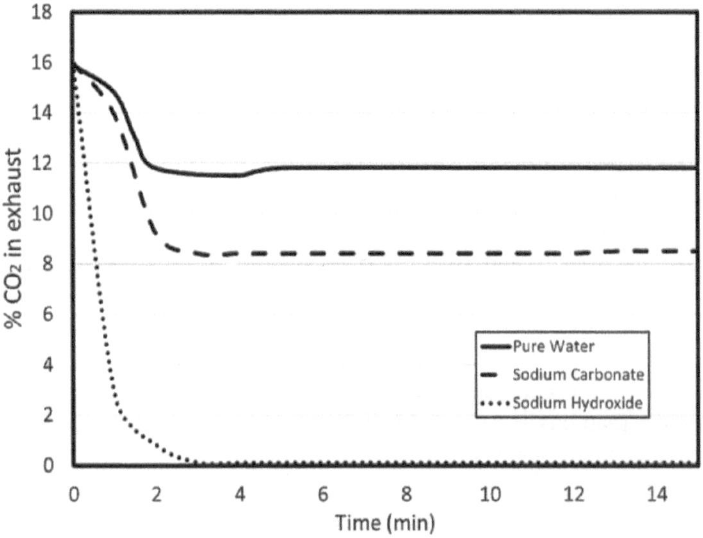

FIGURE 16.7 CO_2 capture from a simulated flue gas using 2 wt% capture solutions of sodium carbonate and hydroxide, compared to distilled water.

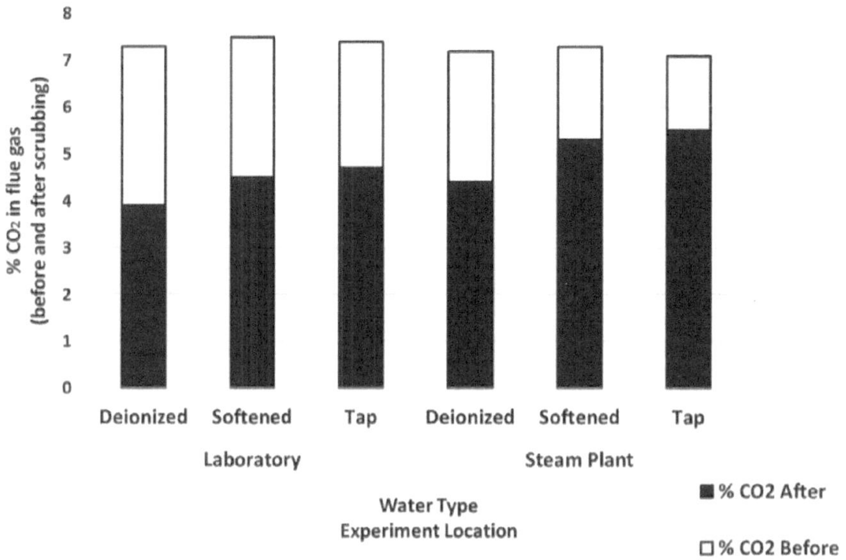

FIGURE 16.8 The purpose of this figure is to show the relative absorption of CO_2 from a flue gas by deionized, softened, and tap water. In the lab, using deionized water, we were able to reduce the CO_2 in the flue gas stream from 7.3% to 3.9%.

In particular, the frothers were intended to decrease the surface tension between the CO_2 and water phases, which should result in smaller bubbles and in turn more overall bubble surface area. The results of adding a frother to 2 wt% sodium carbonate capture solution are presented in Figure 16.9.

As shown in Figure 16.9, the addition of frothers increases the ability of the carbonate solution to capture CO_2. The same principles can likely be applied to other CO_2 scrubbing chemistries, reducing the size of column necessary to capture CO_2. Ultimately, frothers should be selected for compatibility with the scrubbing solution's regeneration cycle, but the overall dosages of frothers required for good performance are not large. Thus, this is a very inexpensive way to ensure the maximum possible CO_2 capture rates.

Table 16.2 summarizes many of the options for post-combustion capture solutions.

16.2.4 Alternative Capture Methods

There are a handful of other capture methods that have received considerable academic attention, but are rarely implemented in industrial scale. These methods are discussed later in this section: mineral capture (Yuen et al., 2016), capture by alkaline process wastes (Pan et al., 2012), adsorption capture via activated carbon (Zhang et al., 2013), or capture with metal-organic frameworks (MOFs) (Sumida et al., 2012). It is worth noting that the vast majority of capture facilities operate amine or hydroxide scrubbers or pre-combustion systems.

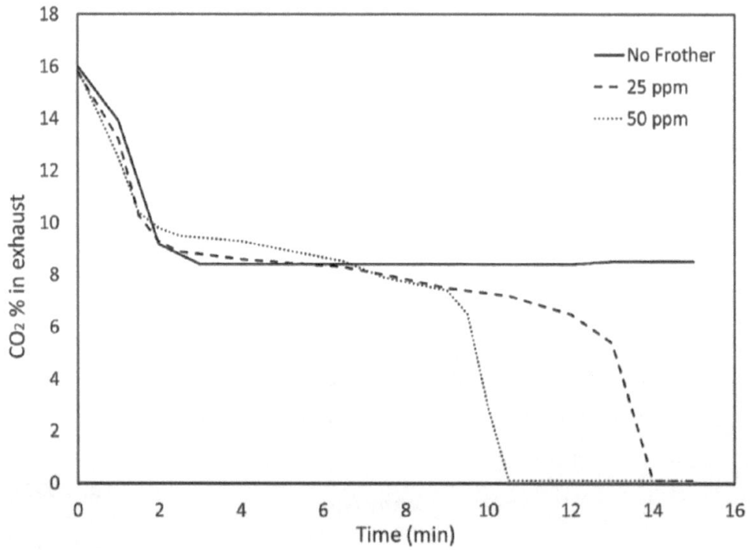

FIGURE 16.9 Frother was added at 25 and 50 ppm to a 2 wt% sodium carbonate solution. The CO_2 passing through the column was measured as a function of time. The time it took for the CO_2 to drop to zero was largely dependent on the development of a stable froth throughout the column. This frother was a polypropylene glycol frother with a molecular weight near 250, but all tested frothers showed good capture performance once the froth evolved.

TABLE 16.2
Comparison of Post-Combustion Capture Reagents for CO_2 Capture

Reagent	Pros	Cons
Amines (MEA, PZ, KS-1, etc.)	Well-studied, good absorption, and regeneration characteristics	Toxic, corrosive, subject to thermal degradation
Alkaline hydroxides (NaOH, KOH)	Very good absorption kinetics	Very corrosive, somewhat difficult regeneration
Carbonates	Good regeneration characteristics, nontoxic, noncorrosive	Slow absorption kinetics
Carbonates + Frothers	Good regeneration characteristics, very good absorption kinetics, noncorrosive	Frothers may thermally degrade or present some toxicity

However, in some cases certain regional characteristics or specific scenarios may imply that far less expensive options may be available. The availability of alkaline minerals or alkaline waste products can provide significant opportunities for CO_2 capture and utilization.

Additionally, some techniques are designed for high efficiency or minimum moving parts, but have not yet received widespread industrial adoption for a variety of reasons. These include techniques such as capture via activated carbon or MOFs for capturing CO_2.

16.2.4.1 Mineral Capture

Because CO_2 is an acid gas, alkaline minerals (hydroxides and many silicates) can typically react with it to form carbonates. The key criterion for this reaction to proceed is that carbonic acid must be a stronger acid than the mineral bicarbonate to be formed. The mechanism is shown in Equation 16.5 below, where X represents a mineral complex which is sufficiently alkaline to have a surface hydroxide group:

$$CO_2 + XOH \rightarrow XOCO_2H \tag{16.5}$$

Any water present can catalyze the reaction by hydrating the mineral surface or converting the CO_2 into carbonic acid. The resulting carbonates and bicarbonates tend to be very stable in the absence of extreme heat or acidic conditions. In the case where X is a monovalent metal cation like sodium, the carbonates tend to be soluble in water, while divalent and trivalent cations tend to form insoluble carbonate species. While the uses of many carbonate compounds are somewhat limited in scope, many common carbonates do possess large-scale uses in smelting, cooking, and cleaning. Additionally, calcium carbonate can be used as a feedstock for making cement or directly as a chalk for sidewalks or blackboards.

Some example minerals that possess sufficient alkalinity to capture CO_2 include quicklime (calcium oxide), serpentinite (magnesium silicate hydroxides), and olivine (magnesium iron silicate). Recent studies indicate that hydrotalcites from serpentinites also provide a significant quantity of CO_2 sequestration (Turvey et al., 2018). These minerals can capture CO_2 from the atmosphere given exposed surfaces and time, but these processes are only naturally fast when considered on geological time scales. Sequestration can be made more practical by transforming the mineral into a more active chemical group first by the addition of a catalyst or pseudo-catalyst (Wang and Maroto-Valer, 2011; Yuen et al., 2016). In this context, pseudo-catalysts are reagents which decrease the activation energy required for the carbonation of the minerals to proceed, but are not automatically regenerated as a part of the reaction steps. In most cases, these catalysts are acidic compounds such as ammonium chloride, acetic acid, hydrochloric acid, or organic acids. In each case, these pseudo-catalysts solubilize some of the capturing ions (almost always magnesium and calcium) from the mineral species which are then precipitated by the addition of carbon dioxide under alkaline conditions (Yuen et al., 2016). For the highest reaction rates and energy efficiency, the aqueous phase can be eliminated entirely and the entire process can be run under substantially elevated pressures instead – the pseudo-catalytic compounds still activate the minerals by releasing the active ions, but the reaction is driven by the presence of high-pressure CO_2 rather than aqueous equilibria (Yuen et al., 2016). The selectivity of these high-pressure processes is lower, however, so if the formation of specific and separated carbonate species is desired, they are less suitable (Yuen et al., 2016).

The difficulty of these pseudo-catalytic processes is that there is an energy cost associated with regenerating the reagents. However, most of the used-up reagents are no longer particularly active, so it is difficult to devise regeneration processes which are sufficiently energetically efficient (Yuen et al., 2016). The restrictions on energy efficiency stem from the fact that this is a CO_2 capture process: if the regeneration step requires the generation of more CO_2 than is actually captured initially, there is no point.

Mineral capture is practical because the resulting carbonates are stable and nonvolatile. However, aside from the regeneration of CO_2 from sodium carbonate, many of these compounds are a bit too stable to practically use to generate a pure CO_2 stream from. Thus, mineral capture does not typically fall into the same category of capture technique as the amine and alkaline scrubber systems – rather, the carbonates from mineral capture are usually the goal, not the method. In this way, mineral capture is the first sequestration technique discussed in this chapter.

16.2.4.2 Alkaline Process Wastes

In mineral processing, paper processing, and around the chemical industry, there are numerous examples of wastes that cannot be easily disposed of and present significant safety and environmental hazards due to their alkalinity (Pan et al., 2012). The alkalinity of these wastes can typically be partially or completely neutralized by the addition of CO_2. Again, the reaction mechanism is based on the acidic behavior of CO_2 gas, so the overall mechanism is the same as previously discussed.

However, in the case of waste neutralization the final end product of reacting CO_2 with these wastes is typically of no concern. In the case of amine and alkaline scrubbing, the idea was to create a purified CO_2 stream that could be used to separate CO_2 from other gases. With mineral carbonation, it may occasionally be ideal to ensure that the resulting carbonates are separable or sufficiently pure to sell. In the case of CO_2 being used for waste management, however, the form that the CO_2 ultimately ends up in is typically of no concern so long as it is nonvolatile and the waste being treated is at least partially neutralized. In truth, in all of these cases the CO_2 will end up as carbonates, but which carbonates is often of comparatively little concern.

These processes are particularly attractive when there are hazards, economic or environmental, presented by both the release of the CO_2 and the alkalinity of the waste in question. As the CO_2 can be sequestered as carbonates and the alkalinity can be reduced in the process, this provides a win–win scenario for the owners of both waste streams. There are numerous examples of alkaline wastes in industry. For instance: steel and iron-making slags, air pollution control residues, fly-ashes, bottom ashes, cement wastes, nickel or asbestos tailings, bauxite residue, paper-making pulps, and wastewater sludges (Pan et al., 2012). Most of these wastes are roughly comparable in how much CO_2 can be absorbed per kilogram of waste with the notable exceptions of coal-combustion fly-ashes and bauxite residue which are much lower (Pan et al., 2012).

However, despite the low capacity for CO_2 absorption per ton, bauxite residue is a particularly interesting case. Bauxite residue, also known as red mud, is a highly alkaline byproduct of processing bauxite (a rock containing various aluminum hydroxide compounds) into alumina for smelting. While the sequestration capacity of red mud is only on the order of 45 g of CO_2 per kilogram of red mud, it is a byproduct of essentially all current aluminum refining processes. Roughly 1.5 tons of red mud are produced per ton of alumina produced (Chandra, 1996), and roughly 130 million metric tons of alumina were produced in 2018 (IAI, 2019), meaning that the annual production of red mud is on the order of 200 million tons/year. Thus, there is a potential to sequester roughly 10 million tons of CO_2/year on annually produced red mud alone, even without considering the stockpiles of red mud that have accumulated over the years.

The dangers of red mud are primarily due to its quantity, alkalinity, and traces of embedded toxic metals. In 2010, a retaining wall on a red mud pond in western Hungary failed, releasing a tremendous quantity of red mud into the local ecosystem in what was described in 2014 as the largest release of caustic waste in history (Anton et al., 2014). However, despite the scale of the disaster, the ecosystem showed significant signs of recovery after only a handful of years. Thus, if the red mud's alkalinity could be reduced the threat it would pose to the environment would likely decrease proportionally, correspondingly increasing the ease with which the red mud can be handled. Furthermore, as the toxic metals present within it, such as vanadium, chromium, cobalt, nickel, and arsenic (Anton et al., 2014), have industrial uses, this increased ability to handle red mud may lead to additional processing opportunities to increase the recovery of nonaluminum valuables. Some red muds contain a considerable concentration of rare earth metals, which may also present a valuable extraction opportunity.

The other alkaline waste materials present similar challenges and opportunities.

16.2.4.3 Activated Carbon

Activated carbon and some other surface frameworks can be used to capture CO_2 via adsorption. These systems capture CO_2 by having a similar overall surface chemistry to CO_2 itself. Thus, when an air stream containing a high concentration of CO_2 passes over the activated carbon surface, the CO_2 tends to stick to the solid surface instead of remaining in the gas phase.

However, the adsorption is largely a physical phenomenon – if the surface is exposed to air with a very low concentration of CO_2, the equilibrium will reverse and the CO_2 will begin to detach from the activated carbon. The process can be accelerated by increasing the temperature (Zhang et al., 2013).

In this fashion, activated carbon can be easily regenerated compared to essentially any other CO_2 capture system. However, the capture capacities are not extremely high. Even after modification with alkaline surface groups to improve capture efficiency, the CO_2 capture capacity was only 19 mmol/g, or 0.839 g CO_2/g activated carbon, in an atmosphere of 36 bar CO_2 (Zhang et al., 2013).

In addition to comparatively easy regeneration, these systems have some appeal in many of the same use cases as solid sorbents. These activated carbon systems do not require external power sources to effectively capture CO_2, have no moving parts, and do not contain liquids.

16.2.4.4 Metal-Organic Frameworks (MOFs)

MOF is another adsorption capture system. MOFs are porous solids carefully designed to maintain particular, regular crystal structures, which incorporate large pores. These structures have been used in a variety of fields for many purposes, including catalysis, separations, and drug delivery (Sumida et al., 2012).

By using carefully designed metal-organic structures, CO_2 can be selectively adsorbed from gas streams containing other gases. Like activated carbon systems, MOFs can be desorbed via pressure or temperature swings. Also similar to activated carbon systems, MOFs have no moving parts, do not contain liquid, and do not require external power sources to effectively capture CO_2 (Sumida et al., 2012).

While appropriately chosen MOFs are highly efficient and selective at capturing CO_2, there are some general concerns with them. In particular, their thermodynamic, surface, and mechanical stabilities are relatively limited, as they are ultimately largely porous. Most MOFs are susceptible to partial degradation under high temperature or physical stress. In addition, contact with water may result in reaction with the metal or permanent adsorption of the water, both of which are likely to degrade performance (Sumida et al., 2012).

MOFs are not in use industrially yet, due in part to the challenges above. However, a robust MOF that could be shown to be stable enough for industrial use is expected to have several advantages over an amine absorption system. In particular, MOFs can be designed to be selective towards individual flue gas impurities without being deactivated by the remaining – thus, SO_x, NO_x, and CO_2 can be handled separately in any order as desired. The capture capacity of MOFs is considerable: at 35 bar, MOF-177 reached a storage density of $320\,cm^3$(STP)/cm^3 for CO_2, compared to $35\,cm^3$(STP)/cm^3 stored freely at the same pressure (Sumida et al., 2012).

Additionally, MOFs have much lower heat capacities than many liquids, requiring significantly less energy to initiate thermal regeneration than amine scrubbing systems. In addition, high thermal conductivities can be maintained to reduce the time required for temperature swing desorption.

Overall though, MOFs are an exciting technology that is expected to lead to significant improvements in post-combustion capture and in preparing gas streams for oxy-combustion and pre-combustion capture processes. However, the field still requires further development to work out the remaining issues of water control, determining the specific effects of impurities on capture behavior and stability, and selectivity with real flue gases (Lin et al., 2017).

MOFs are the most promising technology listed here for small-scale applications. For example, if one wished to install a CO_2 scrubber into the exhaust of an automobile, then MOFs would be the most space-efficient technology available to do so. Although it is likely impossible to include a regeneration setup in the same volume constraints, the high storage capacity of MOFs means that they could be removed and separately cleaned periodically.

16.3 METHODS OF CO_2 SEQUESTRATION

The primary methods by which the considerable quantity of CO_2 generated by human activities can be sequestered are geological sequestration and oceanic sequestration. Geological sequestration places the CO_2 into underground reservoirs that can contain the CO_2, while oceanic sequestration captures the CO_2 in the ocean by taking advantage of its alkaline nature. In either case, there are considerable additional expenses and risks associated with sequestration. The coal industry is particularly uninterested in the additional cost. The obvious question then becomes: what can be done with CO_2 that provides some inherent profits? If a scheme can be devised by which CO_2 capture and sequestration are actively profitable, then it is expected that capture and sequestration will become far more widespread.

16.3.1 Geological Sequestration

Geological sequestration is the injection of CO_2 into deep underground rock formations that can hold gaseous CO_2 for the indefinite future. This may be due to the geology simply being appropriate to hold gases or due to various geological phenomena which can reduce the mobility of CO_2. The main features targeted for geological sequestration are hydrocarbon reservoirs (fully or partially depleted), deep saline aquifers, and coal beds (Bachu, 2015).

The vast majority of large-scale CO_2 sequestration projects are geological sequestration projects, representing about 31 million tons of CO_2 sequestered per year as of 2017. Of these, approximately 27 million tons of CO_2 per year go towards EOR projects and the remaining 4 million tons are sequestration-only projects (Global CCS Institute, 2017).

The main advantage of geological sequestration is that it presents one of the only potential sinks for CO_2 which can keep up with human CO_2 emissions. The primary disadvantage is that in most cases it is simply a sink – there is no inherent economic incentive to pump CO_2 underground, and it is actually quite expensive. The energy cost of pumping CO_2 underground is almost equivalent to the energy cost of capturing and purifying CO_2 from a flue gas stream in the first place (Rochelle, 2009). In this sense, the popularity of EOR is due to the direct economic incentive it provides. EOR is discussed in more detail in Chapter 17, but it also serves to sequester CO_2.

EOR is naturally only possible in hydrocarbon reservoirs, as there is no oil to recover in deep saline aquifers. Enhanced coal bed methane recovery is a more fledgling technology that can apply to coal beds, but coal beds present additional challenges compared to the typical EOR operation. Deep saline aquifers are suitable for sequestration-only operations.

The amount of CO_2 that can be stored in a hydrocarbon reservoir is relatively easy to calculate as a ratio of densities of the material being pushed in versus what has previously been removed. So long as the pressures involved do not exceed the structural capabilities of the reservoir, this means that an equivalent volume of material can be placed in the reservoir as what has previously been taken out (Bachu, 2015). Storage of CO_2 in coal beds is largely the same idea, but those reservoirs are typically not as resilient against high CO_2 pressures as natural gas reservoirs.

In comparison, CO_2 storage efficiencies in deep saline aquifers are much more complex. As the CO_2 is lighter than water, but the mixture of CO_2 in water is denser than water, the injection of CO_2 will result in the formation of convective currents within the aquifer. However, as the aquifer geometry may be extremely complicated and the effective porosity may vary significantly depending on the direction of flow, it may not be possible for these convective currents to adequately mix the CO_2 being injected. Overall, however, several factors have been found to significantly impact the effective storage capacity of a deep saline aquifer, including (Bachu, 2015)

- Conditions within the aquifer, including pressure, temperature, and composition within the aquifer.
- The resulting flow parameters of the phases within the aquifer, particularly the density, viscosity, and interfacial tension.

- The resulting mixing parameters of the phases within the aquifer, particularly the overall solubilities and permeabilities of the fluids.
- The geometry and geology of the aquifer, particularly as it concerns the variation of porosity and compressibility over location and direction.
- The CO_2 injection mechanism, particularly rate, duration, number, and spacing, and overall strategy such as whether and how much water is added as well.
- Regulatory constraints, such as the maximum allowed bottom-hole injection parameter, scale at which the project is being assessed, the time of assessment, and the actual footprint of the aquifer in question.

The same parameters likely have a significant impact on the storage capacity a hydrocarbon reservoir or coal bed, but as a far simpler methodology already exists for determining storage capacity for those cases, there is less emphasis on applying these factors there.

Another important aspect of CO_2 sequestration is the management of risks inherent to the project. Risks can be divided into four major categories (Pawar et al., 2015):

- Risks to site performance and suitability, including uncertainty in the capability of the site to receive CO_2 either as a total quantity or at a given injection rate. Risks of this sort raise the possibility that investment into a site may end up being fruitless, as unforeseen issues may result in the site being incapable of sustaining a given injection rate or capacity.
- Risks to CO_2 containment, including potential leaks, potential causes of new leaks whether avoidable or unavoidable. This includes considerations such as well integrity or the potential for seismic activity. Significant research has been performed to analyze the integrity of wells into the CO_2 reservoirs, which are typically the largest risk to CO_2 containment.
- Risks to the project based on public perception.
- Risks based on market performance.

Risks based on public perception and market performance vary significantly, both based on location and over time. For this reason, this chapter will not dwell on them much. However, the risks to site performance and CO_2 containment are largely physical problems. The risks of site performance can be mitigated to an extent by appropriately characterizing the reservoir, as discussed previously.

The risks to CO_2 containment however are rather pressing, as they feed into the risks to public perception and market performance, along with potentially releasing as much or more CO_2 than the project sequesters. There are two major sources of potential containment failure for CO_2 reservoirs: well-integrity failures and caprock failures. Well-integrity failures are typically associated with wells which were never built with the intention of being used for CO_2 injection or in the area of CO_2 injection. Newer wells are typically relatively resilient, but especially on old oil fields there may be wells which, even assuming they had been constructed properly and maintained fully since then, were never designed to withstand the forces that can result from CO_2 injection into the reservoir. The rate of well-related incidents vary significantly from field to field and from well to well – one natural gas reservoir had an estimated gas discharge incident rate of 1/50,000 years per well, but less extreme sustained casing pressure incidents were as common as 75% in some fields (Pawar et al., 2015). There has been a significant concern that the high-pressure CO_2 being injected into the wells can react significantly with common cements and other construction materials, but it has not been made clear that these potential reactions lead to an actual weakening of the materials (Pawar et al., 2015).

Caprock failures refers to fractures, whether naturally occurring or as a result of CO_2 injection, in the rock above the CO_2 reservoir by which the CO_2 can escape. Most rock formations which have been considered suitable at all for use as a CO_2 reservoir, particularly shale and evaporates, are understood to be fully up to the task of containing CO_2 (Pawar et al., 2015). However, it is not necessarily the case that these rock formations are actually present across the entirety of any given reservoir. The laboratory determination of caprock quality largely depends on the permeability and capillary pressures: low permeability and high capillary entry pressures are correlated with high-quality caprock (Pawar et al., 2015). However, practical field concerns require determining whether or not the caprock is actually intact, based on the ductility of the rock and previous seismic history, along with the overall geometry of the reservoir (Pawar et al., 2015). Based on these parameters, risk assessment models can be developed.

In summary, geological sequestration is the primary manner by which CO_2 is sequestered. Most geological sequestration projects are EOR projects, wherein CO_2 is used to displace oil from partially depleted oil wells while simultaneously trapping the CO_2. These projects, whether EOR or not, represent significant financial and technological undertakings and carry a significant amount of risk. However, the parameters affecting these risks are largely understood and can be mitigated.

However, by itself geological sequestration also represents a significant money sink – there is no inherent return on investment on pumping CO_2 into the ground. Without some additional incentive, whether that be to avoid punitive fines and taxes or by the production of a saleable product, the pace of adoption can be expected to remain relatively slow.

16.3.2 Oceanic Sequestration

Oceanic sequestration again takes advantage of the fact that CO_2 is an acid gas to mix it with an alkaline solution to form carbonates. In this case, the alkaline solution is the ocean, which maintains a mildly basic pH by dissolving carbonates from sediments at its base. The ocean's capacity for CO_2 is tremendous – it is the largest regulating factor in the carbon cycle. The thought then is that by injecting CO_2 into the ocean, CO_2 can be sequestered at a more rapid pace than simply allowing the ocean to equilibriate with the atmosphere.

However, there are significant concerns regarding the impact such CO_2 injection is likely to have on the ocean's ecosystem and the rate at which CO_2 can be absorbed. In addition, there are also concerns regarding the short potential storage lifespan of CO_2 in the ocean, the high costs associated with transporting CO_2 to the ocean, and the operating costs of an injection site in the ocean (De Silva et al., 2015).

The CO_2 sequestration capacity of the deep ocean is roughly 1,000–10,000 gigatons of carbon. However, the ocean is not a closed system – depending on where the CO_2 is injected, as much as 20% of the CO_2 may be returned to the atmosphere within a 300-year time scale (Parson and Keith, 1998).

With regard to biological life, even a slight shift in the ocean's pH can have a tremendous impact on the survival of many species. It would be best if the impacts of CO_2 injection on these organisms could be investigated in controlled, small-scale experiments before any large-scale oceanic sequestration process took place. It would be comparatively easy to study the sequestration kinetics, the pH changes, and the potential effects on biological life by testing these mechanisms in swimming pools or similar vessels.

Without a major scientific breakthrough, oceanic sequestration does not currently present many advantages over the other CO_2 sequestration options (Pawar et al., 2015).

16.3.3 Biomass Storage

There has been a considerable amount of research into the biological processing of CO_2 into nonvolatile, potentially useful chemicals. Several opportunities are easily possible, such as planting trees and other plants, which absorb CO_2 from the atmosphere via photosynthesis to construct themselves, but these processes take a considerable amount of time and land to achieve modest reductions in CO_2. Higher atmospheric CO_2 concentrations are directly correlated with increased plant growth, a property that is often used in greenhouses where the environment can be controlled to such an extent (Kimball, 1983).

To scale this to levels that are relevant for industrial-scale CO_2 capture in the short term, carefully chosen or engineered microalgae can be used in appropriately sized bioreactors (Huang and Tan, 2014). Some microalgae are capable of doubling their biomass in less than a day given appropriate support. Every ton of microalgae produced in such a fashion uses 1.8 tons of CO_2 as part of their growth. Most of these microalgae increase the alkalinity of their environment, which leads to the precipitation of alkaline carbonates such as $CaCO_3$, further increasing the CO_2 storage potential (Huang and Tan, 2014).

The downside to these bioreactors is just how much space is required. Current understanding suggests that these bioreactors require on the order of 3,000 m^3 of space per ton of CO_2 capture capacity per day (Global CCS Institute, 2010; Huang and Tan, 2014). These bioreactors need to be able to supply the microalgae with sunlight and other nutrients, while also controlling the solution pH to ensure continued growth. Furthermore, the quality of CO_2 used for this process may need to be carefully controlled – many impurities in flue gas can have a detrimental effect on the solution pH and the microalgae in general.

The product of this capture is a large quantity of relatively generic biomass, typically intended for use as biofuel or animal feed. While this does not strictly conform with the other forms of sequestration, it is a potentially carbon-neutral fuel source.

16.4 SUMMARY

There are several highly efficient methods to capture CO_2. The current industry standard for post-combustion process is the amine scrubber, but sorption processes and alternative scrubbing liquids have been investigated thoroughly as well. At Michigan Technological University, a carbonate-based post-combustion capture system has been designed with the (US7919064 as Kawatra et al., 2011) intention of reducing corrosion and toxicity compared to amine or hydroxide scrubbing systems.

Significant efficiency gains can be made by considering the capture of CO_2 early in the design of a plant. The most efficient technologies for CO_2 capture require special design considerations from the very beginning of the plant's life, such as oxy-combustion, pre-combustion capture, and chemical looping capture. However, there are many options available for capturing CO_2 even without redesigning the plant, and post-combustion capture processes have become more efficient over time.

Metal oxide frameworks are probably the most exciting technological development in the area of CO_2 capture and storage, though significant work still needs to be done before they can be used on an industrial scale.

Sequestration of CO_2 has remained a more challenging prospect. There are many factors to take into consideration when designing a geological sequestration project, but it is doable and several such facilities are in operation, in the process of being built, or being planned out to be built in the near future. Most geological sequestration projects are also EOR projects, as that provides a significant financial incentive to push for the project.

Other sequestration locations are relatively limited in usage and investigation. Oceanic sequestration is no longer considered a strong contender among sequestration options, while biomass storage has always required a significant land investment for practically any option.

The one major alternative that does seem pragmatic overall is the use of CO_2 to treat other waste products, as there are several alkaline waste products in industry which have concerning environmental properties. If CO_2 can be sequestered at the same time as another waste product is treated, then multiple problems can be solved simultaneously at a net gain for everyone involved.

The next chapter focuses on the alternatives to sequestration: uses of CO_2 which are intended to provide more practical incentives for industries to capture CO_2 from their processes.

REFERENCES

Aaron, D. and Tsouris, C. (2005), "Separation of CO_2 from Flue Gas: A Review," *Separation Science and Technology*, Vol. 40, pp. 321–348.

Anton, Á.D., Klebercz, O., Magyar, Á., Burke, I.T., Jarvis, A.P., Gruiz, K. and Mayes, W.M. (2014), "Geochemical Recovery of the Torna-Marcal River System after the Ajka Red Mud Spill, Hungary," *Environmental Science: Processes Impacts*, Vol. 16, pp. 2677–2685.

Bachu, S. (2015), "Review of CO_2 Storage Efficiency in Deep Saline Aquifers," *International Journal of Greenhouse Gas Control*, Vol. 40, pp. 188–202.

Boot-Handford, M.E., Abanades, J.C., Anthony, E.J., Blunt, M.J., Brandani, S., Mac Dowell, N., Fernández, J.R., Ferrari, M.-C., Gross, R., Hallett, J.P., Haszeldine, R.S., Heptonstall, P., Lyngfelt, A., Makuch, Z., Mangano, E., Porter, R.T.J., Pourkashanian, M., Rochelle, G.T., Shah, N., Yao, J.G. and Fennell, P.S. (2014), "Carbon Capture and Storage Update," *Energy Environmental Science*, Vol. 7, pp. 130–189.

Chandra, S. (1996), "Red Mud Utilization," *Waste Materials Used in Concrete Manufacturing* (Satish, ed.), Noyes Publications, Westwood, NJ, pp. 292–295.

De Silva, G.P.D., Ranjith, P.G. and Perera, M.S.A. (2015), "Geochemical aspects of CO_2 sequestration in deep saline aquifers: A review," *Fuel*, Vol. 155, pp. 128–143.

EIA (2018), "What Are the Energy-Related Carbon Dioxide Emissions from Fossil Fuels for the United States and the World?" *Energy Information Administration*. Accessed at www.eia.gov/tools/faqs/faq.php?id=79&t=11.

EIA (2019), "Monthly Energy Review, March 2019," *Energy Information Administration*, pp. 205–209.

Global CCS Institute (2010), "Accelerating the Uptake of CCS: Industrial Uses of Captured Carbon Dioxide," *Global CCS Institute*, pp. 16–19.

Global CCS Institute (2017), "Large-Scale CCS Facilities," *Global CCS Institute*. Accessed at //www.globalccsinstitute.com/projects/large-scale-ccs-projects (database updated October-27-2017, accessed January-4-2018).

Huang, C.H. and Tan, C.S. (2014), "A Review: CO_2 Utilization", *Aerosol and Air Quality Research*, Vol. 14, pp. 480–499.

IAI (2019), "Alumina Production," *World Aluminum, International Aluminum Institute*, web page, available at www.world-aluminium.org/statistics/alumina-production/ (Accessed on April-17-2019).

Janssens-Maenhout, G., Crippa, M., Guizzardi, D., Muntean, M., Schaaf, E., Olivier, J.G.J., Peters, J.A.H.W. and Schure, K.M. (2017), Fossil CO_2 and GHG Emissions of All World Countries, EUR 28766 EN, Publications Office of the European Union, Luxembourg, ISBN 978-92-79-73207-2, doi:10.2760/709792, JRC107877. Accessed at http://edgar.jrc.ec.europa.eu/overview.php?v=CO2andGHG1970-2016.

Kawatra, S.K. and Eisele, T.C. (2001), *Coal Desulfurization: High-Efficiency Preparation Methods*, Taylor & Francis, Ann Arbor, MI.

Kawatra, S.K., Eisele, T.C. and Simmons, J.J. (2011), "Capture and Sequestration of Carbon Dioxide in Flue Gases," U.S. Patent 7,919,064.

Kimball, B.A. (1983), "Carbon Dioxide and Agricultural Yield: An Assemblage and Analysis of 430 Prior Observations," *Agronomy Journal*, Vol. 75, No. 5, pp. 779–788.

Li, B., Duan, Y., Luebke, D. and Morreale, B. (2013), "Advances in CO_2 Capture Technology: A Patent Review," *Applied Energy*, Vol. 102, pp. 1439–1447.

Lin, Y., Kong, C., Zhang, Q. and Chen, L. (2017), "Metal-Organic Frameworks for Carbon Dioxide Capture and Methane Storage," *Advanced Energy Materials*, Vol. 7, pp. 1–29.

Lower, S.K. (1999), "Carbonate Equilibria in Natural Waters," *Simon Fraser University*.

MacDonald, M. (2008), "CO_2 Capture in the Cement Industry," *IEA GHG*, Report 3, pp. 3–8 to 3–9.

Oyenekan, B.A. and Rochelle, G.T. (2007), "Alternative Stripper Configurations for CO_2 Capture by Aqueous Amines," *AIChE Journal*, Vol. 53, No. 12, pp. 3144–3154.

Pan, S.Y., Chang, E.E. and Chiang, P.C. (2012), "CO_2 Capture by Accelerated Carbonation of Alkaline Wastes: A Review on Its Principles and Applications," *Aerosol and Air Quality Research*, Vol. 12, pp. 770–791.

Parson, E.A. and Keith, D.W. (1998), "Fossil Fuels without CO_2 Emissions," *Science*, Vol. 282, No. 5391, pp. 1053–1054.

Pawar, R.J., Bromhal, G.S., Carey, J.W., Foxall, W., Korre, A., Ringrose, P.S., Tucker, O., Watson, M.N. and White, J.A. (2015), "Recent Advances in Risk Assessment and Risk Management of CO_2 Geologic Storage," *International Journal of Greenhouse Gas Control*, Vol. 40, pp. 292–311.

Rochelle, G.T. (2009), "Amine Scrubbing for CO_2 Capture," *Science*, Vol. 325, No. 5948, pp. 1652–1654.

Rochelle, G.T. (2014), "Postcombustion Capture Amine Scrubbing", *IEA GHG*.

Satyapal, S., Filburn, T., Trela, J. and Strange, J. (2001), "Performance and Properties of a Solid Amine Sorbent for Carbon Dioxide Removal in Space Life Support Applications," *Energy & Fuels*, Vol. 15, pp. 250–255.

Siriwardane, R.V., Robinson, C., Shen, M. and Simonyi, T. (2007), "Novel Regenerable Sodium-Based Sorbents for CO_2 Capture at Warm Gas Temperatures," *Energy & Fuels*, Vol. 21, pp. 2088–2097.

Spigarelli, B.P., Hagadone, P.C. and Kawatra, S.K. (2013), "Increased Carbon Dioxide Absorption Rates in Carbonate Solutions through Surfactant Addition," *Mining, Metallurgy & Exploration*, Vol. 30, No. 2, pp. 95–99.

Spigarelli, B.P. and Kawatra, S.K. (2013), "Opportunities and Challenges in Carbon Dioxide Capture," *Journal of CO_2 Utilization*, Vol. 1, pp. 69–87.

Stolaroff, J.K., Keith, D.W. and Lowry, G.V. (2008), "Carbon Dioxide Capture from Atmospheric Air Using Sodium Hydroxide Spray," *Environmental Science & Technology*, Vol. 42, No. 8, pp. 2728–2735.

Sumida, K., Rogow, D.L., Mason, J.A., McDonald, T.M., Bloch, E.D., Herm, Z.R., Bae, T-H. and Long, J.R. (2012), "Carbon Dioxide Capture in Metal-Organic Frameworks," *Chemical Reviews*, Vol. 112, pp. 724–781.

Turvey, C.C., Wilson, S.A., Hamilton, J.L., Tait, A.W., McCutcheon, J., Beinlich, A., Fallon, S.J., Dipple, G.M. and Southam, G. (2018), "Hydrotalcites and Hydrated Mg-Carbonates as Carbon Sinks in serpentinite mineral wastes from the Woodsreef chrysotile mine, New South Wales, Australia: Controls on carbonate minerology and efficiency of CO_2 air capture in mine tailings," *International Journal of Greenhouse Gas Control*, Vol. 79, pp. 38–60.

Wang, X. and Maroto-Valer, M.M. (2011), "Integration of CO_2 Capture and Mineral Carbonation by Using Recyclable Ammonium Salts," *ChemSusChem*, Vol. 4, pp. 1291–1300.

Yeh, J.T., Resnik, K.P., Rygle, K. and Pennline, H.W. (2005), "Semi-Batch Adsorption and Regeneration Studies for CO_2 Capture by Aqueous Ammonia," *Fuel Processing Technology*, Vol. 86, pp. 1533–1546.

Yuen, Y.T., Sharratt, P.N. and Jie, B. (2016), "Carbon Dioxide Mineralization Process Design and Evaluation: Concepts, Case Studies, and Considerations," *Environmental Science and Pollution Research*, Vol. 23, pp. 22309–22330.

Zhang, C., Song, W., Sun, G., Xie, L., Wang, J., Li, K., Sun, C., Liu, H., Snape, C.E. and Drage, T. (2013), "CO_2 Capture with Activated Carbon Grafted by Nitrogenous Functional Groups," *Energy & Fuels*, Vol. 27, pp. 4818–4823.

17 CO_2 Utilization

Capturing CO_2 is only half of the challenge of handling CO_2 emissions. Once CO_2 is isolated, it needs to be sequestered or otherwise permanently stored. While geological sequestration options provide a sink for the captured CO_2, sequestration alone also presents a remarkable expense. However, if CO_2 can be used to serve some useful purpose, the expense of capture and sequestration can be mitigated or even turned into a profit. The goal of CO_2 utilization is to provide an inherent financial incentive to CO_2 capture.

CO_2 has many uses stemming from its physical properties. However, the sheer quantity of CO_2 available via capture far outstrips any existing markets for it. The uses of pure CO_2 will be covered in this chapter, but know that most of these uses are already completely saturated by existing supply.

More interestingly, it is also possible to transform CO_2 into other organic substances. Though energy intensive, as CO_2 is a very thermodynamically stable compound, CO_2 can be transformed into a wide variety of useful substances. Many of these conversions will be covered in depth, but as CO_2, water, and air allow for access to almost all of organic chemistry, a truly exhaustive list of potential conversions is essentially impossible. Instead, several key chemicals and processes will be discussed that lead into comparatively large potential markets or comparatively extreme profit margins.

CO_2 utilization presents a particular challenge for coal, in that the enthalpy difference between pure carbon and CO_2 is smaller than for most other fuels. Thus, the conversion of captured CO_2 uses more of the usable energy of coal than it would for other fuels. This fundamental limitation would suggest that conversion efforts for coal should be focused primarily on the lowest energy compounds that can be formed from CO_2.

Coal-fired facilities have been reluctant to implement CO_2 capture systems because there is no profit to be had in pure geological or oceanic sequestration. The idea of CO_2 utilization is to find ways to sell these captured pollutants at a profit, whether due to direct uses of CO_2 or further down the line as derivative products.

Thus, this chapter hopes to answer questions like: Can we produce ethanol from CO_2? Can we produce methanol from CO_2? Can we produce oxalic acid from CO_2? Which of these substances can be produced profitably from which power sources?

17.1 PROPERTIES OF CO_2

Pure CO_2 is a gas at ambient temperatures and pressures. It can be liquefied at pressures above 5.13 atm, with the triple point occurring at –56.4°C and the same pressure. It can be solidified below –78.5°C at atmospheric pressure. CO_2's critical point is 31°C and 72.9 atm, above which it exists as a supercritical fluid. A phase diagram is provided in Figure 17.1.

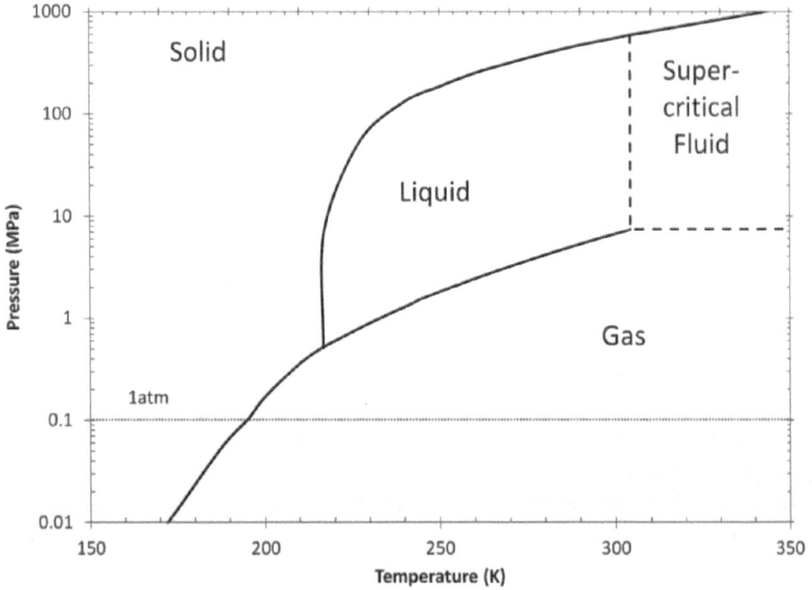

FIGURE 17.1 A phase diagram of CO_2, showing the triple point at 216.7 K and 0.513 MPa and the critical point at 304.19 K and 7.39 MPa. The critical point is the point at which the liquid and gas phases become indistinguishable and the material behaves like a dense, expanding fluid. Based on Witkowski et al. (2014), Lange's *Handbook of Chemistry and Physics* (10th Edition, p. 1463), and the *CRC Handbook of Chemistry and Physics* (44th Edition, pp. 2560–2561).

As a gas, CO_2 is colorless, odorless, and heavier than air. Since CO_2 is used as a signal in human respiration, it can cause hyperventilation and presents a hazard as an asphyxiating gas more readily than other more inert gases. Since it is heavier than air, it can readily become trapped in confined spaces, presenting a significant asphyxiation risk. At pressures below what is required to displace oxygen or trigger hyperventilation, CO_2 can be handled safely with relative ease.

As a liquid, CO_2 is colorless. CO_2 only exists as a liquid at elevated pressures, so piping and storage tanks for liquid CO_2 must be designed with that in mind. CO_2 liquid can be handled relatively easily, but at low pressures it must also be at fairly low temperatures. Liquid CO_2, especially during boiling, can easily present cryogenic hazards such as frostbite.

In the presence of water, liquid CO_2 may present a corrosive hazard to equipment, as it will form carbonic acid. The concentration of carbonic acid which can be achieved by dilute water in pure CO_2 far exceeds the typical concentrations of carbonic acid which can be maintained in aqueous solution. Care should be taken when designing equipment for liquid CO_2, as aqueous pH values below 3.5 from carbonic acid alone are possible at elevated pressures (Toews et al., 1995).

Solid CO_2, typically referred to as dry ice, is a white, crystalline solid. While hard, dry ice is not exceptionally durable. Solid CO_2 is a cryogenic substance, meaning that it typically only exists at unusually low temperatures. In particular solid CO_2 at atmospheric pressure maintains a surface temperature of −78.5°C, and thus presents a thermal hazard to those who work with it. Solid CO_2 maintains this low temperature by sublimating, much like how the liquid phase of boiling water remains near the boiling point regardless of the addition of heat. Due to its low thermal conductivity, however, this cryogenic hazard is easily mitigated with personal protective equipment. Solid CO_2 can be stored short term in cardboard containers, as this provides a level of thermal insulation from ambient conditions. Solid CO_2 can sublimate quickly if heat is transferred into it efficiently, such as by dropping it in water – this produces a fog that has often seen use in practical effects.

CO_2's chemistry is extremely stable – it is not subject to further oxidation reactions, and partakes in few other reaction types. Because the central carbon atom is electron deprived by the comparatively electronegative oxygen atoms it is attached to, CO_2 can be attacked by compounds that donate electrons readily. This includes most bases and most strong nucleophiles – examples of these compounds include sodium hydroxide, lithium aluminum hydroxide, and many metal-organic compounds such as Grignard reagents. Reduction can also be achieved via electrolysis. Essentially every reaction that CO_2 partakes in can be described as a reduction reaction. Altogether this means that the methods for processing CO_2 are relatively limited.

Aside from the respiratory hazards that CO_2 can provide when present in large quantities, it is essentially completely nontoxic. CO_2 is food safe, so long as no unsafe oils or greases are introduced by any pneumatics involved.

17.2 USES OF PURE CO_2

CO_2 has several uses, even not considering the opportunities involving additional chemical processing. The uses of CO_2 are largely divided by phase. Solid CO_2 is used primarily as a cryogenic solid, which is easily handled and readily formed. Liquid and supercritical CO_2 is usually used for extraction, heat transfer, and work transfer. Gaseous CO_2 is primarily used for its properties as a mild acidic gas, or otherwise as a substitute for another inert gas.

17.2.1 Solid CO_2 (Dry Ice)

Dry ice is formed by releasing pressure on liquefied CO_2 and allowing it to evaporate. The evaporative cooling results in the formation of a dry ice "snow," which can be compacted into pellets or blocks as necessary. This is a very simple process when compared to other extremely cold compounds, and can be completed in a single pressurization-depressurization cycle – liquid nitrogen, in comparison, requires a series of gas depressurizations to liquefy nitrogen specifically. Dry ice is fairly simple to handle, as opposed to most cryogenic gases – dry ice can be safely manipulated with gloves or in cardboard boxes, rather than requiring specialized Dewar flasks. Since dry ice is not thermally conductive it sublimates relatively slowly, and typically poses less of an asphyxiation hazard than most other cryogenic gases (excepting oxygen, which has different hazards) (CCPI, 2009; UIGI, 2016; CCPI, 2018).

The storage of dry ice depends largely on the surface area of the piece in question. Large blocks can be stored for quite some time in nothing more than a cardboard box, while fine dry ice dusts are significantly harder to maintain. The longevity of dry ice, as with all cryogenics, is based on the rate at which heat can transfer into the bulk. Dry ice left in an enclosed container which is warmer than it will sublimate over time, potentially leading to a significant pressure buildup (CCPI, 2009; UIGI, 2016).

Dry ice is almost always used in cryogenic applications. As dry ice converts from a solid directly to a gas, a process called sublimation, it absorbs heat from its surroundings. Due to this heat of sublimation, dry ice maintains a surface temperature at an atmospheric pressure of –78.5°C. This makes it a convenient source for mild applications of extreme cold. While dry ice not nearly as cold as liquid nitrogen or other common cryogenic materials, it is sufficient for flash-freezing food (CCPI, 2009; UIGI, 2016).

A common application for dry ice is noncyclic refrigeration. This is opposed to cyclic refrigeration, which pumps heat out of an area by cycling a refrigerant between a heat exchanger and a radiator. Since cyclic refrigeration requires a significant quantity of electricity, it is not always possible to use in some places. Noncyclic refrigeration is often used on food carts and airplanes, and these scenarios typically use dry ice to provide cooling. It can also be used in traveling coolers, such as for storing and transporting donated blood (CCPI, 2009; UIGI, 2016).

Dry ice used for food storage or in food should be of food grade. Dry ice which is not rated as food grade may contain various impurities, typically oils, which could be unfit for human consumption (CCPI, 2009; UIGI, 2016).

Dry ice will sublimate completely and without a trace if there are no impurities present. This makes it practical for use as a sand-blasting agent, as it is hard enough as a crystalline solid to function for sand-blasting, but does not require cleanup. It does require an alternative air supply, as the sublimating CO_2 will displace the oxygen near any such sand-blasting operation, but this is typically advisable regardless for sand-blasting (CCPI, 2009; UIGI, 2016).

17.2.2 Liquid and Supercritical CO_2

Liquid CO_2 can only exist at elevated pressures, above the triple point. Supercritical CO_2 can only exist at temperatures and pressures which are both above the critical point. Liquid and supercritical CO_2 are both powerful nonpolar solvents. Both can exist at temperatures below the thermal decomposition temperatures of most organic compounds, and can be used to selectively extract compounds from various sources. Supercritical CO_2 in particular can also carry a tremendous amount of potential energy, making it a viable working fluid for many applications.

Due to its density and stability at increased temperatures, CO_2 is typically transported in pipelines as a supercritical fluid. Care must be taken to ensure that it does not undergo phase change within a pipe however, as turning into either a liquid or a gas would cause a large and unwanted change in volume. However, maximum transport efficiency, due to density, is achieved in the liquid phase. This typically requires periodic refrigeration to maintain in warmer climates (Zhang et al., 2006).

Note that despite being nonpolar, it is not entirely accurate to say that CO_2 is immiscible with water. Liquid and supercritical CO_2 will be readily attacked by liquid water to form highly concentrated carbonic acid. This process is capable of reaching pH values as low as 2.8, which is well beyond the acidity typically associated with carbonic acid (Toews et al., 1995).

17.2.2.1 Overview of Supercritical Fluids

A liquid takes the shape of its container but maintains its own volume. A gas takes the shape of its container along with the volume of the container. In this respect, a supercritical fluid acts like a gas. However, the density of a supercritical fluid is more like the density of a liquid than of a gas.

A supercritical fluid exists when the difference between gaseous and liquid states vanishes. Liquids have a heat of vaporization that denotes how much energy it takes to transform a unit of that liquid into its gaseous state. This heat of vaporization decreases as the pressure increases, making it easier for liquids and gases to interconvert until, at the critical pressure, the heat of vaporization drops to zero. The liquid phase can remain stable below the critical temperature, but above it the gas and liquid phases freely interconvert. This is the supercritical fluid, which has the properties of both the liquid and gas phases – and often some additional quirks which are not easily ascribed to either.

Many supercritical fluids have solubility properties that do not match their behaviors below the critical point. Solubilities in supercritical fluids usually vary substantially over even relatively small changes in temperature and pressure. This allows the design of highly efficient extraction processes which can selectively move specifically chosen materials using sharp solubility differences to deposit the chosen material at a given location. Typically, supercritical fluids are capable of dissolving more substances than the liquid phase normally would, occasionally to an extreme extent. For example, supercritical water is capable of dissolving glass (Karásek et al., 2013).

Due to the high pressures, temperatures, and relative densities involved, supercritical fluids can also carry a tremendous amount of energy efficiently. This typically makes them excellent working fluids, so long as their pressures, temperatures, and the occasional bit of unusual behavior does not compromise the equipment being used with them.

17.2.2.2 Chemical Extraction

Liquid and supercritical CO_2 are effective nonpolar solvents, meaning that they dissolve oily substances well but water-like substances poorly. One of the primary concerns with typical extraction processes is that thermal decomposition will occur when the solvent needs to be evaporated. However, supercritical CO_2 can exist at temperatures as low as 31°C, which is quite gentle for most organic compounds. The pressures involved in forming liquid or supercritical CO_2 are large compared to atmospheric, but are also small enough to be insignificant when discussing the decomposition of organic compounds. Thus, liquid and supercritical CO_2 provide a relatively gentle polar solvent for dissolving organic compounds (Global CCS Institute, 2010; UIGI, 2016).

Liquid CO_2 has been used as a solvent for separating caffeine from coffee and for dry cleaning. Supercritical CO_2 is also usable for these applications. It is worth re-iterating that CO_2 is food safe so long as it is pure, so these processes need not introduce any harmful additives even if used to prepare chemicals for human consumption (CCPI, 2009; Global CCS Institute, 2010; UIGI, 2016).

With respect to supercritical CO_2 specifically, almost every organic compound becomes soluble in it at sufficiently high pressures. However, all of these solubility curves tend to be relatively unique to the solute in question. Since the pressure can be easily reduced, some interesting separations are possible with supercritical CO_2 (Lou et al., 1997).

17.2.2.3 Supercritical Working Fluid

Supercritical fluids can store a tremendous amount of energy as pressure and heat. However, they do not readily undergo phase changes – within the supercritical regime, they will not condense. Furthermore, while solidification can take place above the critical point, it should be clear from Figure 17.1 that this is not a significant issue for the design of a thermodynamic work cycle – solidification at the critical temperature occurs at roughly 6,000 atm of pressure. Since supercritical fluids flow more like gases than liquids but have much higher densities, the total stored energy per unit volume is much higher than any subcritical gas-based work cycle.

All in all, this allows for relatively simple equipment and cycle design, aside from the pressure and temperature requirements. For CO_2, 72.9 atm and 31°C are not out of reach. Supercritical steam cycles have been investigated, and those require constructing for more than 300 atm pressure and over 600°C temperatures. The wider temperature gaps available to supercritical fluids allow for the construction of more efficient Rankine cycles. Considering steam and coal specifically, subcritical cycles achieve efficiencies near 38% while supercritical cycles can achieve efficiencies above 45%. This corresponds to an 18% increase in energy production from the same coal supply. Similarly, supercritical CO_2 is hoped to improve Rankine efficiencies past subcritical steam cycles (Franco and Diaz, 2009).

Supercritical CO_2 turbines may be a practical replacement for many steam turbine applications. Supercritical CO_2 systems have a much smaller footprint than subcritical steam turbines – some studies have indicated that the turbine footprint of nuclear reactors could be reduced by 75% by replacing existing steam systems with an equivalent supercritical CO_2 system (Ahn et al., 2015).

Supercritical CO_2 turbines may also be practical for geothermal power, where supercritical steam cannot be used. Since supercritical steam dissolves silica, and it is nearly impossible to avoid silica underground, using supercritical steam as the working fluid in geothermal power is ill-advised. After all, what happens if the dissolved silica precipitates out inside of the turbine? However, supercritical CO_2 typically does not dissolve polar compounds, and thus may be practical for use with geothermal energy sources (Brown, 2000; Ahn et al., 2015).

Another application of CO_2 as a working fluid is specifically in cyclic refrigeration. As a refrigerant, CO_2 is known by the code name R744. Advantages of CO_2 over current refrigerants are multitudinous: it is not a chlorofluorocarbon (CFC), the associated environmental and personal health issues are much lower than for CFCs, it does not combust readily, it is less expensive than almost any other refrigerant, it is far more compatible with lubricants than most CFCs, and its thermal properties in the supercritical phase are very good for refrigeration uses. However, CO_2's subcritical thermal properties are less ideal. Furthermore, it is impossible to use supercritical CO_2 alone to cool something below the critical temperature, 31°C. Thus, it is not very useful for air conditioning in homes or cars, or the refrigeration of foods (Aprea et al., 2013).

17.2.2.4 Enhanced Oil Recovery (EOR)

Enhanced oil recovery (EOR) is a technique to improve the recovery of oil from partially depleted oil reservoirs. As oil is pumped out of the ground the hydrostatic pressure within the reservoir drops, which makes it more difficult to pump additional oil out. This can be counteracted by making the oil easier to move, or putting more pressure onto the reservoir itself. EOR takes advantage of various compounds, including CO_2, to achieve these ends. Enhanced gas recovery or enhanced coal bed methane recovery are similar techniques which can be used with gas reservoirs and coal beds to improve product recovery in those conditions as well.

EOR is the most developed of these techniques, with the earliest dedicated CO_2 EOR capture operations dating back to 1972 with the Terrell Natural Gas Processing Plant, TX, USA then known as Val Verde Natural Gas Processing Plant. The CO_2 captured at this facility was then used in the Scurry Area Canyon Reef Operators Committee (SACROC) flood in Scurry County and the North Crossett flood in Crane and Upton counties (Melzer, 2012; Global CCS Institute, 2017).

EOR using CO_2 relies on two primary effects: one, the highly pressurized CO_2 will impart some of its pressure onto the oil remaining in the field; and two, the oil will be partially dissolved by the CO_2, improving its flow properties to further enhance recovery. The mixing of CO_2 with the oil will often significantly lower the viscosity of the oil. In both cases, the goal of EOR with CO_2 is to improve the flow of oil from the reservoir to the surface. Often CO_2 injection is alternated with water injection, wherein the water drives the oil–CO_2 mixture towards the extraction sites. This process is shown in Figure 17.2 (Global CCS Institute, 2010).

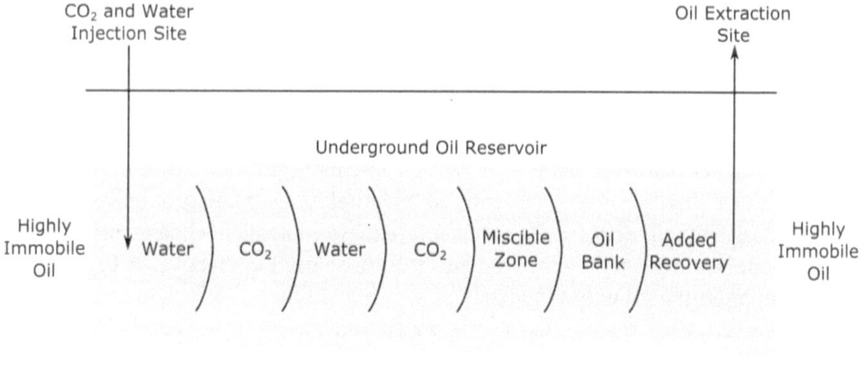

FIGURE 17.2 CO_2-EOR works by injecting alternating phases of CO_2 and driving water to increase the mobility of the otherwise immobile oil phase of a depleted oil reservoir. At the front of the forward CO_2 zone, the oil is mixed with high-pressure CO_2. The CO_2 helps decrease the viscosity of the oil while putting pressure on it so that it can be moved. Water may be used to help force the CO_2–oil mixtures further forward. The oil is then extracted from the extraction site, increasing the overall recovery.

EOR with CO_2 is typically presented as a form of sequestration, as the CO_2 pumped into the reservoir cannot readily escape. However, it is not entirely clear that EOR results in a net decrease in CO_2 emissions, as the oil that is pulled back up obviously goes towards some uses and causes further emissions of CO_2. Thus, EOR with CO_2 should probably be considered as cost effective and profitable more so than environmentally friendly – but, it is still a fact that there is more CO_2 proper in the ground after EOR than there was prior to EOR (Global CCS Institute, 2010, 2017).

In some cases, a much stronger argument for EOR can be made. For example, at the Shute Creek, Labarge, WY, USA facility, natural gas is harvested from a natural gas reservoir composed of approximately two-thirds of CO_2 naturally. This CO_2 has to be separated during processing, and while it could potentially be recycled back to the natural gas reservoir, doing so would incrementally increase operating costs as the CO_2 fraction slowly crept up over time. Using this CO_2 for EOR in nearby oil facilities represents a considerable improvement in profit and CO_2 utilization over simply venting it to atmosphere or not harvesting the gas at all (Thomas, 2009).

It is worth noting that EOR does not need to imply the usage of CO_2. In some cases, depending on the oil chemistry and the underlying geology, CO_2 is not as effective as some other EOR approaches. In this case, amines, surfactants, or other chemicals may be chosen instead, serving different purposes than just applying additional pressure to the reservoir (Global CCS Institute, 2010).

EOR is the one particular use of pure CO_2 that is worth calling out as a large demand – as of 2014, approximately 50 million tons of CO_2 were used for EOR per year. It is predicted that if the supply were available, CO_2 usage for EOR could rise as high as 300 million tons of CO_2 per year. If desired, CO_2 used for EOR can be stored permanently in the oil reservoirs. It is one of only two large-scale industrial CO_2 uses that can permanently store CO_2, alongside mineral carbonation (Global CCS Institute, 2010, 2017; Boot-Handford et al., 2014).

17.2.3 Gaseous and Aqueous CO_2

Gaseous CO_2 has comparatively few uses by itself. It is, after all, a mostly inert gas. It can be used as such – in fact, many handheld fire extinguishers and other fire suppression systems use CO_2 precisely for that. These systems rely on the inability of CO_2 to be further oxidized, but CO_2 is not the only inert gas that can be used for that. CO_2 is heavier than air, and thus tends to accumulate near the ground rather than floating away immediately – this is an advantage over nitrogen when using it for fire suppression, but arguably a disadvantage from a safety perspective. It can also be used to suppress the growth of bacteria in a similar fashion.

Gaseous CO_2 does readily dissolve in water, and can be subsequently hydrolyzed to form carbonic acid. Carbonic acid is a fairly mild acid, but due to the abundance of CO_2 and the safety of carbonic acid, it is frequently used in industrial processes for pH control. Relatedly, CO_2 is also used to carbonate beverages.

17.2.3.1 Inert Gas

An inert gas is one that undergoes comparatively few reactions. Typically this is used to refer to gases that are neither oxidized nor contribute to oxidation, in the sense of fire suppression. The most typical example of an inert gas would be nitrogen, which can displace oxygen and is itself resistant to oxidation, but CO_2 can fill a similar role.

CO_2 resists all common oxidizers and several uncommon oxidizers. Even oxidizing agents as potent as dioxygen difluoride fail to react with CO_2, for which CO_2 is one of the very few compatible solvents. This combined with the fact that CO_2 is a bit heavier than air, which means that it can play an effective role in fire suppression where displacing oxygen is effective (Streng, 1963).

CO_2 will do next to nothing to stop fires which are self-oxidizing, as displacing oxygen is irrelevant there. If a liquid oxidizing agent, such as nitric acid, is present, CO_2 may sometimes be worse than useless if it ends up spreading the oxidizing agent around the blaze. CO_2 is not nearly as resilient against reducing agents – magnesium, for example, can combust in the presence of CO_2 to form magnesium carbonate without any other oxidizing agent. In this case, CO_2 is acting as the oxidizing agent, due to magnesium's extremely strong tendency to give away its electrons. For this reason, CO_2 is not suitable for all metal fires. CO_2 can be used as a shielding gas in some welding operations, though it may not be as effective as noble gas mixtures. In the case of strongly reducing environments, nitrogen gas is likely preferable.

Since CO_2 is mostly inert, it can be used anywhere that does not specifically need a reactive gas. This could include use as an instrument or as a substitute for compressed air. Such applications typically depend on the abundance of CO_2 being used – if a sufficiently pressurized CO_2 line is already available and cool enough to use, then there is no problem with using it for such purposes. Often, compressed air is simply more readily available regardless.

17.2.3.2 pH Control

Since CO_2 reacts with water to form carbonic acid, it can be used to lower the pH of an alkaline solution. The minimum pH which can be achieved is dependent on temperature and composition of the solution, and the pressure of the CO_2 phase it is in contact with. Supercritical CO_2 can achieve exceedingly low pH reading in water, down to 2.8, but atmospheric pressures will reach equilibrium at 5.6–5.7 (Toews et al., 1995; UIGI, 2016).

Carbonic acid is one of very few truly food-safe acids, and is one of the most reliable food-safe weak acids. As such, it is often used in food processing to adjust the pH towards acidic, as needed. pH control in food processes is a significant portion of the 8 million tons of CO_2 used per year by the food industry (Global CCS Institute, 2010; Boot-Handford et al., 2014).

CO_2 is also used during the re-mineralization of water which has been processed via reverse osmosis. As carbonic acid is a weak acid, and dissociates in water, it provides a bare minimum of ionization to the purified water even if no metallic cations are present (Global CCS Institute, 2010).

For the same reason, deionized water exposed to the atmosphere will not remain perfectly deionized, and the pH of natural water will tend to drift towards acidic. Similarly, water will tend to form carbonate salts from any metal cations in it as it evaporates, since it will pick up CO_2 from the atmosphere over time.

Carbonic acid can also be used in cases where extreme quantities of acid are required, without much regard for the fact that it is a weak acid. This can include neutralization of strongly alkaline wastes, or for slightly lowering the pH of high flow-rate streams. The former may be applicable in the remediation and reprocessing of bauxite process red mud, which maintains an exceptionally high pH for most of its lifespan otherwise. The latter has been used in iron ore processing at various facilities.

17.2.3.3 Iron Ore Filtration

As a particular use of CO_2 in pH control, it can be used to control filtration processes. In one particular case study, CO_2 was injected into the filtration feed slurry in an iron ore concentrator to improve filtration rate. The iron ore concentrator processed the concentrate at a high pH value to optimize the flotation grade and recovery. The same conditions lead to a highly dispersing solution which is slow to filter. This dispersion is largely due to the high pH value, which leads to a strong negative charge on all the solid surfaces present in the slurry. While the negative charge on silica is largely unavoidable, as the pH must be dropped to around 2 to avoid that, the negative charge on hematite and magnetite can be mitigated much earlier. Thus, by reducing the pH, the filtration rate can be improved.

Multiple different acids were tested for use in an iron ore concentrator in the Lake Superior iron ore district (Ripke and Kawatra, 2003; Ripke et al., 2004). These included sulfuric acid, hydrochloric acid, and carbonic acid (as CO_2). The strong acids were found to have very little effect in reasonable quantities, and it is suspected that they could not be effectively dispersed through the slurry in an economical fashion. Furthermore, hydrochloric acid is capable of blinding the filter bags used in the iron ore filtration, which was undesirable. CO_2, on the other hand, was inexpensive and easily mixed into the slurry.

Filtration rates were increased by up to 23.7% with the addition of CO_2 to the filter slurry (Ripke and Kawatra, 2003; Ripke et al., 2004), and laboratory tests suggest that even higher filtration rates should be possible. The CO_2 used in this form is partially captured as carbonates that capture magnesium and calcium from the slurry's water hardness and the concentrate's adsorbed moisture, but some will not be permanently captured in this fashion. CO_2 was found to be far more cost effective for improving filtration efficiency than mineral acids, while also presenting much less risk to the filtration equipment. It is also believed that the filter quality may ultimately improve slightly if higher concentrations of CO_2 can be used.

17.2.3.4 Food Preservation and Preparation

CO_2's primary action in food preservation is to displace oxygen, particularly during canning. As many bacteria of concern rely on oxygen to survive and grow, this is often quite effective at extending the shelf life of foods (UIGI, 2016).

Furthermore, many bacteria can only grow and reproduce at reasonable pressures. High pressure is relatively gentle on many organic compounds compared to thermal preservation methods. As such, exposing foods to extremely high pressures is likely to have minimal effect on taste, smell, or appearance. As CO_2 is food safe and readily available, it can be used to inactivate bacteria in food. This involves raising the pressure of a CO_2 atmosphere around the food to approximately 10,000 bar, but inactivates most bacteria and prevents them from growing or reproducing in the food (Cheftel, 1995; UIGI, 2016).

High pressures of CO_2 are also used in beverage carbonation – though much lower pressures than in preservation. Carbonated beverages have been supersaturated with CO_2 above the partial pressure of CO_2 in the atmosphere, and the nucleation of the supersaturated CO_2 provides a distinctive fizziness to the beverage as it is consumed. Once the beverage's container is opened, the CO_2 will slowly begin to escape to the atmosphere – though if enough nucleation potential (e.g. due to shaking) or CO_2 pressure is present, it may do so quite a bit more quickly. Eventually the CO_2 in a carbonated beverage will reach equilibrium with the atmosphere, and the beverage will become flat. This is a major use of CO_2 in the food industry (Gardner et al., 2014).

Pure CO_2 is sometimes used elsewhere in food preparation as well. The most direct and widespread example is likely the "Pop Rocks" candy. The solid candy is formed by crystallizing sugar under a high-pressure CO_2 atmosphere. This causes small bubbles of high pressure CO_2 to be trapped in the candy until the sugar is disrupted. When eaten, this causes the namesake pop.

17.3 CONVERSION OF CO_2 TO USEFUL COMPOUNDS

CO_2's uses as a pure compound are limited compared to the uses of all carbon compounds. Thus, if CO_2 could be efficiently converted to other carbon compounds, unlocking the rest of organic chemistry, perhaps it would be more valuable overall to do that instead of attempting to use it directly.

Table 17.1 considers the theoretical energy cost of creating several basic organic compounds from CO_2 and water on a carbon basis. If graphite is considered to be

TABLE 17.1
Energy Cost to Form Select Compounds Compared to Geological Sequestration

Compound	Net Energy Cost to Create from CO_2 and H_2O (kJ/mol C)	Additional Carbon Needed to Supply Electrolysis Energy (mol C)
Geological sequestration	*Approximately 87.1*	*N/A*
Oxalic acid	121.5	0.000
Formic acid	253.8	0.897
Carbon monoxide	283.2	1.117
Graphite	393.5	1.941
Formaldehyde	570.8	3.267
Ethanol	683.1	4.106
Methanol	725.7	4.424
Methane	890.3	5.654

Source: Adapted from data based on higher heating values available from P.J. Linstrom and W.G. Mallard, Eds., *NIST Chemistry WebBook*, NIST Standard Reference Database Number 69, National Institute of Standards and Technology, Gaithersburg MD, 20899, doi: 10/18434/T4D303, (retrieved 08/17/17).

Additional carbon needed based on conversion to 1 mole of carbon as the compound from graphite with 34% electricity efficiency and only capturing the original mole of CO_2.

representative of a high-quality coal, then only compounds which cost less energy than it can be reasonably formed during CO_2 capture from coal. If the conversion is to be performed via electrolysis, then the electrical energy efficiency of coal needs to be considered as well. Table 17.1 considers a 34% electrical efficiency, which is a common efficiency for coal-fired power plants. After accounting for efficiency, the amount of energy needed for electrolysis can be directly compared to the electrical energy which can be captured from graphite: save for geological sequestration and oxalic acid, there are no organic compounds which can be formed from burning a mole of coal without needing to burn more coal to make up the energy cost.

If hydrogen is readily available however, the thermodynamics change drastically. These compounds can all be created exothermically from hydrogen and CO_2, rather than requiring additional fuel to be burned to supply the necessary energy. These numbers are shown in Table 17.2. In short, energy-efficient use of CO_2 can be achieved immediately if hydrogen can be sourced in an energy-efficient fashion, and subsequently be used as the reducing agent.

However, presently hydrogen is created by the gasification of organic fuels, the water gas shift reaction which uses carbon monoxide to reduce water to hydrogen gas, or the direct electrolysis of water. None of these methods are extraordinarily energy efficient, though the gasification of fuels has been previously discussed as an option to efficiently capture CO_2.

The values in Table 17.1 have been calculated directly from the enthalpies of their components, which paints a somewhat depressing picture. By Hess's law, which

TABLE 17.2
The Energy Required to Form a Given Compound from CO_2 and Hydrogen Gas

Compound	Net Energy Cost to Create from CO_2 and Hydrogen (kJ/mol C)
Ethane	–440.3
Ethanol	–346.7
Methane	–253.0
Graphite	–178.2
Methanol	–130.7
Oxalic acid	–41.9
Formic acid	–31.6
Formaldehyde	–8.2
Carbon monoxide	–2.9

Source: Adapted from data for heats of formation available from P.J. Linstrom and W.G. Mallard, Eds., *NIST Chemistry WebBook*, NIST Standard Reference Database Number 69, National Institute of Standards and Technology, Gaitherburg MD, 20899, doi: 10/18434/T4D303, (retrieved 08/17/17).

Negative values indicate exothermic reactions.

notes that enthalpy is a state function, this means that there is absolutely no clever way to bypass the energy cost of forming these compounds from only water and CO_2. No combination of catalysis or electrolysis will ever reduce the energy costs below those listed in Table 17.1.

If all of the carbon of a coal-fired facility is to be captured, and hydrogen is not readily and cheaply available, the only possible conversion is oxalic acid. The question then becomes: Is it better to convert the CO_2 to oxalic acid, even inefficiently, than to sell the CO_2 as is? Considering that the CO_2 market is exceptionally saturated – possible supply exceeds current demand by over a hundred fold – the price of CO_2 is not very high. The value of CO_2 will be somewhere around the cost of capturing it, so not much more than \$100/ton based on 2006 estimates. The price of oxalic acid in 2018 is somewhere near \$400/ton, and by stoichiometry it can be determined that one can make 2.6 tons of oxalic acid ($H_2C_2O_4$) from 1 ton of CO_2. At those market prices, there is a leeway of nearly \$900/ton of CO_2 for processing costs. This sort of back of the envelope investigation should show that there is some potential to be had in the conversion of CO_2 to useful compounds, though which chemical is optimal is likely to change in the future (Rochelle, 2009).

The other observation to be made from Table 17.1 is that the formation of the higher energy fuels from coal-fired facilities is an energy sink. Unless they can be created with a process with 100% efficiency and subsequently used at 100% efficiency, there is a net energy loss from the conversion. While the creation of these compounds from CO_2 and water will be discussed in this section, creating them from coal may be counterproductive overall. These compounds could be used as precursors in further chemical processes, or as a means of storing otherwise free power.

17.3.1 Reaction Chemistry of CO_2

CO_2 does not react with oxidizing agents, but it can be reduced. Reduction is the general form of all chemistry that CO_2 undergoes. Reduction refers to the addition of electrons. In CO_2's case, the electrons are added to the central carbon in reduction, as it experiences a slight positive charge due to the neighboring oxygen atoms. In its purest form, this electron addition results in the CO_2 anion radical as shown in Equation 17.1.

$$CO_2 + e^- \rightarrow CO_2^{\bullet -} \tag{17.1}$$

In the presence of hydrogen or some other more substantial reducing agent, it is likely that this radical will immediately attack the remainder of the reducing agent to form a compound immediately. Equation 17.2 presents a possible reaction, also known as the reverse water gas shift reaction, which can happen at elevated temperatures.

$$CO_2 + H_2 \rightarrow CO + H_2O \tag{17.2}$$

However, in the absence of an attackable reducing agent, such as in electrolysis, the CO_2 anion radical has only a handful of options. It can dimerize with itself, it can attack an unreduced CO_2 molecule, or it can attack any other readily available molecule (e.g. a solvent, if applicable). The dimerization leads to the formation of the oxalate anion, shown in Equation 17.3 (Fischer et al., 1981).

$$CO_2^{\bullet-} + CO_2^{\bullet-} \rightarrow C_2O_4^{2-} \quad (17.3)$$

This oxalate anion is a crucial step towards creating oxalic acid from CO_2. This reaction is second order in the CO_2 anion radical, and thus is favored by strongly reducing setups – but disfavored by strongly reducing setups that end up preferring other reduction reactions. Unfortunately, this includes the reduction of water into hydrogen, or the reduction of CO_2 directly into carbon monoxide in some cases (Fischer et al., 1981).

Alternatively, it can attack a CO_2 molecule and gain an additional electron to form carbonate and carbon monoxide, as shown in Equation 17.4 (Fischer et al., 1981).

$$CO_2^{\bullet-} + CO_2 + e^- \rightarrow CO_3^{2-} + CO \quad (17.4)$$

In electrolysis the carbonate formed in Equation 17.4 is a bit of a dead end. Since it carries a strongly negative charge, it is nearly impossible to get it to accept any further electrons from the cathode. For this reason, it is preferable to carry out the electrolysis of CO_2 at lower pH values (below 9) (Oloman and Li, 2008).

If protons are available, the CO_2 anion radical can also proceed to form formate, as shown in Equation 17.5 (Fischer et al., 1981).

$$CO_2^{\bullet-} + H^+ + e^- \rightarrow HCOO^- \quad (17.5)$$

It is possible to further reduce formate, which leads towards the production of methanol and methane. Furthermore, it is occasionally possible for the CO_2 anion radical to dimerize with another organic compound containing an electrophilic carbon atom. This leads to the production of heavier hydrocarbons.

17.3.2 Carbon Monoxide

Carbon monoxide (CO) is one of the simplest reduction products that can be formed from CO_2. It can be formed by directly reacting CO_2 with hydrogen as in Equation 17.2, or via disproportionation during electrolysis as in Equation 17.4. The reaction with hydrogen is known as the reverse water gas shift reaction, which is mildly endothermic (+41 kJ/mol) (Jansen et al., 2015).

In electrolysis, carbon monoxide is a common product on most catalytic surfaces. More noble metals in particular tend to form carbon monoxide in larger quantities. Palladium and platinum, for example, form almost exclusively carbon monoxide during electrolysis in nonaqueous electrolytes. Copper, silver, and gold trigger the disproportionation reaction in Equation 17.4, yielding primarily carbon monoxide in aqueous or nonaqueous electrolytes. Most metals will result in the formation of some quantity of carbon monoxide during the reduction of CO_2 (Jitaru et al., 1997).

Carbon monoxide is a useful chemical in many chemical syntheses, owing to its own reducing capabilities. Carbon monoxide can be subsequently transformed into phosgene by reacting with chlorine, or into methanol and higher hydrocarbons such as via the Fischer–Tropsch process. From there, it can be said that carbon monoxide can be used as a precursor to essentially all industrial chemicals (Leitner, 1995; Hietala et al., 2016).

The forward water gas shift reaction, which converts carbon monoxide and water into hydrogen, is used as a source of hydrogen in pre-combustion CO_2 capture to be used as a carbon-free fuel. However, note that pre-combustion capture starts with a fuel – starting from CO_2 would lead to a net energy deficit by the conservation of energy. Thus, converting CO_2 to carbon monoxide for the sake of making hydrogen to use as fuel is counterproductive. Rather, in pre-combustion capture the hydrogen is a useful byproduct of the carbon separation from the original fuel (Jansen et al., 2015).

17.3.3 Elemental Carbon

Reducing carbon monoxide once more by removing the other oxygen would result in the formation of elemental carbon (C). Practically speaking, this is a bit more difficult to achieve – electrolysis on metal cathodes simply does not typically result in this level of reduction. Reduction with hydrogen will result in the formation of methanol or formaldehyde instead of graphite (Jitaru et al., 1997).

Rather, the reduction of carbon compounds all the way to elemental carbon is achieved primarily with extremely high voltages. In the event that carbon does form during electrolysis, its usual role would be to foul the electrode, and is thus perhaps undesirable over all (Ebbesen and Mogensen, 2009).

It is potentially possible via photochemical methods to split the carbon out of CO_2 directly, as in Equation 17.6 below, where γ represents an appropriately high-energy photon. This reaction pathway has been theoretically investigated and is plausible at vacuum ultraviolet frequencies, such as those experienced in space from the sun (Lu et al., 2014).

$$CO_2 + \gamma \rightarrow C + O_2 \tag{17.6}$$

17.3.4 Formic Acid

Industrially, formic acid (H_2CO_2) is presently produced primarily by the hydrolysis of methyl formate into formic acid and methanol as shown in Equation 17.7. Formate salts may also be reacted with acid to form formic acid (Hietala et al., 2016).

$$HCO_2CH_3 + H_2O \rightarrow H_2CO_2 + CH_4O \tag{17.7}$$

Methyl formate is in turn typically produced industrially by the reaction of methanol with carbon monoxide. Thus, producing formic acid from the reduction of CO_2 can be achieved by two primary routes: the direct reduction of CO_2 in the presence of hydrogen as shown in Equation 17.8, or indirectly by the formation of a syngas containing CO and the formation of methanol. Syngas, or synthesis gas, is a mixture of gases containing predominantly carbon monoxide and hydrogen and perhaps a small portion of CO_2. The direct reduction shown in Equation 17.8 is typically unfavorable, as the reactants are far more entropic than the products – the formation of methyl formate as an intermediate product makes the reaction far more likely to proceed in full (Leitner, 1995; Hietala et al., 2016).

$$CO_2 + H_2 \rightarrow CH_2O_2 \tag{17.8}$$

Due to the necessity of hydrogen in Equation 17.8, the electrolytic conversion of CO_2 to formic acid is most easily achieved in an aqueous solution. The electrolytic conversion prefers the use of lead, mercury, indium, or tin as the cathode material, but other metals can achieve modest current efficiencies as well (Jitaru et al., 1997; Oloman and Li, 2008).

Metal complexes have also been explored for the electrolytic conversion of CO_2 to formic acid. These catalysts tend to be extremely sensitive to the precise reaction conditions, demonstrating sensitivity to many solution properties, including the precise composition of the solvent and pH. These catalysts are typically metal-amine complexes based on transition metals, including iron, ruthenium, osmium, rhodium, and cobalt (Leitner, 1995).

Similar catalysts have also been investigated for the photochemical reduction of CO_2 to formic acid. The overall mechanism remains largely the same, except that the creation of the CO_2 anion radical is triggered by the acquisition of an electron liberated by high-energy photons. Some ruthenium-based catalysts are effective, albeit slowly, even with lower-energy photons including visible light (Leitner, 1995).

In the absence of water, the addition of hydrogen to CO_2 can be catalyzed by other metals. Ruthenium complexes in particular have been successful in promoting the hydrogenation of CO_2 to formic acid in alcohol solutions. Unfortunately, these same catalysts tend to catalyze the decomposition of formic acid during the subsequent distillation steps required to purify it. This can be solved by reversibly deactivating the catalyst, but this introduces another complexity into the process with exact requirements – even a tiny trace of active catalyst can significantly lower formic acid yields (Hietala et al., 2016).

Formic acid's direct uses are primarily in the realm of silage for animal feed or products, or for the processing of leather and textiles. Its primary action in these processes tends to be its acidity, though in leather working its activity as a solvent is also valuable. Formic acid can also be used as the precursor to many industrial chemicals. The preparation of formamides or formates from formic acid is direct, and these have considerable uses as solvents and reagents (Hietala et al., 2016).

In particular formaldehyde can be formed by the further hydrogenation and dehydration of formic acid, as shown in Equation 17.9. This runs directly contrary to the typical industrial synthesis of formaldehyde via the catalytic oxidation of methanol. Like formic acid, formaldehyde can be used as precursor chemical to many organic syntheses, and is additionally useful for preservation or occasionally as a solvent. This can be catalyzed electrolytically with ruthenium under some conditions (Jitaru et al., 1997).

$$H_2CO_2 + H_2 \rightarrow CH_2O + H_2O \tag{17.9}$$

17.3.5 Methanol

Methanol (CH_3OH) is the next step after formaldehyde in the reduction of CO_2. It is formed by the hydrogenation of formaldehyde as shown in Equation 17.10, or the triple hydrogenation and single dehydration of CO_2 as shown in Equation 17.11 (Leitner, 1995).

$$CH_2O + H_2 \rightarrow CH_3OH \tag{17.10}$$

$$CO_2 + 3H_2 \rightarrow CH_3OH + H_2O \tag{17.11}$$

Industrially, methanol is made from syngas – a mixture of carbon monoxide and hydrogen. This reaction is shown in Equation 17.12, and the carbon monoxide leading up to it can be formed via Equation 17.2 or 17.4.

$$CO + 2H_2 \rightarrow CH_3OH \tag{17.12}$$

The primary use of methanol in the chemical industry is to be converted to formaldehyde for further synthesis opportunities. In this case, the formation of methanol from CO_2 is perhaps not ideal, as that implies that more reduction has been performed than strictly necessary. Methanol is also often used as a fuel for storing energy.

Starting from CO_2, converting it to methanol may be able to store a meaningful amount of energy for later use. However, because energy is lost during the generation of methanol and during the combustion of methanol combined with the high starting energy cost to create methanol from CO_2, making methanol using fossil fuels does not reduce carbon emissions.

Arriving at methanol from electrolysis is not exceedingly difficult, but is energy intensive. Molybdenum has been found to be an effective cathode material for catalyzing the formation of methanol in aqueous solutions. Many other metals also show some formation of methanol under mildly acidic conditions in aqueous solutions, but carbon monoxide, hydrogen, and formic acid are common byproducts (Jitaru et al., 1997).

17.3.6 Methane

Methane (CH_4) is the simplest and smallest hydrocarbon, and the most reduced product which can be formed from CO_2. It is formed by removing the oxygen from methanol by the addition of hydrogen, as in Equation 17.13 (Leitner, 1995).

$$CH_4O + H_2 \rightarrow CH_4 + H_2O \tag{17.13}$$

Methane's primary use is as a fuel, such as in natural gas or otherwise. Chemical processing of methane typically involves reforming it into carbon monoxide with high-temperature steam first. Methane is largely nonreactive except to oxidation, and is not easy to directly convert to other compounds.

For the same reason that forming fuel methanol is not an ideal use of CO_2, forming fuel methane is also not ideal from CO_2. The products that could be made via methane can be made more directly from the other CO_2 reduction products. However, for the sake of completion, the electrolytic catalysts leading to methane are primarily ruthenium and nickel in aqueous solution. Nickel has the downside of resulting in unavoidable hydrogen production and that the carbon monoxide which is formed strongly bonds to the metal surface, inhibiting further reaction (Jitaru et al., 1997).

17.3.7 Oxalic Acid

Oxalic acid ($H_2C_2O_4$) can be used to leach rare earths from minerals. The rare earths, a national security resource, are used in the production of advanced technology ranging from smartphones and defense to medicine and complex chemical catalysts. Oxalic acid is formed from the dimerization of CO_2 in the presence of hydrogen, as in Equation 17.14. Oxalic acid is one of the strongest organic acids, and has intriguing ligand properties for selectively solubilizing and insolubilizing metal cations. Oxalic acid shares many of the same uses as formic acid, but is also used in rare earth leaching (Sawada and Murakami, 2000).

$$2CO + H_2 \rightarrow H_2C_2O_4 \tag{17.14}$$

Industrial synthesis of oxalic acid today involves the oxidation of various carbohydrates, ethylene glycol, or propylene. Typically the oxidizing agent in these reactions is nitric acid (Sawada and Murakami, 2000).

However as shown in Table 17.1 and implied by the relatively simple dimerization, oxalic acid is not energetically expensive to form from CO_2 if a modest process efficiency can be attained. Oxalic acid is a chemical produced in bulk quantities for the textile and leather industries, and economic analyses suggest that it is one of the most profitable compounds that could be made from CO_2 (Otto et al., 2015).

The direct preparation of oxalic acid from CO_2 and hydrogen is typically impractical, though may be possible with an appropriate catalyst. The reaction can be performed with electrolysis, but only in nonaqueous solutions. Typical cathode materials include stainless steel, lead, mercury, thallium, tin, zinc, molybdenum, and others. Lead and mercury are typically considered to be the most effective flat cathodes for the formation of oxalate (Jitaru et al., 1997).

Nonaqueous electrolytes for this purpose are based on an organic aprotic solvent phase like dimethylformamide or acetonitrile, and a soluble organic salt such as tetraethylammonium bromide or tetrabutylammonium perchlorate. The reaction can be homogenously catalyzed by the presence of an aromatic nitrile, such as o-tolunitrile (Fischer et al., 1981; Gennaro et al., 1996; Jitaru et al., 1997).

A major difficulty of creating oxalate is the separation of the oxalate anion from the electrolytic solution. One solution to this is to use a sacrificial anode that can achieve a +2 ionization, such as zinc, which will promote the immediate precipitation of the oxalate. This precipitate can then be filtered continuously and collected (Fischer et al., 1981).

We have repeated these experiments in our own work, and found that using sacrificial zinc electrodes at appropriately high current densities, zinc oxalate purities of 95% are quite achievable. The major limitations on the rate of oxalic acid formation are the power available, the comparatively low conductivity of the non-aqueous electrolyte as compared to water, and the relatively slow reaction kinetics in laboratory-sized cells. Our best results were achieved using lead and zinc as the cathodes, though with zinc a significant amount of zincite (ZnO) was also formed. Conveniently, fine zincite is actually rather valuable itself, so this may not be so much of an issue.

We compared these results to membrane electrolysis cells. Overall, the membrane electrolysis cell seems to be at an overall disadvantage compared to the sacrificial electrode approach. In particular, the poor conductivity of the nonaqueous electrolyte is no longer an issue, as the membrane's conductivity is typically even worse. Furthermore, the material compatibility of the membrane is a bit of an issue – we were unsuccessful in finding a membrane chemistry that could simultaneously survive the alkaline conditions of the anode chamber and the nonaqueous electrolyte of the cathode chamber. That is not to say that these challenges could not be overcome, but that they do need to be considered before heading down that route in the future.

Oxalic acid is a bit of a dead end for further chemical processing however – while it can be converted to glycolic or glyoxylic acid by further reduction, and it can be reacted to some extent as other carboxylic acids may, there are few reactions it is better suited for than formic acid.

Rather, one of the main draws of oxalic acid is instead its metal chelating properties. In short, oxalic acid forms insoluble complexes with most divalent metal cations but remains soluble with most other metal cations. In the case of rare earth acid leaching processes, most rare metals are divalent metals, so the addition of oxalic acid can allow these metals to precipitate out preferentially (Sawada and Murakami, 2000).

At the time of writing, the market price of oxalic acid far exceeds the market price of an equivalent carbon's worth of coal, allowing for a tremendous profit margin. The estimated value of oxalic acid which can be produced from approximately \$60 worth of coal is nearly \$1.000 of market value. Though this does not include processing expenses or market saturation effects, it is certainly noteworthy that oxalic acid is worth so much comparatively (Otto et al., 2015; EIA, 2017).

17.3.8 Fischer–Tropsch Process

The direct reduction products of CO_2 are all relatively small compounds. While there are a handful of catalysts that result in the formation of small quantities of longer chained carbon compounds directly from CO_2, this is not the norm. Despite that, these longer chain compounds are useful so: how can CO_2 be converted into longer chain molecules? (Jitaru et al., 1997).

The answer is the Fischer–Tropsch process. The Fischer–Tropsch process converts hydrogen and carbon monoxide into hydrocarbon products approximately along the lines of Equation 17.15.

$$nCO + (2n+1)H_2 \rightarrow C_nH_{2n+2} + nH_2O \qquad (17.15)$$

These reactions take place at moderately high temperatures of 250°C–350°C over iron or cobalt catalysts. Since the reverse water gas shift reaction can convert additional CO_2 into carbon monoxide, it is possible to achieve considerable conversion of carbon-containing compounds to higher hydrocarbons. Approximately 1.07 mol of hydrogen are required for every 2 mol of carbon monoxide and every 3 mol of CO_2 present to achieve near complete conversion (Dry, 2002).

These reactions are typically performed in fluidized-bed reactors, allowing for excellent contact between the catalyst and the reacting gases. Metal-based catalysts are used in this reaction – particularly: iron, nickel, cobalt, and ruthenium. Fixed fluidized-bed reactors are preferred, as the catalyst material tends to be quite abrasive and in circulating fluidized-bed reactors more catalyst is needed as only a portion of the catalyst can be in the reactor at once (Dry, 2002).

The Fischer–Tropsch process is not exceedingly selective. Rather, a large variety of products are formed, primarily consisting of fully saturated potentially oxidized products (such as paraffins or alcohols) or moderately unsaturated potentially oxidized products (such as olefins, aldehydes, ketones, and organic acids). Higher operating temperatures tend to result in shorter carbon chains, as does the increased presence of carbon monoxide in the gas stream (Dry, 2002).

Often carbon compounds containing between 10 and 20 carbons are targeted, but it is possible to form several varieties of compounds. Nickel catalysts form primarily methane, while hard, long chain waxes can be formed under high carbon monoxide conditions. These compounds must then be separated into saleable products, typically in a similar fashion to how a crude oil must be processed (Dry, 2002).

The Fischer–Tropsch process provides a methodology by which to produce almost every petroleum product from CO_2 given only energy, the ability to reduce CO_2 into carbon monoxide, and a source of hydrogen.

At present the Fischer–Tropsch process is used primarily to supplement the production of petroleum products and to use excess methane which would otherwise be flared into the atmosphere. The hydrogen is usually formed by reforming methane with steam, which is also the source of carbon monoxide (Dry, 2005).

17.4 SUMMARY

CO_2 produced from coal can be used in many ways to help offset the cost of capturing it. These include using the CO_2 as working fluid or a solvent, for refrigeration, or for conversion to useful chemicals.

While there are many practical uses for pure CO_2, there are very few which actually require tremendous amounts of it. For example, the demands for dry ice or for CO_2 in food preservation are already well met and relatively constant. However, CO_2 for EOR is a growing field, as is the conversion of CO_2 to other chemicals.

Of the many chemical conversions mentioned here, perhaps the most exciting is oxalic acid. It can be used in a number of fields, but it is especially useful in the processing of rare earth minerals. As it is particularly beneficial for its curious ligand properties, which are largely unique to its molecular geometry, and it is a comparatively simple compound it stands to reason that there may never be a practical competitor to it in this particular use case. In addition, oxalic acid's market value is consistently higher than a carbon equivalent's worth of coal, even though it is already produced in bulk around the world. Furthermore, the theoretical energy costs of transforming CO_2 into oxalic acid are very small.

While there is a significant emphasis in literature on the conversion of CO_2 to fuel compounds such as methanol, it is definitely worth remembering that these technologies do not really apply to refining CO_2 with coal power. After all, the resulting energy cost is high enough that more additional CO_2 will be produced than is actually captured. That is not to say that these technologies do not have their uses: if the energy source does not emit CO_2, then there is no immediate problem with using that energy to form any compound that is desired.

CO_2 utilization may not strictly be preferable to CO_2 sequestration, but transforming CO_2 into any nonvolatile compounds should still help to mitigate the atmospheric danger of global climate change. If the potential value presented by CO_2 utilization leads more facilities into developing the infrastructure required to capture CO_2, it is still a win for the environment.

REFERENCES

Ahn, Y., Bae, S.J., Kim, M., Cho, S.K., Baik, S., Lee, J.I. and Cha, J.E. (2015), "Review of Supercritical CO_2 Power Cycle Technology and Current Status of Research and Development," *Nuclear Engineering Technology*, Vol. 47, pp. 647–661.

Aprea, C., Greco, A. and Maiorino, A. (2013), "The Substitution of R134a with R744: An Exergetic Analysis Based on Experimental Data," *International Journal of Refrigeration*, Vol. 36, pp. 2148–2159.

Boot-Handford, M.E., Abanades, J.C., Anthony, E.J., Blunt, M.J., Brandani, S., Mac Dowell, N., Fernández, J.R., Ferrari, M.-C., Gross, R., Hallett, J.P., Haszeldine, R.S., Heptonstall, P., Lyngfelt, A., Makuch, Z., Mangano, E., Porter, R.T.J., Pourkashanian, M., Rochelle, G.T., Shah, N., Yao, J.G. and Fennell, P.S. (2014), "Carbon Capture and Storage Update," *Energy Environmental Science*, Vol. 7, pp. 130–189.

Brown, D.W. (2000), "A Hot Dry Rock Geothermal Energy Concept Utilizing Supercritical CO_2 Instead of Water," *Proceedings of the 25th workshop on geothermal reservoir engineering, Stanford University*, Stanford, CA, pp. 233–238.

CCPI (2009), "What is Dry Ice?" *Continental Carbonic Products, Inc.* Accessible at https://web.archive.org/web/20090727073832/; www.continentalcarbonic.com/dryice/ (archived on July-27-2009).

CCPI (2018), "FAQs – Consumer Purchases," *Continental Carbonic Products, Inc.* Accessible at www.continentalcarbonic.com/faqs.html (accessed on January-7-2018).

Cheftel, J.C. (1995), "Review: High-Pressure, Microbial Inactivation and Food Preservation," *Food Science and Technology International*, Vol. 1, No. 2–3, pp. 75–90.

Dry, M.E. (2002), "The Fischer–Tropsch Process: 1950–2000," *Catalysis Today*, Vol. 71, pp. 227–241.

Ebbesen, S.D. and Mogensen, M. (2009), "Electrolysis of Carbon Dioxide in Solid Oxide Electrolysis Cells," *Journal of Power Sources*, Vol. 193, pp. 349–358.

EIA (2017), "Coal Prices and Outlook," *Energy Information Administration*. Accessed at www.eia.gov/energyexplained/index.php?page=coal_prices.

Fischer, J., Lehmann, Th. and Heitz, E. (1981), "The Production of Oxalic Acid from CO_2 and H_2O," *Journal of Applied Electrochemistry*, Vol. 11, pp. 743–750.

Franco, A. and Diaz, A.R. (2009), "The Future Challenges for 'Clean Coal Technologies': Joining Efficiency Increase and Pollutant Emission Control," *Energy*, Vol. 34, pp. 348–354.

Gardner, D.E., Patel, B.R., Hernandez, V.K., Clark, D., Sorensen, S., Lester, K., Solis, Y., Tapster, D., Savage, A., Hyneman, J. and Dukes, A.D. (2014), "Investigations of the Mechanism of the Diet Soda Geyser Reaction," *The Chemical Educator*, Vol. 19, pp. 358–362.

Gennaro, A., Isse, A.A., Savéant, J.-M., Severin, M.-G. and Vianello, E. (1996), "Homogenous Electron Transfer Catalysis of the Electrochemical Reduction of Carbon Dioxide. Do Aromatic Anion Radicals React in an Outer-Sphere Manner?", *Journal of the American Chemical Society*, Vol. 118, pp. 7190–7196.

Global CCS Institute (2010), "Accelerating the Uptake of CCS: Industrial Uses of Captured Carbon Dioxide," *Global CCS Institute.*

Global CCS Institute (2017), "Large-Scale CCS Facilities," *Global CCS Institute.* Accessed at www.globalccsinstitute.com/projects/large-scale-ccs-projects (database updated October-27-2017, accessed January-4-2018).

Hietala, J., Vuori, A., Johnsson, P., Pollari, I., Reutemann, W. and Kieczka, H. (2016), "Formic Acid," *Ullmann's Encyclopedia of Industrial Chemistry.*

Jansen, D., Gazzani, M., Manzolini, G., van Dijk, E. and Carbo, M. (2015), "Pre-Combustion CO_2 Capture," *International Journal of Greenhouse Gas Control*, Vol. 40, pp. 167–187.

Jitaru, M., Lowy, D.A., Toma, M., Toma, B.C. and Oniciu, L. (1997), "Electrochemical Reduction of Carbon Dioxide on Flat Metallic Cathodes," *Journal of Applied Electrochemistry*, Vol. 27, pp. 875–889.

Karásek, P., Šťavíková, L., Planeta, J., Hohnová, B. and Roth, M., 2013. "Solubility of Fused Silica in Sub- and Supercritical Water: Estimation from a Thermodynamic Model," *The Journal of Supercritical Fluids*, Vol. 83, pp. 72–77.

Leitner, W. (1995), "Carbon Dioxide as a Raw Material: The Synthesis of Formic Acid and Its Derivatives from CO_2," *Angewandte Chemie International Edition*, Vol. 34, pp. 2207–2221.

Lou, X., Jannsen, H.G. and Cramers, C.A. (1997), "Temperature and Pressure Effects on Solubility in Supercritical Carbon Dioxide and Retention in Supercritical Fluid Chromatography," *Journal of Chromatography A*, Vol 785. No. 1–2, pp. 57–64.

Lu, Z., Chang, Y.C., Yin, Q., Ng, C.Y. and Jackson, W.M. (2014), "Evidence for Direct Molecular Oxygen Production in CO_2 Photodissociation," *Science*, Vol. 346, pp. 61–64.

Melzer, L.S. (2012), "Carbon dioxide Enhanced Oil Recovery (CO_2 EOR): Factors Involved in Adding Carbon Capture, Utilization and Storage (CCUS) to Enhanced Oil Recovery," Center for Climate and Energy Solutions, pp. 1–17.

Oloman, C. and Li, H. (2008), "Electrochemical Processing of Carbon Dioxide," *ChemSusChem*, Vol. 1, pp. 385–391.

Otto, A., Grube, T., Schiebahn, S. and Stolten, D. (2015), "Closing the Loop: Captured CO_2 as a Feedstock in the Chemical Industry," *Energy & Environmental Science*, Vol. 8, pp. 3283–3297.

Ripke, S.J. and Kawatra, S.K. (2003), "Effects of Cations on Unfired Magnetite Pellet Strength," *Minerals & Metallurgical Processing*, Vol. 20, pp. 153–159.

Ripke, S.J., Eisele, T.C. and Kawatra, S.K. (2004), "Effects of Retained Calcium Ions in Iron Ore Filtration and Pelletization Performance," *Particle Size Enlargement in Mineral Processing*-Proceedings of the 5th UBC-McGill International Symposium, (Laskowski, ed.) Metallurgical Society of Canadian Institute of Mining and Metallurgy, Hamilton, pp. 384–391.

Rochelle, G.T. (2009), "Amine Scrubbing for CO_2 Capture," *Science*, Vol. 325, No. 5948, pp. 1652–1654.

Sawada, H. and Murakami, T. (2000), "Oxalic Acid," *Kirk-Othmer Encyclopedia of Chemical Technology.*

Streng, A.G. (1963), "The Chemical Properties of Dioxygen Difluoride," *Journal of the American Chemical Society*, Vol. 85, pp. 1380–1385.

Thomas, S. (2009), "LaBarge Field & Shute Creek Facility," *The Wyoming Enhanced Oil Recovery Institute.*

Toews, K.L., Shroll, R.M., Wai, C.M. and Smart, N.G. (1995), "pH-Defining Equilibrium between Water and Supercritical CO_2. Influence of SFE of Organics and Metal Chelates," *Analytical Chemistry*, Vol. 67, No. 22, pp. 4040–4043.

UIGI (2016), "Carbon Dioxide (CO_2) Properties, Uses, Applications: CO_2 Gas and Liquid Carbon Dioxide," *Universal Industrial Gases, Inc.* Accessible at www.uigi.com/carbondioxide.html.

Witkowski, A., Majkut, M. and Rulik, S. (2014), "Analysis of Pipeline Transportation Systems for Carbon Dioxide Sequestration," *Archives of Thermodynamics*, Vol. 35, No. 1, pp. 117–140.

Zhang, Z.X., Wang, G.X., Massarotto, P. and Rudolph, V. (2006), "Optimization of Pipeline Transport for CO_2 Sequestration," *Energy Conversion and Management*, Vol. 47, pp. 702–715.

18 Coal Dust

Coal dust is a tremendously dangerous air pollutant. Not only is coal dust responsible for serious respiratory conditions such as "black lung," it is also responsible for deadly coal mine explosions. Even in general, fine particulate matter (PM) with diameters less than 10 µm (PM_{10}) and especially less than 2.5 µm ($PM_{2.5}$) have been found to be dangerous to human health regardless of composition. As of 2012, the Environmental Protection Agency (EPA) standards for $PM_{2.5}$ were 12 µg/m^3 annual average, and no more than 35 µg/m^3 as the 98th percentile of 24-h averages. PM_{10} restrictions are only slightly looser, allowing up to 150 µg/m^3 as a 24-h average once per year.

For these reasons significant work has been done to examine how coal dust can be effectively controlled. Water sprays are a natural response to controlling coal dust, since water is cheap and readily available. However, it has been reported that water sprays were ineffective as a coal dust suppressant when sprayed at a rate of 95 L/min (25 gpm) on freshly mined coal (Kobrick, 1970). It has been argued that this water spray was ineffective as a dust control reagent due to "incomplete coating of the water droplets by dust particles" (Mohal, 1988).

Surfactants have been studied to examine how effective they are in improving the wetting characteristics of various coals. Surfactants have been used to reduce the contact angles and improve how well the suppressants engulf fine coal particles (Mohal and Chander, 1986; Chander et al., 1987; Kilau, 1993; Kim and Tien, 1993, 1994). Studies have also shown that if wetting reagents are used, the effectiveness of coal dust scrubbers is significantly improved (Kim and Tien, 1993). However, it should be mentioned that the purpose of a scrubber is to treat contaminated air and remove the dust. This is a different type of operation from suppression, which is trying to prevent material from ever becoming airborne in the first place. A dust suppression study in 1993 showed that wetting reagents that improved fine particle engulfment of coal were only marginally effective in dust suppression at a coal mine (Kilau, 1993).

There are fundamental questions which have not been thoroughly explored in previous studies, which focused on improving the wetting characteristics of coal. A correlation between effective fine particle engulfment and significant reductions in airborne coal dust has not been established. Furthermore, studies on iron ore have demonstrated that improvements to the wetting capabilities of a given dust suppressant did not translate to reductions in airborne dust (Copeland and Kawatra, 2005). Such conclusions were the result of controlled experimentation utilizing test methods that examine how well a dust suppressant reduced airborne dust. This type of approach proved vital in the selection of a suitable dust suppressant.

18.1 DUST SUPPRESSANT CHARACTERIZATION METHODS

There are multiple methods which can be utilized to measure the effectiveness of dust suppressants. Contact angle analysis is one of the most common methods used to examine how effectively a dust suppressant wets the surface of the material. A large contact angle would imply that the liquid only poorly wets the surface. The second most common method is the Walker Sink Test, which uses a submerged microbalance to examine the rate at which dust particles can become engulfed in a liquid. Each of these tests measures the ability of the dust suppressant to wet the material. Based on observations from Kobrik in 1970, it was believed that the poor performance of water sprays was due to ineffective wetting of the coal dust. Subsequent studies were therefore conducted to examine how well a given dust suppressant wetted the surface of coal dusts.

18.1.1 Contact Angles

The contact angle is a measure of the resultant interfacial tensions between the liquid droplet, the solid surface, and the air. The contact angle tends to be large if the surfaces are dissimilar, while the contact angle is small if liquid completely wets the surface. Chemical additives which improve wetting characteristics tend to alter the balance of these forces to encourage wetting of the surface. There are two primary methods for measuring contact angles. The first method is the sessile drop method in which a liquid droplet is placed on the solid surface. The second technique is known as the captive bubble method in which a vapor bubble is attached to a submerged solid surface. In each test method, the lower the contact angle is, the more hydrophilic the solid surface behaves. In general, if the measured contact angle is less than 90°, the solid behaves as a hydrophilic surface. The two primary contact angle measurement methods are demonstrated in Figure 18.1.

18.1.2 Particle Engulfment

The Walker Sink Test has been commonly used as another method for evaluating how effectively a dust suppressant will engulf fine dust particles (Mohal and

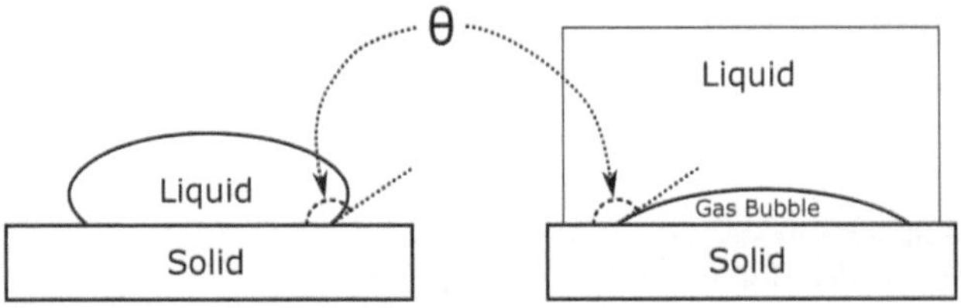

FIGURE 18.1 Depiction of sessile drop method with a liquid drop on a solid in air (left figure) and captive bubble method, where the bubble is caught on a solid surface under the liquid phase (right figure). Contact angle (θ) can be related to the interfacial tensions by Young's equation (Equation 18.1). The liquid demonstrated in this figure is showing poor wetting characteristics, as its contact angle is rather large.

Chander, 1986; Kilau, 1993; Kim and Tien, 1994; Copeland and Kawatra, 2005, 2007; Copeland et al., 2009). The term engulf refers to the ability of the liquid to completely coat the particle surface, replacing the particle–air surface entirely with the particle–liquid surface.

Fine particle engulfment becomes difficult as the material shows hydrophobic behavior and as the size of the particle becomes smaller. Because of these reasons, it was believed that dust suppressant effectiveness could be correlated with rapid particle engulfment. The procedure for measuring sink rates involves using an ultra-sensitive microbalance suspended into a petri dish containing the dust suppressant. The material being tested is then loaded onto the surface of the suppressant, and the accumulation of mass on the balance is measured with respect to time.

The key factors influencing the settling rate are first understood by Young's equation (shown in Equation 18.1). To achieve maximum wetting, the contact angle in Equation 18.1 should be as small as possible.

$$\gamma_{SG} = \gamma_{SL} + \gamma_{LG}\cos(\theta) \tag{18.1}$$

where

γ_{SG} = interfacial tension between solid and gas (N/m)
γ_{SL} = interfacial tension between solid and liquid (N/m)
γ_{LG} = interfacial tension between liquid and gas (N/m)
θ = contact angle within the liquid between the solid and gas phases (radians).

If the liquid–vapor interfacial tension is decreased, and/or the liquid–solid interfacial tension is increased wetting can occur more rapidly. Furthermore, the liquid–vapor interfacial tension is a primary barrier to prevent a particle from penetrating the liquid surface. When a dust suppressant is applied as a spray, the liquid droplet must capture the airborne dust, engulf the dust particle (due to favorable interfacial tension balance), and agglomerate it with neighboring particles.

For the Walker Sink Test, a rapid settling rate indicates that the suppressant can effectively engulf the particles, and a slow settling rate indicates the suppressant cannot effectively engulf the particles. Since coal dust shows hydrophobic behavior, the use of surfactants greatly improves the sink rate, and it is expected that surfactants will likewise improve the ability to suppress coal dust. The Walker Sink Test is demonstrated in Figure 18.2.

The Walker Sink Test directly evaluates the ability of a dust suppressant to engulf a particle of dust. The process of contact between a liquid droplet (as a spray) and capture of an airborne dust particle has been examined (Kim, 1995). The contact process is heavily dictated by appropriate mixing of the airborne dust and the dust suppressant spray, the particle size of the airborne dust and the liquid spray droplets, the impact velocity and direction, and the ability of the dust suppressant to engulf the particle. In the Walker Sink Test, the only force acting upon the dust particle is driven by the resultant interfacial tension and gravity.

Mohal (1988) examined both the initial settling rate as well as the final settling time to characterize how effective a given dust suppressant was in wetting coal. A dust suppressant that shows a high initial settling rate and a low final settling time has

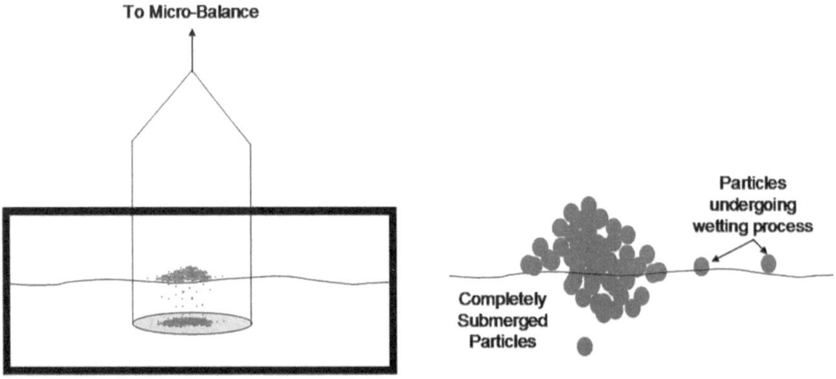

FIGURE 18.2 The Walker Sink Test (after Copeland and Kawatra, 2005, 2007; Copeland et al. 2009). The dust particles are placed on the surface of the suppressant solution. A faster settling rate is considered to be associated with a higher degree of dust suppression.

demonstrated superior ability to wet and engulf fine coal dust particles. In contrast, as the initial settling rate decreases and the settling time increases, it is concluded that the dust suppressant is not wetting and engulfing the fine coal dust particles well. Three relative settling curves are presented in Figure 18.3. Curve A demonstrates a material with a rapid initial settling rate and a low settling time. Curves B and C show materials with lower initial settling rates and longer settling times.

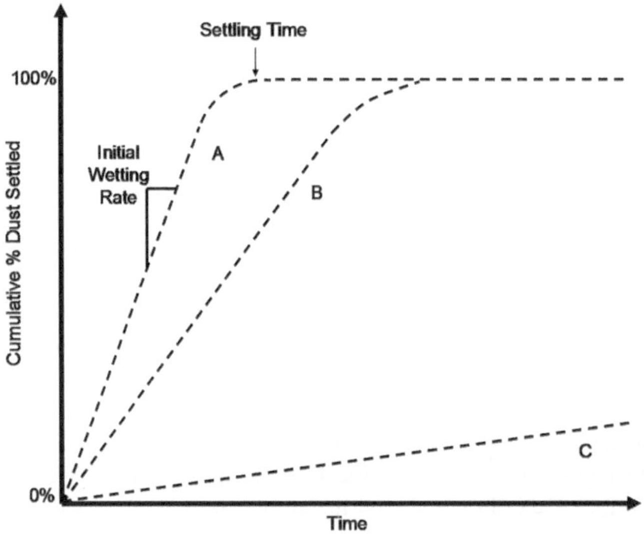

FIGURE 18.3 Hypothetical cumulative percent settling curves using the Walker Sink Test method (after Mohal, 1988). In this figure, liquid A achieves the highest settling rates, which should lead to the highest degree of dust suppression, whereas liquid B performs less well and liquid C performs the worst.

18.1.3 Soak Tests

It is also useful to consider the rate at which the dust suppressant can soak into a bed of dust particles. This is an area that has received particular attention in the literature, and it is expected that favorable wetting phenomena would be demonstrated by a dust suppressant which soaks into the dust bed rapidly. However, many factors affect the rate at which a suppressant can soak into the dust bed, including particle characteristics (e.g. size, shape, interfacial tensions), bed geometry (e.g. pore space, pore diameter), and the characteristics of the suppressant itself (e.g. liquid surface tension and viscosity).

The five stages of a liquid penetrating into the particle bed are shown in Figure 18.4. The first three stages are controlled largely by how the liquid is added and the liquid's own physical properties, but stages 4 and 5 are dominated by the interaction between the particle and the dust. In the worst case, the poorest choices for dust suppressants may never proceed beyond stage 3, and in the best cases stage 3 may be effectively skipped entirely. Stages 1–3 suggest that an ideal dust suppressant should not have too high surface tension. Stages 3 and 4 suggest that an ideal dust suppressant needs to effectively wet the particle bed. Stage 5 suggests that an ideal dust suppressant must be efficiently pulled into the particle bed by some combination of gravity and capillary interactions, which also suggest that the viscosity should not be too high.

There are multiple perspectives regarding the time required for the liquid droplet to penetrate into the bed of particles. First of all, there is debate as to what factors are driving this process. It has been suggested that surface tension and viscosity of the droplet (Popovich et al., 1999; Rocha, et al., 2005), effective particle bed porosity (Popovich et al., 1999; Hapgood et al., 2002), and droplet particle interaction (Larson, 2002) are critical factors that influence the liquid droplet penetration rate. There is debate regarding the relative importance of these factors. Studies on

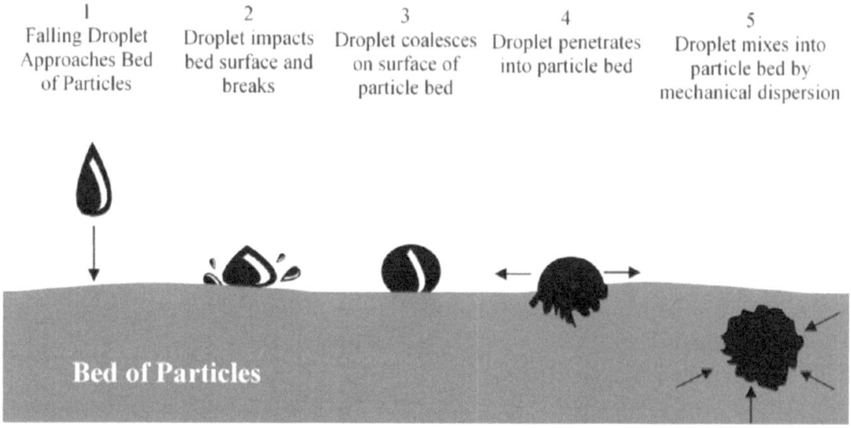

FIGURE 18.4 Five stages of liquid droplet penetration into a bed of particles (after Hapgood et al., 2002). If a poor depressant is used, it may not proceed beyond stage 3.

the penetration of coal-tar into a petroleum coke showed that surface tension and viscosity properties of the coal-tar (liquid phase) were influencing the penetration time (Rocha et al., 2005). However, another study working with carbon black found that there was no correlation between surface tension and penetration times and that the viscosity did not correlate to droplet penetration time as theory would predict (Popovich et al., 1999). The effects of liquid droplet surface tension and viscosity are given in Equations 18.2 and 18.3. These relationships describe the various aspects of liquid droplet penetration into particle beds.

$$Ca = \frac{\eta u}{\gamma_{lv}} \tag{18.2}$$

where

Ca = capillary number (dimensionless)
η = liquid droplet dynamic viscosity (Pa*s)
u = characteristic velocity of liquid droplet (m/s)
γ_{lv} = liquid droplet surface tension (N/m).

The capillary number (Ca) is the ratio of the viscous forces resisting the flow of the droplet into the bed and the surface tension forces driving the penetration. This relationship is given in Equation 18.2. A high capillary number means the penetration time will be high or that the penetration rate will be slow. From this relationship, it is clear that a high liquid viscosity or a low liquid surface tension would increase the penetration time. However, Popovich et al., working with carbon black found that higher viscosity liquids had lower penetration times. They concluded that "... the infiltration process (of a liquid droplet into a bed of carbon black) was governed by more than a balance of capillary (liquid surface tension) and viscous forces" (Popovich et al., 1999).

$$t_0 = \frac{\eta \cdot V_{\text{drop}}}{\gamma_{lv} \cdot A_{\text{drop}} \cdot \varepsilon \cdot S} \tag{18.3}$$

where

t_0 = characteristic time required for liquid penetration (s)
η = liquid droplet dynamic viscosity (Pa*s)
V_{drop} = volume of liquid droplet (m^3)
A_{drop} = area of contact between liquid droplet and particle bed surface (m^2)
γ_{lv} = liquid–vapor interfacial tension (N/m)
ε = particle bed porosity (dimensionless)
S = degree of liquid droplet saturation (volume of fluid divided by total volume of pores).

The characteristic time required for liquid penetration is determined in Equation 18.3 in terms of many of the same variables that go into the capillary number. This number allows for the comparison of the effects of viscosity across multiple liquids, where it was found that the effects of viscosity both decreased the penetration time in some situations and that its effects typically exceeded that which was afforded to it by the capillary number relation (Popovich et al., 1999).

It has also been suggested that a viscous liquid can penetrate more easily because it would tend to widen pores between particles more effectively. This would increase the effective porosity of the particle bed, thus increasing the liquid droplet penetration rate (Popovich et al., 1999). Equation 18.4 accounts for bed porosity more rigorously than Equation 18.3. This model can be used to calculate the droplet penetration time, provided the area of contact (A_{drop}) is constant (Hapgood et al., 2002). Similar models have also been presented in the literature (Denesuk et al., 1993; Middleman, 1995).

$$\tau_{CDA} = 1.35 \left(\frac{V_0^{2/3}}{\varepsilon^2 R_{\text{pore}}} \right) \left(\frac{\eta}{\gamma_{lv} \cos\theta_d} \right) \tag{18.4}$$

where

τ_{CDA} = time required for liquid penetration (s)
V_0 = liquid droplet volume (m^3)
R_{pore} = average pore diameter (m)
ε = particle bed porosity (dimensionless)
γ_{lv} = liquid droplet surface tension (N/m)
η = liquid droplet dynamic viscosity (Pa*s)
θ = interface contact angle (radians).

Further modifications of Equation 18.4 have been proposed to more accurately account for porosity effects in the particle bed. Here, the average pore diameter is treated as an effective pore diameter that is a function of the porosity of the bed, a shape factor, and the surface mean particle diameter (Link and Schlunder, 1996; Hapgood et al., 2002).

$$R_{\text{pore}} = \frac{\phi d_{32}}{3} \left(\frac{\varepsilon}{1 - \varepsilon} \right) \tag{18.5}$$

where

R_{pore} = average pore diameter (m)
Φ = shape factor (dimensionless)
d_{32} = surface mean particle diameter (m)
ε = particle bed porosity (dimensionless).

Interactions between the liquid droplet and the particle surfaces have also been suggested as dominant in influencing liquid droplet penetration rates (Denesuk et al., 1993; Larson, 2002). One study suggested that the liquid droplet penetration rate is primarily due to the adhesion tension (solid–liquid interface energy) and not the liquid–vapor (surface tension of suppressant) surface energy (Larson, 2002). The draw of a liquid into a bed of particles can be modeled by calculating the capillary pressure (or penetration pressure). This pressure drop provides the necessary driving force to draw the liquid into the pores of a bed of particles. This phenomena is described by Equation 18.6 (Denesuk et al., 1993; Larson, 2002), which makes clear that favorable mineral–liquid interactions (γ_{sl}) will provide the necessary pressure drop to drive the wetting process. It has been suggested that as long as the solid–liquid interface is preferred over the solid–vapor interface (i.e. $\gamma_{sl} < \gamma_{sv}$) the liquid droplet will be drawn into the particle bed (Denesuk et al., 1993).

$$\Delta P = \frac{2 \cdot (\gamma_{sv} - \gamma_{sl})}{r} \tag{18.6}$$

where

ΔP = capillary pressure drop (Pa)
γ_{sv} = solid–vapor interfacial tension (N/m)
γ_{sl} = solid–liquid interfacial tension (N/m)
r = capillary diameter (m).

18.2 DUST TOWERS

Dust towers provide a direct measurement of a dust suppressant's ability to reduce airborne dust. Material can be dropped down a dust tower to subject it to a consistent set of mechanical forces. A countercurrent stream of air flowing up the dust tower will catch any generated fine particles and carry them to a side port at the top, which can be used to catch the dust. The dust can then be measured as practical, such as with a PM analyzer. The idea is to present a consistent test bed that can be used to compare the effectiveness of different dust suppressants.

The purpose of a dust tower is to subject the material (either treated or untreated) to the types of impacts, transport, falls, and air currents that generate airborne dust. The dust-laden air can be filtered, calculated (based on mass balances), or directly measured to determine loading and concentration. These experiments are in contrast with wetting enhancement tests which only measure how the suppressant behaves when brought into contact with a material.

The process of removing a sample of airborne dust for direct measurements must account for the sampling efficiency. When airborne dust is sampled, it can either be concentrated or diluted depending on the airflow rates, sampling device face velocities, and the characteristics of the dust particle such as density, size, and shape. Improper accounting of these factors can lead to improper measurement results of the airborne dust concentrations.

18.2.1 Dust Tower Design

The purpose of the dust tower is to expose the sample material to the types of mechanical and impact forces which result in the generation of airborne dust. There are multiple dust towers that have been developed for this purpose. A complete summary of these different dust towers and a detailed discussion on design considerations have been documented in literature (Copeland and Kawatra, 2011).

Two primary variables must be controlled in order to generate a systematic and repeatable test for a given material. The amount of airborne dust that can be generated irrespective of the properties of the material are mechanical and airflow related.

The mechanical aspects involve factors such as drop height, number of impacts, and the properties of the surface the sample impacts (dust tower material of construction). Airflow factors include face velocities and direction of flow. An effective dust tower provides a means to control these factors in a fixed manner so that differences in dust suppressant capabilities can be detected. Mechanical factors are managed through the design of the dust tower itself, which can be made to be fixed. Airflow factors are controlled by using fixed equipment such as a vacuum source to create air upflow that can be verified before a test is run. A vane anemometer or a simple pressure drop measurement can be used to ensure that the face velocities of the air stream are consistent from test to test.

The dust tower test involves taking a fixed amount of treated or untreated material, and dropping it from a fixed height through a fixed countercurrent airflow stream. This process is demonstrated in Figure 18.5. As the material falls, it contacts several mechanical surfaces which abrade the material (external collisions) in addition to internal collisions within the material itself. The result of both external and internal collisions results in the generation of dust which is entrained in the air. The countercurrent airflow stream then suspends the dust creating airborne dust, which then transports the airborne dust away from the sample. This airborne dust is either

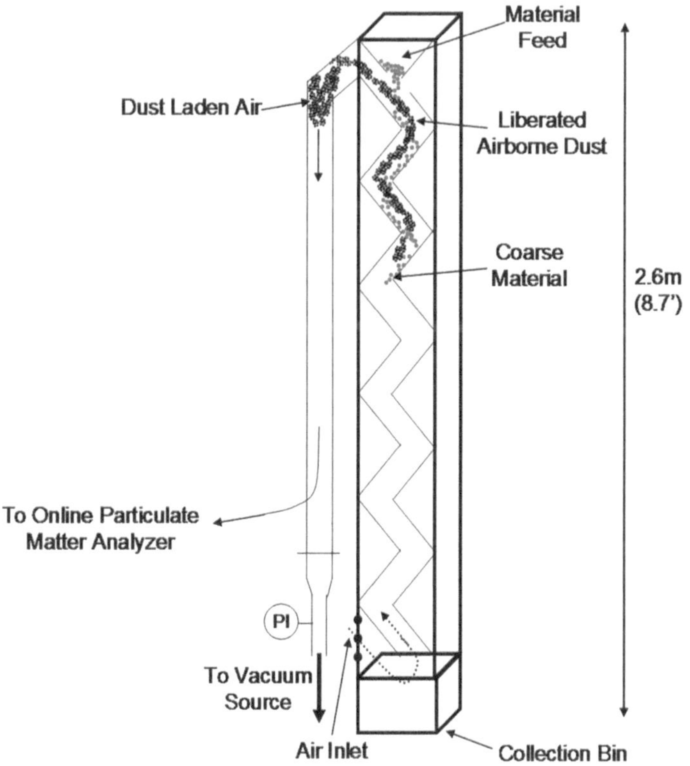

FIGURE 18.5 Dust tower apparatus (after Copeland and Kawatra, 2011). For clarity, the air inlet is the dashed line which is pointed to at the bottom of the tower.

filtered through an air filter or extracted by a sampling device to determine concentrations. These direct measurements are then used to determine the effectiveness of a given dust suppressant on the material being tested.

18.2.2 Airborne Dust Sampling

The process of collecting a sample of airborne dust for measurement and detection follows a three-step process; aspiration, transmission, and transportation (Baron and Willeke, 2001). The three-stage sampling process is demonstrated in Figure 18.6. Each stage has an associated efficiency which is a function of the airflow velocities and the characteristics of the airborne dust particles such as density, size, and shape. Airflow velocities for both the dust-laden air and the velocities inside the sample tube impact the tendency to either concentrate or dilute the concentration of dust. The density, size, and shape of the dust particle will influence how the particle trajectory is changed based upon the airflow contours around the sample inlet plane. The inlet efficiency is calculated as the product of the aspiration and transmission efficiencies.

Once the particle is successfully aspirated and transmitted to the sample tube, it then must be transported. During transport, particles can be lost to the sample walls due to impacts, gravitational settling, and electrostatic attraction. Each of these phenomena must be accounted for in designing a sampling system and for correcting measured values to account for the overall sampling efficiency. A description of the three-stage sampling process is provided below. A more detailed look at sampling efficiency calculations is provided in previous works (Copeland and Kawatra, 2011; Baron and Willeke, 2001).

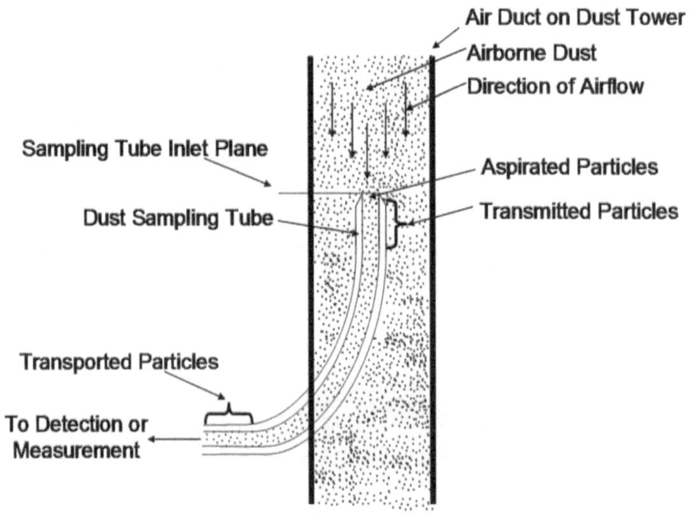

FIGURE 18.6 Air sampling in a dust tower. Airborne dust must be aspirated at the inlet plane of the sample tube, transmitted to the sample tube, then transported through the sample tube to be detected and measured (after Copeland and Kawatra, 2011).

The aspiration efficiency is the result of a balance of both the velocity of the approaching dust particle in the dust-laden air and the air velocity inside the dust sampling tube. An aspirated dust particle is a particle that has crossed the sample tube inlet plane. The most efficient sampling occurs when the velocity of the approaching dust particle is the same as the air velocity inside the dust sampling tube. If the approach velocity of the dust particle is less than the air velocity inside the dust sampling tube, the dust particles will tend to get diluted. In contrast if the approach velocity of the dust particle is greater than the air velocity inside the dust sampling tube, the dust particles will tend to get concentrated. This concentration and/or dilution effect is primarily influenced by the diameter of the sampling nozzle and the face velocity at the inlet plane (see Figure 18.6). As such, careful consideration and calculation of the sample tube diameter must be taken with respect to the expected face velocities of the airborne dust to ensure that most accurate sampling will occur.

Once the particle has been aspirated into the tube, the air pressure inside of the tube will tend to push it back out. The transmission efficiency is the probability that the particle overcomes this hurdle. The transmission efficiency is again a strong function of both the approach velocity of the dust particle and the face velocity inside the sample tube. Like the aspiration efficiency, transmission is most efficient when the approach velocity of the dust particle is equal to the face velocity of the air inside the sample tube. As the balance between the approach velocity of the particle and the face velocity of air deviates, the transmission efficiency deviates from ideal.

When a sampled dust particle has been successfully aspirated and transmitted, it must then be transported through the sample tube to be detected. The three primary causes for poor transportation efficiencies are particle impacts on tube walls, gravitational deposition along horizontal runs, and electrostatic attraction between the dust particle and the sample tube wall. Losses due to impacts are best managed by avoiding turbulent flow regimes and instead ensuring a laminar airflow inside the sample tube. Transport losses due to impacts are also minimized if 90° bends are avoided in the sample tube. Gravitational losses are unavoidable if any section of the sample tube has a horizontal component. As such, the design of a sampling system should minimize horizontal runs. Electrostatic losses can be managed by using a conductive surface such as copper that is electrically grounded to avoid building up a charge. In the dust tower system, gravitational losses account for the greatest loss in transport efficiency (Copeland and Kawatra, 2011).

18.2.3 Application to PM Measurements

The airborne dust generated from coal contains particle of different size classes. Each of the sampling stages, including aspiration, transmission, and transportation efficiencies, is a strong function of the size of the particle. A western sub-bituminous coal with a density of 1.2 g/cm^3 resulted in the sampling efficiency curves shown in Figure 18.7 (after Copeland and Kawatra, 2011). The sampling efficiency is the product of inlet efficiency (product of the aspiration and transmission efficiencies) and transport efficiency.

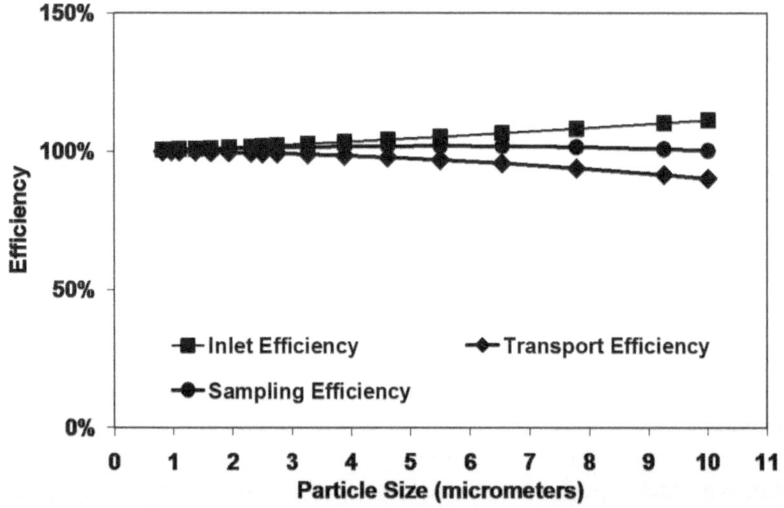

FIGURE 18.7 Coal particle inlet, transport, and sampling efficiencies in a dust tower sampling system. The approach velocity of the coal dust particle was 0.53 m/s and the average face velocity inside the sample tube was 2.0 m/s.

When making PM measurements such as PM_{10}, it is important to account for the different size classes present in the airborne dust. PM_{10} is defined as the concentration of airborne dust particles with an aerodynamic particle diameter of 10 μm and less. Since PM measurements report the concentration of airborne dust in units of mass per volume, a weighted average approach can be used to calculate the overall sampling efficiency when making PM_{10} measurements (after Copeland and Kawatra, 2011). This is demonstrated in Table 18.1, where the weighted average sampling

TABLE 18.1
Overall Sampling Efficiency for PM_{10} Measurement of a Western Sub-Bituminous Coal

$d_{particle}$(μm)	η_{sample}(%)	Individual Size Fractions (weight %)	Weighted Sampling Efficiency Per Size (%)
0.97	100	0.0	0.0
1.38	101	0.0	0.0
1.94	101	0.0	0.0
2.5	101	8.7	8.8
2.75	101	11.0	11.2
3.89	102	10.6	10.8
5.50	102	26.0	26.5
7.78	102	26.0	26.4
10.0	100	29.5	29.61
PM_{10} sampling efficiency			113

efficiency for PM_{10} using a western sub-bituminous coal was 113%. This result means that the measured concentration of PM_{10} with this coal sample is 13% higher than the actual. As such, measured results require to be divided by 1.13 to obtain the corrected values (Copeland and Kawatra, 2011; Baron and Willeke, 2001).

18.3 CASE STUDIES

A western sub-bituminous coal sample was obtained from a utility plant located in Northern Michigan of the United States. The proximate analysis is given in Table 18.2. This sample was a mixture of two western sub-bituminous coals containing 55% Wyoming coal and 45% Montana coal. The sample was received in six 19-L (5 gallon) buckets, each containing approximately 18 kg (40 lbs) of coal. First, the sample (all six buckets) was spread out onto large pans for a period of 4 days for drying. This was done in a laboratory at ambient temperatures of 25°C. After air drying, the sample was screened at 8 mesh (Tyler series) to remove fines. This was done to minimize the extent of oxidized material for experiments. After screening, the sample was passed through a jaw crusher and a gyratory crusher for light grinding and then rotary split using a 12-segment rotary sample splitter. During the rotary split stage, the 12 incremental splits were recombined to the original six buckets which had been triple lined with plastic bags. The samples were then immediately transported for storage in a freezer. Each of the six containers now contained approximately 9 kg (20 lbs) of crushed coal. As the experiments progressed, the following procedure was used to prepare a coal sample for a given test:

1. Obtained previously prepared 9 kg sample of crushed coal
2. Passed 9 kg sample through cone crusher and rotary split into 500 g samples

TABLE 18.2
Proximate Analysis of Coal Sample Used in This Study (Sample Was 55% Wyoming and 45% Montana Coal)

Location	Wyoming	Montana
Seam	Anderson	Anderson/Dietz
Rank	Sub-bituminous	Sub-bituminous
% of total	55%	45%
Moisture	26.5%	25.0%
Ash for contract	5.25%	4.50%
BTU/lb	8,800	9,350
Volatiles	31.15%	32.5%
Fixed carbon	37.0%	38.25%
Sulfur	0.20%	0.34%
#SO_2/MBTU	0.45	0.73
#Ash/MBTU	5.97	4.81
Hardgrove	62	56

3. 500 g samples placed into freezer bags and labeled
4. Freezer bags placed into freezer and stored
5. Frozen coal (500 g samples) removed from freezer no earlier than 4 h prior to testing.

One set of dust tower studies utilized coal which had been ground further. For these tests, the previously frozen 500 g samples were placed in an attritor. Three kilograms of 1/8" steel beads were added as grinding media, and the attritor was engaged for 1 min at a rate of 260 rpm. This procedure was done several times to accumulate the number of samples needed for the dust studies. Once sufficient sample had been produced in this manner, it was rotary split into 500 g samples, placed in freezer bags and stored in a freezer until 4 h prior to experiment trials. An acetylenic glycol surfactant, alkanolamide-oil fluid, tall oil pitch (tall oil is a byproduct from the paper and pulp industry, which was further processed and emulsified in water), and No. 2 fuel oil were examined in this study.

The experimental work was divided into two phases. The purpose of the first phase was to examine how the wetting characteristics could be enhanced using an effective surfactant. These experiments included contact angle analysis, the Walker Sink Test, and suppressant soakability methods. The second phase of the experimental studies involved systematic dust studies with crushed coal to examine which types of suppressants were effective in reducing PM_{10} as the material fell through a novel dust tower (Copeland and Kawatra, 2005).

18.3.1 Wetting Improvements to Coal Dust

18.3.1.1 Contact Angle Improvement

Acetylenic glycol was extremely effective for a western sub-bituminous coal. The natural contact angle of this material was around 60°, and the acetylenic glycol surfactant reduced the contact angle to less than 5°. These results are shown in Figure 18.8.

18.3.1.2 Coal Dust Particle Engulfment

The amount of time required for the particles to wet on the surface, and subsequently settle onto the balance pan was the wetting time. A low wetting time meant that the particles settled onto the pan rapidly (low wetting time means a rapid wetting rate). It has been argued that effectiveness of dust control is maximized when the suppressant engulfs the particles rapidly. These experiments were carried out to fully characterize this coal sample and to prove that acetylenic glycol improved particle engulfment. For these experiments, a Thermo Electron Cahn C35 Ultra-Microbalance which was accurate within 0.1 μg was used.

The use of the acetylenic glycol surfactant significantly reduced the wetting time for the western sub-bituminous coal. These results are given Figure 18.9. Furthermore, if the acetylenic glycol surfactant was not used, the coal particles did not sink in the Walker Sink Test.

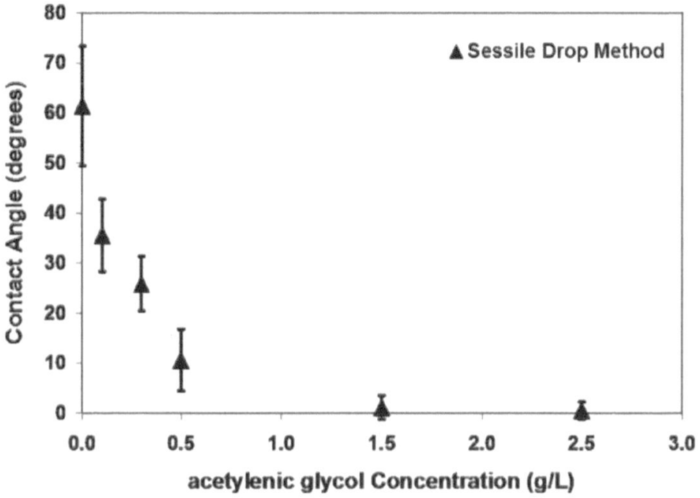

FIGURE 18.8 Effects of acetylenic glycol surfactant on sessile contact angle for western sub-bituminous coal. Each data point is a composite of nine individual samples with contact angle (θ) measured four times with error bars = $\pm 1\sigma$ (standard deviation of the mean). Acetylenic glycol reduced the contact angle to less than 5° for this mineral.

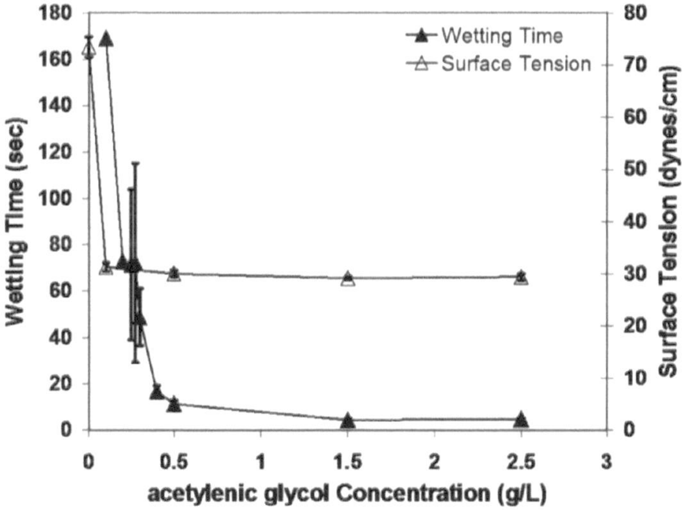

FIGURE 18.9 Effects of acetylenic glycol surfactant on the wetting time of crushed western sub-bituminous coal sample. Walker Sink Wetting Time for 20 mg of crushed coal (−400 mesh) (P_{90} = 51 μm, P_{80} = 40 μm, P_{10} = 4 μm). At concentrations greater than 0.25 g/L, each data point is a composite of four measurements with error bars = $\pm 1\sigma$ (standard deviation of the mean). Surface tensions were determined using the DuNouy Ring Method. Each surface tension data point is a composite of four measurements with error bars = $\pm 1\sigma$ (standard deviation of the mean). This reagent significantly decreased the wetting time for the crushed coal sample. If no surfactant was used, the crushed coal sample did not wet.

18.3.1.3 Coal Dust Particle Soakability

The last set of wetting experiments involved a soakability test that measured how quickly a liquid droplet soaked into a bed of dry particles. This type of method has been previously used to determine how well a suppressant engulfed fine coal particles for enhanced coal dust control (Kilau, 1993). The particle bed soak test is outlined in Figure 18.10. The soak time was designated using the classifications given in Table 18.3. For this experiment, a flat bed of particles was placed into a 25 mL Petri Dish. A hemispherical depression with a diameter of 4 cm was created, and 200 μL of suppressant was added to the depression. Following the addition of the suppressant, the behavior was observed, and the total time required for the droplet to soak into the bed of particles was recorded.

Initial studies were conducted to examine how well various suppressants behaved in this test. The objective was to discover a reagent which soaked into the bed of coal particles with relative ease. These results are given in Figure 18.11. For these tests, the exact quantity of coal (weight) was not measured, as well as the amount of suppressant (purpose was simply to examine overall behavior). The distilled water suppressant did not soak in at all. This result was expected due to the hydrophobic behavior of the material. The acetylenic glycol behavior was of particular interest. Although this suppressant was previously shown to reduce the contact angle to less than 5° and tremendously enhance the engulfment capabilities of fine coal particles, it did not penetrate the bed of coal. Instead, it engulfed nearby particles to some extent. This type of

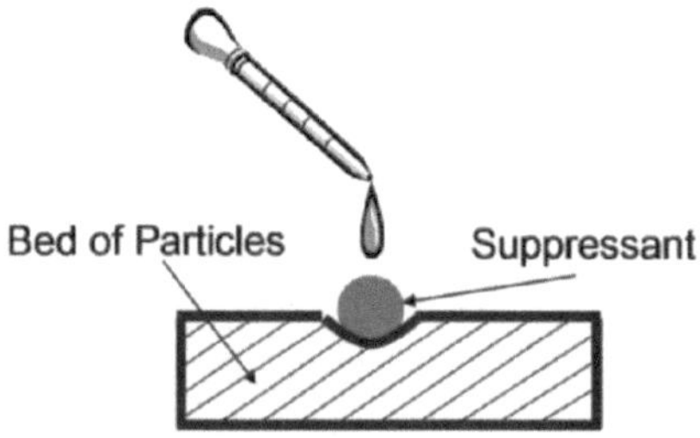

FIGURE 18.10 Suppressant soakability experiment.

TABLE 18.3
Wetting Designations Based on Observation Time

Observation	Wetting Designation
Soaked in <3 s	Instant wetting
Soaked in 3 s < x < 10 s	Fast wetting
Soaked in 10 s < x < 30 min	Intermediate wetting
Soaked in 30 min < x < 2 h	Slow wetting
Did not soak in	No wetting

Recall That Faster Wetting Is Preferable.

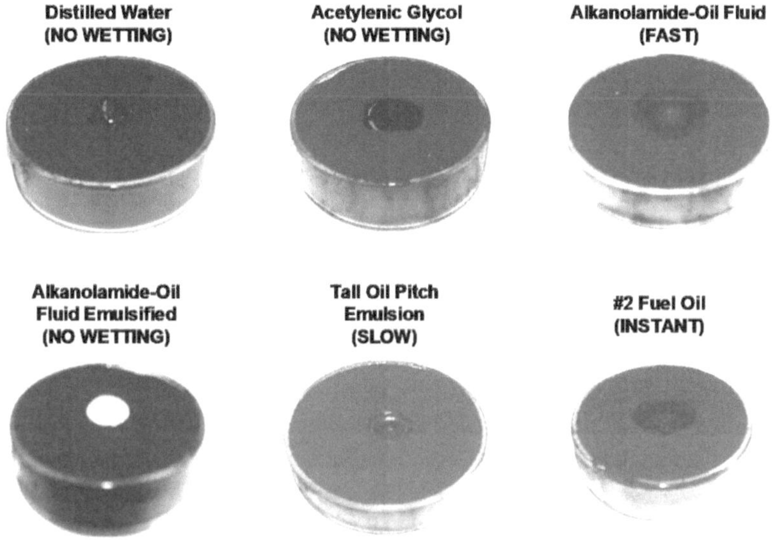

FIGURE 18.11 Soakability of various reagents on western sub-bituminous coal. Distilled water, acetylenic glycol, and emulsified alkanolamide-oil reagents did not soak in. Alkanolamide-oil fluid, tall oil, and #2 fuel oil were soaked in, where #2 fuel oil soaked in most rapidly. This would suggest that #2 fuel oil would be the best dust suppressant.

behavior has been reported in previous studies (Kilau, 1993). Other reagents, such as an alkanolamide-oil fluid and tall oil pitch, were evaluated. These reagents are oils, and it was expected they would have higher tendencies to soak into the coal since they were hydrophobic. The alkanolamide-oil soaked in with relative ease (far superior to acetylenic glycol) but the tall oil did not soak in as well (it took much longer). One possibility was that the tall oil was an emulsion with water. To further illustrate this, the alkanolamide-oil fluid was emulsified with water, and this emulsion did not soak in. This again likely occurred because of the presence of water.

A fifth reagent, No. 2 fuel oil, was also examined on its ability to soak into a bed of western sub-bituminous coal. The No. 2 fuel oil was the best reagent in that it soaked into the bed of coal particles within 3 s. Therefore, the No. 2 fuel oil was examined further in dust suppression studies to see if the soak test provided a better indication than contact angle or Walker Sink results for predicting how effective a reagent will be in controlling fugitive dust. It should be pointed out that using a No. 2 fuel oil for coal dust control in an industrial setting is not recommended due to flammability concerns. The purpose of these studies was to determine whether how well the suppressant soaks into a bed of particles should be considered in selecting a suitable dust suppressant.

After the initial soakability experiments were conducted, further studies were completed using the exact procedures outlined in this section. The objective here was to evaluate reproducibility and establish a standard test. For these tests, distilled water, acetylenic glycol, and No. 2 fuel oil were examined. The distilled water did not soak into the bed, the acetylenic glycol only engulfed neighboring particles, and the No. 2 fuel oil completely soaked into the bed within 1 s. These results are given in Figure 18.12. From these results, it was clear that the surface tension was not the dominant factor in bed soakability. The No. 2 fuel oil and acetylenic glycol reagents both have surface tensions around 30 dynes/cm. Yet, their behavior in the soakability experiment was completely different. It is possible that the fuel oil was interacting with the coal particles more favorably. In fact, sessile drop contact angles were verified to be zero. This could have caused a higher capillary pressure drop leading to more effective penetration.

The difference in soakability between the acetylenic glycol (2.5 g/L) and the No. 2 fuel oil was of particular interest. Properties for these two reagents are summarized in Table 18.4. From these results, it was clear that simply lowering the surface tension was not sufficient to allow wetting to occur in the soakability experiment. Second, the No. 2 fuel oil was the best reagent in these experiments yet it was significantly more viscous. Although theory would predict that No. 2 fuel oil would penetrate more slowly, it actually wetted the best. Furthermore, the creation of surface cracks in the particle bed was not observed. This suggested that viscosity was not impacting the wetting ability in this case, as it was in previous studies (i.e. lower viscosity wetting better) (Rocha et al., 2005) and higher viscosity wetting better (Popovich et al., 1999). Therefore, the most likely explanation is that it is the difference in interface energies which is important in driving the wetting process. In this case, it is more likely that since the solid–liquid interface is more favored (i.e. low γ_{sl}); the #2 fuel oil was able to penetrate into the coal bed more easily than the other reagents.

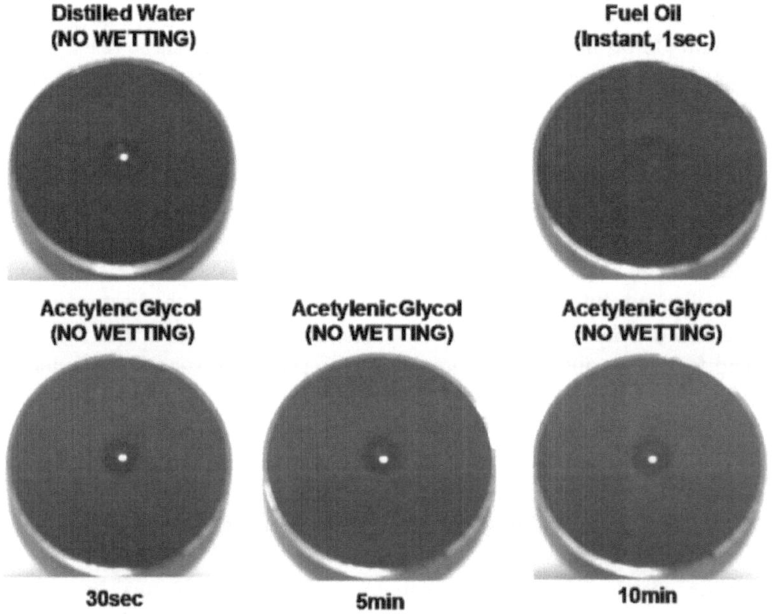

FIGURE 18.12 Soakability of various reagents on western sub-bituminous coal. Distilled water and acetylenic glycol did not soak in, and #2 fuel oil soaked within 1 s, which would suggest that it would be the best dust suppressant.

TABLE 18.4
Summary of Viscosity and Surface Tension for Acetylenic Glycol and #2 Fuel Oil Viscosity Measured Using a #50 Cannon-Fenske Viscometer in a 70°F (21°C) Constant Temperature Bath

Reagent	Relative Viscosity (η/η_0) @ 21°C (70°F)	Surface Tension (γ_{lv}) dynes/cm
Distilled water	1.00 ± 0.004	73.5 ± 1.9
Acetylenic glycol (2.5 g/L)	0.98 ± 0.012	29.4 ± 0.6
#2 fuel oil	4.16 ± 0.084	34.1 ± 0.1

Surface tension measured using the DuNouy Ring Method. Error bars are = ±1σ(standard deviation of the mean).

The viscosities and surface tensions shown in Table 18.4 do not necessarily correlate to how effective these dust suppressants are. However, the high surface tension of distilled water in air would tend to imply that it would poorly wet coal. The relative viscosities would suggest that either acetylenic glycol or #2 fuel oil is superior, based on contradictory models as to whether low or high viscosity is superior. The Walker Sink Test results would suggest that #2 fuel oil is the best suppressant of these three. Perhaps the most important takeaway is that these tests, while interesting and insightful, are insufficient to completely characterize the dust suppression behavior of these reagents.

18.3.2 Coal Dust Suppression Dust Tower Studies

To examine how effective various reagents were in reducing respirable dust from coal, a novel dust tower was used (see Figure 18.5). This tower simulated materials handling and made direct PM_{10} measurements. This was done to determine which type of wetting behavior was needed to effectively reduce PM_{10} from the coal.

Once the material was treated and cured, the material was dropped in the dust tower. As the material fell through the column, it struck several flat plates (simulates multiple impacts) and was passed through a countercurrent air stream that carried the airborne dust through the column, into a duct for analysis. An isokinetic nozzle located inside the air duct of the tower extracted a sample of the airborne dust and transported it to a TSI Dust Trak® that determines the concentrations of particulates finer than 10 μm (PM_{10}) in milligrams per cubic meter. The maximum (peak) PM_{10} value which was detected was reported as the peak concentration. The objective was to examine which reagents were most effective in reducing the peak PM_{10} from coal.

Studies conducted on the western sub-bituminous coal showed that the acetylenic glycol surfactant was only marginally effective in suppressing PM_{10} from this material. The data presented in Figure 18.13 is for a sample of the coal which was lightly ground using a short head crusher. It was expected that acetylenic glycol suppressant would be significantly more effective than distilled water because

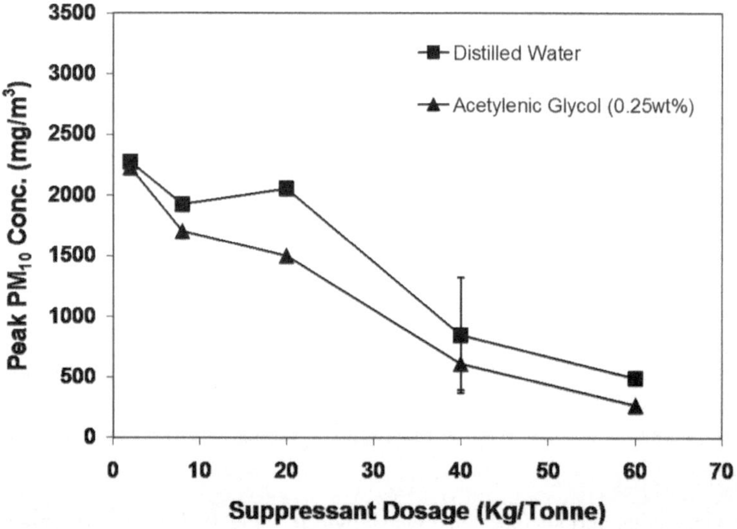

FIGURE 18.13 Effects of distilled water and acetylenic glycol on the peak PM_{10} in the dust tower dust analyzer from western sub-bituminous coal. Five hundred grams of cone crushed western sub-bituminous coal ($P_{90} = 2{,}900\,\mu m$, $P_{80} = 2{,}387\,\mu m$, $P_{10} = 148\,\mu m$). Five minutes cure time, 9 L/min flow rate on tower. Experiments were repeated three times and averaged (where error bars shown). Error bars are $\pm 1\sigma$(standard deviation of the mean). The acetylenic glycol was not significantly more effective than distilled water in reducing PM_{10} from this material.

of the Walker Sink Test and contact angle results. The acetylenic glycol reagent clearly enhanced wetting in these experiments, yet was only marginally more effective than distilled water in suppression of PM_{10} from the western sub-bituminous coal. Another important point was that as the amount of suppressant increased, the peak PM_{10} decreased. Previous studies have shown that although coal particles are hydrophobic and do not penetrate water droplets without wetting reagents, they still tend to stick to the surfaces of the liquid droplets (Stulov et al., 1978; Kilau, 1993). This behavior was observed when the soakability experiment was conducted with the distilled water suppressant. It was likely that when the suppressant dosage was increased, more PM_{10} was stuck to the surface of the suppressant droplets, leading to the observed decrease in PM_{10} during dust tests.

One possibility for the poor performance of the acetylenic glycol suppressant in the lightly ground coal tests was that the material did not produce enough PM_{10} to observe a benefit. Therefore, the coal was ground further using an attritor, producing a material that was much dustier. Again, the acetylenic glycol suppressant did not show a significant benefit in suppression of PM_{10} from the western sub-bituminous coal. These results are given in Figure 18.14.

Recall that the soakability test showed that although the acetylenic glycol reagent did improve coal particle imbibition, it did not allow the suppressant to effectively soak into the material. However, it was determined that #2 fuel oil soaked into the coal very effectively. Dust tower studies showed that the #2 fuel oil was extremely

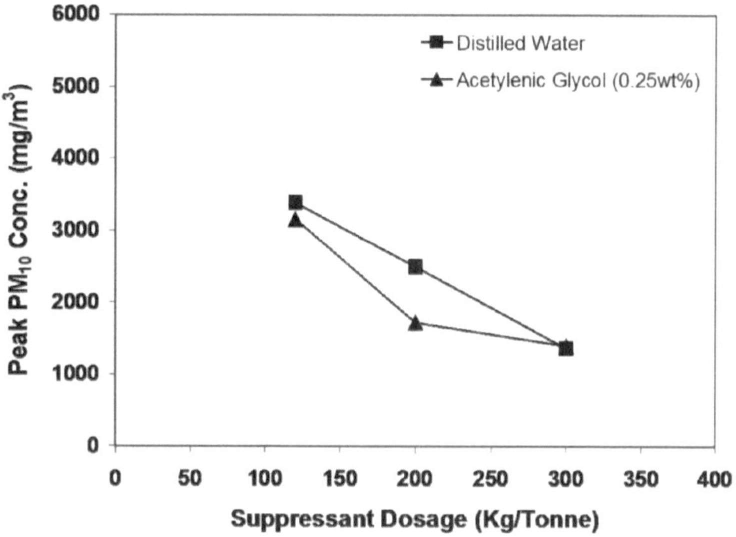

FIGURE 18.14 Effects of distilled water and acetylenic glycol on the peak PM_{10} from crushed western sub-bituminous coal. Hundred grams of cone crushed and pulverized coal ($P_{90} = 170\,\mu m$, $P_{80} = 136\,\mu m$, $P_{10} = 19\,\mu m$). Five minutes cure time, 9 L/min flow rate on tower. The acetylenic glycol was not significantly more effective than distilled water in reducing PM_{10} from this material.

effective in reducing PM_{10} from the western sub-bituminous coal, as shown in Figure 18.15. Effective dust suppression was achieved at a treatment of 9 kg/tonne, which was tremendously lower than distilled water treatments (>100 kg/tonne) used for the same material (see Figure 18.16 for comparison). Although this practice is strongly discouraged due to potential fire hazards, it illustrates a strong point that the ability of the suppressant to soak into the material is a stronger predictor of suppressant effectiveness than traditional wetting experiments. There are paraffinic solvents such as ISOPAR® which have had the light volatiles removed, making them less flammable than fuel oil. Further studies should be conducted to test the feasibility of using these types of reagents for dust control.

18.3.3 Other Dust Suppression Strategies

So far, a considerable amount of focus has been placed on long-chain surfactants or long-chain hydrocarbons. However, dust suppression studies in iron ore have found that hygroscopic tendencies may also lead to very effective dust suppression. This is largely orthogonal to the thinking provided previously in this chapter, and further suggests that the surface chemistry approach is incomplete.

These results were acquired by testing dust suppressants with iron ore pellets from a Midwest United States iron ore concentrator. The primary constituents of iron ore are magnetite and hematite, which in contrast to coal are both highly ionic species. Thus, iron ore is naturally quite wettable by water. Soak tests and surface

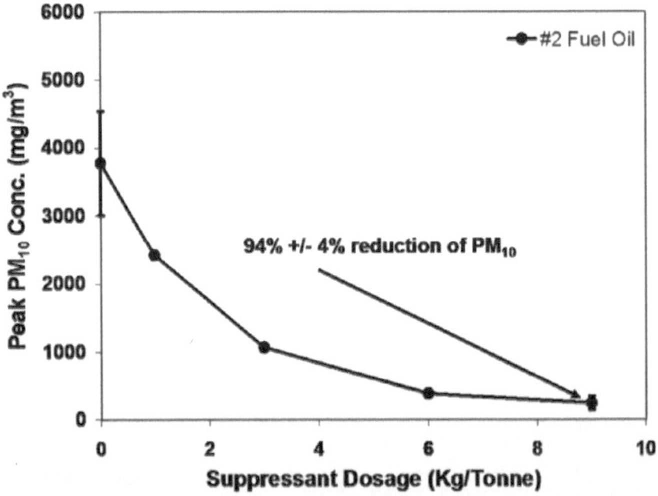

FIGURE 18.15 Effects of #2 fuel oil on the peak PM_{10} from crushed western sub-bituminous coal. Hundred grams of cone crushed and pulverized coal (P_{90} = 170 µm, P_{80} = 136 µm, P_{10} = 19 µm). Five minutes cure time, 9 L/min flow rate on tower. Experiments repeated three times and averaged (where error bars shown). Error bars are ±1σ(standard deviation of the mean). This reagent significantly reduced airborne dust from this coal because it adequately soaked into and wetted the material. It is not recommended to use this reagent in practice because of flammability concerns; however, alternative reagents which are much less flammable should be considered in future studies.

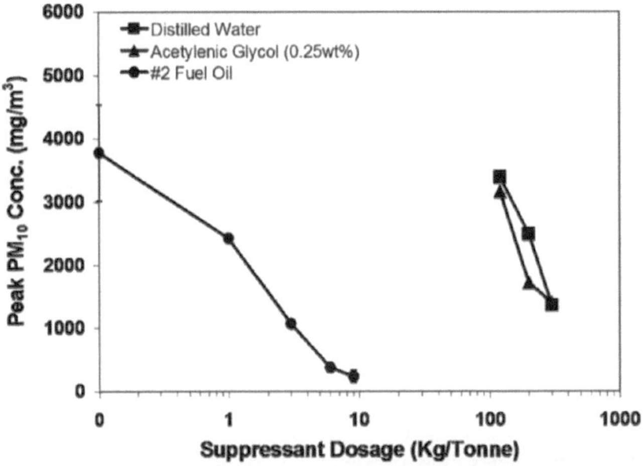

FIGURE 18.16 Summary of distilled water, acetylenic glycol (0.25 wt%), and #2 fuel oil on PM_{10} from a pulverized western sub-bituminous coal sample. Hundred grams of cone crushed and pulverized coal (P_{90} = 170 µm, P_{80} = 136 µm, P_{10} = 19 µm). Five minutes cure time, 9 L/min flow rate on tower. Experiments repeated three times and averaged (where error bars shown). Error bars are ±1σ(standard deviation of the mean). A treatment of around 300 kg/tonne of distilled water was equivalent to a treatment around 3 kg/tonne of #2 fuel oil.

tension tests with iron ore strongly favor acetylenic glycol as a candidate for dust suppression. However, dust tower tests with iron ore find that acetylenic glycol is, once again, no more effective than water, as shown in Figure 18.17.

Dust tower results using a calcium chloride solution are shown in Figure 18.18. In contrast to acetylenic glycol, this solution shows a very considerable decrease in

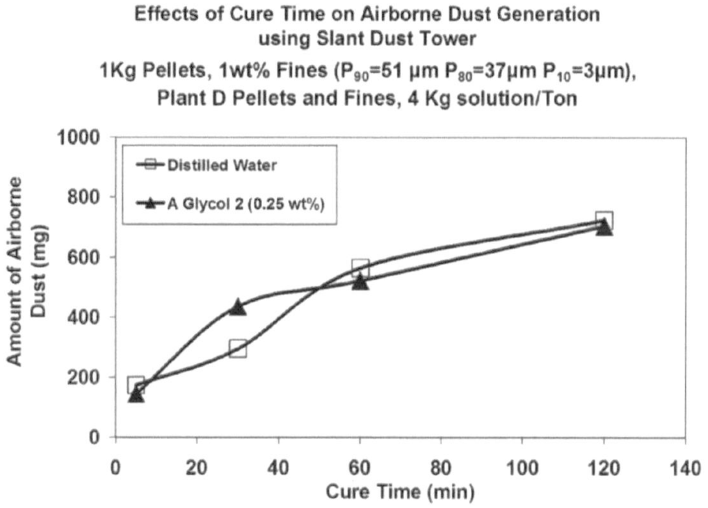

FIGURE 18.17 Acetylenic glycol as a dust suppressant on iron ore pellets over various cure times. There is little difference between the suppression achieved by acetylenic glycol and water.

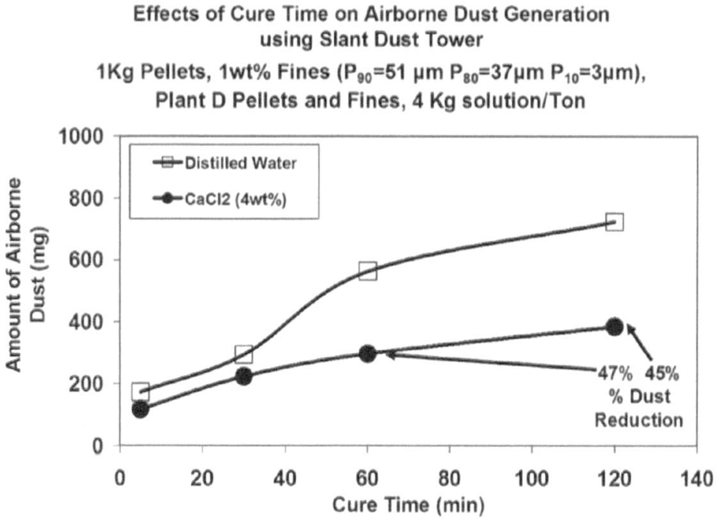

FIGURE 18.18 Calcium chloride as a dust suppressant on iron ore pellets over various cure times. Once curing is achieved, there is a significant reduction in dust generation.

dust production. While not shown here, calcium acetate ($Ca(C_2H_3O_2)_2$) had a similar effect in decreasing dust production in the dust tower.

It is difficult to predict whether or not calcium chloride would also be effective on coal, as coal's natural surface chemistry is considerably different. It may be that the increased ionic character of the calcium chloride solution is effective for binding the water phase to the iron ore, but would be ill-suited for binding to the typically more hydrophobic coal phases.

18.4 SUMMARY

Systematic dust studies have shown that an acetylenic glycol surfactant which significantly improved fine particle engulfment was only marginally more effective than water in controlling PM_{10} generations from a western sub-bituminous coal. This means that utilizing reagents which lower contact angles and improve particle engulfment may not necessarily lead to effective dust control. Further studies showed that the reason the acetylenic glycol surfactant was not very effective as a dust control reagent was that it did not soak into the bed of coal particles, sufficiently coating them.

The dust tower procedure outlined in this chapter has been used to great effect in increasing the effectiveness of dust suppression in an iron ore plant. The laboratory work on iron ore found that acetylenic glycol had favorable properties for suppression and that calcium chloride worked very effectively as a dust suppressant. Calcium chloride was also tested and found to be effective at the plant scale. It is unclear that these specific results are applicable to coal, but it was found that for coal #2 fuel oil is an effective dust suppressant. Although using #2 fuel oil is not recommended in practice due to flammability concerns, this result illustrates a point that effective dust control reagents should soak into the material rather than engulf the fine particles.

Our understanding is that the hygroscopic calcium chloride ($CaCl_2$) and calcium acetate ($Ca(C_2H_3O_2)_2$) can bond strongly with the iron ore which is an ionic compound, making the entire material more hygroscopic and less prone to dustiness. It might even be that the selective adsorption of calcium onto the iron ore's surface is the largest factor in these compounds' effectiveness as dust suppressants.

Further experiments should be conducted to determine the ideal depressant for coal. Paraffinic solvents such as ISOPAR® may be effective, which have had light volatiles removed and may be successful in achieving dust control without added concerns of flammability. Additionally, other variables of interest may be identified to lead to alternative approaches, such as hygroscopicity.

REFERENCES

Baron, P.A. and Willeke, K. (2001), *Aerosol Measurement Principles, Techniques, and Applications*, 2nd Edition, Springer, Berlin, pp. 143–195.

Chander, S., Mohal, B.R. and Aplan, F.F. (1987), "Wetting Behavior of Coal in the Presence of Some Nonionic Surfactants," *Colloids and Surfaces*, Vol. 26, pp. 205–216.

Copeland, C.R. and Kawatra, S.K. (2005), "Dust Suppression in Iron Ore Processing Plants," *Minerals and Metallurgical Processing*, Vol. 22, No. 4, pp. 177–191.

Copeland, C.R. and Kawatra, S.K. (2007), "Are Surfactants Effective for Iron Ore Dust Control?" *International Journal of Materials and Structural Integrity*, Vol. 1, No. 1–3, pp. 259–276.

Copeland, C.R., Eisele, T.C. and Kawatra, S.K. (2009) "Suppression of Airborne Articulates in Iron Ore Processing Facilities," *International Journal of Mineral Processing*, Vol. 93, pp. 232–238.

Copeland, C.R. and Kawatra, S.K. (2011), "Design of a Dust Tower for Suppression on Airborne Particulates for Iron Making," *Minerals Engineering*, Vol. 24, pp. 1459–1466.

Denesuk, M., Smith, G.L., Zelinski, B.J.J., Kreidl, N.J. and Uhlmann, D.R. (1993), "Capillary Penetration of Liquid Droplets into Porous Materials," *Journal of Colloid and Interface Science*, Vol. 158, No. 1, pp. 114–120.

Hapgood, K.P., Litster, J.D., Biggs, S.R. and Howes, T. (2002), "Drop Penetration into Porous Powder Beds," *Journal of Colloid and Interface Science*, Vol. 253, No. 2, pp. 353–366.

Kilau, H.W. (1993), "The Wettability of Coal and Its Relevance to the Control of Dust During Coal Mining," *Journal of Adhesion Science and Technology*, Vol. 7, No. 6, pp. 649–667.

Kim, J. and Tien, J.C. (1993), "Enhanced Dust Suppression Using Surfactants," *Proceedings of the 6th US Mine Ventilation Symposium*, June 21–23, Salt Lake City, Utah, pp. 523–528.

Kim, J. and Tien, J.C. (1994), "The Effect of Added Base on Coal Wetting Ability of Nonionic Surfactant Solutions Used for Dust Control," *Mining Engineer (London)*, Vol. 154, No. 399, pp. 151–155

Kim, J. (1995), "*Respirable Coal Dust Control using Surfactants–With Special Emphasis on a Liquid Spray System*," PhD Thesis, University of Missouri, Rolla.

Kobrick, T. (1970), "Water as a Control Method State of the Art Sprays and Wetting Agents," *U.S. Bureau of Mines. Information Circular*, Vol. 8458, pp. 123–131.

Larson, B. (2002), "Study of the Factors Affecting the Sensitivity of Liquid Penetrant Inspections: Review of Literature Published from 1970 to 1998," U.S. Department of Transportation Federal Aviation Administration Report Number DOT/FAA/AR-01/95. January. Document available through the National Technical Information Service (NTIS) Springfield, Virginia 22161, p. 5.

Link, K.C. and Schulender, E.-U. (1996), "New Method for the Characterization of the Wettability of Powders," *Chemical Engineering & Technology*, Vol. 19, No. 5, pp. 432–437.

Middleman, S. (1995), *Modeling Axisymmetric Flows: Dynamics of Films, Jets, and Drops.* Academic Press, San Diego, CA, p. 232

Mohal, B.R. and Chander, S. (1986), "A New Technique to Determine Wettability of Powders-Imbibition Time Measurements," *Colloids and Surfaces*, Vol. 21, pp. 193–203

Mohal, B.R. (1988), "Enhancement of the Wettability of Coal Powders Using Surfactants," Ph.D. Thesis. The Pennsylvania State University. University Park, Pennsylvania.

Popovich, L.L., Feke, D.L., and Manas-Zloczower, I. (1999), Influence of Physical and Interfacial Characteristics on the Wetting and Spreading of Fluids on Powders," *Powder Technology*, Vol. 104, No. 1, pp. 68–74.

Rocha, V.G., Blanco, C., Santamaria, R., Diestre, E.I., Menendez, R. and Granda, M. (2005), "Pitch/Coke Wetting Behaviour," *Fuel*, Vol. 84, No. 12–13, pp. 1550–1556.

Stulov, L.D., Murashkevich, F.I. and Fuchs, N. (1978), "Efficiency of Collision of Solid Aerosol Particles with Water Surfaces," *Journal of Aerosol Science*, Vol. 9, No. 1, pp. 1–6.

19 Rare Earths

The world of the coal industry is facing tough times, but there are two silver linings:

1. CO_2 from coal can potentially be used as a feedstock into other chemical processes, as discussed in Chapter 17.
2. Coal ash has higher than typical concentrations of the rare earth elements (REE).

19.1 RARE EARTH ELEMENTS

REEs are a set of 17 elements on the periodic table covering the 15 lanthanides plus scandium and yttrium. These elements play an essential role in many technologies, including but not limited to catalysts, permanent magnets, lamp phosphors, and rechargeable batteries. As such the REEs are becoming increasingly important as a strategic resource. With China currently exporting more than 90% of the global REE output and its increasingly tight export regulations, the rest of the world faces a risk to their ability to access REEs (Binnemans et al., 2013). While the United States has significant reserves of rare earth minerals, it is necessary to develop efficient processing technologies to extract these elements at low cost.

In truth, aside from promethium, the REEs are not actually exceptionally rare throughout the earth's crust. The primary issue with extracting rare earths is not their rarity, but how evenly distributed they are. Whereas many elements are concentrated into nearly pure mineral ores in which they are a major component, such as iron into hematite and magnetite, the rare earths are typically only found as occasional substituents into other mineral species. This means that the rare earths must be extracted from minerals where they are extremely sparse compared to most traditional mineral processing operations. For the sake of completeness, promethium is instead derived through radioactive decay processes, as it is in fact extremely rare.

19.2 RARE EARTHS IN COAL

There are coal deposits containing elevated concentrations of germanium, gallium, selenium, lithium, and REEs (including yttrium). These deposits are of considerable interest, as when these coals are burned, the rare earths will concentrate in the resulting ashes. This concentration makes it easier to recover these elements when compared to directly extracting them from the coal seams.

In addition to the traditional REEs described in Table 19.1, the other elements mentioned above have significant uses in several industries including defense (Seredin et al., 2013). Germanium in particular is used in the production of solar cells, as it

TABLE 19.1
The 17 REEs and Several of Their Uses (Eyring et al., 2002; Hammond, 2009; Krishnamurthy and Gupta, 2015; Tekumalla et al, 2015; Paul et al., 2016; Barbi et al., 2018; Gimaev et al., 2018)

Atomic Number	Symbol	Name	Applications/Uses	Abundance, ppm
21	Sc	Scandium	Aerospace components, mercury-vapor lamps, and as a tracer in oil refineries.	22
39	Y	Yttrium	Yttrium aluminum garnet (YAG) lasers, television red phosphor, yttrium barium copper oxide (YBCO) high-temperature superconductors, yttria-stabilized zirconia (YSZ), yttrium iron garnet (YIG) microwave filters, energy-efficient light bulbs, spark plugs, gas mantles, steel additive, and cancer treatments.	33
57	La	Lanthanum	Telescope lenses, hydrogen storage, alkali-resistant glass, flint, rechargeable batteries, camera lenses, and as a fluid catalytic cracking catalyst for oil.	39
58	Ce	Cerium	Oxidation of CO emissions from vehicles, polishing powder, yellow colors in glass and ceramics, catalyst for self-cleaning ovens, cracking catalyst for oil, ferrocerium flints for lighters. Also alloyed with aluminum, iron, and magnesium.	66.5
59	Pr	Praseodymium	Powerful rare earth magnets, carbon arc lighting, as a colorant in glass and enamels, and in didymium glass used in welding goggles.	9.2
60	Nd	Neodymium	Rare earth magnets, lasers, as a violet color additive in glass and ceramics, ceramic capacitors, and electric motors for electric automobiles.	41.5
61	Pm	Promethium	Luminous paint and nuclear batteries	10^{-15}
62	Sm	Samarium	Neutron absorber for control rods in nuclear power plants, rare earth magnets, lasers, and masers.	7.05
63	Eu	Europium	Europium oxide is used for red and blue phosphor in TV screens and fluorescent lamps. Europium is also used in lasers, mercury-vapor lamps, and as an Nuclear Magnetic Resonance (NMR) relaxation agent.	2

(*Continued*)

TABLE 19.1 (*Continued*)
The 17 REEs and Several of Their Uses (Eyring et al., 2002; Hammond, 2009; Krishnamurthy and Gupta, 2015; Tekumalla et al, 2015; Paul et al., 2016; Barbi et al., 2018; Gimaev et al., 2018)

Atomic Number	Symbol	Name	Applications/Uses	Abundance, ppm
64	Gd	Gadolinium	High refractive index glass, lasers, X-ray tubes, computer memory, neutron shielding for nuclear reactor control rods, magnetic resonance imaging (MRI) contrast agent, and in NMR spectroscopy. Alloyed with some metals to improve their workability and resistance.	6.2
65	Tb	Terbium	Rare earth magnets, biological probes, naval sonar systems and other sensors, and a stabilizer in fuel cells.	1.2
66	Dy	Dysprosium	High-temperature rare earth magnets, lasers, magnetostrictive alloys such as terfenol-D, hard disk drives.	5.2
67	Ho	Holmium	Wavelength calibration standards for optical spectrophotometers, rare earth magnets.	1.3
68	Er	Erbium	Infrared lasers and fiber optics. Alloyed into vanadium steel.	3.5
69	Tm	Thulium	Portable X-ray machines, lasers, and metal halide lamps.	0.52
70	Yb	Ytterbium	Portable X-ray machines, nuclear medicine, infrared lasers, stress gauges, earthquake monitoring equipment, decoy flares, and as an alloying component in stainless steel.	3.2
71	Lu	Lutetium	Positron emission tomography detectors, high refractive index glass, meteorite dating, catalysts, and LED light bulbs.	0.8

has excellent photoelectric conversion efficiency. Germanium is also used in energy-saving Light-Emitting Diodes (LEDs). Germanium has already been successfully extracted from coal in addition to other mineral reserves.

Franus et al. (2015) have carried out extensive research with coal ashes, and their results are summarized in Table 19.2. Lanthanum contents were reported by the US Geological Survey in several coal benches in US coal production regions, including the Northern Great Plains, Eastern Interior, Appalachian, and various regions in the Interior Basin (Rozelle et al., 2016).

TABLE 19.2
Summary of Rare Earth Concentrations Measured from 12 Coal Fly-Ashes

Element	Minimum Abundance, ppm	Median Abundance, ppm	Mean Abundance, ppm	Maximum Abundance, ppm	Natural Abundance, ppm
La	**15.5**	**60.9**	**58.5**	**81.7**	**39**
Ce	30.7	**124.5**	**117.3**	**172.5**	66.5
Pr	3.3	**14**	**13.5**	**20.5**	9.2
Nd	12.7	**54.3**	**52.4**	**81.3**	41.5
Sm	2.8	**11.1**	**10.7**	**17**	7.05
Eu	0.56	**2.5**	**2.4**	**3.8**	2
Gd	2.85	**9.0**	**9.0**	**14.7**	6.2
Tb	0.45	**1.6**	**1.5**	**2.4**	1.2
Dy	2.61	**8.4**	**8.1**	**12.2**	5.2
Ho	0.59	**1.7**	**1.6**	**2.6**	1.3
Er	1.79	**4.7**	**4.6**	**7.4**	3.5
Tm	0.27	**0.8**	**0.7**	**1.1**	0.52
Yb	1.8	**4.5**	**4.4**	**6.7**	3.2
Lu	0.3	0.70	0.67	**1.0**	0.8
Y	17.9	**48.8**	**48.4**	**73.2**	33
Sc	2	**24**	**24**	**45**	22

Source: From Franus et al. (2015).
Values higher than the natural abundance are bold. Most coal fly-ashes are more concentrated in rare earths than the natural abundance.

19.3 EXTRACTION OF RARE EARTHS FROM COAL

It is clear from Table 19.3 especially that the REEs are significantly concentrated in the coal ashes. A typical coal will contain on the order of 60 ppm of REEs, while a typical power plant fly-ash will contain 470 ppm of total REEs. King et al. (2018) leached REEs from various ashes and found for Powder River Basin ashes that the best leaching performance was achieved with HCl. These leaching results are very encouraging. The subsequent processing of the rare earths which are concentrated in coal ashes or from any other source can be accomplished by four primary methods:

- Flotation, whether in conventional flotation cells or via column flotation.
- Gravity separation
- Magnetic separation
- Electrostatic separation

All of these methods have been previously covered in this book. If the rare earths are to be extracted from fly-ash or bottom ash, it is much better to use dry separation methods.

TABLE 19.3
Concentration of REEs in Coal Bottom Ashes after Combustion

Element	Raw Coal Concentration, mg/kg	Coal Bottom Ash Concentration, mg/kg	Concentration Factor	Natural Abundance, ppm
La	9.1	259.8	28.6	39
Ce	20.9	468.8	22.4	66.5
Pr	4.8	59.0	12.3	9.2
Nd	8.5	236.0	27.8	41.5
Eu	0.3	7.6	27.3	2
Tb	0.5	10.3	19.1	1.2
Dy	2.1	61.5	29.4	5.2
Y	8.2	408.3	49.9	33

Source: From Mayfield and Lewis (2013).

The National Energy Technology Center funded a research program at the University of Kentucky, Lexington, USA and the West Virginia University, Morgan Town, USA. They found that froth flotation was the only technique that could be employed for the extraction of rare earths. Gravity-based separation techniques were ineffective for extracting rare earths due to the ultrafine grain sizes. If rare earths are to be extracted from the ashes of the utility plants, it is best to use dry separations as described in Chapter 6.

However, the findings of Jordens et al. (2014) were quite different. They studied flotation, gravity, magnetic, and electrostatic techniques, and found that gravity separation and magnetic separation performed the best. The group of Kentucky and West Virginia used raw material from Kentucky, whereas Jordens et al. (2014) used raw material from the Nechalacho deposit in the northwest territories in Canada.

19.3.1 Hydrometallurgical Extraction of Rare Earths

Battsengel et al. (2018) have studied the hydrometallurgical route for the extraction of rare earths from apatite. Such ore deposits are present in Mongolia. It seems likely that the hydrometallurgical process will form the core of any extraction of rare earths from coal, but there is not yet enough research in the area to be certain. It is also likely that the optimal process for each raw material depends strongly on the material itself.

Rozelle et al. (2016) have studied the use of ionic liquids to leach REEs from coal. An ionic liquid is a liquid that consists primarily of codissolved anions and cations. This highly ionic nature is similar to an extremely polar solvent, so for the same reason, water is better at dissolving salts than oils, and ionic liquids should be better at dissolving salts than water.

The roughest outline of using ionic liquids for extraction is that the rare earth oxides will have significantly more ionic character than coal, and thus will be preferentially leached into an ionic liquid as opposed to coal, which will be largely rejected from such a system. An ionic liquid and a deep eutectic solvent system were compared to an ammonium sulfate leaching solution, but were not found to have significantly improved performance. Additionally, the ammonium sulfate solution did achieve a high leaching efficiency, so the additional expense of ionic liquids may not be warranted for the extraction of rare earths from coal sources. This is compounded by the relatively unknown environmental and economic concerns of using ionic liquids on an industrial scale, as they have not been widely applied for these sorts of operations.

Typically, the extraction of rare earths from mineral sources proceeds via comminution followed by concentration by flotation, magnetic, or gravity separation methods. These steps are only intended to concentrate the rare earth bearing sections of the ore however. In each case, the actual metal is recovered by a hydrometallurgical process (Xie et al., 2014). Both alkaline and acidic leaching approaches can be used to extract rare earths, though in bastenites, alkaline leaching may lead to radioactives such as thorium reporting to both the concentrate and the tailings (Xie et al., 2014). There are some ores that can be leached effectively with only acids – those where the rare earths are primarily present as adsorbed ions rather than crystal substitutions or otherwise (Xie et al., 2014).

A typical rare earth extraction circuit from minerals goes something like: concentration (comminution and physical separation steps), roasting (possibly in the presence of strong acid or base), leaching (in acid or base), neutralization, and general precipitation (to produce mixed rare earths) or element-specific separations (to recovery pure rare earths) (Xie et al., 2014). Each of these steps may be composed of several stages, and especially, the individual element separation steps may represent dozens of stages of processing. It is not unusual for the complete rare earth extraction process to add up to hundreds of stages (Xie et al., 2014).

19.4 SUMMARY

There are traces of REEs and other strategic metals in coal. When the coal is burned, these elements are concentrated in the coal ashes. These ashes may serve as a valuable source of these very important metals, and as such the extraction of these rare earths from these ashes is of significant interest to the coal mining industry. However, the technology to extract these rare earths from coal ashes is still in development, and only germanium has been successfully extracted on a significant scale to date. Despite this, the REEs are invaluable to a modern lifestyle and to national security, so it is essentially certain that work on these extraction processes will continue.

REFERENCES

Barbi, S., Mugoni, C., Montorsi, M., Affatigato, M., Gatto, C. and Siligardi, C. (2018), "Structural and Optical Properties of Rare-Earths Doped Barium Bismuth Borate Glasses," *Journal of Non-Crystalline Solids*, Vol. 481, pp. 239–247.

Battsengel, A., Batnasan, A., Narankhuu, A., Haga, K., Watanabe, W. and Shibayama, A. (2018), "Recovery of Light and Heavy Rare Earth Elements from Apatite Ore Using Sulphuric Acid Leaching, Solvent Extraction and Precipitation," *Hydrometallurgy*, Vol. 179, pp. 100–109.

Binnemans, K., Jones, P.T., Blanpain, B., Van Gerven, T., Yang, Y., Walton, A. and Buchert, M. (2013), "Recycling of Rare Earths: A Critical Review," *Journal of Cleaner Production*, Vol. 51, pp. 1–22.

Eyring, L., Gschneidner, K.A. and Lander, G.H., eds. (2002), *Handbook on the Physics and Chemistries of Rare Earths*, Vol. 32, Elsevier, North Holland.

Franus, W., Wiatros-Motyka, M.M. and Wdowin, M. (2015), "Coal Fly Ash as a Resource for Rare Earth Elements," *Environmental Sciences and Pollution Research*, Vol. 22, No. 12, pp. 9464–9474.

Gimaev, R., Kopeliovich, D., Spichkin, Y. and Tishin, A. (2018), "Measurements of the Magnetic and Magnetothermal Properties of Heavy Rare Earths," *Journal of Magnetism and Magnetic Materials*, Vol. 459, pp. 215–220.

Hammond, C.R. (2009), "Section 4; The Elements," *CRC Handbook of Chemistry and Physics* (Lide, ed.), 89th Edition, CRC Press/Taylor and Francis, Boca Raton, FL.

Jordens, A., Sheridan, R.S., Rowson, N.A. and Waters, K.E. (2014), "Processing a Rare Earth Mineral Deposit Using Gravity and Magnetic Separation," *Minerals Engineering*, Vol. 62, pp. 9–18.

King, J.F., Taggar, R.K., Smith, R.C., Hower, J.C. and Hsu-Kim, H. (2018), "Aqueous Acid and Alkaline Extraction of Rare Earth Elements from Coal Combustion Ash," *International Journal of Coal Geology*, Vol. 195, pp. 75–83.

Krishnamurthy, N. and Gupta, C.K. (2015), *Extractive Metallurgy of Rare Earths*, CRC Press, Boca Raton, FL.

Mayfield, D.B. and Lewis, A.S. (2013), "Environmental Review of Coal Ash as a Resource for Rare Earth and Strategic Elements," *Proceedings of the 2013 World of Coal Ash (WOCA) Conference*, Lexington, KY, Vol. 2013, pp. 22–25.

Paul, M.C., Bysakh, S., Das, S., Dhar, A., Pal, M., Bhadra, S.K., Sahu, J.K., Kir'yanov, A.V. and d'Acapito, F. (2016), "Recent Developments in Rare-Earths Doped Nano-Engineered Glass Based Optical Fibers for High Power Fiber Lasers," *Transactions of the Indian Ceramic Society*, Vol. 75, No. 4, pp. 195–208.

Rozelle, P.L., Khadilkar, A.B., Pulati, N., Soundarrajan, N., Klima, M.S., Mosser, M.M., Miller, C.E. and Pisupati, S.V. (2016), "A Study on Removal of Rare Earth Elements from U.S. Coal Byproducts by Ion Exchange," *Metallurgical and Materials Transactions E*, Vol. 3, No. 1, pp. 6–17.

Seredin, V.V., Dai, S., Sun, Y. and Chekryzhov, I.Y. (2013), "Coal Deposits as Promising Sources of Rare Metals for Alternative Power and Energy-Efficient Technologies," *Applied Geochemistry*, Vol. 31, pp. 1–11.

Tekumalla, S., Seetharaman, S., Almajid, A. and Gupta, M. (2015), "Mechanical Properties of Magnesium-Rare Earth Alloy Systems: A Review," *Metals*, Vol. 5, No. 1, pp. 1–39.

Xie, F., Zhang, T.A., Dresinger, D. and Doyle, F. (2014), "A Critical Review on Solvent Extraction of Rare Earths from Aqueous Solutions," *Minerals Engineering*, Vol. 56, pp. 10–18.

Appendix 1

Physical Characteristics of Microscopic Pyrite (Hucko and Gala, 1990)

Pyrite Type	Crystal Size, µm	Unit Size, µm	Shape	Internal Structure	Time and Mode of Origin
Massive syngenetic	10 or larger	1–1,000	Irregular, or filling cell cavities	Contain cell wall materials as submicron organic layers or cell wall shaped vitrinite	Syngenetic: formed within the peat before compression
Massive dendritic	5–1,000	5–1,000	Radial dendritic form; several forms often coalesce	Pyrite unstructured, but coal inclusions common	Syngenetic to diagenetic: formed in gelification stage of coalification
Massive bedding plane	1–1,000	1–1,000	Tabular to lenticular	Rarer interlocking crystal faces, or relic voids between crystals that formed patches	Epigenetic or diagenetic: bedding plane fracture fillings, or diagenetically modified patches
Massive cell filling	0.1–1,000	0.1–1,000	Shape determined by the internal shape of the cell cavity	Crystal faces are rarely apparent	Epigenetic or syngenetic: after deposition of fusinite or semifusinite in the peat, through epigenetic filling of these cavities within the solid coal.
Massive breccia matrix	10–1,000	10–1,000 µm	Irregular, often not evident in microblock	Breccia of fragmented vitrinite, fusinite, or semifusinite	Epigenetic or syngenetic: most probably epigenetic

(*Continued*)

Pyrite Type	Crystal Size, µm	Unit Size, µm	Shape	Internal Structure	Time and Mode of Origin
Massive fracture filling	0.1–1,000	0.1–1,000	Shape determined by shape of the fracture	Rare interlocking crystal faces	Epigenetic: fills fractures formed within the solid coal. Rarely is syngenetic and fills shrinkage cracks formed in coalification
Patches	0.1–10	2–100	Lenticular to irregular	Contain subhedral to euhedral crystals; sometimes patterned or aligned	Syngenetic: formed within the peat or early in the coalification stage
Framboids	0.1–10	4–100	Spherical	Contain euhedral to anhedral crystals; sometimes arranged concentrically	Syngenetic: formed within the peat or early in the coalification stage
Isolated euhedral crystals	0.1–10	0.1–10	Octahedral or cubic	None	Syngenetic to epigenetic: probably formed in the peat stage, but can form during any stage of coalification

REFERENCE

Hucko, R.E. and Gala, H.B. (1990), "Surface Science of Coal Preparation," U.S. Department of Energy, Pittsburgh Energy Technology Center, DOE/PETC/TR-90/2 (DE90007161).

Appendix 2

Comparison of SO_2 Emission Reduction Techniques (Vernon and Jones, 1993)

Process	Applicability	Benefits	Disadvantages
Switch to natural gas	New or dual-fired plants (large or small)	Minimal SO_2 emissions, few other emissions, low capital costs, no wastes	Possible future price increases, limited reserves in many countries
Switch to low-sulfur oil	New or dual-fired plants (large or small)	Allows some emissions regulations to be met without major capital expense	Limited availability due to refinery configurations, low-sulfur oils significantly more expensive
Switch to nuclear power	New plants (large)	No SO_2, NO_x, or CO_2 emissions	High capital costs, major political difficulties
Switch to renewable sources	New plants (small except for hydropower)	No SO_2 emissions, no fuel costs	Mostly small scale except for hydropower, increasing concerns over impacts of hydropower
Switch to low-sulfur coal	All users	Allows some emissions regulations to be met without major capital expense	Insufficient to meet some regulations, price likely to increase in the future
Conventional coal cleaning	All users	Reduces ash content, improves coal heating value	Limited sulfur removal ability, unable to remove organic sulfur
Advanced physical coal cleaning	All users	Potential to remove 80%–90% of inorganic sulfur	Requires fine grinding of coal, cannot remove organic sulfur
Chemical coal cleaning	All users	Potential to remove significant fraction of all sulfur	Still under development, potentially high costs
Biological coal cleaning	All users	Potential to remove significant fraction of all sulfur	Still under development, potentially high costs

(*Continued*)

Process	Applicability	Benefits	Disadvantages
Atmospheric fluidized bed combustion	New plants (large or small)	High SO_2 removal, emission control integrated with process, considerable practical experience	Large amounts of waste residue to be disposed of, SO_2 removal requires large amounts of absorbent
Circulating fluidized bed combustion	New plants (large or small)	High SO_2 removal, long residence time favors efficient absorbent use	Large amounts of waste residue to be disposed of, limited commercial experience
Pressurized fluidized bed combustion	New plants (large)	High SO_2 removal, high combustion efficiency, small size	Limited commercial experience, complex technology, large amounts of waste residue
Integrated gasification combined cycle	New plants (large)	Very high sulfur removal, high efficiency, adaptable to a wide range of fuels	Demonstration scale only, costs uncertain
Dry sulfur absorbents	New plants or retrofit (large or small)	Simple process, low capital cost, rapid installation	SO_2 removal limited, may cause boiler fouling, potential problems with waste
Spray-dry scrubbers	New plants or retrofit (large/medium)	High SO_2 removal, established technology, flexible operation, low capital cost	Higher operating cost than wet scrubbers, potential for fouling, large amounts of waste for disposal
Wet scrubbers	New plants or retrofit (large)	Widely used technology, high SO_2 removal, potential to produce marketable byproducts	Higher capital costs than spray-dry, high space and water requirements, wet residue difficult to handle
Regenerative scrubbers	New plants or retrofit (large)	High SO_2 removal, potential for marketable byproducts, low absorbent demand	High capital costs, complex process, only economical if byproducts are sold locally
Combined SO_2/NO_x scrubbers	New plants or retrofit (large)	High SO_2 removal, controls two major pollutants	Limited commercial experience, complex process

REFERENCE

Vernon, J.L. and Jones, T. (1993), *Sulphur and Coal*, IEA Coal Research, London, Report No. IEACR/57.

Index

N

O

P

R

S

T

W

X

Z